Ecological Studies, Vol. 206

Analysis and Synthesis

Ecological Studies

Further volumes can be found at springer.com

Volume 189
Ecology of Harmful Algae (2006)
E. Granéli and J.T. Turner (Eds.)

Volume 190
Wetlands and Natural Resource Management (2006)
J.T.A. Verhoeven, B. Beltman, R. Bobbink, and D.F. Whigham (Eds.)

Volume 191
Wetlands: Functioning, Biodiversity Conservation, and Restoration (2006)
R. Bobbink, B. Beltman, J.T.A. Verhoeven, and D.F. Whigham (Eds.)

Volume 192
Geological Approaches to Coral Reef Ecology (2007)
R.B. Aronson (Ed.)

Volume 193
Biological Invasions (2007)
W. Nentwig (Ed.)

Volume 194
***Clusia*: A Woody Neotropical Genus of Remarkable Plasticity and Diversity** (2007)
U. Lüttge (Ed.)

Volume 195
The Ecology of Browsing and Grazing (2008)
I.J. Gordon and H.H.T. Prins (Eds.)

Volume 196
Western North American *Juniperus* Communites: A Dynamic Vegetation Type (2008)
O. Van Auken (Ed.)

Volume 197
Ecology of Baltic Coastal Waters (2008)
U. Schiewer (Ed.)

Volume 198
Gradients in a Tropical Mountain Ecosystem of Ecuador (2008)
E. Beck, J. Bendix, I. Kottke, F. Makeschin, R. Mosandl (Eds.)

Volume 199
Hydrological and Biological Responses to Forest Practices: The Alsea Watershed Study (2008)
J.D. Stednick (Ed.)

Volume 200
Arid Dune Ecosystems: The Nizzana Sands in the Negev Desert (2008)
S.-W. Breckle, A. Yair, and M. Veste (Eds.)

Volume 201
The Everglades Experiments: Lessons for Ecosystem Restoration (2008)
C. Richardson (Ed.)

Volume 202
Ecosystem Organization of a Complex Landscape: Long-Term Research in the Bornhöved Lake District, Germany (2008)
O. Fränzle, L. Kappen, H.-P. Blume, and K. Dierssen (Eds.)

Volume 203
The Continental-Scale Greenhouse Gas Balance of Europe (2008)
H. Dolman, R. Valentini, and A. Freibauer (Eds.)

Volume 204
Biological Invasions in Marine Ecosystems: Ecological, Management, and Geographic Perspectives (2009)
G. Rilov and J.A. Crooks (Eds.)

Volume 205
Coral Bleaching: Patterns, Processes, Causes and Consequences (2009)
M.J.H. van Oppen and J.M. Lough (Eds.)

Volume 206
Marine Hard Bottom Communities: Patterns, Dynamics, Diversity, and Change (2009)
M. Wahl (Ed.)

Martin Wahl
Editor

Marine Hard Bottom Communities

Patterns, Dynamics, Diversity, and Change

Editor
Prof. Dr. Martin Wahl
IFM-GEOMAR
Düsternbrookerweg 20
24105 Kiel, Germany
mwahl@ifm-geomar.de

ISBN 978-3-540-92703-7 e-ISBN 978-3-540-92704-4
DOI 10.1007/978-3-540-92704-4
Springer Dordrecht Heidelberg London New York

Ecological Studies ISSN 0070-8356

Library of Congress Control Number: 200992151

Cover illustrations: The book cover shows a small selection of typical hard bottom community species of the temperate North Atlantic, with a tropical gorgonian in the background. While the two cephalopods are occasional visitors to hard bottom systems, the star fish and the mussel tend to prefer hard over soft bottom habitats, and the remaining forms are exclusively found on natural hard bottom. The species shown represent the important trophic functional types of predators, grazers and suspension feeders.

Cover design: WMXDesign GmbH, Heidelberg, Germany

Printed on acid-free paper

Springer is part of Springer Science+Business Media (www.springer.com)

Preface

Hard-bottom communities represent some of the most productive and diverse ecosystems on our planet. They include very dissimilar types of assemblages such as hydrothermal vent communities, kelp forests, tropical coral reefs, fouling on man-made structures or epibioses on living surfaces. These communities are composed of microbial and macrobial organisms which are sessile, sedentary or motile, auto-, mixo- or heterotrophic, solitary or colonial, and represent most of the marine phyla. Due to their impressive taxonomical and functional diversity, as well as their internal dynamics, these communities provide irreplaceable ecosystem services such as nutrient cycling, water purification, benthic–pelagic coupling, or nursery grounds for juveniles. Besides this, they represent an evolutionary heritage which in species richness is comparable, but in functional diversity is superior to tropical rainforests.

As a large proportion of the worldwide hard-bottom communities are located in shallow and near-shore regions, they are particularly prone to the impacts of human activities and global change. In the course of the ongoing shift of abiotic variables associated with climate change, and in combination with more directly man-induced pressures like over-fishing, eutrophication, invasions or the restructuring of coastlines, it can be expected that marine hard-bottom communities will structurally reorganize over the coming decades. If this change in species composition is associated with a change in functional group richness, then shifts in ecosystem services will ensue.

In an effort to improve our understanding of the value, dynamics, diversity and change of marine hard-bottom communities, in this book more than 50 marine ecologists report on various aspects of these systems from a large number of biogeographic regions.

The result is a comprehensive picture of the particularity of these communities, the dynamics of their reproduction, recruitment and interactions, and the patterns of their taxonomic and functional diversity at a variety of spatial and temporal scales. The role and value of diversity regarding community stability and ecosystem services, and the present challenges to this diversity and the associated risks are highlighted. Finally, suitable measures to protect or restore community diversity are presented and discussed. The book ends with some recommendations on new and appropriate research tools.

We hope that this work will help scientists, decision makers and students—as the coming responsible generation—to recognize, understand and hopefully master the challenges associated with global change.

To whatever extent this books fulfils these expectations, its value reflects the cumulated input of numerous colleagues and, particularly, the efficient and competent effort of the parts coordinators. I deeply appreciate all the enthusiastic and insightful contributions, comments, suggestions and discussions. The highly competent editorial assistance by Nina Blöcher greatly facilitated the task. All in all, it was a highly rewarding experience.

February 2009

Martin Wahl
Kiel, Germany

Contents

Contributors

L. Airoldi
Dipartimento di Biologia Evoluzionistica Sperimentale and Centro Interdipartimentale di Ricerca per le Scienze Ambientali, Università di Bologna, Via S. Alberto 163, 48100 Ravenna, Italy
laura.airoldi@unibo.it

F. Arenas
Laboratory of Coastal Biodiversity, Centro Interdisciplinar de Investigação Marinha e Ambiental–CIIMAR, Universidade do Porto, Rua dos Bragas 289, 4050-123 Porto, Portugal
farenas@ciimar.up.pt

G. Bavestrello
Dipartimento di Scienze Mare, Università Politecnica delle Marche, Via Brecce Bianche, 60131 Ancona, Italy
g.bavestrello@univpm.it

M.W. Beck
The Nature Conservancy and Institute of Marine Sciences, 100 Shaffer Road-LML, University of California, Santa Cruz, CA 95060, USA
mbeck@tnc.org

L. Benedetti-Cecchi
Dipartimento di Biologia, University of Pisa, Via A. Volta 6, 56126 Pisa, Italy
lbenedetti@biologia.unipi.it

D. Blockley
Centre for Research on Ecological Impacts of Coastal Cities, Marine Ecology Laboratories A11, University of Sydney, NSW 2006, Australia
d.blockley@usyd.edu.au

M.E.S. Bracken
Marine Science Center, Northeastern University, 430 Nahant Road, Nahant, MA 01908, USA
m.bracken@neu.edu

J. Canning-Clode
Leibniz Institute of Marine Sciences at the University of Kiel, Duesternbrooker Weg 20, 24105 Kiel, Germany;
University of Madeira, Centre of Macaronesian Studies, Marine Biology Station of Funchal, 9000-107 Funchal, Madeira, Portugal
jcanning-clode@ifm-geomar.de, jcanning-clode@daad-alumni.de

C. Cerrano
Dipartimento per lo studio del Territorio e delle sue Risorse, Università di Genova, Corso Europa 26, 16132 Genova, Italy
cerrano@dipteris.unige.it

M.G. Chapman
Centre for Research on Ecological Impacts of Coastal Cities, Marine Ecology Laboratories A11, University of Sydney, NSW 2006, Australia
gee@eicc.bio.usyd.edu.au

B.G. Clynick
Centre for Research on Ecological Impacts of Coastal Cities, Marine Ecology Laboratories A11, University of Sydney, NSW 2006, Australia
bgclynick@yahoo.com.au

S.D. Connell
Southern Seas Ecology Laboratories, DP 418, University of Adelaide, SA 5005, Australia
sean.connell@adelaide.edu.au

T.P. Crowe
School of Biology and Environmental Science, University College Dublin, Belfield, Dublin 4, Ireland
tasman.crowe@ucd.ie

M. Cusson
Département des Sciences Fondamentales, Université du Québec à Chicoutimi, 555 Boulevard de l'Université, Chicoutimi, Québec G7H 2B1, Canada
mathieu.cusson@uqac.ca

B.A.P. da Gama
Departamento & Programa de Pós-Graduação em Biologia Marinha, Universidade Federal Fluminense, C.P. 100644, 24001-970 Niterói, Rio de Janeiro, Brazil
bapgama@pq.cnpq.br

A.R. Davis
School of Biological Sciences, University of Wollongong, NSW 2522, Australia
adavis@uow.edu.au

P.W. Fofonoff
Smithsonian Environmental Research Center, P.O. Box 28, Edgewater, MD 21037, USA
fofonoffp@si.edu

S. Fraschetti
Laboratorio di Zoologia e Biologia Marina, Dipartimento di Scienze e Technologie Biologiche ed Ambientale, Università del Salento, CoNISMa, 73100 Lecce, Italy
simona.fraschetti@unile.it

A.L. Freestone
Smithsonian Environmental Research Center, P.O. Box 28, Edgewater, MD 21037, USA
freestonea@si.edu

L. Gamfeldt
Department of Marine Ecology, Göteborg, University of Gothenburg, Box 461, 40530 Göteborg, Sweden
Lars.Gamfeldt@marecol.gu.se

J.-M. Gili
Departamento Biología Marina y Oceanografía, Institut de Ciències del Mar (CSIC), Passeig Maritim de la Barceloneta 37–49, 08003 Barcelona, Spain
gili@icm.csic.es

S.J. Goldstien
Marine Ecology Research Group, School of Biological Sciences, University of Canterbury, Private Bag 4800, 8140 Christchurch, New Zealand
sharyn.goldstien@canterbury.ac.nz

P.J. Goodsell
Centre for Research on Ecological Impacts of Coastal Cities, Marine Ecology Laboratories A11, University of Sydney, NSW 2006, Australia
pgoodsell@eicc.bio.usyd.edu.au

J.N. Griffin
The Marine Biological Association of the UK, Citadel Hill, Plymouth PL1 2PB, UK;
Marine Biology and Ecology Research Centre, Marine Institute, School of Biological Sciences, University of Plymouth, Plymouth PL4 8AA, UK
john.griffin@plymouth.ac.uk

C.D.G. Harley
Department of Zoology, University of British Columbia, 6270 University Blvd., Vancouver, BC V6T1Z4, Canada
harley@zoology.ubc.ca

J. Havenhand
Department of Marine Ecology, Tjärnö, University of Gothenburg, 45296 Stromstad, Sweden
jon.havenhand@marecol.gu.se

S.J. Hawkins
The Marine Biological Association of the UK, Citadel Hill, Plymouth
PL1 2PB, UK;
School of Ocean Sciences, Bangor University, Menai Bridge, Anglesey
LL59 5AB, UK
s.hawkins@bangor.ac.uk

A.C. Jackson
Centre for Research on Ecological Impacts of Coastal Cities, Marine Ecology
Laboratories A11, University of Sydney, NSW 2006, Australia;
Present address: Environmental Research Institute, North Highland College,
UHI Millennium Institute, Castle Street, Thurso, Caithness KW14 7JD, UK
Angus.Jackson@thurso.uhi.ac.uk

S.R. Jenkins
School of Ocean Sciences, Bangor University, Menai Bridge, Anglesey LL59
5AB, UK
s.jenkins@bangor.ac.uk

J. Kotta
Estonian Marine Institute, University of Tartu, Mäealuse 10a, 12618 Tallinn,
Estonia
jonne.kotta@sea.ee

D.J. Marshall
School of Biological Sciences, University of Queensland, 4072 Queensland,
Australia
d.marshall1@uq.edu.au

C.D. McQuaid
Department of Zoology and Entomology, Rhodes University, Grahamstown 6140,
South Africa
C.McQuaid@ru.ac.za

M. Molis
Biologische Anstalt Helgoland, Alfred-Wegener-Institute for Polar and Marine
Research, Kurpromenade 201, 27498 Helgoland, Germany
markus.molis@awi.de

P.S. Moschella
The Marine Biological Association of the UK, Citadel Hill, Plymouth
PL1 2PB, UK;
Commission Internationale pour l'Exploration Scientifique de la Mer
Méditerranée, 16 Bd de Suisse, MC 98000, Monaco,
pmoschella@ciesm.org

S. Navarrete
Estación Costera de Investigaciones Marinas & Center for Advanced Studies in Ecology and Biodiversity, Pontificia Universidad Católica de Chile, Casilla 114-D, Alameda 340, Santiago, Chile
snavarrete@bio.puc.cl

L.M.-L.J. Noël
The Marine Biological Association of the UK, Citadel Hill, Plymouth PL1 2PB, UK;
Marine Biology and Ecology Research Centre, Marine Institute, School of Biological Sciences, University of Plymouth, Plymouth PL4 8AA, UK
noelaure@gmail.com

P.S. Petraitis
Department of Biology, University of Pennsylvania, Philadelphia, PA 19104-6018, USA
ppetrait@sas.upenn.edu

G.M. Ruiz
Smithsonian Environmental Research Center, P.O. Box 28, Edgewater, MD 21037, USA
ruizg@si.edu

R. Russell
CIESIN, The Earth Institute, Columbia University, 2910 Broadway, MC 3277, New York, NY 10025, USA;
The Sandhill Institute, 5800 Edwards Road, Grand Forks, BC V0H 1H9, Canada
roly@fulbrightweb.org

D.R. Schiel
Marine Ecology Research Group, School of Biological Sciences, University of Canterbury, Private Bag 4800, 8140 Christchurch, New Zealand
david.schiel@canterbury.ac.nz

E.A. Serrão
CCMAR-CIMAR, Universidade do Algarve, Gambelas, 8005-139 Faro, Portugal
eserrao@ualg.pt

C. Simkanin
Department of Biology, University of Victoria, P.O. Box 3020 STN CSC, Victoria, BC V8W 3N5, Canada
simkanin@uvic.ca

C. Styan
RPS Environment, Level 2, 47 Ord St, West Perth WA 6005, P.O. Box 465, Subiaco WA 6904
StyanC@rpsgroup.com.au

H.E. Sugden
School of Ocean Sciences, Bangor University, Menai Bridge, Anglesey LL59 5AB, UK
h.sugden@bangor.ac.uk

A. Terlizzi
Dipartimento di Scienze e Tecnologie Biologiche ed Ambientali, Università del Salento, CoNISMa, 73100 Lecce, Italy
antonio.terlizzi@unile.it

R.C. Thompson
Marine Biology and Ecology Research Centre, Marine Institute, School of Biological Sciences, University of Plymouth, Plymouth PL4 8AA, UK
R.C.Thompson@plymouth.ac.uk

M. Wahl
IFM-GEOMAR, Duesternbrookerweg 20, 24105 Kiel, Germany
mwahl@ifm-geomar.de

J.D. Witman
Department of Ecology and Evolutionary Biology, Box G-W, Brown University, Providence, RI 02912, USA
Jon_Witman@brown.edu

J.T. Wootton
Department of Ecology & Evolution, The University of Chicago, 1101 East 57th Street, Chicago, IL 60637, USA
twootton@uchicago.edu

A.J. Underwood
Centre for Research on Ecological Impacts of Coastal Cities, Marine Ecology Laboratories A11, University of Sydney, NSW 2006, Australia
aju@bio.usyd.edu.au

C.M. Young
Oregon Institute of Marine Biology, University of Oregon, P.O. Box 5389, Charleston, OR 97420, USA
cmyoung@uoregon.edu

Part I
Habitat, Substrata and Communities

Coordinated by Andrew R. Davis

Introduction

Andrew R. Davis

In terrestrial systems, the substratum is acknowledged as playing a highly significant role in community organisation; rock type determines soil composition as well as nutrient availability and, in turn, the floral and faunal communities that will be supported. In the marine realm, in contrast, organisms adhere to surfaces but rarely do organisms derive benefit from these (see Chisholm et al. 1996 for an intriguing exception). While it is well known that the orientation and slope of surfaces in the subtidal zone will influence incident light and sedimentation and, therefore, the resultant community, it is assumed that the nature of the substratum will be of minor importance. The chapters in this opening Part I of the book explore this notion in a range of subtidal systems.

In the opening chapter, Wahl first examines the unique features of the subtidal zone, providing a foundation for discussion of hard bottom communities. He then applies the concept of functional groups to these communities across an unprecedented scale, with data from a global experiment using identical substrata—PVC panels. Given the hard-won nature of data from subtidal communities, his examination of alternative approaches to exploring ecological pattern and process is laudable. Indeed, it is useful not to be constrained by taxonomy, particularly as the taxonomic identity of some subtidal organisms—in particular, sponges and colonial ascidians—can be difficult to determine. This is further clouded by the presence of species' complexes or cryptic species (Davis et al. 1999). Although Wahl's dataset is from a relatively short time period and the panels small, it is nevertheless a starting point and underscores the power of adopting the same methodology across a broad geographic scale; it is rare to have direct comparisons across distinct fauna in disparate coastal zones. Overall, the functional group approach is worthy of closer scrutiny.

In the second chapter, Davis contrasts the communities that develop on mineral, living and artificial substrata in shallow water (i.e. accessible by SCUBA). Artificial surfaces have proven particularly useful in understanding patterns of settlement and recruitment but the small spatial and temporal scales over which most such experiments are done questions their utility in understanding general ecological pattern and process. Moreover, panels are often at some distance from natural processes affecting the benthos. Living or biogenic surfaces represent a special case as substrata because strong selection pressures are expected to be operating, and their ability to attract,

M. Wahl (ed.), *Marine Hard Bottom Communities*, Ecological Studies 206,
DOI: 10.1007/978-3-540-92704-4_I, © Springer-Verlag Berlin Heidelberg 2009

deter or remove colonisers is well established. The recent finding that surface microtopography may be an important settlement deterrent is particularly exciting. Microtexture has become apparent for a variety of organisms, including on bivalve shells, molluscan egg masses and cetaceans. In relation to mineral surfaces, it is becoming clear that surface mineralogy may play an important structuring role for subtidal communities. First, the degree to which surfaces are pitted or their surface rugosity modified due to mineral dissolution may affect patterns of propagule settlement and recruitment. Second, the cracks and crevices associated with certain rock types may be an important ecological driver in some circumstances. Third, surprising evidence is mounting that the mineralogy of surfaces has direct effects on the settlement and survivorship of propagules even over broad scales, although this notion needs further examination with well-replicated experiments.

The next two chapters treat hard bottom communities in special environments: the deep sea and on living surfaces. In Chapter 3, Young extends our understanding of patterns and processes for hard bottom communities into the deep ocean. Although this habitat is usually considered a ‘sea of mud’, Young reveals that hard surfaces can be extensive in the deep ocean and that deep coral reefs may exceed in area that of their shallow-water tropical counterparts. Striking features of his contribution include the variety of hard surfaces that are available for colonisation, the specialisations of some of the colonisers and the significant progress that has already been made in understanding patterns and processes in this environment. Given the difficulties in working at depth, progress on this latter point is particularly impressive. In addition, ecological work in this environment provides an opportunity to test the generality of ecological theory already developed in less logistically (and financially) challenging environments (Lawton 1996, 1999). On a sobering note, these environments may be no more immune to anthropogenic effects than are shallow-water reefs; anticipated changes in the calcite and aragonite saturation horizons stemming from fossil fuel-generated CO_2 threaten the dissolution of key elements in these habitats (Guinotte et al. 2006).

In the final chapter in this book part, Wahl brings us up to date with recent developments in the field of epibiosis (fouling). Historically, this area has stood on the interface of natural products chemistry, invertebrate ecology and microbial biology. Increasingly, particularly with the advent of cultureless microbial techniques, molecular biologists are becoming involved. Wahl’s contribution emphasises the increasing complexity being revealed in the ‘arms race’ between the host (basibiont) and the fouler (epibiont). The responses of the key players and the outcomes of their interactions can be astoundingly variable.

Conclusion

To conclude, these chapters have highlighted recent developments in these disparate habitats; they have also outlined some of the challenges we face in moving our field forward. In my view, these chapters underscore two important points and, although, my observations are not particularly novel, they are worth emphasising. First, there is the overarching importance of an experimental approach (sensu Hurlbert 1984) in seeking to better understand natural systems. Even the seemingly intractable deep sea offers possibilities in this regard. My second point is the power of using identical methodology to explore pattern and process in disparate locations. The application of both of these approaches to subtidal communities can only strengthen our field.

References

Chisholm JRM, Dauga C, Ageron E, Grimont PAD, Jaubert JM (1996) 'Roots' in mixotrophic algae. Nature 381:382

Davis AR, Roberts DE, Ayre DJ (1999) Conservation of sessile marine invertebrates: you don't know what you've got 'til it's gone. In: Ponder W, Lunney D (eds) The other 99%: the conservation and biodiversity of invertebrates. Mosman, Transactions of the Royal Society of New South Wales, pp 325–329

Guinotte JM, Orr J, Cairns S, Freiwald A, Morgan L, George R (2006) Will human-induced changes in seawater chemistry alter the distribution of deep-sea scleractinian corals? Frontiers Ecol Environ 4:141–146

Hurlbert SH (1984) Pseudoreplication and the design of ecological field experiments. Ecol Monogr 54:187–211

Lawton J (1996) Patterns in ecology. Oikos 75:145–147

Lawton J (1999) Are there general laws in ecology. Oikos 84:177–192

Chapter 1
Habitat Characteristics and Typical Functional Groups

Martin Wahl

1.1 Particularities of the Aquatic Medium

In the aquatic environment, evolution has produced a number of life forms which are rare or missing in terrestrial ecosystems. These include sessile organisms, i.e. microbes, plants and animals living attached to hard substrata without trophically depending on these substrata. These life forms constitute the bulk of the hard bottom communities treated in this book. Consequently, hard bottom communities are typically aquatic, and reach their highest diversity and largest biomasses in the marine realm. Important examples are coral reefs and kelp forests.

The reason why many of the functional groups which compose marine hard bottom communities are missing on land lies in the fundamental differences between the media, air and salt water, with regard to a number of physicochemical properties with biological relevance. In the following, I will briefly review some of these and their ecological consequences (Fig. 1.1).

Water is denser and more viscous than air, and the ratios of these parameters (approx. 80:1 and 100:1 respectively) change with temperature, salinity and pressure. Additionally, the dipole nature of water molecules makes this fluid the 'universal solvent' for an extremely wide range of elements and molecules. As a further consequence of the dipole nature, water molecules interlinking by hydrogen bonds form clusters. The existence of such clusters is the basis for the high viscosity of water. Cluster forming also is the reason for the remarkable heat capacity of water as compared to other liquids: much solar energy is used to break up hydrogen bonds which link the molecules in a cluster, rather than raising water temperature. This stored energy is released during cooling when the clusters form. The released heat slows down the cooling. A further difference between the terrestrial and the aquatic environment concerns the availability of light for photosynthesis or optical orientation. Water molecules, and particulate or dissolved matter in the water, absorb light much more efficiently than does air, or the low concentration of molecules or particles in air. Absorption is strongest in the ultraviolet, yellow, red and infrared wavelengths. As a consequence, light changes with water depth both in quantity and quality. The role of solar radiation as a source of energy and information is virtually

M. Wahl (ed.), *Marine Hard Bottom Communities*, Ecological Studies 206,
DOI: 10.1007/978-3-540-92704-4_1, © Springer-Verlag Berlin Heidelberg 2009

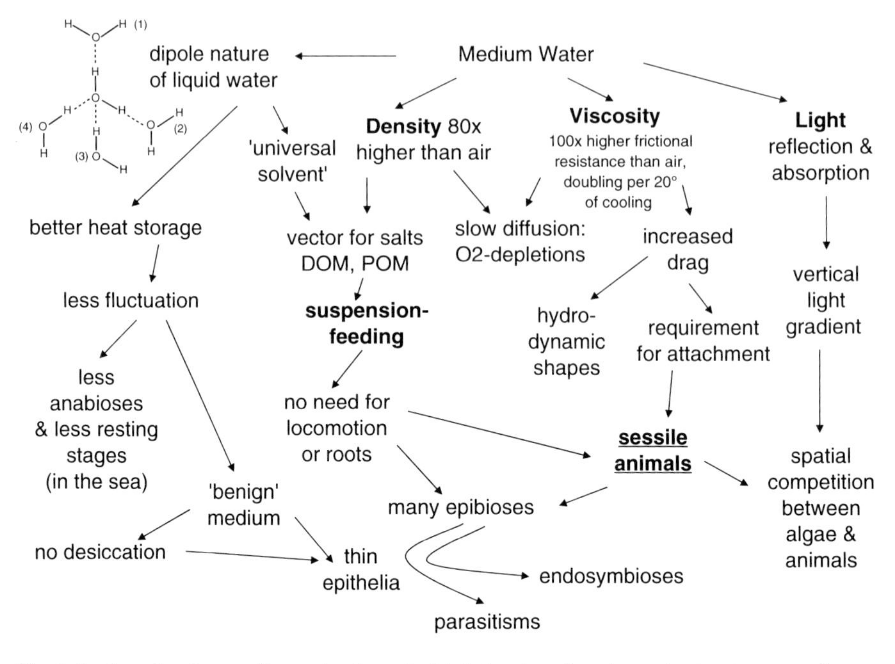

Fig. 1.1 Causal pathways illustrating how the high density, viscosity and solvent power of water enable the existence of the functional group of sessile suspension feeders, one of the most important components of hard bottom communities

nil below the first couple of hundred meters. On the other hand, the aquatic medium is better suited than air as a vector for acoustic, electric and chemical information.

These physicochemical properties, singly or in their interaction, have consequences which strongly affect marine life, enabling the evolution of life forms which have never made their way onto land. The high density of water reduces the relative weight of aquatic organisms. The proportions of heavy mineral skeletons, almost neutral organic tissue and buoyant gaseous or lipid inclusions determine the net weight of submerged organisms. This will always be lower than for a similarly built organism in air. One consequence of this is that floating in water consumes less energy, and aqueous nekton and plankton are incomparably richer in biomass and diversity than their aerial counterparts. The other side of the low-weight coin is a dramatic reduction in friction between bottom-dwelling organisms and the substratum. Reduced friction in conjunction with the elevated viscosity of the medium poses a challenge to 'staying put' in a water current. To avoid being entrained, aquatic organisms must swim or attach. Permanent attachment on land usually is possible only when the substratum also serves as a source for food, e.g. soil for plants, and hosts for parasites. Since water acts as a vector for a rich load of suspended organic material (seston, plankton, nekton), attached heterotrophic organisms may acquire all energy they need by filter feeding or by capturing deposited particles.

Animals usually require more food than is available within the immediate reach of their teeth or tentacles. They must exploit larger areas or volumes. The relative movement between consumer and 'food space' is assured by locomotion of terrestrial and motile aquatic animals, and by currents around sessile animals in the sea. In addition, the by-flowing 'universal solvent' provides gases (O_2, CO_2), nutrients used by algae, and organic compounds, and it eliminates excreted waste products. Many of these solutes may be taken up as additional energy sources (lipids, sugars, peptides), may enable intraspecific communication (various infochemicals), or may drive other interactions (interspecific cues, defence metabolites). Gametes and propagules may also be disseminated by the flowing medium. While the attached mode of living is energetically beneficial, one shortcoming is the inability to escape local biotic or abiotic stress. Consumption and overgrowth may be limited by the evolution of structural or chemical defences. Adverse abiotic conditions are more difficult to avoid. If the abiotic conditions in the sea were as variable as they are on land, then permanently attached animals would have to be extremely tolerant, capable of homoeostasis, or able to pass stressful phases in a state of reduced activity (e.g. anabiosis). However, the underwater 'climate' in a given subtidal location varies much less in time than is the case for the terrestrial climate. Due to the high heat capacity, temperature is virtually constant below 1,000m, and at shallow depths of polar and tropical regions. Even in shallow (<20m) temperate seas, water temperatures typically vary by less than 1°C on a diurnal scale and by less than 20°C seasonally. Viscosity, inertia and internal friction buffer the temporal variability of currents, except at depths affected by surface waves and tidal currents. The chemical composition of salt water is extremely stable, with the exception of the compounds metabolized by organisms (e.g. nitrate, phosphate, silicate). One further notable exception to the general stability of abiotic variables in the oceans is oxygen. Because diffusion in water is much slower than in air, oxygen concentrations may range from anoxia to supersaturation, driven by local heterotrophic and autotrophic processes. This imbalance is enhanced when the relative movement between the water and organisms is reduced. Thus, anoxia is common in stagnant bottom waters, and hyperoxia may occur in plankton blooms.

Overall, aquatic habitats may be considered physicochemically more benign than terrestrial ones, because environmental variables vary less in the former. As a consequence, a menacing departure from the physiological optimum is less likely, and costly protective adaptations are less urgently required.

Across habitats, and particularly on a large spatial scale, however, the underwater climate exhibits notable gradients. Vertically, light decreases to biologically unusable levels within the upper tens or hundreds of meters, depending on the load of dissolved organic material and suspended particles; temperature outside the polar regions decreases to an almost constant 2–4°C below 1,000m; pressure increases by 1bar per 10m depth, leading to a slight increase of density and a stronger increase of gas solubility with depth. In addition, temporal variability of all parameters generally decreases with depth. Horizontally, salinity increases towards regions of intense evaporation and little precipitation (e.g. the Red Sea), and decreases towards zones of high precipitation and low evaporation (e.g. the Baltic Sea). Nutrients tend to decrease from the sediment upwards, and from upwelling cells or estuaries outwards.

Despite strong large-scale gradients, at a given subtidal location conditions are usually relatively stable and predictable. A sessile organism experiences abrupt and strong changes only when water pockets of a different physicochemical nature drift by—a situation usually restricted to certain shallow coastal zones. For most sessile organisms, the challenge is to settle in a suitable (micro-) habitat.

In conclusion, the high density and viscosity of seawater, together with its solvent and heat capacity, have enabled the evolution of a typically aquatic functional group, the sessile suspension feeder, which is characteristic for and often dominant in marine hard bottom communities. In nutrient-rich and euphotic habitats, however, suspension feeding animals have to compete with macroalgae for substratum, a potentially limiting resource.

1.2 Life Forms in Hard Bottom Communities

Hard bottom communities around the world, also termed fouling communities (mostly used for assemblages on manmade substrata) and epibioses (on living substrata), harbour tens of thousands of species. In the following, I will concentrate on the sessile species, since many of the motile forms are not restricted to this substratum type. All major macroalgal groups are present in hard bottom communities, as long as sufficient light is available. Sessile forms of most animal classes and phyla share hard substratum with algae at light-exposed sites, and exclusively occupy these when light is low or absent: all sponges, most cnidarians, a few bivalves and sessile gastropods, most bryozoans and phoronids, a few boring urchins, many tube-building annelids, and all ascidians. Many sessile representatives of these phyla do not strictly differentiate between rocky, artificial or living substratum (Wahl and Mark 1999). Large groups with rare or no sessile representatives in hard bottom communities are the platyhelminths, nematodes, echinoderms and vertebrates. They do, however, contribute to the motile components of these communities.

In view of the enormous number of species and the apparent lack of phylogenetic predilection for hard substrata, we need an alternative method of classification of community components. One which is ecologically more meaningful than phylogenetics and which has seen a revival during the past 15 years is the concept of functional groups (e.g. Steneck and Dethier 1994; Bengtsson 1998; Petchey and Gaston 2002; Bremner et al. 2003; Micheli and Halpern 2005; Wright et al. 2006; Halpern and Floeter 2008). Functional groups are suites of species which play a similar ecological role (Petchey and Gaston 2002; Blondel 2003), i.e. have similar requirements with regard to one or more resources (e.g. light, food particles, prey, substratum) and/or provide similar services (e.g. oxygen production, denitrification, shelter). More traits may be added to characterize a functional group: e.g. degree of motility, adult body size, reproductive mode, larval feeding mode and dispersal, longevity and coloniality. The more dimensions are used to define a functional group, the more similar (in an ecological sense) are its members and the more intensively they may compete if one or more of the shared resources is or are limited.

By evolutionary convergence, phylogenetically very different species may share a functional group. Sponges and encrusting colonial ascidians, for instance, are all non-motile, require hard substratum for larval attachment, metamorphosis and colony growth, feed on similar size ranges of plankton, settle in similar habitats, may exploit still waters because they create their own filtering current, may grow at comparable rates, reach similar body sizes and live to similar ages. Such a degree of functional similarity should lead to pronounced competition where these forms co-occur. In contrast, different ontogenetic stages of the same species, like the meroplanktonic larva of barnacles and the sessile adult, belong to distinct functional groups and do not compete in the least.

The concept of functional groups is useful in two regards at least. When asking questions such as 'which are typical components of hard bottom communities?' or 'what determines the stability of hard bottom communities?', species identities are not informative at all. Species found on hard substrata vary from site to site for evolutionary and ecological reasons and do not provide a general pattern. In contrast, their functionality, i.e. their membership to a given functional group, or the degree to which species overlap with regard to a given functional trait, is highly relevant (Duffy et al. 2007). Functional diversity of a community, i.e. the number of different functional groups, is related to resource use and productivity (Naeem and Li 1997; Cardinale et al. 2002). Functional redundancy (Loreau 2004), i.e. the number of species in a community which play similar ecological roles, seems to be relevant for ecological and structural stability (Fonseca and Ganade 2001; Britton-Simmons 2006; Stachowicz et al. 2007).

Functional groups generally are defined as the suite of those species sharing a number of biological traits (but see Wright et al. 2006 for a critical review of classification schemes). Which and how many traits should be considered has been a matter of debate for some time (e.g. Petchey and Gaston 2006). The main issues of conflict have been (1) which life history traits are ecologically relevant, (2) how many traits are required to characterize the functionality of a species and (3) whether traits should be used in a categorical (either 'suspension feeder' or 'deposit feeder') or a continuous way ('mostly suspension feeder, occasionally switching to deposit feeding') (e.g. Bremner et al. 2003). Because certain traits are interrelated to some extent (e.g. longevity, size, trophic level), body size has been suggested to be the one most important trait, and suitable as a proxy or sum parameter for most ecologically relevant characteristics of a species (e.g. Woodward et al. 2005). A less extreme approach was taken in two studies on redundancy in functional diversity using the two traits motility and diet (Micheli and Halpern 2005), and on latitudinal diversity patterns emphasizing the traits diet and dispersal (the latter including larval dispersal and adult motility; Hillebrand 2004). Possibly the most sophisticated approach to functional group definition was suggested by Bremner et al. (2003), who used nine biological traits subdivided further into a total of 34 categories which again contained three levels each of applicability to a given species. However, the resolution of this system is so fine that it characterizes individual species ecologically, rather than grouping functionally similar species into guilds. Functional diversity concepts are treated in more detail by Crowe and Russell in Chapter 5 (5.2).

In the following, I provide an example of characterizing hard bottom communities using functional groups. For a study on community stability, we exposed panels (15×15cm, roughened PVC) to natural colonization for 9 months in different biogeographic regions in the North Sea, Mediterranean Sea, South East Pacific, West Atlantic, Central East Atlantic, Tasmanian Sea and temperate West Pacific. The colonization of the panels occurred in the respective summer seasons of the year 2005, at the same depth (1m) and inclination (vertical), and the resulting communities were analyzed according to the same taxonomic and functional criteria. (It should be kept in mind that the communities studied were relatively young and had established on spatially restricted substrata.)

Ideally, before combining species into functional groups the functional role of each species should be identified experimentally (Wright et al. 2006). However, this is not realistic for species-rich communities, especially in poorly studied regions (Halpern and Floeter 2008). Consequently, for this global approach we employed the commonly used a priori definition of functional role. We initially chose an intermediate level of resolution by selecting eight traits of ecological relevance: size, longevity, growth form, growth rate, trophic type, modularity, dispersal and motility, each subdivided into a number of levels (Table 1.1). These traits were identified for the adult individual. We recognize that earlier ontogenetic stages may belong to different functional groups but, for the majority of the sessile species considered here, their meroplanktonic propagules do not belong to the fouling community, and juveniles constituted only a minor biomass in these mid-successional assemblages. Because motile species often switch between habitats, we considered only sessile species for the analysis of hard bottom community properties relevant for structural stability. After eliminating 'impossible' combinations like 'size XX'×'longevity S', the number of plausible combinations of seven traits was 1,484. Following the suggestion that some of the traits are related (Woodward et al. 2005), we also tested a coarser approach and used only those four traits considered the least connected and the most important for characterizing structure and dynamics of a hard bottom community: size, growth form, trophic type and modularity. With this reduced set of dimensions, the number of plausible combinations was 114.

A total of 356 multicellular species were identified in the 9-month-old fouling communities in ten biogeographic regions. Only rarely were species encountered in more than one region. Exceptions to this regional segregation were the invasives *Mytilus galloprovincialis*, *Bugula neritina* and *Enteromorpha* spp. These 356 taxonomic identities grouped into 146 functional groups when using seven functional traits for the ecological characterization and into 57 functional groups when using only four traits. Thus, roughly 10% of the plausible 7-trait groups but 50% of the plausible 4-trait groups were encountered. Unexpectedly, despite the more than tenfold higher resolution of the 7-trait approach as compared to the coarser 4-trait description, the functional diversity levels yielded by the two approaches hardly differed at the site or at the panel level (Fig. 1.2). Apparently, the three traits omitted in the coarser approach (longevity, growth form, dispersal) did not vary independently of the four others. Defining functional groups of sessile hard bottom species using the traits size, growth form, trophic type and modularity apparently provides sufficient information to characterize

Table 1.1 Life history traits as used in the global experiment on the structure of young fouling communities. 'Motility' was not considered in the present analysis. The remaining traits were used in the 7-trait approach, those in italics in the 4-trait approach

Size	Longevity	*Growth form*	*Trophic type*	*Modularity*	Dispersal	Growth rate	Motility
<1mm (S)	<1month (S)	Encrusting (E)	Autotroph (A)	Solitary (S)	10sm (L)	>x2/week (F)	Attached (A)
1–10 mm (M)	1 month-1 year (M)	Massive (M)	Predator (P)	Colonial (C)	100sm (M)	~x2/month (M)	Crawling (C)
10–100 mm (L)	>1year (L)	Bushy (B)	Suspension feeder (S)		1,000sm (R)	<x2/month (S)	Swimming (S)
100–1,000 mm (X)		Filamentous (F)	Deposit feeder (D)				Drifting (D)
>1,000mm (XX)			Grazer (G)				Burrowing (B)

the functional composition of this kind of community and to evaluate functional redundancy or the potential for competition.

Functional redundancy expressed as the number of species within a given functional group also was similar at the two resolution levels: per panel, 7-trait functional groups on average contained 1.22 (SE 0.05) species, and 4-trait panels 1.33 (SE 0.05) species. Redundancy seems to decrease with spatial scale considered, i.e. with presumed intensification of interactions. This is illustrated by the increasing similarity of the species-, FG7- and FG4-richness values from the global, over the site and to the panel scale (Fig. 1.2). Thus, maximum potential competition, i.e. functional similarity in all traits, seems to be restricted to few interactors at the small scale of a panel. Naturally, competition along single traits is more common. This applies to all sessile species when the resource competed for is substratum. It should be noted, however, that at this functional resolution not all 'suspension feeders', for instance, necessarily compete with each other. A finer subdivision of the trophic type (e.g. Bremner et al. 2003) might reveal that co-occurring species exploit different particle sizes, or filter in different strata or at different times of the day. Also, in relation to their body mass, encrusting forms require more attachment substratum than do bushy forms. Consequently, the former are more sensitive to a scarcity of substratum but, on the other hand, may pre-empt this resource more

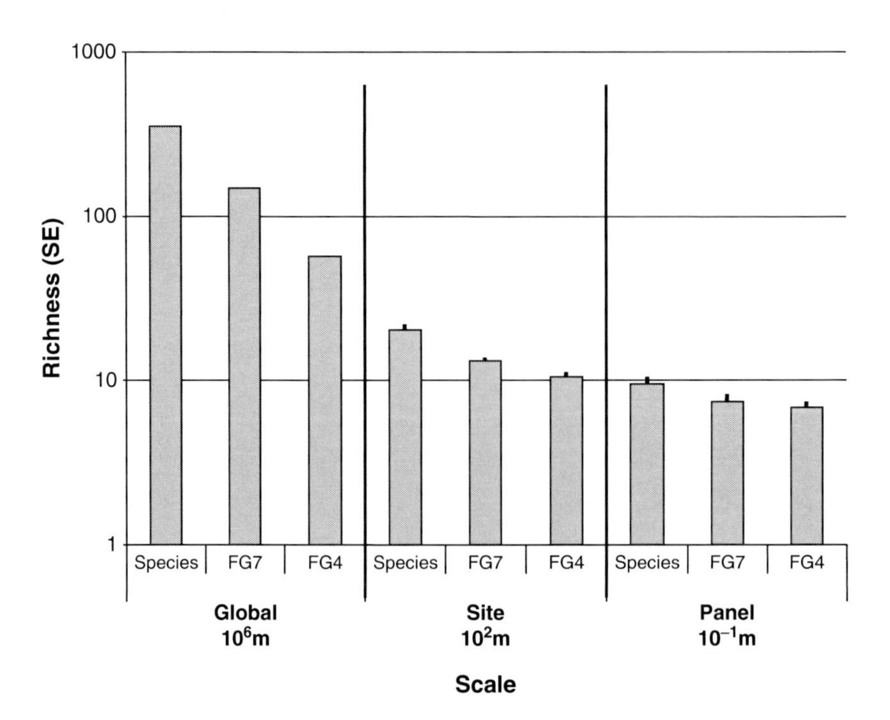

Fig. 1.2 Richness of species numbers, 7-trait and 4-trait functional groups at global, site and panel scale as quantified on 15×15 cm settlement panels exposed for several months on the coasts of New Zealand, Australia, Malaysia, Japan, Brazil, Chile, England, Egypt, Germany and Finland

extensively than do the latter. Clearly, to optimize the usefulness of the functional group approach, the identity and resolution of the life history traits used must be adapted to the questions asked.

Which now are the single functional traits most commonly found in the young shallow-water hard bottom communities investigated? Among the 4-trait functional groups across all panels exposed to colonization in the ten biogeographic regions, the most prevalent properties of the sessile hard bottom species encountered were large size (75% of all functional groups identified) plus massive or encrusting growth form (75%) plus suspension feeding (63%) or autotrophy (33%) plus either colonial (58%) or solitary (42%) life history (Fig. 1.3). The lack of the largest size class may, however, be a bias introduced by limited panel size and duration of the experiment. The lack of predators or grazers is due to the initial decision to consider only sessile species. These most prevalent life history traits are, for instance, encountered in sponges, colonial ascidians, many bryozoans and sea anemones. The combination, however, of common and less common traits makes up the observed considerable functional diversity.

Only few functional groups represented most species and the bulk of the biomass in most regions. Among the commonest were large or medium/massive/suspension-feeding/solitary organisms like mussels or solitary ascidians, large/encrusting or bushy/suspension feeding/colonial species like bryozoans or hydrozoans, or large/bushy or filamentous/autotroph/solitary forms as encountered in numerous algae. Similar patterns were found in more mature communities on rocky substratum (Witman et al. 2004).

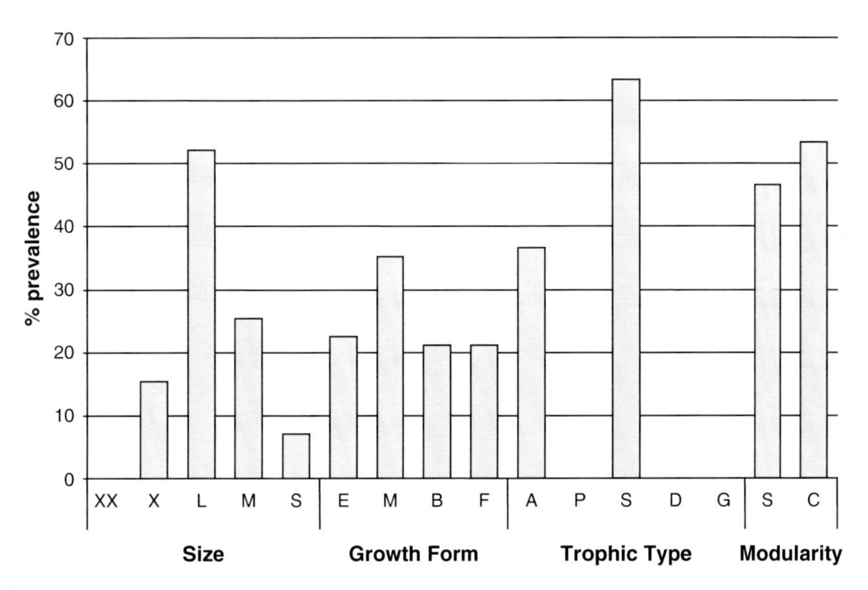

Fig. 1.3 Prevalence of the life history traits encountered in the 356 hard bottom species examined (abbreviations as in Table 1.1)

So, which are the advantages of the functional group approach? In the preceding example, it enabled us to characterize typical life forms of young shallow-water hard bottom communities worldwide, despite extreme taxonomic heterogeneity between the sites. The approach could prove useful to investigate the relative contribution of genotypic (i.e. taxonomic) and phenotypic (i.e. functional) diversity to the structural or functional stability of a community in the face of environmental change. Also, the long-lasting debate about the relationship between local and regional richness (e.g. Cornell et al. 2008) may progress when the ecologically more relevant functional richness is considered at the local scale. Finally, it may be rewarding to verify whether the latitudinal diversity gradient as described for taxonomic richness (e.g. Mittelbach et al. 2007) applies also for the diversity of life history traits and their combinations into functional groups.

References

Bengtsson J (1998) Which species? What kind of diversity? Which ecosystem function? Some problems in studies of relations between biodiversity and ecosystem function. Appl Soil Ecol 10:191–199
Blondel J (2003) Guilds or functional groups: does it matter? Oikos 100:223–231
Bremner J, Rogers SI, Frid CLJ (2003) Assessing functional diversity in marine benthic ecosystems: a comparison of approaches. Mar Ecol Prog Ser 254:11–25
Britton-Simmons KH (2006) Functional group diversity, resource preemption and the genesis of invasion resistance in a community of marine algae. Oikos 113:395–401
Cardinale BJ, Palmer MA, Collins SL (2002) Species diversity enhances ecosystem functioning through interspecific facilitation. Nature 415:426–429
Cornell HV, Karlson RH, Hughes TP (2008) Local-regional species richness relationships are linear at very small to large scales in west-central Pacific corals. Coral Reefs 27:145–151
Duffy JE, Carinale BJ, France KE, McIntyre PB, Thebault E, Loreau M (2007) The functional role of biodiversity in ecosystems: incorporating trophic complexity. Ecol Lett 10:522–538
Fonseca CR, Ganade G (2001) Species functional redundancy, random extinctions and the stability of ecosystems. J Ecol 89:118–125
Halpern BS, Floeter SR (2008) Functional diversity responses to changing species richness in reef fish communities. Mar Ecol Prog Ser 364:147–156
Hillebrand H (2004) On the generality of the latitudinal diversity gradient. Am Nat 163:192–211
Loreau M (2004) Does functional redundancy exist? Oikos 104:606–611
Micheli F, Halpern BS (2005) Low functional redundancy in coastal marine assemblages. Ecol Lett 8:391–400
Mittelbach GG, Schemske DW, Cornell HV, Allen AP, Brown JM, Bush MB, Harrison SP, Hurlbert AH, Knowlton N, Lessios HA, McCain CM, McCune AR, McDade LA, McPeek MA, Near TJ, Price TD, Ricklefs RE, Roy K, Sax DF, Schluter D, Sobel JM, Turelli M (2007) Evolution and the latitudinal diversity gradient: speciation, extinction and biogeography. Ecol Lett 10:315–331
Naeem S, Li S (1997) Biodiversity enhances ecosystem reliability. Nature 390:507–509
Petchey OL, Gaston KJ (2002) Functional diversity (FD), species richness and community composition. Ecol Lett 5:402–411
Petchey OL, Gaston KJ (2006) Functional diversity: back to basics and looking forward. Ecol Lett 9:741–758
Stachowicz JJ, Bruno JF, Duffy JE (2007) Understanding the effects of marine biodiversity on communities and ecosystems. Annu Rev Ecol Evol Syst 38:739–766

Steneck RS, Dethier MN (1994) A functional group approach to the structure of algal-dominated communities Oikos 69(3):476–498
Wahl M, Mark O (1999) The predominantly facultative nature of epibiosis: experimental and observational evidence. Mar Ecol Prog Ser 187:59–66
Witman JD, Etter RJ, Smith F (2004) The relationship between regional and local species diversity in marine benthic communities: a global perspective. Proc Natl Acad Sci 101:15664–15669
Woodward G, Ebenman B, Ernmerson M, Montoya JM, Olesen JM, Valido A, Warren PH (2005) Body size in ecological networks. Trends Ecol Evol 20:402–409
Wright JP, Naeem S, Hector A, Lehman C, Reich PB, Schmid B, Tilman D (2006) Conventional functional classification schemes underestimate the relationship with ecosystem functioning. Ecol Lett 9:111–120

Chapter 2
The Role of Mineral, Living and Artificial Substrata in the Development of Subtidal Assemblages

Andrew R. Davis

2.1 Patterns on Temperate Hard Substrata

Representatives of almost every phylum can be found on or adjacent to hard substrata in the marine environment (Barnes et al. 2001). Certain taxa dominate in this habitat, depending on key environmental gradients including depth, wave exposure, surface orientation and latitude (Witman and Dayton 2001). In the temperate zone, which is the primary focus of this chapter, algae dominate horizontal surfaces, while vertical faces and overhangs are largely the preserve of sessile invertebrates. These systems are often space limited and sessile invertebrates, such as sponges, ascidians and cnidarians, engage in competitive interactions for this limited resource (Jackson 1977; see Fig. 2.1a). A number of sources attest to these patterns (Earl and Erwin 1983; Moore and Seed 1985; Hammond and Synnot 1994; Underwood and Chapman 1995) and it is not my intention to summarise further this literature. In this chapter, I explore how the nature of the substratum affects patterns of colonisation and the development of assemblages. I will restrict this treatment to field-based experiments. My primary focus is the subtidal zone but, where appropriate or where there is a paucity of information on the subtidal zone, I may draw on studies from the rocky intertidal zone.

The development of hard substratum assemblages in the subtidal zone may occur on three types of substratum—mineral (natural rock or surfaces of organic origin: e.g. valves of dead clams), artificial surfaces of anthropogenic origin, and biogenic (living) surfaces. To provide context for this chapter, I start by examining trends in the subtidal literature over the last 25 years (1980–2005)—specifically, I examine temporal trends in (1) the number of publications relating to each substratum type in 5-year increments, (2) the number of publications focussing on algae relative to sessile invertebrates over the same time increments and (3) the geographic locations at which the research was done over the entire 25-year period. Two databases were searched, 'Biosis' and 'Biological Abstracts', and more detail on the search criteria can be found in the caption of Fig. 2.2. Overall, four patterns

M. Wahl (ed.), *Marine Hard Bottom Communities*, Ecological Studies 206,
DOI: 10.1007/978-3-540-92704-4_2, © Springer-Verlag Berlin Heidelberg 2009

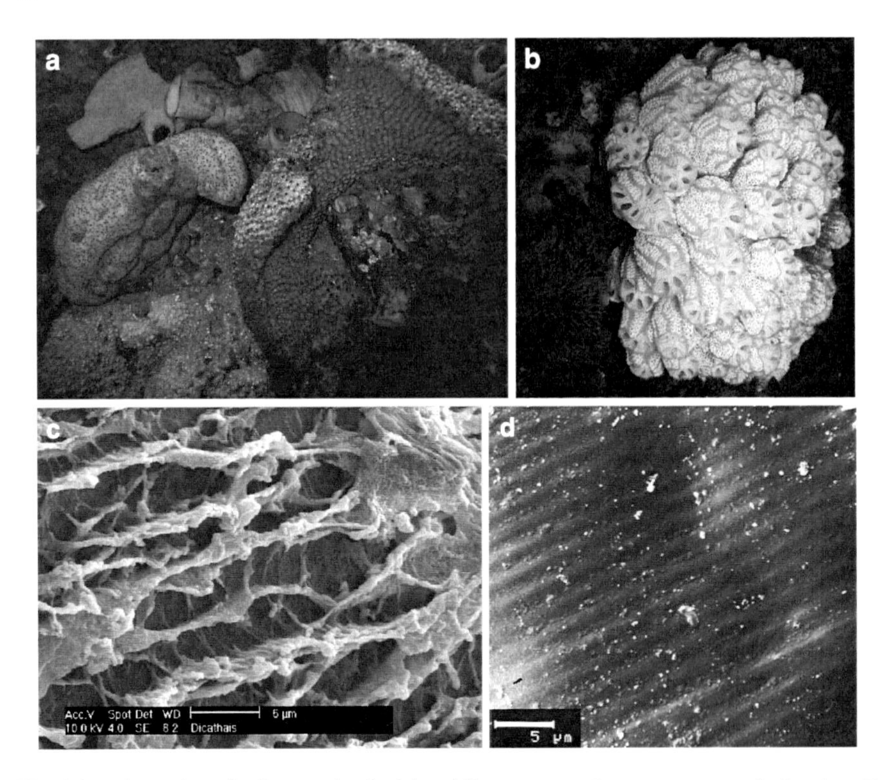

Fig. 2.1 **a** A number of solitary and colonial ascidians compete for space on a vertical rock wall in southern New South Wales, Australia. Species include *Polyandrocarpa lapidosa*, *Phallusia obesa*, *Cnemidocarpa pedata*, *Herdmania grandis*. Photo A.R. Davis. **b** A mature colony of the ascidian *Botrylloides magnicoecum* displaying no evidence of fouling organisms on its surface. Photo S. Schulz. **c** Micro-topography on the surface of egg masses of the muricid gastropod *Dicathais orbita*. Photo K. Benkendorff. **d** Microstructuring on the periostracum of the mytilid *Mytilus* sp. Photo A.V. Bers

emerge—first, the majority of work has focussed on natural rocky substrata across all time increments, with the relative attention on rocky substrata increasing markedly in the last time increment (Fig. 2.2a). Second, there has been an apparent surge in interest in colonisation of biogenic substrata in the last time increment (2001–2005), reaching almost 25% of all studies in that time increment (Fig. 2.2a). Third, overwhelmingly attention has been focussed on sessile invertebrates across all time periods (Fig. 2.2b). Finally, patterns among the three substratum types within each geographic region were very similar, with two exceptions: artificial substrata were poorly represented in North American studies but were overrepresented in Asia (Fig. 2.2c). What these data cannot reveal is the relative impact of research in these regions, because many of the conceptual models and the data that inspired or supported these are derived from the well-studied coasts of North America and Europe.

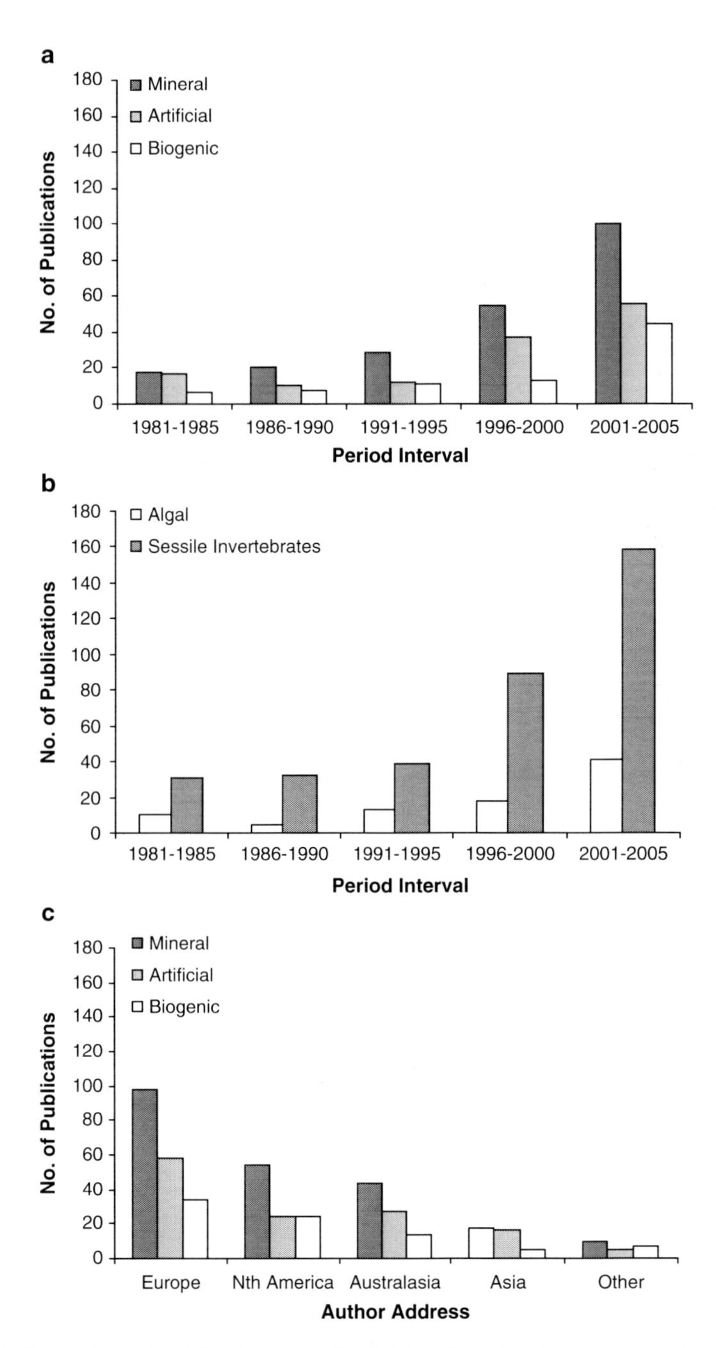

Fig. 2.2 Comparison of publication output for field studies of hard bottom assemblages in the temperate subtidal zone for **a** three types of substrata—mineral (natural reef), artificial surfaces and biogenic; **b** algal vs. sessile invertebrates and **c** geographic location. Data were sourced worldwide for a 25-year period (1981–2005) from *Biological Abstracts* and *Biosis Previews*. Information was gleaned from the abstracts of more than 400 articles obtained in the searches. If an article fell into more than one category (e.g. natural and artificial substrata), then each category was given 0.5 publications. In this way, publications were only 'recorded' once

2.2 The Colonisation Process

New surfaces placed in the marine environment rapidly develop a biofilm, which is then followed by the attachment, metamorphosis and growth of algal and invertebrate taxa (see reviews by Davis et al. 1989; Wahl 1989). The colonisation process can be highly unpredictable and is usually characterised by significant spatial and temporal variation (e.g. Sutherland and Karlson 1977; Keough 1984; Underwood and Anderson 1994). The development of assemblages on hard substrata is underpinned by a combination of pre- and post-settlement processes. I consider settlement and post-settlement interactions in the next section but it is appropriate to start by focussing on pre-settlement determinants of patterns of colonisation.

Compelling experimental evidence from the intertidal region indicates that the number of larvae encountering a suitable surface will determine the number of settlers and that larval supply is one of the key determinants of colonisation pattern (e.g. Grosberg 1982; Gaines and Roughgarden 1985; Gaines et al. 1985; Minchinton and Scheibling 1991). As far as I am aware, there are no subtidal zone examples linking larval supply to patterns of colonisation but some information on patterns of larval distribution is available for the subtidal zone. Larval availability will clearly be a function of the number of competent larvae in the water column. It will also be a function of the rate at which larvae encounter surfaces and, thus, of the volume of water 'sampled' by a given substratum. Larvae present at relatively low densities may still encounter surfaces frequently in areas of high flow (Keough 1983). As many propagules show a negative exponential decay with distance from the parent (Harper 1977), combined with the observation that some invertebrate larvae disperse surprisingly short distances prior to settlement (Grosberg 1987; Davis and Butler 1989), an elevated colonisation rate would be anticipated with proximity to the larval pool, i.e. the parental habitat (MacArthur and Wilson 1967). Invertebrate larvae show this pattern near subtidal rock walls, with larval densities of many philopatric species declining dramatically even quite small distances away from the faces of rock walls (Graham and Sebens 1996). In addition, in a challenging project requiring the identification of sponge larvae to species level, Mariani et al. (2005) observed that the density of sponge larvae in the water column immediately above the substratum was a positive function of the cover of each sponge species.

2.3 The Role of Substrata in the Colonisation and Development of Assemblages

The small size of most larvae and the difficulties associated with examining these in situ make realistic examinations of their interactions with surfaces challenging, prompting Young (1990) to liken the process to "cracking a black box". Consequently, much of our understanding of larval responses to surfaces is derived from (1) laboratory observations, which may modify the very behaviour we are seeking to examine, or (2) the use of settlement panels deployed for short periods of time in the field. There are inherent risks in these approaches, given that larvae are likely to be responding to a variety of cues (Crisp 1976; Pawlik 1992), many of which are poorly understood and difficult to simulate in the laboratory. Nevertheless, behavioural observations at very small scales have underscored the complexity of the settlement process (Walters et al. 1999).

I now focus on the outcome of field experiments that examine the colonisation and development of assemblages on different substrata. I will take each substratum type in turn.

2.3.1 *Artificial Substrata*

Artificial substrata have played a key role in advancing our understanding of ecological pattern and process. At small scales, settlement panels are convenient and consistent surfaces on which to examine the responses and timing of larval settlement into subtidal systems, while larger surfaces such as pier pilings and breakwaters present an opportunity to explore pattern and process at larger spatial and temporal scales. Here, I use the definition of artificial substrata advocated by Svane and Petersen (2001) in their review "Submerged man-made structures susceptible to fouling". The relative consistency of PVC, ceramic or glass surfaces has meant that much of the variation due to the nature of the substratum could be removed from examinations of spatial and temporal variation in settlement and recruitment. This has proven to be a powerful technique when comparisons are drawn within a substratum type (see Brown 2005).

A clear outcome of this settlement panel research is the important role that surface rugosity plays in the settlement or post-settlement survivorship of algae and invertebrates (e.g. Harlin and Lindbergh 1977; Keough and Downes 1982; Hills and Thomason 1996). The propensity of larvae to settle in concavities is particularly well documented for barnacles, with smooth surfaces receiving the lowest number of settlers (e.g. Crisp and Barnes 1954; Wethey 1986; Hills and Thomason 1996). Importantly, concavities of a scale similar to that of settling organisms provide settlement cues for a range of phyla (Chabot and Bourget 1988; Bourget et al. 1994; Walters and Wethey 1996; Hills and Thomason 1998). Furthermore, the development of settlement panels that are accurate to the micron scale, identical and can be produced in sufficient numbers to give good levels of replication has enabled the interactions among settlers and the structure of the surfaces they contact to be explored more fully (Hills et al. 1999; Köhler et al. 1999). Laboratory and field experiments using surfaces of standardised rugosity (glass beads of a range of diameters embedded into a latex lacquer) in a relatively impoverished Baltic Sea assemblage have revealed highly complex outcomes that are extremely difficult to predict (Wahl and Hoppe 2002). Surface rugosity (bead diameter) interacted with settlement density of invertebrates and the activities of grazers. Although the complexity of these interactions may be daunting in a natural setting, a better understanding of settlement responses and of how they are influenced by hydrodynamics is crucial as we move away from broad-spectrum, highly toxic antifouling paints to the development of low-toxicity or non-toxic antifoulants (Andersson et al. 1999; Köhler et al. 1999).

Clearly, research on settlement panels does have limitations. Research is usually focussed on single species at small spatial and temporal scales. Indeed, it is rare for studies to be on panels larger than 25×25 cm, and for panels to be deployed for periods longer than 1 year or at more than a couple of locations. There are a number of noteworthy exceptions (Kay and Keough 1981; Caffey 1985; Hughes et al. 1999). In general, assemblages appear to take much longer to develop than the timeframe over which panels are usually deployed; Butler and Connolly (1999) quantified the establishment of a fouling community on a large steel pier—they were not convinced that the assemblage had reached a climax after 13.5 years. Davis and Ward (2009) contrasted experimental plots (0.18 m^2) with natural areas of rock wall and noted marked differences in the cover of invertebrates and statistically significant differences in the composition of the assemblage after almost 5 years of monitoring. The timeframe of settlement studies may be sufficient to detect patterns in the timing of recruitment and the interactions among recruits but timescales for the development of climax assemblages are significantly longer.

Artificial reefs offer an opportunity to explore ecological issues and the development of assemblages at a much broader scale. Although artificial reefs have been used extensively for more than three decades, they are attracting much more interest as they are seen as a means of increasing fisheries production, restoring degraded habitat, improving water quality and maintaining biodiversity (Baine 2001; Svane and Petersen 2001). Their potential utility as experimental models has also been emphasised (Butler and Connolly 1999; Miller 2002).

Larger-scale patterns of colonisation are usually examined on piers and breakwaters (e.g. Knott et al. 2004). Sunken ships are another means of examining relatively large-scale pattern, although the likely confounding effects of the antifouling applied to the lower sections of their hulls while the vessel was in service need to be considered. While most of the attention on artificial reefs is directed at fish or fished species, a small number of studies have explored patterns of algal and invertebrate colonisation (Carter et al. 1985; Bailey-Brock 1989; Baynes and Szmant 1989; Relini et al. 1994). The assemblages that develop on artificial reefs are for the most part quite distinct from those in natural areas of reef. This is not unexpected, given that they are of different ages, composed of different materials and are often physically arranged quite differently. The heterogeneity associated with artificial reefs not only provides suitable habitat for a range of sessile organisms but may also modify patterns of distribution and abundance of their consumers. For example, artificial reefs have higher fish biomass than do natural reefs (reviewed by Bohnsack 1991) and teleosts may exert a strong influence on the assemblage that develops (Russ 1980; Keough 1984).

2.3.2 Mineral Substrata

The colonisation of natural mineral surfaces will generally be much more complex than that occurring on settlement panels, as larvae and propagules must contend with an existing assemblage of macro-organisms, some of which will be competitors or predators (e.g. Dean and Hurd 1980). The focus of most studies examining subtidal assemblages has been on seasonal and spatial patterns (e.g. bathymetry) (e.g. Ojeda and Dearborn 1989; Newton et al. 2007). It is generally assumed that rock type has little impact on hard bottom assemblages and, consequently, this question has received scant attention. A small number of well-replicated field studies have examined this question in the intertidal zone, with contradictory outcomes. Caffey (1982) focussed on the responses of the barnacle *Tesseropora rosea* to four rock types (gabbro, mudstone, sandstone and shale) and failed to detect consistent differences in settlement and survival. McGuinness (1989), in reinterpreting Caffey's findings, noted that rock type had an effect but that this was simply variable in space and time. In contrast to Caffey's (1982) findings, Raimondi (1988) reported enhanced settlement and survivorship of *Chthamalus anispoma* on natural granite and pitted basalt relative to natural basalt and cut (smooth) granite. He concluded that the differences in survivorship were due to the contrasting thermal properties of the light-coloured granite versus dark-coloured basalt, resulting in an elevated upper limit of this barnacle on the granite shore. McGuinness and Underwood (1986) contrasted the inhabitants of sandstone and shale boulders following a reciprocal transplant between two shores, and emphasised the importance of microhabitats in each rock type in providing refugia for consumers that then modified the structure of the assemblages. All of these workers and more recent studies examining rock type (e.g. Herbert and Hawkins 2006) emphasised the potential for surface 'texture' (defined by Crisp 1974 as small-scale surface relief) to affect settlement; generally, the more pitted surfaces were preferred. McGuinness and Underwood (1986) also emphasised the importance of surface contour (defined by Crisp 1974 as surface features that are large relative to the scale of larvae, such as cracks and holes) in providing refugia for grazers that, in turn, modified the structure of the assemblage. It appears then that where rock type has an effect in the intertidal zone, it is due largely to the amelioration of abiotic stressors. These stressors do not operate in the subtidal zone and, hence, the effect of rock type on patterns of colonisation and the development of assemblages should be limited, although this has only recently been examined formally (see below). There are instances where subtidal rocks may affect the development of assemblages; Huxley et al. (1984) report enhanced settlement on shale that contained settlement inducers, while storms have been reported to remove organisms from friable mudstones, periodically creating new patches of habitable space (Foster 1982). However, such examples appear rare.

Field experiments indicate that microflora or bioorganic films can play an important role in modifying patterns of settlement of sessile invertebrates (Todd and Keough 1994; Keough and Raimondi 1995, 1996) and there is some evidence that biofilms may be specific to biogenic substrata (e.g. Gil-Turnes et al. 1989).

However, comparisons among natural rock substrata in the marine environment have not detected significant differences in biofilm composition (Faimali et al. 2004; Totti et al. 2007). Faimali et al. (2004) reported statistically significant differences in the biofilm on glass relative to three other substrata. Importantly, they did not detect differences between the two natural substrata tested: marble and quartz. Totti et al. (2007) failed to detect differences in the colonisation of epilithic diatoms on quartzite, slate and marble that had been polished to remove the likely confounding effect of surface roughness.

Given the rapidity of biofilm formation, it has always been assumed that larva will come into contact with surface biofilms, rather than make direct contact with natural rock surfaces (Wahl 1989). Consequently, the crystallographic structure of rock has not been considered an important structuring agent in subtidal assemblages—surface roughness has always taken centre stage. Indeed, comparisons between artificial substrata (concrete) and sandstone, although restricted to sites in SE Australia, failed to detect meaningful differences between substrata (Connell 2000; Glasby 2000; Knott et al. 2004). Evidence is mounting, however, that mineralogy may indeed play a structuring role in subtidal epibenthic assemblages (Cerrano et al. 1999). Oyster spat settle more heavily on calcium-rich limestone than on quartzite (metamorphosed sandstone) and, although Soniat and Burton (2005) consider this "likely" to be due to the chemical composition of the substratum, they did not rule out differences in the presence or composition of biofilms between the two substrata. Indeed, Zobell (1939) argued that biofilms might promote invertebrate settlement by increasing alkalinity at the surface of the substratum. Bavestrello et al. (2000) provide evidence that quartz-rich rocks and sands (associated with granitic rocks) have negative effects on species richness and the cover of assemblages at three locations on the Italian and Sardinian coast. Distinct assemblages formed on the quartzite rocks relative to sandstones, marl limestone and puddingstone; the latter three lack significant quartz components. Laboratory experiments revealed that planulae of a hydroid avoided quartzose sands relative to marble sands; however, no differences in metamorphosis were apparent. It contrast to the assertions of Bavestrello et al. (2000), negative effects of quartz were not apparent in several laboratory and field studies. Field settlement of the barnacle *Balanus balanoides* examined daily over 14 days was significantly higher on quartz than on most of the 14 other rock types tested by Holmes et al. (1997). The formation of laboratory biofilms and the subsequent settlement of the barnacle *Balanus amphitrite* (Faimali et al. 2004) or the colonisation by epilithic diatoms (Totti et al. 2007) was not depressed on quartz. Groppelli et al. (2003) reported a positive effect of quartz in the laboratory settlement and area of attachment for the ascidian *Phallusia mammillata.* Schiaparelli et al. (2003) demonstrated that the vermetid gastropod *Dendropomapetraeum* was significantly more abundant on granite (quartz-rich) shores than on those composed of limestone. Another vermetid species, *Vermetus triquetrus*, showed no such pattern.

Guidetti et al. (2004) provided further correlative support for the suggestion that mineralogy affects community structure, with a focus on fish, algal and sessile invertebrate assemblages on granite and limestone outcrops on the Sardinian coast. Their nMDS ordinations revealed striking (although not statistically significant) differences between fish and epibenthic assemblages on each rock type. Univariate analyses confirmed statistically significant differences for several taxa. In exploring the mechanisms underlying these patterns, they did not detect differences in slope, rugosity and boulder size for each rock type at their study sites, and argued that differences were due rather to the formation of silicon-based and OH radicals in the surrounding water. Cattaneo-Vietti et al. (2002) extended this work to include non-granitic shores. Although rather poorly replicated, they detected statistically significant differences between epibenthic guilds, with total abundance highest for sandstone and basalt relative to serpentine rocks. These Italian studies, although often with contradictory outcomes, argue that the effect of rock type on benthic community structure deserves experimental attention.

Clearly, there are difficulties disentangling the features of natural substrata that act to determine the structure of benthic communities. Surface roughness, surface texture and the crystallographic structure of rocks are likely to be confounded. In turn, these features of the substratum may also affect the bioorganic film that develops. Consequently, effects of the substratum on settling larvae or algal propagules may be direct or indirect, and how they respond to these cues across a variety of spatial and temporal scales constitutes a significant challenge to marine ecologists.

It is clear that rock type plays a larger role than the literature attests in structuring subtidal assemblages. The surface texture and contour (as defined above) of the substratum are related to rock type. Rock type affects the degree to which cracks and crevices may form from fracturing, as well as the size of resultant cobbles and boulders. Igneous rocks (basalt) usually form numerous cracks from natural joint sets, and crevices at the 10–50 cm scale are common, producing cobbles and boulders of these dimensions (L. Moore, personal communication). Igneous rock also commonly produces featureless expanses of substratum (low-profile reef; personal observation). Granites tend to form orthogonal jointing at the scale of 1–3 m, associated with rounded boulders of these dimensions. In addition, minerals such as feldspar are weathered from the surface of granite, leaving a pockmarked, etched surface texture at the scale of mm, comprised of quartz (Twidale 1982). Bedded sediments such as sandstone commonly form undercut walls and caves (Young and Young 1992). The fracturing of the substratum will contribute to the heterogeneity on a reef, and has the potential to provide refuge for consumers that, in turn, may modify the assemblage that forms. For example, the SE Australia urchin barrens are dominated by the large diadematid sea urchin *Centrostephanus rodgersii*. This urchin is strongly dependent on shelter during daylight hours; urchins removed from crevices or irregularities in the rock are attacked and rapidly consumed by the large wrasse *Achoerodus viridis* (personal observation). The nocturnal foraging activities of these urchins modify the algal and invertebrate assemblages in close proximity to crevices, producing pavements of grazer-resistant crustose coralline algae (Fletcher 1987; Davis et al. 2003). In an impressive demonstration of the significance of substratum heterogeneity in these trophic interactions, involving the use of ‘lift bags’, Andrew (1993) introduced large boulders into kelp forests, providing shelter for urchins and producing a shift from a kelp-dominated assemblage to one dominated by crustose coralline algae.

Rock type may also affect the likelihood that vertical faces will form; fine-grained sedimentary rock (sandstone) and hard metamorphic rock such as granite are more likely to support vertical walls (Young and Young 1992; Huggett 2007). Vertical surfaces are associated with elevated covers and diversity of sessile invertebrates in part because sediment tends not to accumulate on vertical surfaces (Moore 1977). Vertical surfaces also represent a significant impediment to grazing sea urchins (Sebens 1985). In SE Australia, low walls (<2 m) are grazed extensively by large urchins and these remove a high proportion of invertebrates and algae, leaving surfaces dominated by grazer-resistant crustose coralline algae (Davis et al. 2003). In contrast, high walls are dominated by sessile invertebrates, usually exceeding 95% cover. It appears that there is considerable biogeographic variation in this pattern, as using the same sampling methodology in Mediterranean Spain revealed no such disparity between the assemblages on short and tall walls (Davis et al. 2003).

The other mineral surfaces that larvae often colonise are the mineralised external shells of bivalves and crustaceans. Although these organisms are alive, their surfaces are inorganic, which is why I have chosen to discuss these in this section. The small patch sizes of these surfaces means they are more likely to be colonised by abundant larvae such as those of bryozoans, barnacles and tubeworms (Kay and Keough 1981; Keough 1984) but, occasionally, patches are colonised and subsequently dominated by competitively superior colonial ascidians or sponges (Keough 1984). The provision of these small patches of hard substrata may enable species normally associated with hard substrata to persist in areas dominated by unconsolidated sediments (Kay and Keough 1981) or may prove to be a focus for the development of assemblages normally associated with hard substrata. In some instances, the host (basibiont) derives a benefit from the presence of the fouler (Pitcher and Butler 1987) but, usually, the effects are negative (e.g. Wahl 1997; Branch and Steffani 2004). Mineralised surfaces may not always form isolated patches; oyster reefs form in estuaries, and the physical arrangement of the reef can affect flow and, in turn, the mortality of reef-associated organisms as well as the individual performance of oyster recruits (Lenihan and Peterson 1998; Lenihan 1999).

2.3.3 Biogenic Living Surfaces

Unless protected, hard surfaces exposed in the marine environment rapidly become fouled. Living surfaces can form an important focus for a variety of sessile organisms and, although rates of epibiosis are usually low (e.g. Davis and White 1994; see Fig. 2.1b), in some circumstances assemblages of sessile flora and fauna may form (Farnsworth and Ellison 1996). Generally, there are numerous disadvantages for the organism being settled upon—the basibiont. These may include the restriction of photosynthesis, the occlusion of feeding structures or an increased likelihood of dislodgement under high flow or shear conditions (Wahl 1989; Davis et al. 1989). Using chemical means, basibionts often seek to restrict foulers (see reviews by Wahl 1989; Davis et al. 1989; Pawlik 1992; Davis and Bremner 1999), although alternative approaches such as the sloughing of fouled tissue (reviewed by Davis et al. 1989) and the pre-emptive establishment of benign foulers such as the thin investing sponge on the stalked ascidian *Pyura spinifera* (Davis et al. 1996) have also been reported.

Topographic complexity also affects establishment on biogenic substrata but the outcome appears to be determined by the growth form of the settlers. Walters and Wethey (1996) predicted that settling larvae of species with limited attachment to the substratum would settle in topographically complex refuges—creases in the test of the solitary ascidian *Styela plicata*—while settlers with large areas of attachment would not. Their predictions were borne out but more recent examination of interactions between settlers and the complexity of biogenic surfaces has revealed an unexpected relationship. It is now apparent that a variety of organisms, including mussels, echinoderms, molluscs (egg cases) and even cetaceans, possess microstructure at the micron scale and that this micro-topography dissuades the settlement of micro- and macrofoulers (Baum et al. 2003; Bers and Wahl 2004; Bers et al. 2006; Lim et al. 2007; see Fig. 2.1c, d). In the case of mytilids, the microstructuring of the periostracum is apparent across genera and has proven effective against barnacle larvae (Bers et al. 2006).

Given that most biogenic surfaces possess some means of defence, direct fouling of surfaces is much less significant to the structure of hard substratum assemblages than is the role played by organisms as ecosystem engineers (sensu Jones et al. 1994). The physical modification of habitats by organisms, such as the presence of clump-forming mussels, beds of turfing algae or groups of solitary ascidians, provides important structure that acts to reduce wave shock, provide predictable escapes in space and time from predators as well as to ameliorate physiologically stressful conditions (Witman 1985; Ojeda and Dearborn 1989; Bertness and Leonard 1997). The removal or exploitation of these habitat-forming species has important consequences for biodiversity (Coleman and Williams 2002). Aggregations of organisms will also increase the structural heterogeneity of habitats (Bell et al. 1991). By providing structure around which drift algae and other detritus may accumulate or become entangled (e.g. Bell et al. 1995), additional resources may be offered to distinct guilds of organisms, e.g. detritivores.

The presence of organisms may also modify patterns of settlement and survivorship for subsequent colonists (Dean 1981; Wethey 1986; Petraitis 1990). Usually, the effects of an established fauna on later colonists are negative; in some instances, larvae may avoid settling at locations where superior competitors or predators have already established (Grosberg 1981; Young and Chia 1981; Young 1989a) or, alternately, larvae may be consumed by the established fauna (e.g. Olson and McPherson 1987; Young and Gotelli 1988; Davis and Butler 1989; but see Young 1989b). These negative effects are underscored by the interactive model of species accumulation; under this model, colonisation rates are depressed and, indeed, pre-existing occupants may pre-empt space (e.g. Underwood and Anderson 1994). Similarly, in a developing fouling assemblage Dean and Hurd (1980) reported that "inhibitory interactions appeared to be much more prevalent than facilitative interactions". In general, positive interactions among fauna have not been viewed as playing a major role in community organisation (Karieva and Bertness 1997); yet, it is clear that they occur in the subtidal zone. The well-characterised positive effects of barnacles on settling conspecifics is rather a special case, given the reproductive imperative of cross-fertilisation. The attraction of mussels to filamentous structures sees enhanced settlement on the byssal threads of conspecifics (de Blok and Geelen 1958).

Positive interactions among allo-specific species are less common. Davis (1996) interpreted the reliance of the stalked ascidian *Pyura spinifera* on the unstalked *Cnemidocarpa pedata* as a cue for settling at locations relatively free of grazing urchins. He provided experimental field evidence that *P. spinifera* settled only on *C. pedata*, and used the observation that *C. pedata* was not observed in urchin-grazed barrens to imply that *P. spinifera* was seeking 'enemy free space'. Some organisms may also provide associational defence where predators avoid species such as chemically defended sessile invertebrates that, in turn, provide a degree of protection for neighbouring species (Littler et al. 1986). All of these interactions with living organisms have the potential to influence the development of an assemblage.

2.4 Future Focus

For reasons of cost and convenience, most attention in the subtidal zone has been focussed on relatively shallow water, generally <20 m depth but often less than 10 m. Do the patterns observed in the shallows and the processes that underpin these hold in deeper water? Patterns in deepwater assemblages have received scant attention, let alone the examination of process (but see Witman and Sebens 1990; Maldonado and Young 1998). Increasingly, technology is being used to explore ecological pattern and process at depths below those considered safe for SCUBA. Examining patterns is an important precursor to understanding process (Lawton 1996) and, for hard substrata, this is relatively easily and cheaply achieved remotely by means of 'jump' cameras (e.g. Roberts et al. 1994; Roberts 1996) or acoustic surveys (Jordan et al. 2005). However, establishing experimental manipulations to explore process is significantly more complex and expensive, usually requiring remotely operated vehicles (ROVs) and, more recently, automated underwater vehicles (AUVs). One viable option is to explore patterns in deep water and then conduct experiments in the shallows (Roberts et al. 2006), although clearly these outcomes need to be interpreted with caution. The challenge for the future will be to develop means of exploring process in deeper water.

There also may be risks in drawing general conclusions from a limited number of studies. For example, Witman and Sebens (1990) examined patch dynamics in deep water and noted that patches of substratum were slow to close, owing to the slow growth of the dominant sponge *Hymedesmia* sp. Some temperate zone sponges in shallow water also grow slowly (Ayling 1983) but this is not a general outcome, as some temperate species can effect patch closure very rapidly (Kay and Keough 1981).

Determining the appropriate ecological scale at which to work represents a significant challenge to ecologists. As I have emphasised, a large body of evidence from experiments with settlement panels attests to the importance of surface texture in enhancing settlement and survivorship in pits and crevices but it remains unclear whether these small-scale effects are as significant in structuring communities on a broad scale. Indeed, the outcomes of large-scale experiments are often difficult to predict from small-scale ones. These emergent properties represent a significant challenge to our understanding, and two recent examples underscore this. Large-scale fish exclosures on a near-shore tropical reef produced an unprecedented profusion of tropical algae standing several metres high (Hughes et al. 2007), a pattern that would not have been observed in small exclosures. In a temperate example, Babcock et al. (1999) describe a trophic cascade following the closure of a 5-km stretch of the northern New Zealand coastline to human collecting in the mid-1970s. This was unexpected because small-scale manipulative experiments provided no suggestion that this would occur (Andrew and Choat 1982). Clearly, neither of these outcomes was or could be predicted from the use of small-scale experiments, arguing that as ecologists we need to think big, as do funding agencies!

Acknowledgements I thank Martin Wahl for the opportunity to contribute this chapter and his patience when it was slow to materialise. He, Todd Minchinton and Kirsten Benkendorff provided insightful comments that substantially improved an early draft. Discussions with several geologists were useful in improving my understanding of how organisms may interact with rock type—special thanks to Leah Moore and Colin Murray-Wallace in this regard. Without the able assistance of Allison Broad this chapter would have been much more difficult to assemble, although any omissions or errors are my own. This represents contribution number 283 from the Ecology and Genetics Group at the University of Wollongong, Australia.

References

Andersson M, Berntsson K, Jonsson P, Gatenholm P (1999) Microtextured surfaces: towards macrofouling resistant coatings. Biofouling 14:167–180

Andrew NL (1993) Spatial heterogeneity, sea urchin grazing, and habitat structure on reefs in temperate Australia. Ecology 74:292–302

Andrew NL, Choat JH (1982) The influence of predation and conspecific adults on the abundance of juvenile *Evechinus chloroticus* (Echinoidea: Echinodermata). Oecologia 54:80–87

Ayling AL (1983) Growth and regeneration rates in thinly encrusting demospongiae from temperate waters. Biol Bull 165:343–352

Babcock RC, Kelly S, Shears NT, Walker JW, Willis TJ (1999) Large-scale habitat change in a temperate marine reserve. Mar Ecol Prog Ser 189:125–134

Bailey-Brock JH (1989) Fouling community development on an artificial reef in Hawaiian waters. Bull Mar Sci 44:580–591

Baine M (2001) Artificial reefs: a review of their design, application, management and performance. Ocean Coastal Manag 44:241–259

Barnes RSK, Calow P, Olive PJW, Golding DW, Spicer JI (2001) The invertebrates: a synthesis, 3rd edn. Blackwell, Oxford

Baum C, Simon F, Meyer W, Fleischer LG, Siebers D, Kacza J, Seeger J (2003) Surface properties of the skin of the pilot whale *Globicephala melas*. Biofouling suppl 19:181–186

Bavestrello G, Bianchi CN, Calcinai B, Cattaneo - Vietti, Cerrano C, Morri C, Puce S, Sara M (2000) Bio-mineralogy as a structuring factor for marine epibenthic communities. Mar Ecol Prog Ser 193:241–219
Baynes TW, Szmant AM (1989) Effects of current on the sessile benthic community structure of an artificial reef. Bull Mar Sci 44:546–566
Bell SS, McCoy ED, Mushinsky HR (1991) Habitat structure: the physical arrangement of objects in space. Chapman and Hall, London
Bell SS, Hall MO, Robbins BD (1995) Toward a landscape approach in seagrass beds: using macroalgal accumulation to address questions of scale. Oecologia 104:163–168
Bers AV, Wahl M (2004) The influence of natural surface microtopographies on fouling. Biofouling 20:43–51
Bers AV, Prendergast GS, Zurn CM, Hansson L, Head RM, Thomason JC (2006) A comparative study of the anti-settlement properties of mytilid shells. Biol Lett 2:88–91
Bertness MD, Leonard GH (1997) The role of positive interactions in communities: lessons from intertidal habitats. Ecology 78:1976–1989
Bohnsack JA (1991) Habitat structure and the design of artificial reefs. In: Bell SS, McCoy ED Mushinsky HR(eds) Habitat structure: the physical arrangement of objects in space. Chapman and Hall, London, pp 412–426
Bourget E, DeGuise J, Daigle G (1994) Scales of substratum heterogeneity, structural complexity, and the early establishment of a marine epibenthic community. J Exp Mar Biol Ecol 181:31–51
Branch GM, Steffani CN (2004) Can we predict the effects of alien species? A case-history of the invasion of South Africa by *Mytilus galloprovincialis* (Lamarck). J Exp Mar Biol Ecol 300:189–215
Brown CJ (2005) Epifaunal colonization of the Loch Linnhe artificial reef: influence of substratum on epifaunal assemblage structure. Biofouling 21:73–85
Butler AJ, Connolly RM (1999) Assemblages of sessile marine invertebrates: still changing after all these years? Mar Ecol Prog Ser 182:109–118
Caffey HM (1982) No effect of naturally-occurring rock types on settlement or survival in the intertidal barnacle Tesseropora rosea (Krauss). J Exp Mar Biol Ecol 63:119–132
Caffey HM (1985) Spatial and temporal variation in settlement and recruitment of intertidal barnacles. Ecol Mongr 55:313–332
Carter JW, Carpenter AL, Foster MS, Jessee WN (1985) Benthic succession on an artificial reef designed to support a kelp-reef community. Bull Mar Sci 37:86–113
Cattaneo-Vietti R, Albertelli G, Bavestrello G, Bianchi CN, Cerrano C, Chiantore M, Gaggero L, Morri C, Schiaparelli S (2002) Can rock composition affect sublittoral epibenthic communities? Mar Ecol 23:65–77
Cerrano C, Arillo A, Bavestrello G, Benatti U, Calcinai B, Cattaneo-Vietti R, Cortesogno L, Gaggero L, Giovine M, Puce S, Sara M (1999) Organism-quartz interactions in structuring benthic communities: towards a marine biomineralogy? Ecol Lett 2:1–3
Chabot R, Bourget E (1988) Influence of substratum heterogeneity and settled barnacle density on the settlement of cypris larvae. Mar Biol 97:45–56
Coleman FC, Williams SL (2002) Overexploiting marine ecosystem engineers: potential consequences for biodiversity. Trends Ecol Evol 17:40–44
Connell SD (2000) Floating pontoons create novel habitats for subtidal epibiota. J Exp Mar Biol Ecol 247:183–194
Crisp DJ (1974) Factors influencing the settlement of marine invertebrate larvae. In: Grant PT, Mackie AN(eds) Chemoreception in marine organisms. Academic Press, London, pp 177–215
Crisp DJ (1976) Settlement responses in marine organisms. In: Newell RC(ed) Adaptation to environment: essays on the physiology of marine animals. Butterworth, London, pp 83–124
Crisp DJ, Barnes H (1954) The orientation and distribution of barnacles at settlement with particular reference to surface contour. J Anim Ecol 23:142–162

Davis AR (1996) Association among ascidians: facilitation of recruitment in *Pyura spinifera*. Mar Biol 126:35–41

Davis AR, Bremner J (1999) Potential antifouling natural products from ascidians: a review. In: Fingerman M, Nagabhushanam R, Thompson MF(eds) Recent Advances in Marine Biotechnology, vol III. Biofilms, bioadhesion, corrosion and biofouling. Science Publishers, Enfield, NH, pp 259–308

Davis AR, Butler AJ (1989) Direct observations of larval dispersal in the colonial ascidian *Podoclavella moluccensis* Sluiter: evidence for closed populations. J Exp Mar Biol Ecol 127:189–203

Davis AR, Ward DW (2009) The establishment and persistence of species rich patches in a species poor landscape: role of a structure-forming subtidal barnacle. Mar Ecol Prog Ser (in press)

Davis AR, White GA (1994) Epibiosis in a guild of sessile subtidal invertebrates in south-eastern Australia: a quantitative survey. J Exp Mar Biol Ecol 177:1–14

Davis AR, Targett NM, McConnell OJ, Young CM (1989) Epibiosis of marine algae and benthic invertebrates: natural products chemistry and other mechanisms inhibiting settlement and overgrowth. In: Scheuer PJ (ed) Bioorganic Marine Chemistry, vol 3. Springer, Berlin Heidelberg New York, pp 85–114

Davis AR, Ayre DJ, Billingham MR, Styan CA, White GA (1996) The encrusting sponge *Halisarca laxus*: population genetics and association with the ascidian Pyura spinifera. Mar Biol 126:27–33

Davis AR, Fyfe SK, Turon X, Uriz MJ (2003) Size matters sometimes: wall height and the structure of subtidal benthic invertebrate assemblages in southeastern Australia and Mediterranean Spain. J Biogeogr 30:1797–1807

Dean TA (1981) Structural aspects of sessile invertebrates as organizing forces in a fouling community. J Exp Mar Biol Ecol 53:163–180

Dean TA, Hurd LE (1980) Development in an estuarine fouling community: the influence of early colonists on later arrivals. Oecologia 46:295–301

de Blok JW, Geelen HJ (1958) The substratum required for the settling of mussels (Mytilus edulis L.). Arch Neerl Zool jubilee vol, pp 446–460

Earl R, Erwin DG(eds) (1983) Sublittoral ecology: the ecology of the shallow sublittoral benthos. Clarendon Press, Oxford

Faimali M, Garaventa F, Terlizzi A, Chiantore M, Cattaneo-Vietti R (2004) The interplay of substrate nature and biofilm formation in regulating *Balanus amphitrite* Darwin, 1854 larval settlement. J Exp Mar Biol Ecol 306:37–50

Farnsworth EJ, Ellison AM (1996) Scale-dependent spatial and temporal variability in biogeography of mangrove-root epibiont communities. Ecol Monogr 66:45–66

Fletcher WJ (1987) Interactions among subtidal Australian sea urchins gastropods and algae: effects of experimental removals. Ecol Monogr 57:89–109

Foster MS (1982) The regulation of macroalgal associations in kelp forests. In: Srivastava L (ed) Synthetic and degradative processes in marine macrophytes. W. de Gruyter, Berlin, pp 185–205

Gaines SD, Roughgarden J (1985) Larval settlement rate: a leading determinant of structure in an ecological community. Proc Natl Acad Sci USA 82:3707–3711

Gaines SD, Brown S, Roughgarden J (1985) Spatial variation in larval concentrations as a cause of spatial variation in settlement for the barnacle, *Balanus glandula*. Oecologia 67:267–273

Gil-Turnes MS, Hay ME, Fenical W (1989) Symbiotic marine bacteria chemically defend crustacean embryos from a pathogenic fungus. Science 246:116–118

Glasby TM (2000) Surface composition and orientation interact to affect subtidal epibiota. J Exp Mar Biol Ecol 248:177–190

Graham KR, Sebens KP (1996) The distribution of marine invertebrate larvae near vertical surfaces in the rocky subtidal zone. Ecology 77:933–949

Groppelli S, Pennati R, Scari G, Sotgia C, De Bernardi F (2003) Observations on the settlement of *Phallusia mammillata* larvae: effects of different lithological substrata. Ital J Zool 70:321–326

Grosberg RK (1981) Competitive ability influences habitat choice in marine invertebrates. Nature 290:700–702
Grosberg RK (1982) Intertidal zonation of barnacles: the influence of planktonic zonation of larvae on vertical distribution of adults. Ecology 63:894–899
Grosberg RK (1987) Limited dispersal and proximity-dependent mating success in the colonial ascidian *Botryllus schlosseri*. Evolution 41:372–384
Guidetti P, Bianchi CN, Chiantore M, Schiaparelli S, Morri C, Cattaneo-Vietti R (2004) Living on the rocks: substrate mineralogy and the structure of subtidal rocky substrate communities in the Mediterranean Sea. Mar Ecol Prog Ser 274:57–68
Hammond LS, Synnot RN(eds) (1994) Marine biology. Longman Cheshire, Melbourne
Harlin MM, Lindbergh JM (1977) Selection of substrata by seaweeds: optimal surface relief. Mar Biol 40:33–40
Harper JL (1977) The population biology of plants. Academic Press, London
Herbert RJH, Hawkins SJ (2006) Effect of rock type on the recruitment and early mortality of the barnacle *Chthamalus montagui*. J Exp Mar Biol Ecol 334:96–108
Hills JM, Thomason JC (1996) A multi-scale analysis of settlement density and pattern dynamics of the barnacle *Semibalanus balanoides*. Mar Ecol Prog Ser 138:103–115
Hills JM, Thomason JC (1998) The effects of scales of surface roughness on the settlement of the barnacle (*Semibalanus balanoides*). Biofouling 12:57–70
Hills JM, Thomason JC, Muhl J (1999) Settlement of barnacle larvae is governed by Euclidean and not fractal surface characteristics. Functional Ecol 13:868–875
Holmes SP, Sturgess CJ, Davies MS (1997) The effect of rock-type on the settlement of *Balanus balanoides* (L.) cyprids. Biofouling 11:137–147
Huggett RJ (2007) Fundamentals of geomorphology. Rutledge, London
Hughes TP, Baird AH, Dinsdale EA, Moltschaniwskyj NA, Pratchett MS, Tanner JE, Willis BL (1999) Patterns of recruitment and abundance of corals along the Great Barrier Reef. Science 397:59–63
Hughes TP, Rodrigues MJ, Bellwood DR, Ceccarelli D, Hoegh-Guldberg O, McCook L, Moltschaniwskyj N, Pratchett MS, Steneck RS, Willis B (2007) Phase shifts, herbivory, and the resilience of coral reefs to climate change. Curr Biol 17:360–365
Huxley R, Holland DL, Crisp DJ (1984) Influence of oil shale on intertidal organisms: effect of oil shale surface roughness on settlement of the barnacle *Balanus balanoides* (L.). J Exp Mar Biol Ecol 82:231–237
Jackson JBC (1977) Competition on marine hard substrata: the adaptive significance of solitary and colonial strategies. Am Nat 111:743–767
Jones CG, Lawton JH, Shachak M (1994) Organisms as ecosystem engineers. Oikos 69:373–386
Jordan A, Lawler M, Halley V, Barrett N (2005) Seabed habitat mapping in the Kent Group of islands and its role in marine protected area planning. Aquat Conserv Mar Freshw Ecosyst 15:51–70
Karieva PM, Bertness MD (1997) Re-examining the role of positive interactions in communities. Ecology 78:1945
Kay AM, Keough MJ (1981) Occupation of patches in the epifaunal communities on pier pilings and the bivalve *Pinna bicolor* at Edithburgh, South Australia. Oecologia 48:123–130
Keough MJ (1983) Patterns of recruitment of sessile invertebrates in two subtidal habitats. J Exp Mar Biol Ecol 66:213–245
Keough MJ (1984) Dynamics of the epifauna of the bivalve *Pinna bicolor*: interactions among recruitment, predation and competition. Ecology 65:677–688
Keough MJ, Downes BJ (1982) Recruitment of marine invertebrates: the role of active larval choices and early mortality. Oecologia 54:348–352
Keough MJ, Raimondi PT (1995) Responses of settling invertebrate larvae to bioorganic films: effects of different types of films. J Exp Mar Biol Ecol 185:235–253
Keough MJ, Raimondi PT (1996) Responses of settling invertebrate larvae to bioorganic films: effects of large-scale variation in films. J Exp Mar Biol Ecol 207:59–78

Knott NA, Underwood AJ, Chapman MG, Glasby TM (2004) Epibiota on vertical and on horizontal surfaces on natural reefs and on artificial structures. J Mar Biol Assoc UK 84:1117–1130
Köhler J, Hansen PD, Wahl M (1999) Colonization patterns at the substratum-water interface: how does surface micro-topography influence recruitment patterns of sessile organisms. Biofouling 14:237–248
Lawton J (1996) Patterns in ecology. Oikos 75:145–147
Lenihan HS (1999) Physical-biological coupling on oyster reefs: how habitat structure influences individual performance. Ecol Monogr 69:251–275
Lenihan HS, Peterson CH (1998) How habitat degradation through fishery disturbance enhances impacts of hypoxia on oyster reefs. Ecol Appl 8:128–140
Lim NSH, Everuss KJ, Goodman AE, Benkendorff K (2007) Comparison of surface microfouling and bacterial attachment on the egg capsules of two molluscan species representing Cephalopoda and Neogastropoda. Aquat Microb Ecol 47:275–287
Littler M, Taylor PR, Littler DS (1986) Plant defence associations in the marine context. Coral Reefs 5:63–71
MacArthur RH, Wilson EO (1967) The theory of island biogeography. Princeton University Press, Princeton, NJ
Maldonado M, Young CM (1998) Limits on the bathymetric distribution of keratose sponges: a field test in deep water. Mar Ecol Prog Ser 174:123–139
Mariani S, Uriz MJ, Turon X (2005) The dynamics of sponge larvae assemblages from northwestern Mediterranean nearshore bottoms. J Plankton Res 27:249–262
McGuinness KA (1989) Effects of some natural and artificial substrata on sessile marine organisms at Galeta Reef, Panama. Mar Ecol Prog Ser 52:201–208
McGuinness KA, Underwood AJ (1986) Habitat structure and the nature of communities on intertidal boulders. J Exp Mar Biol Ecol 104:97–123
Miller MW (2002) Using ecological processes to advance artificial reef goals. ICES J Mar Sci suppl 59:27–31
Minchinton TE, Scheibling RE (1991) The influence of larval supply and settlement on the population structure of barnacles. Ecology 72:1867–1879
Moore PG (1977) Inorganic particulate suspensions in the sea and their effects on marine animals. Oceanogr Mar Biol Annu Rev 15:225–363
Moore PG, Seed R(eds) (1985) The ecology of rocky coasts. Hodder and Stoughton, London
Newton KL, Creese B, Raftos D (2007) Spatial patterns of ascidian assemblages on subtidal rocky reefs in the Port Stephens-Great Lakes Marine Park, New South Wales. Mar Freshw Res 58:843–855
Ojeda FP, Dearborn JH (1989) Community structure of macroinvertebrates inhabiting the rocky subtidal zone in the Gulf of Maine: seasonal and bathymetric distribution. Mar Ecol Prog Ser 57:147–161
Olson RR, McPherson R (1987) Potential vs. realized larval dispersal: fish predation on larvae of the ascidian *Lissoclinum patella* (Gottschaldt). J Exp Mar Biol Ecol 110:245–256
Pawlik JR (1992) Chemical ecology of the settlement of benthic marine invertebrates. Oceanogr Mar Biol Annu Rev 30:273–335
Petraitis PS (1990) Direct and indirect effects of predation, herbivory and surface rugosity on mussel recruitment. Oecologia 83:405–413
Pitcher CR, Butler AJ (1987) Predation by asteroids, escape response, and morphometrics of scallops with epizoic sponges. J Exp Mar Biol Ecol 112:233–249
Raimondi PT (1988) Rock type affects settlement, recruitment, and zonation of the barnacle *Chthamalus anisopoma* Pilsbury. J Exp Mar Biol Ecol 123:253–267
Relini G, Zamboni N, Tixi F, Torchia G (1994) Patterns of sessile macrobenthos community development on an artificial reef in the Gulf of Genoa (Northwest Mediterranean). Bull Mar Sci 55:745–771
Roberts DE (1996) Patterns in subtidal marine assemblages associated with a deep-water sewage outfall. Mar Freshw Res 47:1–9

Roberts DE, Fitzhenry SR, Kennelly SJ (1994) Quantifying subtidal macroinvertebrate assemblages on hard substrata using a jump camera method. J Exp Mar Biol Ecol 177:157–170

Roberts DE, Davis AR, Cummins SP (2006) Experimental manipulation of shade, silt, nutrients and salinity on the temperate reef sponge *Cymbastela concentrica*. Mar Ecol Prog Ser 307:143–154

Russ GR (1980) Effects of predation by fishes, competition, and structural complexity on the establishment of a marine epifaunal community. J Exp Mar Biol Ecol 42:55–69

Schiaparelli S, Guidetti P, Cattaneo-Vietti R (2003) Can mineralogical features affect the distribution patterns of sessile gastropods? The Vermetidae case in the Mediterranean Sea. J Mar Biol Assoc UK 83:1267–1268

Sebens KP (1985) Community ecology of vertical walls in the Gulf of Maine, USA: small scale processes and alternative community states. In: Moore PG, Seed R(eds) The ecology of rocky coasts. Columbia University Press, New York, pp 346–371

Soniat TM, Burton GM (2005) A comparison of the effectiveness of sandstone and limestone as cultch for oysters, *Crassostrea virginica*. J Shellfish Res 24:483–485

Sutherland JP, Karlson RH (1977) Development and stability of the fouling community at Beaufort, North Carolina. Ecol Monogr 47:425–446

Svane I, Petersen JK (2001) On the problems of epibiosis, fouling and artificial reefs, a review. PSZNI Mar Ecol 33:169–188

Todd CD, Keough MJ (1994) Larval settlement in hard substratum epifaunal assemblages: a manipulative field study of the effects of substratum filming and the presence of incumbents. J Exp Mar Biol Ecol 181:159v187

Totti C, Cucchiari E, De Stefano M, Pennesi C, Romagnoli T, Bavestrello G (2007) Seasonal variations of epilithic diatoms on different hard substrates, in the northern Adriatic Sea. J Mar Biol Assoc UK 87:649–658

Twidale CR (1982) Granite landforms. Elsevier, Amsterdam

Underwood AJ, Anderson MJ (1994) Seasonal and temporal aspects of recruitment and succession in an intertidal estuarine fouling assemblage. J Mar Biol Assoc UK 74:563–584

Underwood AJ, Chapman MG(eds) (1995) Coastal marine ecology of temperate Australia. University of New South Wales Press, Sydney

Wahl M (1989) Marine epibiosis. I. Fouling and antifouling: some basic aspects. Mar Ecol Prog Ser 58:75–89

Wahl M (1997) Increased drag reduces growth of snails: comparison of flume and in situ experiments. Mar Ecol Prog Ser 151:291–293

Wahl M, Hoppe K (2002) Interactions between substratum rugosity, colonization density and periwinkle grazing efficiency. Mar Ecol Prog Ser 225:239–249

Walters LJ, Wethey DS (1996) Settlement and early post-settlement survival of sessile marine invertebrates on topographically complex surfaces: the importance of refuge dimensions and adult morphology. Mar Ecol Prog Ser 137:161–171

Walters LJ, Miron G, Bourget E (1999) Endoscopic observations of invertebrate larval substratum exploration and settlement. Mar Ecol Prog Ser 182:95–108

Wethey D (1986) Ranking of settlement cues by barnacle larvae: influence of surface contour. Bull Mar Sci 39:393–400

Witman JD (1985) Refuges, biological disturbance, and rocky subtidal structure in New England. Ecol Monogr 55:421–445

Witman JD, Dayton PK (2001) Rocky subtidal communities. In: Bertness MD, Gaines SD, Hay ME(eds) Marine community ecology. Sinauer, Sunderland, MA, pp 339–366

Witman JD, Sebens KP (1990) Distribution and ecology of sponges at a subtidal rock ledge in the central Gulf of Maine. In: Rutzler K (ed) New perspectives in sponge biology. Smithsonian Institution Press, Washington, DC, pp 391–396

Young CM (1989a) Selection of predator-free settlement sites by larval ascidians. Ophelia 30:131–140

Young CM (1989b) Larval depletion by ascidians has little effect on settlement of epifauna. Mar Biol 102:481–489

Young CM (1990) Larval ecology of marine invertebrates: a sesquicentennial history. Ophelia 32:1–7
Young CM, Chia F-S (1981) Laboratory evidence for delay of larval settlement in response to a dominant competitor. Int J Invert Reprod 3:221–226
Young CM, Gotelli NJ (1988) Larval predation by barnacles: effects on patch colonization in a shallow subtidal community. Ecology 69:624–634
Young R, Young A (1992) Sandstone landforms. Springer, Berlin Heidelberg New York
Zobell CE (1939) The role of bacteria in the fouling of submerged surfaces. Biol Bull 77:302

Chapter 3
Communities on Deep-Sea Hard Bottoms

Craig M. Young

3.1 Islands in a Sea of Mud

Most of the deep-sea floor is covered with muddy sediment up to several km thick, yet the deep ocean also contains a surprising amount of hard substratum, including the largest mountain ranges on earth, and tens or perhaps hundreds of thousands of active and inactive volcanic seamounts (Smith 1991; Kitchingman et al. 2007). Slopes steeper than 22° along continental margins, in submarine canyons, and on island or seamount flanks are generally free of sediment (Smith and Demopoulos 2003), and exposed basalt typically surrounds mid-ocean spreading centers, which are enormous linear features spanning the entire globe (Tunnicliffe et al. 2003). The total area of deep-sea coral reefs is estimated to exceed the area (approximately 284,300 km^2) covered by all shallow-water reefs in the tropics (Freiwald and Roberts 2005). Given the sizes and numbers of these features, it seems likely that the deep sea contains more hard surfaces than do all nearshore intertidal and subtidal habitats combined. Hard substrata in the deep sea are also surprisingly diverse. In addition to volcanic and carbonate rocks, they include surfaces as varied as whale bones, manganese nodules, sponge stalks, ancient coral reefs, and methane ice. This chapter will briefly review (1) the kinds of hard-bottom substrata available for community development in the deep sea, (2) the dominant groups of organisms found on deep-sea hard bottoms, and (3) our current limited understanding of mechanisms that control patterns of spatial and temporal patchiness in deep-sea epifaunal communities.

During most of the 19th and 20th centuries, deep-sea biology focused primarily on the soft bottoms of the ocean basins and abyssal plains, which are among the largest and most biologically diverse habitats on the planet. The dramatic discovery of rich chemosynthetic communities on mid-ocean spreading centers in the late 1970s shifted attention toward animals living on hard bottoms of volcanic origin, including the basalts and sulfides near hydrothermal vents, and the authigenic carbonates and asphalts found in areas of hydrocarbon seepage. More recently, seafloor mapping efforts, lucrative fisheries, and conservation concerns have focused major attention on seamounts, which are nearly always populated by a remarkable fauna of hard-bottom animals including octocorals and reef-forming scleractinian

M. Wahl (ed.), *Marine Hard Bottom Communities*, Ecological Studies 206,
DOI: 10.1007/978-3-540-92704-4_3,

corals. Deep-water corals, and the species associated with these have come into the limelight recently, as they are particularly vulnerable to trawling and other human perturbations. Finally, specialized organisms living on whale bones and submerged wood are receiving renewed interest.

3.2 Types of Hard Substrata in the Deep Sea

The fine sediments of the deep ocean basins originate from atmospheric dust combined with the skeletons of planktonic organisms that sink from the overlying water column. With the exception of drop stones from icebergs (Oschmann 1990) and the bones of dead animals, hard bottoms in the deep sea are virtually all authigenic, originating in the places where they are currently found. These authigenic deep-sea substrata include volcanic basalts and glasses on mid-plate volcanoes and within a few kilometers of spreading centers; ferromanganese crusts and nodules; carbonates of microbial origin near methane seeps; carbonates from ancient and modern reef formations; naturally occurring asphalt; sedimentary rocks; methane hydrate "ice"; and the hard shells or skeletons of living and dead organisms. Because community development is tied closely to the nature and location of the substratum, it is useful to consider the origin and development of these various substratum types.

3.2.1 Substrata Formed by Volcanism

Volcanic lava erupting at mid-ocean spreading centers and volcanoes may solidify in the form of sheets (Gregg and Fink 1995), but more often forms distinctive meandering or ovoid structures known as pillow lavas. The surface consistency of these lavas varies with the rate of cooling. Thus, some new pillows are covered with volcanic glass (obsidian), whereas others are made of basalt with a fine crystalline structure. Older pillow lavas and sheets may be encrusted with ferromanganese layers. Recent work (Tebo et al. 2004, 2005; Templeton et al. 2005) demonstrates that microbial activity can modify the texture and composition of volcanic glasses, facilitating the encrustation of manganese, and presumably changing its attractiveness as a settlement substratum. On the slopes of volcanic islands, sheet flows, broken-up volcanic debris (breccia), and rocks spewed into the sea from above-water eruptions also provide substrata for attachment of animals.

Undersea asphalt pavements have recently been discovered in a region of salt dome formation near the Campeche Knolls in the Gulf of Mexico (McDonald et al. 2004). The hard asphalt, which occurs at approximately 3,000 m depth, appears to have been extruded by seafloor volcanism at temperature higher than that of seawater, and it provides a surface for a variety of chemosynthetic and non-chemosynthetic epifaunal organisms.

Hydrothermal vents transport minerals to the seafloor from deep within the crust. These minerals, particularly anhydrite (calcium sulfate), precipitate around the orifices of vents, creating rapidly growing towers that become permeated with a variety of metals and other minerals (Haymon 1983; Hannington et al. 1995). The resulting chimneys and mounds of metallic sulfides provide substrata for those organisms (notably, polychaetes and molluscs) that can tolerate the toxic conditions and elevated temperatures found there.

3.2.2 Polymetallic Nodules and Manganese Crusts

Polymetallic or manganese are present in all oceans of the world, and are particularly abundant in the equatorial Pacific and the central Indian Ocean between 4,000 and 6,000 m depths. Nodules are roughly spherical concretions formed of concentric layers of iron, manganese hydroxides, and other minerals around a microscopic central core, which may consist of a small shell, rock, shark's tooth, or other hard piece of debris. The nodules form very slowly in the deep sea (on the order of 1 cm in several million years). Using time-lapse photography, Gardner et al. (1984) monitored beds of nodules between 3,000 and 5,000 m over 3 years. During this time, none of the nodules moved, though organisms were seen eating sediments from the surfaces of nodules. Based on these and other observations, it appears that manganese nodules may be stable over very long periods of time, perhaps thousands of years.

At depths ranging from 400 to 4,000 m, but especially between 1,000 and 3,000 m, most exposed rocks eventually become encrusted with ferromanganese layers. These crusts form very slowly, precipitating at a rate of about 2.5 mm per million years (Moore and Clague 2004), but the mechanisms of this encrustation are not fully understood (Verlaan 1992).

From an ecological standpoint, polymetallic nodules may be regarded as tiny islands of hard substratum resting in a sea of soft sediment. Because most nodules are less than 10 cm in diameter, and are often partly buried, they are seldom colonized by large epifauna such as corals, but they do provide a substratum for foraminiferans and other small organisms.

3.2.3 Carbonates

In the deep sea, carbonate substrata from ancient reefs commonly comprise the flanks of coral islands that have subsided (Darwin 1842). Calcium carbonate skeletons of corals, and thalli of calcareous green algae such as *Halimeda* spp. may accumulate in deep water near islands, where they become consolidated and cemented into limestone structures known as bioherms or lithoherms (Neumann et al. 1977; Messing et al. 1990; Paull et al. 2000; Reed 2002).

Authigenic carbonates, which may be aragonite, dolomite or calcite, are formed by consortia of bacteria that oxidize methane and other hydrocarbons, often on continental margins. Where these carbonates form exposed pavements on topographic high points, they are colonized by filter-feeders and other epifaunal organisms, and often form the base for extensive reefs of ahermatypic corals.

3.2.4 Methane Hydrate

Methane seeping from subsea oil deposits combines with seawater under appropriate conditions of low temperature and high pressure to form solid methane hydrates, which resemble ice. Hydrates that extend above the surrounding soft substrata in the Gulf of Mexico are occupied by a specialized hesionid polychaete (Desbruyeres and Toulmond 1998; Fisher et al. 2000) that apparently locates these rare hydrate outcroppings with dispersive planktotrophic larvae (Eckelbarger et al. 2001; Pile and Young 2006a).

3.2.5 Biogenic Surfaces

Many living organisms in the deep sea have hard body structures that provide suitable substrata for epibionts. Such organisms include deep-sea cnidarians (corals and octocorals), hexactinellid sponges, and echinoderms such as stalked crinoids. Many deep-sea sponges have exposed spicules that support the body, often in the form of large, braided glass stalks (Beaulieu 2001a). Specialized epibionts such as zoanthids commonly occupy these sponge stalks. Likewise, the spines of cidaroid sea urchins often support specialized barnacles and zoanthids. Because deep-sea filter-feeders seek elevated surfaces, virtually all tall gorgonians, corals, sponges and echinoderms provide perches for other animals.

3.2.6 Organic Remains from the Upper Ocean

The carcasses of dead whales and other animals sinking to the seafloor have recently received considerable attention as biogenic substrata for community development. Whale bones are unique in providing not only a substratum, but also extensive lipid stores that can be exploited by settling organisms (Smith and Baco 2003). Even small body parts such as sharks teeth and squid beaks are often colonized by benthic animals. Water-logged wood is also found on the deep-sea floor, abundantly in some places, and a number of wood specialists, including (but not limited to) boring bivalves, exploit woods in deep water (Turner 1978; Voigt 2007; Tyler et al. 2007).

3.3 Major Groups of Deep-Sea Organisms

With the discovery of hydrothermal vents and other chemosynthetic ecosystems, the number of known hard-bottom species has climbed at an exponential rate over the past 30 years (Tunnicliffe 1991; Van Dover 2000; Desbruyeres et al. 2006). The dominant groups of organisms at hydrothermal vents include bathymodiolin mussels, vesicomyid clams, vestimentiferan tubeworms, a wide variety of small snails and limpets, and numerous polychaetes, notably the alvinellids and polynoids, many of which live inside other organisms. A systematic account of all described vent species has been provided by Desbruyeres et al. (2006). No comparable global summary has been produced for methane seep organisms, but there are extensive species lists in descriptive accounts of individual seeps. Some of the species that exploit whale bones are also chemosynthetic. The recently discovered bone-worm polychaetes of the genus *Osedax* use internal symbiotic bacteria to exploit substances in the bones of dead whales (Rouse et al. 2004; Braby et al. 2007).

The main groups of non-chemosynthetic hard-bottom epifauna in the deep sea have been listed by Gage and Tyler (1991), and many have been illustrated photographically in Chave and Jones (1991), Tyler and Zibrowius (1992), and Chave and Malahoff (1998). An exhaustive list of species found associated with deep-sea coral reefs has been provided by Rogers (1999). Cnidarians are especially common, and include numerous octocorals (Fig. 3.1e–3.1j), hydrocorals, hard corals such as *Lophelia* (Fig. 3.1d) and *Montipora*, and solitary cup corals such as *Caryophyllia*. Echinoderms are often associated with cnidarians in the deep sea. Among the most prominent are comatulid crinoids (Fig. 3.1c), and ophiacanthid (Fig. 3.1c) and euryalid (Fig. 3.1g) ophiuroids, all of which perch on cnidarians, sponges, boulders, and other high points to capture food from currents. Stalked crinoids (Fig. 3.1i), a major group of deep-sea epifauna, are presently confined to deep water, though they were once common constituents of warm, shallow seas.

Sponges are common on both soft and hard bottoms, and these include specialized demosponges (Fig. 3.1a), such as the stony lithistids, as well as hexactinellid glass sponges (Fig. 3.1b, c, n) in tremendous variety. The latter are found only in relatively deep waters, where they sometimes form extensive reefs.

A variety of mobile epifauna, including sea urchins, asteroids and gastropods, are found associated with deep hard-bottom communities. Many groups of active filter-feeders, including ascidians and barnacles, tend to be much less common in the deep sea than in shallow water, presumably because active pumping is not efficient in the low-food deep-sea environment (Monniot 1979). Hexactinellid sponges represent the major exception to this rule, possibly because they feed on bacteria, an abundant food source in deep water (Pile and Young 2006b). Deep-sea barnacles include the unusual verrucimorphs, and the stalked scalpellids. A few of the deep-sea ascidians resemble their shallow-water counterparts, whereas the exclusively deep-sea Octacnemidae (Fig. 3.1l) have hypertrophied branchial siphons, and no cilia in their branchial sacs (Monniot and Monniot 1978). These latter species are carnivores, relying on ambient water currents to bring their food, with no expenditure of energy.

Fig. 3.1 Hard-bottom animals in the deep sea, from 500 to 2,500 m depth (photo credits in parentheses). **a** Spiny brittle star (Ophiacanthidae) at the apex of a large demosponge protruding from the carbonate slope of a Bahamian island (C.M. Young). **b** A volcanic boulder on the slope of Mauna Loa Volcano, Hawaii, supporting a variety of hexactinellid sponges and cnidarians. The largest sponge is *Walteria* sp., and the sea anemone is probably *Actinoscypha* sp. (C.M. Young). **c** Large hexactinellid sponge (Dactylocalycidae) serving as a perch for comatulid crinoids and ophiacanthid ophiuroids in the northern Bahamas (C.M. Young). **d** Thicket of the deep-sea coral *Lophelia pertussa* off the east coast of Florida (S. Brooke). **e** Large pillow lava supporting an aggregation of the octocoral *Anthomastus* sp. on Vailulu'u Seamount, east of American Samoa

3.4 Population and Community Ecology of Hard-Bottom Deep-Sea Epifauna

Because the deep sea is difficult and expensive to access, much of what we know about deep-sea ecology is descriptive in nature, based on exploration, photography, and sampling. Quantitative tools for descriptive analysis commonly include rarefaction curves, regression analysis, size spectrum analysis, ordination, and other multivariate techniques. Trophic relationships are often inferred from stable isotopes, breeding seasons and reproductive modes are inferred from sequential gonad samples, and connectivity among metapopulations is inferred from population genetics. Manipulative experiments with the potential of disclosing the processes underlying patterns are, unfortunately, rather rare in the deep-sea literature. The present review, which is not intended to be comprehensive, will highlight a few of the descriptive and experimental studies that provide insights on the mechanisms underlying spatial or temporal patterns of abundance. Excellent, more comprehensive general reviews of deep-sea ecology are available in Gage and Tyler (1991), Tyler (1995), and Van Dover (2000).

3.4.1 Chemosynthetic Communities

The ecophysiology of vent and seep organisms, and the mechanisms of obtaining nutrition with the aid of chemosynthetic bacteria, have been extensively studied. Reviews by Fisher (1990, 1996), Fiala-Medioni and Felbeck (1990), Childress and Fisher (1992), and Van Dover (2000) provide convenient portals to this large literature.

Fig. 3.1 (continued) (Staudigel et al. 2006). **f** The spiral chrysogorgiid gorgonian *Iridogorgia* sp. off Samoa (C.M. Young). **g** Erect gorgonian supporting a large euryalid ophiuroid in the Bahamas (C.M. Young). **h** Large bushy gorgonian with a crab in the Bahamas (C.M. Young). **i** Stalked crinoid, *Endoxocrinus parrae*, and crab using a tall gorgonian as a substratum in the Bahamas (C.M. Young). **j** Predatory sea star, *Hippasterias* sp., preying on a bamboo coral (Isidae) off the big island of Hawaii (C.M. Young). **k** Brisingid asteroid, *Novodinea antillensis*, in the Bahamas. This animal uses large pedicellaria lining its arms to capture small crustaceans from the plankton (C.M. Young). **l** Two undescribed species of carnivorous octacnemid ascidians, *Situla* sp. and *Dicopia* sp., on a carbonate slope in the Bahamas (C.M. Young). **m** Hydrothermal vent tubeworms and predators found in areas of high sulfide emission at 9N on the East Pacific Rise. Depth: 2,500 m. The small tubeworms are *Tevnia jericonana*, the pioneer species in the successional sequence. A smaller number of the larger worms, *Riftia pachyptila*, are interspersed amongst the *Tevnia*. Predatory zoarcid fishes (*Thermarces cerberus*) and crabs (*Bythygrea thermodon*) are also present (C.M. Young). **n** Stalked hexactinellid sponge, *Hyalonema* sp.: its rope-like braided stalk of glass spicules provides discrete islands of hard substratum in muddy areas of the deep sea. Zoanthids are visible near the base of this stalk, hydroids line the left side of the stalk, and a crab is perched near the top (C.M. Young). **o** A specialized file-shell bivalve, *Acesta oophaga*, which lives on the tubeworm *Lamellibrachia luymesi* in areas of hydrocarbon seepage in the Gulf of Mexico. The bivalve feeds on lipid-rich zygotes released by the female worm (Järnegren et al. 2005)

Most species with symbionts require particular chemical environments, so their distributions are tied closely to patterns of vent fluid flow. For example, the giant vent clam *Calyptogena magnifica* nestles in areas of broken basalt, and diffuse, warm-temperature hydrothermal flow (Hessler et al. 1985) where it obtains carbon dioxide and hydrogen sulfide with its foot, while drawing oxygen and carbon dioxide through its siphons from the overlying water column. The giant tubeworm *Riftia pachyptila*, by contrast, depends on turbulent flow in the near-vent water column to alternately transport sulfide and oxygen to the anterior plume, which absorbs all of the compounds that the symbionts require (Hessler et al. 1988; Johnson et al. 1988). Thus, although the spatial patterns of chemosynthetic organisms at vents are tightly connected to small-scale concentrations of hydrogen sulfide, different strategies for obtaining sulfide result in very different patterns of distribution. One of many papers drawing this general conclusion is the study by Sarrazin et al. (1997) on the Endeavor segment of the Juan de Fuca Ridge. A large chimney named "Smoke and Mirrors" supports six distinctly different faunal assemblages, each correlated with specific physical and chemical conditions. Over several years of monitoring, it was determined that these complex mosaics of assemblages shifted spatially in response to changes in flow. The work by Fustec et al. (1987), and Desbruyeres (1998) at 13N on the East Pacific Rise also reveals a community comprised of a complex patchwork of assemblages. Changes in distributions at this site can often be traced to precipitation of new sulfides, and major disturbance events caused by tectonic activity.

Patterns of species diversity at biogeographic scales have been extensively documented at vents, and these patterns are regularly reevaluated as new vent fields are explored in the more remote parts of the oceanic ridge system. This extensive literature has been reviewed by Tunnicliffe (1988, 1991), Van Dover (2000), and Van Dover et al. (2002). Of great relevance to the issues of biogeography are genetic and developmental patterns. The former have been studied in most parts of the world, and for many of the dominant hard-bottom groups (reviewed by Vrijenhoek 1997; Vrijenhoek et al. 1998). Gene exchange within ridge systems is extensive for most, but not all, groups that have been studied (e.g., France et al. 1992; Black et al. 1994; Creasey et al. 1996; Vrijenhoek 1997; Vrijenhoek et al. 1998).

Developmental and reproductive studies of vent animals, which are relevant to dispersal and recruitment processes, have been reviewed by Tyler and Young (1999), Van Dover (2000), and Young (2003). Most vent and seep animals that have been examined produce lipid-rich lecithotrophic larvae capable of relatively long dispersal (Marsh et al. 2001; Pradillon et al. 2001, 2005). Plankton collections, larval rearing studies, and modeling of currents indicate that larvae may disperse either in the near-bottom flow, or at the level of the buoyant plume (Kim et al. 1994; Kim and Mullineaux 1998; Marsh et al. 2001), but some species, notably the vent crab *Bythograea thermodon*, can migrate 2,500 m into the surface waters where much greater dispersal distances are possible (Epifanio et al. 1999; Jinks et al. 2002). *Alvinella pompeiana*, a vent polychaete that occupies the hottest habitable portions of chimneys, requires warm water to complete its development (Pradillon et al. 2001, 2005). Embryos arrest development reversibly at normal deep-sea temperatures, presumably permitting very long dispersal with minimal energy expenditure (Pradillon et al. 2001).

The best studied hydrothermal vents are found on the volcanic mid-ocean ridge systems in the eastern Pacific, including 9N on the East Pacific Rise, the Galapagos Rift, the Guaymas Basin, and the Juan de Fuca Ridge. The oft-visited 9N has sites of considerable ecological work, where scientists have observed major disturbance events in the form of volcanic eruptions that devastate entire communities, killing the residents and resetting the colonization clocks. The most famous of these eruptions occurred in April 1991 (Haymon et al. 1993). Subsequent colonization and succession has been documented by Shank et al. (1998), and Lutz et al. (2001, 2002). Succession at this site appeared deterministic, occurring in a predictable sequence that lasted about 4 years, and corresponded with successive reductions in the concentration of sulfide (Shank et al. 1998). The primary colonists were tubeworms, *Tevnia jericonana* (Fig. 3.1m), which were replaced with larger tubeworms, *Riftia pachyptila* (Fig. 3.1m), then the mussel *Bathymodiolus thermodron*, and finally two species of serpulid polychaetes. At another well-studied site, the Endeavor segment of the Juan de Fuca Ridge, Tunnicliffe et al. (1997) characterized colonization following another eruption. In dramatic contrast to the study by Shanks et al. (1998), they interpreted the colonization process as a stochastic lottery, with trajectories of community development being determined by the pools of larvae that happen to be in the water column when new habitats open up.

Using experimental deployments of clean basalt blocks, Mullineaux et al. (2000) tested three alternative hypotheses that might explain the observed successional sequence of *Tevnia* followed by *Riftia* and *Bathymodiolus* on the East Pacific Rise. The experiments showed that *Riftia* and another tubeworm, *Oasisia*, only colonize surfaces already occupied by *Tevnia*. In other deterministic successional systems (e.g., the classic old-field succession model), pioneer species modify the environment in ways that favor the survival of species that eventually outcompete pioneers. There was no evidence for this in the 9N system. Instead, Mullineaux et al. (2000) hypothesize that *Tevnia* facilitates recruitment of the other species by producing chemical cues used at larval settlement.

The role of settlement cues, however, has been difficult to prove in hydrothermal vent systems. In the only successful experiment on metamorphic cues conducted at vents, Rittschoff et al. (1998) demonstrated that vent polychaetes, *Paralvinella* sp., burrow into gelatinous substrata exuding sulfide, whereas they do not colonize identical controls without sulfide. Although the presence of juveniles on adult tubes or shells of various vent animals has frequently been observed, the role of gregarious and associative settlement cues remains largely unresolved at hydrothermal vents. Hunt et al. (2004) attempted to test the settlement cue hypothesis advanced by Mullineaux et al. (2000) by examining the recruitment patterns of tubeworms in the presence of adult tubes and artificial tubes. There was no evidence for selective settlement in any of these experiments.

Mullineaux et al. (2003) also examined successional patterns in other vent species at 9N. They found that both facilitation and inhibition are important, with most sessile species facilitating subsequent colonizers, and most mobile species being inhibitory. The role of predation in community development at 9N was examined by Micheli et al. (2002) by deploying caged and uncaged colonization blocks. Their results showed a persistent pattern of inhibition in which mobile predators such as zoarcid fishes (Fig. 3.1m) reduced the number of small grazing gastropods. This predation had a cascading effect in the system because fewer snails were present to graze off small fauna, and to bulldoze off new colonists. The zoarcids are discriminatory in their choice of food (Micheli et al. 2002; Sancho et al. 2005), so they had qualitative effects on species composition, as well as quantitative effects on overall abundance of the fauna (Micheli et al. 2002). The greatest caging effects were observed in those species (including small snails and amphipods) that were preferred by the predatory fishes.

In the same study, Micheli et al. (2002) showed that large bythograeid crabs (Fig. 3.1m) feed primarily on tubeworms and mussels, rather than on small mobile epifauna. However, because limpets and other small organisms are often more abundant on tubeworms (Governor et al. 2005) and in mussel beds (Van Dover 2003) than on open basalt substrata, predators that influence the large species could have cascading effects on the smaller ones. To determine whether higher densities on tubeworm clumps were caused by some chemical attribute of the tubeworms, or by the physical structure of their tubes, Governor and Fisher (2006) compared community development in clusters of artificial tubes (made of PVC hose) with that of real tubeworm clusters. Most species, but not all, colonized artificial tubeworms as readily as real tubeworm clusters, indicating that physical structure is important in establishing habitat parameters.

Hydrothermal vents often have two-dimensional habitat gradients of primary (chemosynthetic) productivity, sulfide concentration, metal concentrations, and temperature that are similar in many ways to the disturbance and desiccation gradients found in the rocky intertidal. Many of the experiments mentioned above have been deployed at sites along the productivity gradient to determine if the gradient itself interacts with other ecological processes. Thus, for example, Governor and Fisher (2006) found that the physical structure was much more important in areas of high and intermediate vent fluid flow than in the low-flow region. Likewise, the predator exclusion experiments of Micheli et al. (2002) revealed much greater predation effects in the zones of high productivity.

To determine settlement rates along the vent fluid flow gradient, Mullineaux et al. (1998) placed basalt blocks in high-temperature areas near black smokers, and low-temperature areas with diffuse flow. They discovered that most major groups of animals were able to colonize both zones. The bathymodiolin mussels and serpulid polychaetes settled broadly both within and outside the zones where adult conspecifics occur, whereas other species, including tubeworms, scallops and paralvinellid polychaetes, colonized only the zones where adults were found. These results suggested that both pre-settlement selection and post-settlement mortality play roles in community development, but that the relative importance of these two processes varies with species. This same conclusion could be made for most non-chemosynthetic hard-bottom communities in shallow water.

3.4.2 Seamounts, Continental Slopes, and Islands

Recent analyses of global sea-floor topography indicate that there are at least 14,000 seamounts more than 1 km in height (Kitchingman et al. 2007), and estimates of the total number of large and small seamounts in the Pacific Ocean alone range between 600,000 and 1.5 million (Smith 1991; Koslow 2007). Most oceanic islands may be regarded as seamounts with exposed peaks, and the rocky areas of continental slopes and canyons have much in common with hard-bottom seamounts. Because seamounts have economic importance in some parts of the world, the vast preponderance of information relates to seamount fisheries (Keating et al. 1987; Rogers 1994). Studies of benthic communities in these habitats have been largely descriptive, focusing on patterns of diversity, biogeography, and endemism (e.g., Grigg et al.1987; Forges et al. 2000) at the community level, or on genetics, morphology, and systematics of specific taxa. The CENSEAM program of the Census of Marine Life and Seamounts Online http://seamounts.sdsc.edu/, an NSF-supported program at Scripps Institution of Oceanography, currently maintains extensive bibliographies of worldwide seamount research. A careful search of these data bases, which contain several thousand references, reveals very few experimental studies addressing the factors that control the distribution and abundance of seamount benthic invertebrate fauna. Most of the literature on seamount ecology consists of faunal lists, species descriptions, and information on fisheries. This literature has been reviewed by Rogers (1994) and in recent books devoted to the subject (Keating et al. 1987; Pitcher et al. 2007). The present paper will not deal with the voluminous literature on fisheries, or the large literature dealing with the concentration of plankton and nekton above seamounts (e.g., Boehlert 1988); the focus here is on the relatively small body of literature dealing with factors known to influence patterns of faunal abundance in the deep-sea hard-bottom benthos.

Unsedimented portions of seamounts, continental slopes, and island slopes are dominated by suspension feeders, organisms that remove their food from the water column. With the exception of sponges, actively pumping filter-feeders such as bryozoans, barnacles, and ascidians are much less common here than in shallow-water communities, presumably because filter-feeding is a marginal proposition even where plankton concentrations are high (Jorgensen 1955). Deep-water suspension-feeding communities tend to be dominated by octocorals (Fig. 3.1e–j), scleractinian corals (Fig. 3.1d), sea anemones (Fig. 3.1b), hydroids, hexactinellid (Fig.3.1b, c) and lithistid sponges, and suspension-feeding echinoderms including crinoids (Fig. 3.1b, i), brisingid asteroids (Fig. 3.1k), as well as ophiacanthid, gorgonocephalid, and euryalid ophiuroids (Fig. 3.1a, b, g). Such communities are found most abundantly on topographical high points such as ridges and knolls, where faster currents transport more particles per unit time past the feeding structures (Jumars and Gallagher 1982; Genin et al. 1986, 1992; Tyler and Zibrowius 1992; Staudigel et al. 2006). Currents also keep hard surfaces clean of sediments. In a study of sponge distribution on the Bahamian Slope, demosponges and hexactinellids dominated only on vertical surfaces where sediments did not accumulate, and on slopes where currents were fast enough to remove the sediment (Maldonado and Young 1996).

Relationships between flow speed and community development have been documented repeatedly in the deep sea at several different scales. For example, Genin et al. (1992) studied the communities on Blake Spur, a carbonate ridge at the edge of the Blake Plateau in the Northwest Atlantic. Suspension-feeding animals such as lithistid sponges and ophiuroids occurred at high densities (up to 25% cover) relative to the surrounding areas. The authors attributed these high abundances to the Gulf Stream western boundary current that impinges on the slope. Topographic promontories compress streamlines in the benthic boundary layer, resulting in greater velocities, and therefore greater fluxes of plankton. This is consistent with the findings that, on seamounts, the apices invariably support the highest densities of benthic organisms. Genin et al. (1986) undertook a survey of cnidarian distributions on a number of multi-peaked seamounts in the Eastern Pacific. They found that antipatharians and gorgonians were consistently more abundant on the tops of peaks than at comparable depths on the mid-slopes. Corals were abundant on the crests of narrow peaks, but also occurred on the flanks of broader peaks. All of these observations support the hypothesis that organisms concentrate in areas where flows are accelerated. An interesting exception to this pattern was documented on Volcano 7 seamount (Wishner et al. 1990). In this case, fauna were more abundant on the highly oxygenated flanks than on the peak, which lies within a strong hypoxic zone.

On a larger scale, Tyler and Zibrowius (1992) showed that the highest biomass of suspension feeders off SW Ireland was found at the depths (2,100–2,600 m) that are bathed by North Atlantic Deep Water, presumably because of greater plankton flux at these depths. In the same study, the general observation of faunal concentration on rocks and peaks still held true (Tyler and Zibrowius 1992).

In an elegant series of experiments involving field deployments and laboratory flume studies, Mullineaux and Butman (1990) concluded that colonization patterns on millimeter scales are determined by the local patterns of flow on seamounts. This study hints that distributions of seamount fauna can be determined at the settlement or colonization phase. It is noteworthy, however, that no experimental studies have addressed the proximate causes of filter-feeder distributions at other scales. Thus, we do not know if elevated densities of these organisms on topographic highs is caused by post-settlement mortality, settlement choices of larvae, philopatric dispersal, or other factors.

Current patterns near the apices of seamounts, including internal waves, eddies, and Taylor columns, may concentrate larvae, fishes, particles, and plankton (Boehlert and Genin 1987). The potential role of Taylor's caps and Taylor's columns in larval retention was detailed by Mullineaux (1994), and tested in a series of experiments on Fieberling Guyot off California (Mullineaux and Mills 1997). Although there was evidence for larval retention in tidally rectified flows, patterns of larval distribution were not those expected by Taylor-cap retention. Additional work is required to explain how insular seamounts retain their faunas, and indeed develop high numbers of endemic species (De Forges et al. 2000). One answer may come from a study of a shallow seamount off Oregon where Parker and Tunnicliffe (1994) demonstrated that most species have direct development, or short-lived lecithotrophic larvae. Calder (2000) described a similar situation on an Atlantic seamount, where most hydroids have life cycles that lack free-swimming medusoid stages.

An important point in all of these studies is that plankton concentrations are generally too low in the deep sea to support the kinds of active filter-feeders common in shallow water (Monniot 1979). Animals compensate for low particle concentrations by living in areas with high flux, by extending their feeding structures high into the benthic boundary layer, and by using passive filtration mechanisms to minimize energy expenditure (Macurda and Meyer 1974). Gorgonians and other cnidarians tend to adopt erect, planar morphologies in the deep sea, oriented perpendicular to the prevailing flows. Suspension-feeding echinoderms such as crinoids and brisingid asteroids (Fig. 3.1k) employ mutable collagenous tissues to maintain optimal orientations relative to the currents, without the continuous use of energy-demanding muscles (Motokawa 1984; Wilkie and Emson 1988; for deep-sea examples, see Wilkie et al. 1993, 1994; Emson and Young 1994).

Most deep-sea ascidians, including the carnivorous octacnemids, lack cilia in their branchial sacs, and are therefore incapable of active pumping (Monniot 1979). Stalked ascidians such as *Culeolus* spp. use currents to passively reorient the incurrent siphon into the currents, in much the same way as the shallow-water stalked styelids (Young and Braithwaite 1980). Octacnemids (Fig. 3.1l) are sit-and-wait predators, feeding like Venus fly traps, or by exploiting the Bernoulli effect in the benthic boundary layer to entrain prey into their hypertrophied branchial siphons (Young and Vazquez, unpublished data).

Some groups, such as stalked crinoids and hexactinellid sponges, are confined to deeper waters for reasons that remain unclear. Because stalked crinoids were once very common in shallow seas, it has been proposed that they moved into the deep sea with the radiation of shallow-water predators. In support of this hypothesis, a recent study (McClintock et al. 1999) demonstrated that these animals lack chemical defense mechanisms. Hexactinellids may be limited to deep water because of a high demand for dissolved silica to form their very substantial skeletons. Competition with diatoms in shallow water may be the reason for their distributional limits (Maldonado et al. 1999). Dayton et al. (1982) erected and tested eight hypotheses that might explain why an Antarctic barnacle, *Bathylasme corolliforme*, is restricted to deep water. By transplanting individuals into shallower water, they negated the hypothesis that the barnacles were restricted by physiological tolerances, and concluded that these passive suspension feeders do not experience sufficient water movement in the shallow waters of McMurdo Sound.

3.4.3 Deep Coral Reefs

Several species of ahermatypic scleractinian corals are commonly found at bathyal and abyssal depths throughout the world oceans. The loose reefs formed by these highly branching hard corals begin on rock outcrops or authigenic carbonates, but as the polyps senesce and die, the dead portions of the coral colonies often break off to form additional substratum on which the overlying reef can expand (Messing et al. 2008). Consequently, reefs of *Lophelia*, *Madrepora*, and *Oculina* may be many kilometers long, and form mounds up to several hundred meters high (Fig. 3.1d). The reefs provide substratum and habitat for a rich array of epifaunal and infaunal organisms (Rogers 1999; Koslow 2007; Messing et al. 2008). Much, but not all, of this fauna appears to be facultative; the species also occupy non-coral habitats in the deep sea (Rogers 1999). Reefs are patchy on a large scale, and there has been much discussion about the conditions that are suitable for reef formation. Brisk currents seem to play a role, as does the existence of a hard substratum on which coral planulae may initially become established (Messing et al. 2008). Nevertheless, corals are absent at many apparently suitable sites in the deep sea. Recent laboratory experiments have yielded insights on larval development (Brooke and Young 2003), on the current speeds required and tolerated by the polyps, and on the inhibitory effects of sedimentation (Brooke, Holmes and Young, unpublished data). Nevertheless, the causes underlying large-scale patchiness remain enigmatic, and are not likely to be discovered without rigorous in situ experimentation. As in other deep-sea hard-bottom systems, the work to date has been mostly descriptive and autecological.

3.4.4 Ferromanganese Nodules

Apart from large deposit feeders that sweep across their surfaces (Gardner et al. 1984), most of the organisms associated with ferromanganese nodules are very small, with the fauna dominated by foraminiferans (Mullineaux 1987), and small crustaceans such as harpacticoid copepods. Aspects of their micro-distribution have been studied experimentally by Mullineaux (1988, 1989). The smooth upper surfaces of nodules are dominated by suspension-feeding organisms, whereas the sides of the nodules, which tend to be rougher, are dominated by deposit feeders (Mullineaux 1989). In an attempt to unravel the processes underlying this pattern, Mullineaux (1988) deployed defaunated smooth and rough nodules on and above the seafloor. Her conclusion, based on recruitment data over a 7-week period, was that texture alone does not determine the patterns of distribution. Instead, the experimental data (Mullineaux 1988) suggested a three-way interaction among texture, flow, and particle concentration. Particle concentration could partly be predicted from patterns of shear stress around the nodules, which Mullineaux (1989) modeled in flume studies.

3.4.5 Organic Materials from the Upper Ocean

Wood on the deep-sea floor supports specialized boring organisms, including, but not limited to, the opportunistic xylophagid bivalves. These borers appear very quickly in virtually all wood falls, suggesting that there is a large pool of larvae in the water column waiting for recruitment opportunities (Turner 1973). Growth rates of these bivalves are rapid, fecundities are high (Tyler et al. 2007), and they attain reproductive maturity at a small size. Recent work (Voigt 2007) in the NE Pacific, where there are extensive forests exporting wood to the ocean, demonstrates a high diversity of bivalves. Voigt (2007) noted a consistent pattern of species replacement that was related to life-history attributes, and possible competitive interactions. Because this system can be manipulated by deploying artificial habitats, it is ripe for further experimental work on community dynamics, recruitment, and distribution. Another deep-sea group that specializes on organic falls from above is the cocculinid limpets. These small organisms are common on wood falls, squid beaks, whale bones, and other materials that reach the seafloor from the water column (McLean and Harasewych 1995).

Large organic falls such as whale carcasses have recently received much attention in deep-sea ecology. The successional sequence of colonization in these systems, as generalized in a recent review by Smith and Baco (2003), consists of four stages. In the first stage, soft tissues are removed by sharks, crabs, hagfish, and other mobile predators, exposing the bones for hard-bottom organisms. Once the hard bones are exposed, they are colonized by small opportunistic polychaetes and crustaceans, which may occur at up to 40,000 per m2, the highest known densities of any hard-bottom animals in the deep sea. Colonization by opportunists leads to the sulphophilic stage, in which species that tolerate or require hydrogen sulfide colonize the bones. This stage of succession may involve as many as 185 species, the highest local diversity of any known hard-bottom community in the deep sea (Smith and Baco 2003).

Osedax bone worms apparently colonize in the opportunistic or sulphophilic stage (Rouse et al. 2004; Braby et al. 2007). These gutless polychaetes have symbiotic bacteria in their roots that transport and metabolize lipids and other organic materials from the whale bones. The relatively large females extend their plumes into the water column, and the diminutive males reside in the oviducts of the females. Embryos and lecithotrophic larvae have now been reared for several of these species, and the length of larval life has been estimated (Rouse et al. 2009), but factors controlling recruitment and distribution of these worms are mostly unknown.

The transitions among successional stages in whale-bone systems seem to be mediated in part by facilitation effects. Thus, for example, scavengers and bacteria expose bones, facilitating the colonization by epifauna (Smith and Baco 2003). Like deep-sea wood falls, this is a system that lends itself to manipulative experiments. Bones and carcasses, though big and unwieldy, can be placed on the seafloor in various configurations, and at different times. Smith and his colleagues have made excellent use of this system, and various teams are currently focusing on aspects of recruitment and community development (Smith and Baco 2003; Braby et al. 2007).

3.4.6 Epizooism

The common occurrence of mobile epifauna on tall cnidarians and sponges in the deep sea has already been mentioned. Many of these mobile species, including most comatulid crinoids, isocrinid crinoids, and ophiacanthid ophiuroids, are opportunistic; they are as likely to be found on the tops of rocky ridges or boulders as on gorgonians or hexactinellids. The asteroschematid serpent stars, by contrast, are mostly obligate epizoites on gorgonians (Fig. 3.1g), and they tend to specialize on particular species. Although there have been some studies of feeding biology in asteroschematids, the factors controlling their distributions on both geographic and within-host scales remain unknown.

Beaulieu (2001a) studied the fauna of organisms colonizing the rope-like glass spicule stalks of *Hyalonema* spp. (Fig. 3.1n). Although these sponges live on soft bottoms, their stalks provide islands of hard substratum that persist long after the sponges die. Zoanthids were the most common and abundant organisms on these stalks, though the stalks also supported filter-feeding serpulid polychaetes, hydroids, and stalked ascidians, as well as more than 100 species of small metazoans and foraminiferans nestled among the epifauna, and apparently feeding on particulate material trapped on the sponge stalks (Beaulieu 2001a). Although there was no significant correlation between height on the sponge stalk and the number of taxa present, some species of large suspension feeders were concentrated near the tops of the stalks where currents were more than twice as strong as near the bottom (Beaulieu 2001a). Sheet-like colonial zoanthids tended not to be colonized by other organisms, possibly because of chemical defenses. Moreover, their colonial growth form enabled their domination on the most desirable portions of the stalks, probably because of competitive dominance in overgrowth interactions. Beaulieu (2001b) also studied recruitment on defaunated sponge stalks over periods of several months. Many of the dominant species such as the zoanthids did not recruit during this time; recruitment was highest in serpulids and foraminiferans, which were also two of the most numerically dominant groups on the natural stalks.

Zoanthids are also common on the shells of deep-sea hermit crabs, and on the spines of cidaroid sea urchins, which also support specialized veruccimorph and scalpellid barnacles. Many species have been described, but there have been no ecological studies on the factors controlling their distributions, and the larval biology of deep-sea zoanthids remains completely unknown (Ryland et al. 2000).

An interesting case of epizooism has recently been described and analyzed by Järnegren et al. (2005). The bivalve file-shell *Acesta oophaga* lives as an epibiont on cold-seep tubeworms, *Lamellibrachia luymesi*, in the Gulf of Mexico (Fig. 3.1o). The shells of the bivalve are modified to surround the tube of the vestimentiferan worm so as hold the anterior plume within the mantle cavity. Virtually all individuals are on females, rather than males, and isotopic work has revealed the reason: the bivalves feed on the lipid-rich buoyant eggs of their continuously breeding host (Järnegren et al. 2005).

3.5 Conclusions

There is probably more hard-bottom habitat in the deep sea than in shallow water, and the deep-water hard-bottom fauna is much more diverse and common than generally supposed. A considerable literature spanning more than 150 years describes the taxonomy and biogeography of this fauna. However, with the major exception of a few frequently visited hydrothermal vent systems, most ecological work on deep-sea hard bottoms has been correlative and descriptive in nature. Ecological work requiring repeated visits to a site is expensive because of reliance on submersibles and ROVs, yet there are tremendous opportunities to explore community dynamics in these extensive and diverse habitats. A major limitation is imposed by the very long lives of some organisms such as gorgonians (which may live for thousands of years), and the apparently low recruitment and growth rates in some of these systems. However, dynamic communities dominated by opportunistic and short-lived species are also common, and these systems are very amenable to manipulative experimentation.

Acknowledgements Our work on hard bottoms in the deep sea has been supported by grants from the U.S. National Science Foundation (most recently, OCE-0527139), the National Undersea Research Program (NOAA/NURP Hawaii and Wilmington), the NOAA Office of Ocean Exploration, and the Mineral Management Service.

References

Beaulieu SE (2001a) Life on glass houses: sponge stalk communities in the deep sea. Mar Biol 138(4):803–817

Beaulieu SE (2001b) Colonization of habitat islands in the deep sea: recruitment to glass sponge stalks. Deep-Sea Res I 48(4):1121–1137

Black MB, Lutz RA, Vrijenhoek RC (1994) Gene flow among vestimentiferan tube worm (*Riftia pachyptila*) populations from hydrothermal vents in the eastern Pacific. Mar Biol 120:33–39

Boehlert GW (1988) Current-topography interactions at mid-ocean seamounts and the impact on pelagic ecosystems. Geojournal 16(1):45–52

Boehlert GW, Genin A (1987) A review of the effects of seamounts on biological processes. In: Keating BH., Fryer P., Batiza R, Boehlert GW (eds) Seamounts islands and atolls. Geophys Monogr 43:319–334

Braby CE, Rouse GW, Johnson SB, Jones WJ, Vrijenhoek RC (2007) Bathymetric and temporal variation among *Osedax* boneworms and associated megafauna on whale-falls in Monterey Bay, California. Deep-Sea Res I 54:1773–1791

Brooke S, Young CM (2003) Embryogenesis and larval biology of the ahermatypic scleractinian *Oculina varicosa*. Mar Biol 146:665–675

Calder DR (2000) Assemblages of hydroids (Cnidaria) from three seamounts near Bermuda in the Western North Atlantic. Deep-Sea Res 1 Oceanogr Res Pap 47:1125–1139

Chave EH, Jones AT (1991) Deep-water megafauna of the Kohala and Haleakala slopes, Alenuihaha Channel, Hawaii. Deep-Sea Res 38:781–803

Chave EH, Malahoff A (1998) In deeper waters: photographic studies of Hawaiian deep-sea habitats and life forms. University of Hawaii Press

Childress JJ, Fisher CR (1992) The biology of hydrothermal vent animals: physiology, biochemistry, and autotrophic symbioses. Oceanogr Mar Biol Annu Rev 30:61–104

Creasey S, Rogers AD, Tyler PA (1996) Genetic comparison of two populations of the deep-sea vent shrimp *Rimicaris exoculata* (Decapoda: Bresiliidae) from the Mid-Atlantic Ridge. Mar Biol 125:473–482

Darwin C (1842) The structure and distribution of coral reefs. J. Murray, London

Dayton PK, Newman WA, Oliver J (1982) The vertical zonation of the deep-sea Antarctic acorn barnacle, *Bathylasma corolliforme* (Hoek): experimental transplants from the shelf into shallow water. J Biogeogr 9:95–109
De Forges BR, Koslow JA, Poore GCB (2000) Diversity and endemism of the benthic seamount fauna in the southwest Pacific. Nature 405:944–946
Desbruyeres D (1998) Temporal variations in the vent communities on the East Pacific Rise and Galapagos Spreading Centre: a review of present knowledge. Cah Biol Mar 39:241–244
Desbruyeres D, Toulmond A (1998) A new species of hesionid worm, *Hesiocaeca methanicola* sp. nov (Polychaeta: Hesionidae) living in ice-like methane hydrates in the deep Gulf of Mexico. Cah Biol Mar 39:93–98
Desbruyeres D, Segonzak M, Bright M (2006) Handbook of deep-sea hydrothermal vent fauna. Denisia 18:1–544
Eckelbarger KJ, Young CM, Brooke S, Ramirez Llodra E, Tyler PA (2001) Reproduction, gametogenesis and early development in the methane "ice worm" *Hesoicoeca methanicola* from the Louisiana Slope. Mar Biol 138:761–775
Emson RH, Young CM (1994) The feeding mechanism of a brisingid sea star, *Novodinea antillensis*. Mar Biol 118:433–442
Epifanio CE, Perovich G, Dittel AG, Cary SC (1999) Development and behavior of megalopa larvae and juveniles of the hydrothermal vent crab *Bythograea thermydron*. Mar Ecol Prog Ser 185:147–154
Fiala-Medioni A, Felbeck H (1990) Autotrophic processes in invertebrate nutrition: bacterial symbioses in bivalve mollusks. Comp Physiol 5:49–69
Fisher CR (1990) Chemoautotrophic and methanotrophic symbioses in marine invertebrates. Crit Rev Aquat Sci 2:399–436
Fisher CR (1996) Ecophysiology of primary production at deep-sea vents and seeps. Biosyst Ecol Ser 11:313–336
Fisher CR, MacDonald IR, Sassen R, Young CM, Macko SA, Hourdez S, Carney RS, Joye S, McMullin E (2000) Methane ice worms: *Hesiocaeca methanicola* colonizing fossil fuel reserves. Naturwissenschaften 87:184–187
Forges BR, Koslow JA, Poore GCB (2000) Diversity and endemism of the benthic seamount fauna in the southwest Pacific. Nature 405:944–947
France SC, Hessler RR, Vrijenhoek RC (1992) Genetic differentiation between spatially-disjunct populations of the deep-sea hydrothermal vent-endemic *Ventiella sulfuris*. Mar Biol 114:552–556
Freiwald A, Roberts JM (eds) (2005) Cold-water corals and ecosystems. Springer, Berlin Heidelberg New York
Fustec A, Desbruyeres D, Juniper SK (1987) Deep-sea hydrothermal vent communities at 13°N on the East Pacific Rise: microdistribution and temporal variations. Biol Oceanogr 4:121–164
Gage JD, Tyler PA (1991) Deep-sea biology: a natural history of organisms at the deep-sea floor. Cambridge University Press, Cambridge
Gardner WD, Sullivan LG, Thorndike EM (1984) Long-term photographic, current and nephelometer observations of manganese nodule environments in the Pacific. Earth Planet Sci Lett 70:95–109
Genin A, Dayton PK, Lonsdale PF, Speiss FN (1986) Corals on seamount peaks provide evidence of current acceleration over deep-sea topography. Nature 322:59–61
Genin A, Paull CK, Dillon WP (1992) Anomalous abundances of deep-sea fauna on a rocky bottom exposed to strong currents. Deep-Sea Res 39:293–302
Governor B, Fisher CR (2006) Experimental evidence of habitat provision by aggregations of *Riftia pachyptila* at hydrothermal vents on the East Pacific Rise. Mar Ecol 28:3–14
Governor B, Le Bris N, Gollner S, Glanville J, Aperghis A, Hourdez S, Fisher CR (2005) Epifaunal community structure associated with *Riftia pachyptila* aggregations in chemically different hydrothermal vent habitats. Mar Ecol Prog Ser 305:67–77

Gregg TKP, Fink JH (1995) Quantification of lava flow morphologies through analog experiments. Geology 23:73–76

Grigg RW, Malahoff A, Chave EH, Landahl J (1987) Seamount benthic ecology and potential environmental impact from manganese crust mining. In: Keating BH, Fryer P, Batiza R, Boehlert GW (eds) Seamounts, islands and atolls. Geophys Monogr 43:379–390

Hannington MD, Honasson IR, Herzig PM, Petersen S (1995) Physical and chemical processes of seafloor mineralization at mid-ocean ridges. In: Humphries SE, Zierenberg RA, Mullineaux LS, Thomson RE (eds) Seafloor hydrothermal systems: physical, chemical and geochemical interactions. Geophys Monogr 91:115–157

Haymon RM (1983) Growth history of black smoker hydrothermal chimneys. Nature 301:695–698

Haymon RM, Fornari DJ, Von Damm KL, Lilley MD, Perfit MR, Edmond JM, Shanks WC III, Lutz RA, Grebmeier JM, Carbotte S, Wright D, McLaughlin E, Smith M, Beedle N, Olson E (1993) Volcanic eruption of the mid-ocean ridge along the East Pacific Rise crest at 9° 45-52'N: direct submersible observations of sea-floor phenomena associated with an eruption event in April, 1991. Earth Planet Sci Lett 119:85–101

Hessler RR, Smithey WM Jr, Keller CH (1985) Spatial and temporal variation of giant clams, tubeworms and mussels at deep-sea hydrothermal vents. Biol Soc Wash Bull 6:411–428

Hessler RR, Smithey WM, Keller CH (1988) Temporal change in megafauna at the Rose Garden hydrothermal vent (Galapagos Rift: eastern tropical Pacific). Deep-Sea Res 35:1681–1709

Hunt HL, Metaxas A, Jennings RM, Halanych KM, Mullineaux LS (2004) Testing biological control of colonization by vestimentiferan tubeworms at deep-sea hydrothermal vents (East Pacific Rise, 9 degrees 50'N). Deep-Sea Res I 51:225–235

Järnegren J, Tobias CR, Macko SA, Young CM (2005) Egg predation fuels unique species association at deep sea hydrocarbon seeps. Biol Bull 209:87–93

Jinks RN, Markley TL, Taylor EE, Perovich G, Dittel AI, Epifanio CE, Cronin TW (2002) Adaptive visual metamorphosis in a deep-sea hydrothermal vent crab. Nature 420:68–70

Johnson KS, Childress JJ, Hessler RR, Sakamoto-Arnold CM, Beehler CL (1988) Chemical and biological interactions in the Rose Garden hydrothermal vent field, Galapagos spreading center. Deep-Sea Res 35:1723–1744

Jorgensen CB (1955) Quantitative aspects of filter feeding in invertebrates. Biol Rev Camb Philos Soc 30:391–454

Jumars PA, Gallagher ED (1982) Deep-sea community structure: three plays on the benthic proscenium. In: Ernst WG, Morin JG (eds) The environment of the deep sea. Prentice Hall, Englewood Cliffs, NJ, pp 217–285

Keating BH, Fryer P, Batiza R, Boehlert GW (1987) Seamounts, islands and atolls. Geophys Monogr 43

Kim SL, Mullineaux LS (1998) Distribution and near-bottom transport of larvae and other plankton at hydrothermal vents. Deep-Sea Res II 45:423–440

Kim SL, Mullineaux LS, Helfrich KR (1994) Larval dispersal via entrainment into hydrothermal vent plumes. J Geophys Res 99:12,655–12,665

Kitchingman A, Lai S, Morato T, Pauly D (2007) How many seamounts are there and where are they located? In: Pitcher TJ, Morato T, Hart PJB, Clark MR, Haggan N, Santos RS (eds) Seamounts: ecology, conservation and management. Blackwell, Oxford, Fish and Aquatic Resources Series, chap 2, pp 26–40

Koslow T (2007) The silent deep. University of Chicago Press, Chicago, IL

Lutz RA, Shank TM, Evans R (2001) Life after death in the deep sea. Am Sci 89:422–431

Lutz RA, Shank TR, Fornari DJ, Haymon RD, Lilley MD, Von Damm K, Desbruyeres D (2002) Rapid growth at deep-sea vents. Nature 371:663–664

Macurda DB Jr, Meyer DL (1974) Feeding posture of modern stalked crinoids. Nature 247(5440):394–396

Maldonado M, Young CM (1996) Bathymetric patterns of sponge distribution on the Bahamian Slope. Deep-Sea Res I 43:897–915

Maldonado M, Carmona C, Uriz MJ, Cruzado A (1999) Decline in Mesozoic reef-building sponges explained by silicon limitation. Nature 401:785–788
Marsh A, Mullineaux L, Young CM, Manahan DL (2001) Larval dispersal potential of the tube-worm *Riftia pachyptila* along a deep-ocean ridge axis. Nature 411:77–80
McClintock JB, Baker BJ, Baumiller TK, Messing CG (1999) Lack of chemical defense in two species of stalked crinoids: support for the predation hypothesis for Mesozoic bathymetric displacement. J Exp Mar Biol Ecol 232:1–7
McDonald IR, Bohrmann G, Escobar E, Abegg F, Blanchon P, Blinova V, Bruckmann W, Drews M, Eisenhauer A, Han X, Heeschen K, Meier F, Mortera C, Naehr T, Orcutt B, Bernard B, Brooks J, de Farago M (2004) Asphalt volcanism and chemosynthetic life in the Campeche Knolls, Gulf of Mexico. Science 304:999–1002
McLean JH, Harasewych MG (1995) Review of Western Atlantic species of cocculinid and pseudococculinid limpets, with descriptions of new species (Gastropoda: Cocculiniformia). Natural History Museum Los Angeles County Contrib Sci 453:1–33
Messing CG, Neumann AC, Lang JC (1990) Biozonation of deep-water lithoherms and associated hardgrounds in the northeastern Straits of Florida. Palaios 5(1):15–33
Messing CG, Reed JK, Brooke SD, Ross SW (2008) Deep-water coral reefs of the United States. In: Riegl B, Dodge RE (eds) Coral reefs of the USA. Springer, Berlin Heidelberg New York, pp 763–787
Micheli F, Peterson CH, Mullineaux LS, Fisher CR, Mills SA, Sancho G, Johnson GA, Lenihan HS (2002) Predation structures communities at deep-sea hydrothermal vents. Ecol Monogr 72:365–382
Monniot C (1979) Adaptations of benthic filtering animals to the scarcity of suspended particles in deep water. Ambio Spec Rep 6:73–74
Monniot C, Monniot F (1978) Recent work on the deep-sea tunicates. Oceanogr Mar Biol Annu Rev 16:181–228
Moore JG, Clague DA (2004) Hawaiian submarine manganese-iron oxide crusts—a dating tool? Geol Soc Am Bull 115:337–347
Motokawa T (1984) Catch connective tissue in echinoderms. Biol Rev 59:255–270
Mullineaux LS (1987) Organisms living on manganese nodules and crusts: distribution and abundance at three North Pacific sites. Deep-Sea Res 34:165–184
Mullineaux LS (1988) The role of initial settlement in structuring a hard-substratum community in the deep sea. J Exp Mar Biol Ecol 120:247–261
Mullineaux LS (1989) Vertical distributions of the epifauna on manganese nodules: implications for feeding and settlement in flow. Limnol Oceanogr 34(7):1247–1262
Mullineaux LS (1994) Implications of mesoscale flows for dispersal of deep-sea larvae. In: Young CM, Eckelbarger KJ (eds) Reproduction, larval biology and recruitment of the deep-sea benthos. Columbia University Press, Irvington, NY, pp 201–223
Mullineaux LS, Butman CA (1990) Recruitment of encrusting benthic invertebrates in boundary-layer flows: a deep water experiment on Cross Seamount. Limnol Oceanogr 35:409–423
Mullineaux LS, Mills SW (1997) A test of the larval retention hypothesis in seamount-generated flows. Deep-Sea Res 44:745–770
Mullineaux LS, Mills SW, Goldman E (1998) Recruitment variation during a pilot colonization study of hydrothermal vents (9°50'N East Pacific Rise). Deep-Sea Res II 45:441–464
Mullineaux LS, Fisher CR, Peterson CH, Schaeffer SW (2000) Vestimentiferan tubeworm succession at hydrothermal vents: use of biogenic cues to reduce habitat selection error? Oecologia 123:275–284
Mullineaux LS, Peterson CH, Micheli F, Mills SW (2003) Successional mechanism varies along a gradient in hydrothermal fluid flux at deep-sea vents. Ecol Monogr 73:523–542
Neumann AC, Kofoed JW, Keller GH (1977) Lithoherms in the Straits of Florida. Geology 5:4–10
Oschmann W (1990) Dropstones—rocky mini-islands in high-latitude pelagic soft substrate environments. Senckenb marit 21:55–75

Parker T, Tunnicliffe V (1994) Dispersal strategies of the biota on an oceanic seamount: implications for ecology and biogeography. Biol Bull 187:336–345

Paull CK, Neumann AC, am Ende BA, Ussler W, Rodriguez NM (2000) Lithoherms on the Florida Hatteras Slope. Mar Geol 136:83–101

Pile AJ, Young CM (2006a) Consumption of bacteria by larvae of a deep-sea polychaete. Mar Ecol 27:15–19

Pile AJ, Young CM (2006b) The natural diet of a hexactinellid sponge: benthic-pelagic coupling in a deep-sea microbial food web. Deep-Sea Res I 53:1148–1156

Pitcher TJ, Morato T, Hart PJB, Clark MR, Haggan N, Santos RS (2007) Seamounts: ecology, fisheries and conservation. Blackwell, Oxford

Pradillon F, Shillito B, Young CM, Gaill F (2001) Deep-sea ecology: developmental arrest in vent worm embryos. Nature 413:698–699

Pradillon F, Le Bris N, Shillito B, Young CM, Gaill F (2005) Influence of environmental conditions on early development of the hydrothermal vent polychaete *Alvinella pompeijana*. J Exp Biol 208:1551–1561

Reed J (2002) Comparison of deep-water coral reefs and lithoherms off southeastern USA Hydrobiologia 471:57–69

Rittschoff D, Forward RB, Cannon G, Welch JM, McClary M, Holm ER, Clare AS, Conova S, McKelvey LM, Bryan P, Van Dover CL (1998) Cues and context: larval responses to physical and chemical cues. Biofouling 12:31–44

Rogers AD (1994) The biology of seamounts. Adv Mar Biol 30:305–351

Rogers AD (1999) The biology of *Lophelia pertusa* (Linnaeus 1758) and other deep-water reef-forming corals and impacts from human activities. Int Rev Hydrobiol 84:315–406

Rouse GW, Wilson NG, Goffredi SK, Johnson SB, Smart T, Widmer C, Young CM, Vrijenhoek RC (2009) Spawning and development in *Osedax* boneworms (Siboglinidae, Annelida). Mar Biol 156:395–405

Rouse GW, Goffredi SK, Vrijenhoek RC (2004) *Osedax*: bone-eating marine worms with dwarf males. Science 305:668–671

Ryland JS, de Putron S, Scheltema RS, Cimonides PJ, Zhadan DG (2000) Semper's (zoanthid) larvae: pelagic life, parentage and other problems. Hydrobiologia 440:191–198

Sancho G, Fisher CR, Mills S, Micheli F, Johnson GA, Lenihan HS, Peterson CH, Mullineaux LS (2005) Selective predation by the zoarcid fish *Thermarces cerberus* at hydrothermal vents. Deep-Sea Res 152:837–844

Sarrazin J, Robigou V, Juniper SK, Delaney J (1997) Biological and geological dynamics over four years on a high-temperature sulphide structure at the Juan de Fuca Ridge hydrothermal observatory. Mar Ecol Prog Ser 153:5–24

Shank TM, Formari DJ, von Damm KL, Lilley MD, Haymon RM, Lutz RA (1998) Temporal and spatial patterns of biological community development at nascent deep-sea hydrothermal vents (9°50'N, East Pacific Rise). Deep-Sea Res 45:465–515

Smith DK (1991) Seamount abundances and size distributions, and their geographic variations. Adv Mar Biol 5:197–210

Smith CR, Baco A (2003) Ecology of whale falls at the deep-sea floor. Oceanogr Mar Biol Annu Rev 41:311–354

Smith CR, Demopoulos AWJ (2003) The deep Pacific ocean floor. In: Tyler PA (ed) Ecosystems of the world: ecosystems of the deep ocean. Elsevier, Amsterdam, pp 179–218

Staudigel H, Hart S, Pile A, Bailey B, Baker E, Brooke S, Haucke L, Hudson I, Jones D, Koppers A, Konter J, Lee R, Pietsch T, Tebo B, Templeton A, Zierenberg R, Young CM (2006) Vailulu'u Seamount, Samoa: life and death on an active submarine volcano. Proc Natl Acad Sci 103:6448–6453

Tebo BM, Barger JR, Clement BG, Dick GJ, Murray KJ, Parker D, Verity R, Webb SM (2004) Biogenic manganese oxides: properties and mechanisms of formation. Annu Rev Earth Planet Sci 32:287–328

Tebo BM, Templeton AS, Johnson HA, McCarthy J (2005) Geomicrobiology of Mn(II)-biomineralization. Trends Microbiol 13:421–428
Templeton AS, Staudigel H, Tebo BM (2005) Diverse Mn(II)-oxidizing bacteria isolated from submarine basalts at Loihi Seamount. Geomicrobiol J 22:129–137
Tunnicliffe V (1988) Biogeography and evolution of hydrothermal vent fauna in the eastern Pacific Ocean. Proc R Soc Lond B 233:347–366
Tunnicliffe V (1991) The biology of hydrothermal vents: ecology and evolution. Oceanogr Mar Biol Annu Rev 29:319–417
Tunnicliffe V, Embley RW, Holden JF, Butterfield DA, Massoth GJ, Juniper SK (1997) Biological colonization of new hydrothermal vents following an eruption on Juan de Fuca Ridge. Deep-Sea Res 44:1627–1644
Tunnicliffe V, Juniper SK, Sibuet M (2003) Reducing environments of the deep-sea floor. In: Tyler PA (ed) Ecosystems of the world: ecosystems of the deep-sea floor. Elsevier, Amsterdam, pp 81–110
Turner RD (1973) Wood-boring bivalves, opportunistic species in the deep sea. Science 180:1377–1379
Turner RD (1978) Wood mollusks and deep-sea food chains. Am Malacol Bull (1977):13–19
Tyler PA (1995) Conditions for the existence of life at the deep-sea floor: an update. Oceanogr Mar Biol Annu Rev 33:221–244
Tyler PA, Young CM (1999) Reproduction and dispersal at vents and cold seeps. J Mar Biol Assoc UK 79:193–208
Tyler PA, Zibrowius H (1992) Submersible observations of the invertebrate fauna on the continental slope southwest of Ireland (NE Atlantic Ocean). Oceanol Acta 15:211–226
Tyler PA, Young CM, Dove F (2007) Settlement, growth and reproduction in the deep-sea wood-boring bivalve mollusc *Xylophaga depalmai*. Mar Biol 343:151–159
Van Dover CL (2000) The ecology of deep-sea hydrothermal vents. Princeton University Press, Princeton, NJ
Van Dover CL (2003) Variation in community structure within hydrothermal vent mussel beds of the East Pacific Rise. Mar Ecol Prog Ser 253:55–56
Van Dover CL, German CR, Speer KG, Parson LM, Vrijenhoek RC (2002) Evolution and biogeography of deep-sea vent and seep invertebrates. Science 395:1253–1257
Verlaan PA (1992) Benthic recruitment and manganese crust formation on seamounts. Mar Biol 113:171–174
Voigt JR (2007) Experimental deep-sea deployments reveal diverse Northeast Pacific wood-boring bivalves of Xylophagainae (Myoida: Pholadidae). J Mollus Stud 73:377–391
Vrijenhoek RC (1997) Gene flow and genetic diversity in naturally fragmented metapopulations of deep-sea hydrothermal vent animals. J Heredity 88:285–293
Vrijenhoek RC, Shank T, Lutz RA (1998) Gene flow and dispersal in deep-sea hydrothermal vent animals. Cah Biol Mar 39:363–366
Wilkie IC, Emson RH (1988) Mutable collagenous tissues and their significance for echinoderm palaeontology and phylogeny. In: Paul CRC, Smith AB (eds) Echinoderm phylogeny and evolutionary biology. Clarendon Press, Oxford, pp 311–330
Wilkie IC, Emson RH, Young CM (1993) Smart collagen in sea lilies. Nature 366:519–520
Wilkie IC, Emson RH, Young CM (1994) Variable tensility of the ligaments in the stalk of a sea-lily. Comp Biochem Physiol 109A:633–641
Wishner K, Levin LA, Gowing M, Mullineaux L (1990) Involvement of the oxygen minimum in benthic zonation on a deep seamount. Nature 346:57–59
Young CM (2003) Reproduction, development and life-history traits. In: Tyler PA (ed) Ecosystems of the world: ecosystems of the deep oceans. Elsevier, Amsterdam, pp 381–426
Young CM, Braithwaite LF (1980) Orientation and current-induced flow in the stalked ascidian *Styela montereyensis*. Biol Bull 159:428–440

Chapter 4
Epibiosis: Ecology, Effects and Defences

Martin Wahl

4.1 Sessile Mode of Life

The high density and viscosity of water directly and indirectly favour a sessile mode of life, even for animals (for details, see Chap. 1). As a consequence, hard substratum may become a limiting resource in this medium. Competition for settlement substratum may have driven the evolution of a lifestyle which is rare on land but common in and typical for the marine benthos: epibiosis.

In an early paper (Wahl 1989), I have suggested some definitions around this theme which since then have evolved a little: 'epibiosis' is the spatial association between a substrate organism ('basibiont') and a sessile organism ('epibiont') attached to the basibiont's outer surface without trophically depending on it. The built-in restrictions enable epibiosis to be distinguished from ectoparasitism or trophic symbiosis, or from 'visiting' motile animals simply engaging in an ephemeral relationship. Epizoans and epiphytes are animal and alga epibionts respectively. There are as yet no established terms for epibiotic prokaryotes or fungi: epibacteria and epifungi might represent a logical solution.

4.2 Establishment of an Epibiotic Community

'Fouling' is the process of colonization of a solid/fluid interface. Any solid surface, living or non-living, exposed to natural seawater is subject to fouling. The various stages of fouling have been described by successional (Davis et al. 1989) and probabilistic models (Clare et al. 1992; Maki and Mitchell 2002). The entire process is usually composed of four phases (Fig. 4.1): thermodynamic adsorption of macromolecules, bacterial attachment, unicellular colonization, settlement of animal and macroalgae (cf. details in Dexter and Lucas 1985; Wahl 1997b; Railkin 2004). While these phases frequently happen sequentially, this does not prove causality. The sequence could rather represent a gradient of availability and reaction time of the different groups of colonizers: macromolecules are always present and can react quickly; bacterial, protist and fungal densities in the water column do follow some seasonal cycle in mid-latitudes and react slower in cold water but there

M. Wahl (ed.), *Marine Hard Bottom Communities*, Ecological Studies 206,
DOI: 10.1007/978-3-540-92704-4_4, © Springer-Verlag Berlin Heidelberg 2009

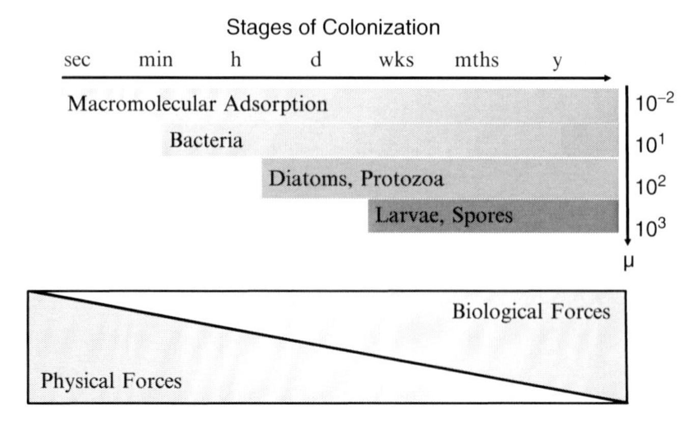

Fig. 4.1 Commonly observed phases of colonization at a solid/liquid interface in the aquatic environment. The forces driving the process shift from physical to biological during the first days

is some heterotrophic microbial colonization occurring at any time of year; autotroph unicellular forms like animal larvae and macroalgal propagules colonize very little during winter.

However, some conditioning of the substratum by colonization undoubtedly takes place (e.g. Railkin 2004). Macromolecular adsorption masks the physicochemical properties of the substratum and leads to an energy accumulation at the interface which is attractive to bacteria. Physicochemical properties of biofilms (bacteria, protists, fungi) and their exudates can attract, repel or not affect further microbial or macrobial colonizers. Settlement of further epibionts (mostly multicellulars) can be induced (Bryan et al. 1997; Harder et al. 2002) or inhibited (Mearns-Spragg et al. 1998; Armstrong et al. 2001; Dobretsov and Qian 2004) by the presence and/or activity of a bacterial biofilm. The main pattern seems to be that the activities of bacteria are strain-specific and may be modulated in a multi-strain assemblage (Boyd et al. 1999; Dobretsov and Qian 2004), and that the response of colonizers is species-specific (Wahl et al. 1994; Maximilien et al. 1998; Egan et al. 2000) and changes with the age of the propagule (Railkin 2004; Gribben et al. 2006). In many biofilms, however, less than 20% of the bacteria may exhibit fouling-modulating activity (Kanagasabhapathy et al. 2006). The most common effect described for biofilms on inert substrata has recently been suggested to be one of facilitation (Wieczorek and Todd 1998), especially with regard to the settlement of animal and alga propagules. However, any generalization is difficult because the interactions between biofilms and multicellular settlers are species- and context-specific (e.g. Wahl 2008).

Multicellular colonizers also produce fouling-modulating metabolites which affect the survival or, more often, the settlement behaviour of conspecifics or other species (e.g. Railkin 2004). They also modify the three-dimensional structure of the substratum. Different scales of rugosity are favourable to different colonizer species (Bourget and Harvey 1998; Wahl and Hoppe 2002). Thus, barnacles and tube-building polychaetes facilitate each other's recruitment by the structure they provide (Thieltges and Buschbaum 2007). Mussel larvae preferentially settle on filamentous structure, a behaviour which may be enhanced by algal exudates and certain biofilms (Dobretsov 1999 but see Davis and Moreno 1995). Numerous settlement inducers have recently been described (Hadfield and Paul 2001), e.g. the chemical substance responsible for the notoriously gregarious settlement of the barnacle *Balanus amphitrite* (Dreanno et al. 2006).

Of course, settlement and attachment are only the first stages of colonization. Direct and indirect interactions among epibionts, interactions with the host, and fluctuating environmental conditions continue to drive the dynamics of the epibiotic community.

4.3 Consequences of Epibioses

Most interactions between an aquatic organism and its biotic and abiotic environment are somehow linked to a set of properties of its outer surface: through this interface gases, dissolved nutrients or ions are taken up or excreted; light energy is absorbed; chemical, acoustic, mechanical and visual signals are perceived and emitted. Colour, texture, wettability, electrical charge and microtopography govern diverse processes such as heat exchange, friction or particle adherence. Together with overall size and shape, surface friction determines the drag an organism experiences. Obviously, some or all of these properties and functions are modulated when a living surface becomes colonized (Table 4.1). Depending on the environmental context and on the composition and density of the epibiotic assemblage, the effect on

Table 4.1 Potential advantages and disadvantages of an epibiotic association for the two partners (for details, see Wahl 1989, 1997a)

Possible beneficial effects	Possible detrimental effects
For epibionts	
Gain of attachment site	Unstable substratum
Expanding substratum	Exuded toxins
Exuded nutrients from basibiont	Risky habitat change
Favourable hydrodynamics	Shared doom
Shock-absorbing substratum	
Favourable irradiation	
Free transport	
Associational resistance	
For basibionts	
Camouflage	Weight increase
Protection	Drag & friction increase
Nutrients or vitamins from epibionts	Decreased flexibility
Drag reduction	Increased brittleness
Improvement of housing	Increased siltation
Associational resistance	Insulation effect
	Mechanical damage
	Chemical damage
	Competition for nutrients with epibionts
	Shared doom
	Increased susceptibility to predation

the basibiont can be specific or general, weak or strong, direct or indirect, beneficial or detrimental (reviewed in detail by Railkin 2004; Wahl 2008).

An adsorbed film of sugars and proteins ('chemical fouling') may modify surface charge and wettability but will not interfere with other functions. An epibacterial assemblage already constitutes a physiological filter which may process molecules exuded by the basibiont or those reaching the basibiont's surface from the surrounding water (e.g. Saroyan 1968). The biofilm may further alter the chemical conditions at the interface through its metabolic activities (concentrations of O_2, CO_2, nutrients, H^+; e.g. Araujo Jorge et al. 1992; Thevanathan et al. 2000). This may benefit or damage the host. Biofilms of a certain thickness and/or containing pigmented components (e.g. diatoms) can be expected to isolate the basibiont's surface from vital resources (e.g. light; Costerton et al. 1987). The insulating effect of larger epibionts is additive to that of the biofilm. One particular aspect of 'insulation' is the hiding of the basibiont from optical, tactile or chemical detection. This may be beneficial when the basibiont is hidden from consumers (see below), detrimental epibionts or pathogens (Gilturnes and Fenical 1992). It is detrimental when recognition by symbionts or mates is hindered. This interference of epibionts with the biotic interactions of the host is particularly well studied for predator-prey relationships. Well-defended epibionts like some bacteria, sponges, hydrozoans, actinians, ascidians and some algae deter predation on their basibiont host by gastropods (Cerrano et al. 2001; Dougherty and Russell 2005; Marin and Belluga 2005), urchins (Wahl and Hay 1995), starfishes (Laudien and Wahl 1999, 2004; Marin and Belluga 2005), crabs (Wahl and Hay 1995; Wahl et al. 1997), fishes (Manning and Lindquist 2003) or birds (Prescott 1990). This beneficial effect is called associational resistance. In contrast, predation may also be enhanced by epibionts ('shared doom'). Epibiotic hydrozoans facilitate the detection of burrowing bivalves by their crab predators (Manning and Lindquist 2003). Barnacles on mussel shells improve the grip of crab claws and, thereby, predation pressure on the basibiont (Enderlein et al. 2003). Epibiotic hydrozoans and boring polychaetes may weaken the shell of molluscs, thereby enhancing the success of crushing predators (Bach et al. 2006; Buschbaum et al. 2006). This effect may also be expected from other boring epibionts like some sponges, green algae, phoronids and actinians. Enhanced apparency by epibionts may also increase predation (Threlkeld and Willey 1993). A given epibiont species may have contrasting effects, i.e. increase *and* reduce predation, depending on which basibiont species it grows on (Wahl and Hay 1995) or which predator species is in the vicinity (Manning and Lindquist 2003). Consequently, the nature of epibiont impact on the basibiont is often context-specific. Epibionts on swimming bivalves may harm their hosts by increasing drag and weight but the repellent effect they have on predators may overcompensate this disadvantage, when the predators are present (Forester 1979; Pitcher and Butler 1987). Then again, the negative effects of epibionts may be additive. Barnacle epibionts have multiple effects on their snail basibionts:

1. They increase drag and, consequently, reduce growth (Wahl 1997a).
2. Reduced growth will keep the snails longer in the window of vulnerability (Buschbaum and Reise 1999) with regard to crab predation.
3. Predatory success may further be enhanced by improved handling due to fouling (Enderlein et al. 2003).

4. At the same time, the presence of epibiotic barnacles favours the settlement of shell-boring polychaetes (Thieltges and Buschbaum 2007). The borers weaken the shell and will further extend the time that snails are vulnerable to crab predation.
5. Finally, barnacle epibionts hinder the reproduction of the snails (Buschbaum and Reise 1999), so that the enhanced mortality by extended predation may not be compensated for.

Other examples for complex, and sometimes contrasting, effects of epibionts include filamentous epiphytes on sea grass which (1) increase drag, (2) compete with the host for light and (3) nutrients, (4) may be more palatable to grazers than their host and, thus, deflect grazing pressure, (5) offer shelter from UV radiation and desiccation during emersion and (6) attract mussel recruits, which may be responsible for a fatal increase in weight.

The consequences of an epibiotic way of life for epibionts are somewhat more straightforward (reviewed in Wahl 1997b). The main benefit may be the acquisition of a settlement substratum. Without such, most larvae of sessile animals and all algal propagules would be doomed. The elevated position offers the double advantage of escaping the boundary layer near the bottom, which often is stagnating and shaded. Temporarily strong water velocity may not constitute a serious problem because usually the basibiont has a flexible build, or avoids strong currents by body contraction or motility. Epibionts may also take nutritional advantage of the exudates or sloppy feeding habits of their host. When the basibiont is well defended against predation, the epibiont is likely to benefit from this shelter (associational resistance). On the other hand, epibionts have to cope with or adapt to a limited life expectancy of their substratum and to particular chemical and mechanical conditions at the living surface. The former is a consequence of the limited longevity of the host, or the mechanical dynamics of its outer surface (mucus secretion, sloughing, moulting). The latter may be due to transcutaneous exchanges linked to the basibiont's primary metabolism or to the production and excretion of secondary defensive metabolites. Finally, epibionts on motile basibionts may be exposed to severe changes of environmental variables when the host switches between epi- and endobenthic or between aquatic and aerial habitats. Also, epibionts may become accidental victims to predators of their basibiont.

Thus, an epibiotic association entails a highly complex suite of advantages and disadvantages for both partners. Species often engaging in this association seem to draw a net benefit from their role of epi- or basibiont. Those which do not will avoid settling on a live substratum or evolve mechanisms deterring potential epibionts.

4.4 Distributional Patterns of Epibioses

Epibiotic species and epibiotic communities ('epibioses') are heterogeneously distributed among taxa, individuals and body parts, habitats and/or in time. Body surfaces of potential basibionts which are important for transcutaneous exchange of photons, gases, solutes or particles, which are engaged in particle trapping (filtering devices), or where friction must be minimized should be kept clean of epibionts.

Likewise, organisms which depend on flexibility or light weight should avoid being colonized by rigid or heavy epibionts.

On the colonizer side, a successful colonizer should not grow larger than the substratum organism and reach reproductive maturity before the basibiont surface sloughs or the basibiont dies. In general, the epibiont must be able to cope with all aspects of the basibiont's lifestyle and surface properties.

Such requirements (described in more detail in Wahl 1997b) are the reason why the functional roles of epi- and basibiont are not equally represented in all phyla (Fig. 4.2), as found when over 2,000 epibiont-basibiont pairings where analysed (Wahl and Mark 1999). Most basibiont species seem to be found in larger hard-shelled molluscs and crustaceans. Epibionts are particularly common among small filamentous algae or hydrozoans, encrusting algae or bryozoans, and unicellular algae (bacteria were not recorded).

Within a given species, we find variability of epibiosis at a number of scales. Quantitatively, the amount of epibionts may vary between ontogenetic stages, different parts of the organism, between seasons or among years, and with the fitness of an individual. Generally, older individuals or older parts of an individual tend to be more heavily covered by epibionts (e.g. Maldonado and Uriz 1992; Warner 1997; Fernandez et al. 1998; Dougherty and Russell 2005). Different epibiont species may occupy different microhabitats on a given basibiont individual (e.g. Gili et al. 1993; Patil and Anil 2000). To avoid competition, epibiotic hydroids may

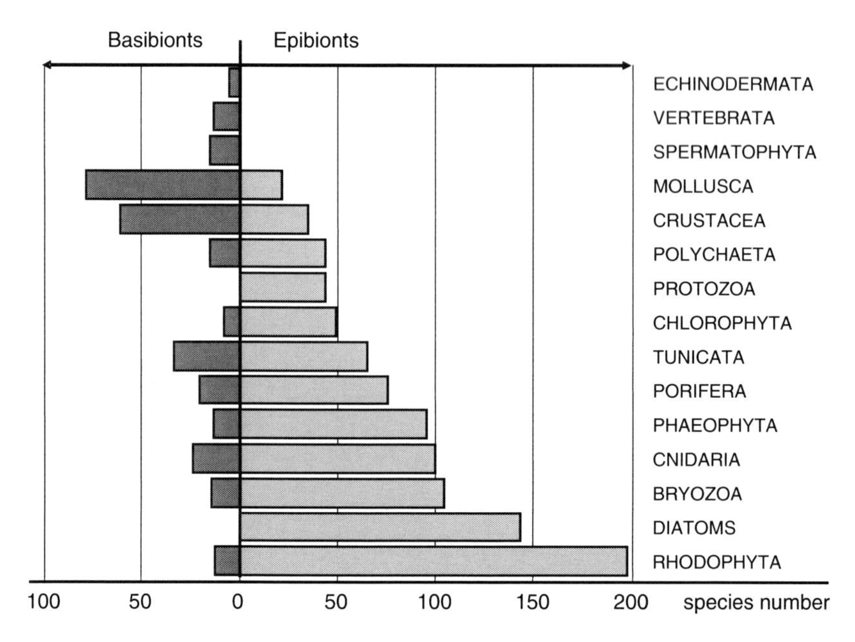

Fig. 4.2 The number of basibiont (*left*) and epibiont (*right*) species described in various phyla (ordered by commonness of epibionts). The qualifications for either functional group tend to be inversely related among the phyla

differentiate trophically (phenotypic plasticity) in response to certain species combinations in a given epibiotic assemblage (Orlov 1997). Also, the composition and quantity of epibioses may differ between sexes within a species (e.g. Maldonado and Uriz 1992; Patil and Anil 2000). Species occurring in different habitats or regions often feature different epibiont species or different degrees of epibiotic coverage (Key et al. 1996; Fernandez et al. 1998; Reiss et al. 2003; Dougherty and Russell 2005). Within the same habitat, different basibiont species may also host different epibiotic communities (e.g. Chiavelli et al. 1993; Davis and White 1994). Specific associations between basibiont and epibionts, however, seem to be the exception. In most surveys, less than 20% of epibionts are restricted to this mode of life, and less than 5% occur exclusively on one basibiont species (Barnes and Clarke 1995; Gutt and Schickan 1998; Cook et al. 1998; Wahl and Mark 1999). With regard to microepibionts, the situation may be slightly different. With the advent of new technologies enabling the identification of non-cultivable strains, an increasing number of papers have reported on host-specific assemblages of epibacteria (e.g. Harder et al. 2003; Lee and Qian 2004; Dobretsov et al. 2006). Seasonal or between-year variability of epibiotic communities is common (e.g. Chiavelli et al. 1993; Davis and White 1994; Fernandez et al. 1998; Dougherty and Russell 2005). One would expect that the susceptibility to epibiosis should also vary among basibiont genotypes and with stress. Genetic causes for variability in epibiosis could reside in different capacities of antifouling defence. Single, or suites of, stressors should jeopardize antifouling defences. However, too few data are published on these two aspects to draw any conclusions.

The conspicuous variability in quantity and quality (i.e. composition) of epibiosis at a number of scales has two main causes. The composition of the pool of locally available potential colonizer species may vary stochastically (e.g. currents), with region for evolutionary reasons, with habitat for ecological or hydrographic reasons, and with season. Which of the available potential epibionts will successfully settle and grow when a substratum becomes available depends on the properties of the basibiont's surface, i.e. its consistency, texture, wettability, the deployment of mechanical, chemical or behavioural defences (see next section), and the presence of previous settlers (e.g. biofilms). All of these may vary at the scales considered.

4.5 Responses of the Host

All aquatic organisms are exposed to fouling pressure. The intensity of this pressure varies with region, habitat and season but some potential settlers are always 'around'. Sometimes, the pressure is so strong that surfaces become completely fouled within 1 week (author's personal observation; Hodson and Burke 1994). Faced with this challenge, potential basibionts may evolve antifouling defences or tolerance of epibionts. Of course, these are only the two extremes of a continuum. Intermediate strategies are differential responses at a spatial (defences of sensitive organs only) or temporal scale (defence when fouling pressure is strong or when the host would be most vulnerable to the effects of epibiosis), a specific defence against the most harmful (i.e. large) epibionts only, or a 'subcontracted' defence (i.e. tolerance of well-defended epibionts like certain bacteria, sponges, hydrozoans, tunicates). Some organisms exhibit a complete and constitutive (permanent) defence, keeping all epibionts (often even including bacteria) at bay (e.g. Wahl and Banaigs 1991; Maximilien et al. 1998; Dobretsov et al. 2004, 2005). Other species tolerate certain epibiont species while rejecting others (e.g. Dobretsov et al. 2004, 2006), which may produce basibiont-specific epibioses. Some examples include bryozoans on kelp (e.g. Manriquez and Cancino 1996; Hurd et al. 2000), sponges on clams (e.g. Marin and Belluga 2005), kamptozoans on ascidians (Wahl and Lafargue 1990), hydrozoans on bivalves (e.g. Manning and Lindquist 2003), and monospecific bacterial communities on lobster eggs (Gilturnes and Fenical 1992) or polyspecific bacterial films on algae (Egan et al. 2000; Dobretsov and Qian 2002) or invertebrates (e.g. Dobretsov and Qian 2004). When only certain epibionts are found on a given organism, speculation is often raised as to whether this is simply due to these particular epibionts' tolerance for the specific conditions (including defences) on the basibiont surfaces, or whether the basibiont actively attracts beneficial epibionts ('pro-fouling') which may reduce drag (e.g. Bernadsky et al. 1993), deter predation (e.g. Cerrano et al. 2001) or inhibit further fouling (e.g. Dobretsov and Qian 2004; Lee et al. 2006).

Epibiosis may be reduced by repelling colonizers before they settle, by inhibiting their attachment, by killing colonizers when they approach or contact the surface, or by removing colonizers periodically (reviewed in Wahl 1989; Clare 1996). Mechanically, settlers may be removed by grooming, sloughing, moulting or abrasion. Inhibition of attachment by a surface wettability in the biocompatible range of 20–30 mN/m has never been demonstrated unequivocally (Becker et al. 2000). Extreme surface pH values (mostly acidic) have been shown in gastropods, echinoderms, ascidians and algae (e.g. Pelletreau and Muller-Parker 2002) but its utility is questionable (Davis and Wright 1989). Numerous studies report on toxic substances killing bacteria, fungi, diatoms, larvae and spores. However, often these compounds were tested at unnatural concentrations and against target organisms not occurring in the natural habitat of the defended species (e.g. Clare 1996; Steinberg et al. 1998), and the admonition that non-toxic but nevertheless repellent chemical activities may be more common than suspected is becoming louder (e.g. Wahl et al. 1994, 1998; Maximilien et al. 1998; Kelly et al. 2003). Lastly, potential basibionts may benefit from some fouling-reducing activity (chemical exudates, filtering activity, grazing) of neighbouring organisms (e.g. Wahl 2001).

In view of the phylogenetic and physiological diversity of potential epibionts, it is not surprising that defences are similarly diverse. In fact, as a general rule, defence systems seem to be complex, i.e. composed of multiple types of defences. Thus, mussels employ a combination of microtopography plus several metabolites plus foot grooming plus cumulative filtration (e.g. Bers et al. 2006). Some ascidians defend their surfaces by sloughing plus numerous different bioactivities plus associative grazing (Wahl and Banaigs 1991). Some algae combine chemical antifouling defences with regular sloughing (Nylund et al. 2005). Not all defence components within a multiple defence system are necessarily expressed at the same time. In ongoing, as yet unpublished studies, we find that at any given time there may be substantial variability of defence activities between conspecifics, and intra-specifically between seasons and habitats. Moreover, defences vary among body parts and in time for a given individual. From the point of view of a potential epibiont, this results in a highly variable, 'flickering' defence landscape represented by the members of a basibiont population expressing different defences to different degrees. The difficulty to adapt to such a multiple-stress situation might explain the scarcity of obligate and specialized epibionts.

Concluding, epibiotic associations are omnipresent, variable and diverse. They have a profound ecological impact, since the presence of epibionts affects the fitness of basibionts directly as well as indirectly, by modulating its interactions with the abiotic and biotic environment (reviewed by Wahl 2008).

References

Araujo Jorge TC, Coutinho CMLM, Aguiar LEV (1992) Sulphate-reducing bacteria associated with biocorrosion—a review. Mem Inst Oswaldo-Cruz 87:329–337

Armstrong E, Yan LM, Boyd KG, Wright PC, Burgess JG (2001) The symbiotic role of marine microbes on living surfaces. Hydrobiologia 461:37–40

Bach CE, Hazlett BA, Rittschof D (2006) Sex-specific differences and the role of predation in the interaction between the hermit crab, *Pagurus longicarpus*, and its epibiont, *Hydractinia symbiolongicarpus*. J Exp Mar Biol Ecol 333:181–189

Barnes DKA, Clarke A (1995) Epibiotic communities on sublittoral macroinvertebrates at Signy Island, Antarctica. J Mar Biol Assoc UK 75:689–703

Becker K, Hormchong T, Wahl M (2000) Relevance of crustacean carapace wettability for fouling. Hydrobiologia 426:193–201

Bernadsky G, Sar N, Rosenberg E (1993) Drag reduction of fish skin mucus: relationship to mode of swimming and size J Fish Biol 42(5):797–800

Bers AV, Prendergast GS, Zurn CM, Hansson L, Head RM, Thomason JC (2006) A comparative study of the anti-settlement properties of mytilid shells. Biol Lett 2:88–91

Bourget E, Harvey M (1998) Spatial analysis of recruitment of marine invertebrates on arborescent substrata. Biofouling 12(1/3):45–55

Boyd KG, Adams DR, Burgess JG (1999) Antibacterial and repellent activities of marine bacteria associated with algal surfaces. Biofouling 14:227–236

Bryan PJ, Rittschof D, Qian PY (1997) Settlement inhibition of bryozoan larvae by bacterial films and aqueous leachates. Bull Mar Sci 61:849–857

Buschbaum C, Reise K (1999) Effects of barnacle epibionts on the periwinkle *Littorina littorea* (L.). Helgoland Mar Res 53:56–61

Buschbaum C, Buschbaum G, Schrey I, Thieltges DW (2006) Shell-boring polychaetes affect gastropod shell strength and crab predation. Mar Ecol Prog Series 329:123–130

Cerrano C, Puce S, Chiantore M, Bavestrello G, Cattaneo-Vietti R (2001) The influence of the epizoic hydroid *Hydractinia angusta* on the recruitment of the Antarctic scallop *Adamussium colbecki*. Polar Biol 24:577–581

Chiavelli DA, Mills EL, Threlkeld ST (1993) Host preference, seasonality, and community interactions of zooplankton epibionts. Limnol Oceanogr 38:574–583

Clare AS (1996) Natural product antifoulants: status and potential. Biofouling 9:211–229

Clare AS, Rittschof D, Gerhart DJ, Maki JS (1992) Molecular approaches to nontoxic antifouling. Invertebr Reprod Dev 22:67–76
Cook JA, Chubb JC, Veltkamp CJ (1998) Epibionts of *Asellus aquaticus* (L.) (Crustacea, Isopoda): an SEM study. Freshwater Biol 39:423–438
Costerton JW, Cheng KJ, Geesey GG, Ladd TI, Nickel JC, Dasgupta M, Marrie TJ (1987) Bacterial biofilms in nature and disease. Annu Rev Microbiol 41:435–464
Davis AR, Moreno CA (1995) Selection of substrata by juvenile *Choromytilus chorus* (Mytilidae)—are chemical cues important. J Exp Mar Biol Ecol 191:167–180
Davis AR, White GA (1994) Epibiosis in a guild of sessile subtidal invertebrates in South-Eastern Australia—a quantitative survey. J Exp Mar Biol Ecol 177:1–14
Davis AR, Wright AE (1989) Interspecific differences in fouling of two congeneric ascidians (*Eudistoma olivaceum* and *E. capsulatum*): is surface acidity an effective defense? Mar Biol 102:491–497
Davis AR, Targett NM, McConnell OJ, Young CM (1989) Epibiosis of marine algae and benthic invertebrates: natural products chemistry and other mechanisms inhibiting settlement and overgrowth. In: Scheuer PJ (ed) BioOrganic Marine Chemistry, vol 3. Springer, Berlin Heidelberg New York, pp 86–114
Dexter SC, Lucas KE (1985) The study of biofilm formation under water by photoacoustic spectroscopy. J Coll Interf Sci 104:15–27
Dobretsov SV (1999) Effects of macroalgae and biofilm on settlement of blue mussel (*Mytilus edulis* L.) larvae. Biofouling 14:153–165
Dobretsov SV, Qian PY (2002) Effect of bacteria associated with the green alga *Ulva reticulata* on marine micro- and macrofouling. Biofouling 18:217–228
Dobretsov S, Qian PY (2004) The role of epibotic bacteria from the surface of the soft coral *Dendronephthya* sp in the inhibition of larval settlement. J Exp Mar Biol Ecol 299:35–50
Dobretsov S, Dahms HU, Qian PY (2004) Antilarval and antimicrobial activity of waterborne metabolites of the sponge *Callyspongia* (Euplacella) *pulvinata*: evidence of allelopathy. Mar Ecol Prog Ser 271:133–146
Dobretsov S, Dahms HU, Qian PY (2005) Antibacterial and anti-diatom activity of Hong Kong sponges. Aquat Microb Ecol 38:191–201
Dobretsov S, Dahms HU, Harder T, Qian PY (2006) Allelochemical defense against epibiosis in the macroalga *Caulerpa racemosa* var. *turbinata*. Mar Ecol Prog Ser 318:165–175
Dougherty JR, Russell MP (2005) The association between the coquina clam *Donax fossor* Say and its epibiotic hydroid *Lovenella gracilis* Clarke. J Shellfish Res 24:35–46
Dreanno C, Matsumura K, Dohnae N, Takio K, Hirota H, Kirby R, Clare AS (2006) An {alpha} 2-macroglobulin-like protein is the cue to gregarious settlement of the barnacle *Balanus amphitrite*. Proc Natl Acad Sci USA 103:14396–14401
Egan S, Thomas T, Holmstrom C, Kjelleberg S (2000) Phylogenetic relationship and antifouling activity of bacterial epiphytes from the marine alga *Ulva lactuca*. Environ Microbiol 2:343–347
Enderlein P, Moorthi S, Rohrscheidt H, Wahl M (2003) Optimal foraging versus shared doom effects: interactive influence of mussel size and epibiosis on predator preference. J Exp Mar Biol Ecol 292:231–242
Fernandez L, Parapar J, Gonzalez-Gurriaran E, Muino R (1998) Epibiosis and ornamental cover patterns of the spider crab *Maja squinado* on the Galician coast, northwestern Spain: influence of behavioral and ecological characteristics of the host. J Crustacean Biol 18:728–737
Forester AJ (1979) The association between the sponge *Halichondria panicea* (Pallas) and the scallop *Chlamys varia* (L.): a commensal–protective mutualism. J Exp Mar Biol Ecol 36:1–10
Gili JM, Abello P, Villanueva R (1993) Epibionts and intermolt duration in the crab *Bathynectes piperitus*. Mar Ecol Prog Ser 98:107–113
Gilturnes MS, Fenical W (1992) Embryos of *Homarus americanus* are protected by epibiotic bacteria. Biol Bull 182:105–108

Gribben PE, Marshall DJ, Steinberg PD (2006) Less inhibited with age? Larval age modifies responses to natural settlement inhibitors. Biofouling 22:101–106
Gutt J, Schickan T (1998) Epibiotic relationships in the Antarctic benthos. Antarctic Sci 10:398–405
Hadfield MG, Paul V (2001) Natural chemical cues for settlement and metamorphosis of marine invertebrate larvae. In: McClintock B, Baker BJ (eds) Marine chemical ecology. CRC Press, Boca Raton, FL, pp 1–610
Harder T, Lam C, Qian PY (2002) Induction of larval settlement in the polychaete *Hydroides elegans* by marine biofilms: an investigation of monospecific diatom films as settlement cues. Mar Ecol Prog Ser 229:105–112
Harder T, Lau SCK, Dobretsov S, Fang TK, Qian PY (2003) A distinctive epibiotic bacterial community on the soft coral *Dendronephthya* sp and antibacterial activity of coral tissue extracts suggest a chemical mechanism against bacterial epibiosis. FEMS Microbiol Ecol 43:337–347
Hodson SL, Burke C (1994) Microfouling of salmon cage netting: a preliminary investigation. Biofouling 8:93–105
Hurd CL, Durante KM, Harrison PJ (2000) Influence of bryozoan colonization on the physiology of the kelp *Macrocystis integrifolia* (Laminariales, Phaeophyta) from nitrogen-rich and -poor sites in Barkley Sound, British Columbia, Canada. Phycologia 39:435–440
Kanagasabhapathy M, Sasaki H, Haldar S, Yamasaki S, Nagata S (2006) Antibacterial activities of marine epibiotic bacteria isolated from brown algae of Japan. Ann Microbiol 56:167–173
Kelly SR, Jensen PR, Henkel TP, Fenical W, Pawlik JR (2003) Effects of Caribbean sponge extracts on bacterial attachment. Aquat Microb Ecol 31:175–182
Key MM, Jeffries WB, Voris HK, Yang CM (1996) Epizoic bryozoans, horseshoe crabs, and other mobile benthic substrates. Bull Mar Sci 58:368–384
Laudien J, Wahl M (1999) Indirect effects of epibiosis on host mortality: seastar predation on differently fouled mussels. PSZNI Mar Ecol 20:35–47
Laudien J, Wahl M (2004) Associational resistance of fouled blue mussels (*Mytilus edulis*) against starfish (*Asterias rubens*) predation: relative importance of structural and chemical properties of the epibionts. Helgoland Mar Res 58:162–167
Lee OO, Qian PY (2004) Potential control of bacterial epibiosis on the surface of the sponge *Mycale adhaerens*. Aquat Microb Ecol 34:11–21
Lee OO, Lau SCK, Qian PY (2006) Defense against epibiosis in the sponge *Mycale adhaerens*: modulating the bacterial community associated with its surface. Aquat Microb Ecol 43:55–65
Maki JS, Mitchell K (2002) Biofouling in the marine environment. In: Bitton G (ed) Encyclopedia of environmental microbiology. Wiley, New York, pp 610–619
Maldonado M, Uriz MJ (1992) Relationships between sponges and crabs—patterns of epibiosis on *Inachus aguiarii* (Decapoda, Majidae). Mar Biol 113:281–286
Manning LM, Lindquist N (2003) Helpful habitant or pernicious passenger: interactions between an infaunal bivalve, an epifaunal hydroid and three potential predators. Oecologia 134:415–422
Manriquez PH, Cancino JM (1996) Bryozoan-macroalgal interactions: do epibionts benefit? Mar Ecol Prog Ser 138:189–197
Marin A, Belluga MDL (2005) Sponge coating decreases predation on the bivalve *Arca noae*. J Mollus Stud 71:1–6
Maximilien R, de Nys R, Holmstrom C, Gram L, Givskov M, Crass K, Kjelleberg S, Steinberg PD (1998) Chemical mediation of bacterial surface colonisation by secondary metabolites from the red alga *Delisea pulchra*. Aquat Microb Ecol 15:233–246
Mearns-Spragg A, Bregu M, Boyd KG, Burgess JG (1998) Cross-species induction and enhancement of antimicrobial activity produced by epibiotic bacteria from marine algae and invertebrates, after exposure to terrestrial bacteria. Lett Appl Microbiol 27:142–146
Nylund GM, Cervin G, Hermansson M, Pavia H (2005) Chemical inhibition of bacterial colonization by the red alga *Bonnemaisonia hamifera*. Mar Ecol Prog Ser 302:27–36
Orlov D (1997) Epizoic associations among the white sea hydroids. Sci Mar 61:17–26

Patil JS, Anil AC (2000) Epibiotic community of the horseshoe crab *Tachypleus gigas*. Mar Biol 136:699–713
Pelletreau KN, Muller-Parker G (2002) Sulfuric acid in the phaeophyte alga *Desmarestia munda* deters feeding by the sea urchin *Strongylocentrotus droebachiensis*. Mar Biol 141:1–9
Pitcher CR, Butler AJ (1987) Predation by asteroids, escape response, and morphometrics of scallops with epizoic sponges. J Exp Mar Biol Ecol 112:233–249
Prescott RC (1990) Sources of predatory mortality in the bay scallop *Argopecten irradians* (Lamarck)—interactions with seagrass and epibiotic coverage. J Exp Mar Biol Ecol 144:63–83
Railkin AI (2004) Marine biofouling: colonization processes and defenses. CRC Press, Boca Raton, FL
Reiss H, Knauper S, Kröncke I (2003) Invertebrate associations with gastropod shells inhabited by *Pagurus bernhardus* (Paguridae)—secondary hard substrate increasing biodiversity in North Sea soft-bottom communities. Sarsia 88:404–414
Saroyan JR (1968) Marine biology in antifouling paints. J Paint Technol 41:285–303
Steinberg PD, De Nys R, Kjelleberg S (1998) Chemical inhibition of epibiota by Australian seaweeds. Biofouling 12:227–244
Thevanathan R, Nirmala N, Manoharan A, Gangadharan A, Rajarajan R, Dhamotharan R, Selvaraj S (2000) On the occurrence of nitrogen fixing bacteria as epibacterial flora of some marine green algae. Seaweed Res Utiln 22:189–197
Thieltges DW, Buschbaum C (2007) Vicious circle in the intertidal: facilitation between barnacle epibionts, a shell boring polychaete and trematode parasites in the periwinkle *Littorina littorea*. J Exp Mar Biol Ecol 340:90–95
Threlkeld ST, Willey RL (1993) Colonization, interaction, and organization of Cladoceran epibiont communities. Limnol Oceanogr 38:584–591
Wahl M (1989) Marine epibiosis. 1. Fouling and antifouling—some basic aspects. Mar Ecol Prog Ser 58:175–189
Wahl M (1997a) Increased drag reduces growth of snails: comparison of flume and in situ experiments. Mar Ecol Prog Ser 151:291–293
Wahl M (1997b) Living attached: aufwuchs, fouling, epibiosis. In: Nagabhushanam R, Thompson MF (eds) Fouling organisms of the Indian Ocean: biology and control technology. Oxford & IBH, New Delhi
Wahl M (2001) Small scale variability of benthic assemblages: biogenic neighborhood effects. J Exp Mar Biol Ecol 258:101–114
Wahl M (2008) Ecological lever and interface ecology: epibiosis modulates the interactions between host and environment. Biofouling 24:427–438
Wahl M, Banaigs B (1991) Marine epibiosis. 3. Possible antifouling defense adaptations in Polysyncraton lacazei (Giard) (Didemnidae, Ascidiacea). J Exp Mar Biol Ecol 145:49–63
Wahl M, Hay ME (1995) Associational resistance and shared doom—effects of epibiosis on herbivory. Oecologia 102:329–340
Wahl M, Hoppe K (2002) Interactions between substratum rugosity, colonization density, and periwinkle grazing efficiency. Mar Ecol Prog Ser 225:239–249
Wahl M, Lafargue F (1990) Marine epibiosis. 2. Reduced fouling on *Polysyncraton lacazei* (Didemnidae, Tunicata) and proposal of an antifouling potential index. Oecologia 82:275–282
Wahl M, Mark O (1999) The predominantly facultative nature of epibiosis: experimental and observational evidence. Mar Ecol Prog Ser 187:59–66
Wahl M, Jensen PR, Fenical W (1994) Chemical control of bacterial epibiosis on ascidians. Mar Ecol Prog Ser 110:45–57
Wahl M, Hay ME, Enderlein P (1997) Effects of epibiosis on consumer-prey interactions. Hydrobiologia 355:49–59
Wahl M, Kroger K, Lenz M (1998) Non-toxic protection against epibiosis. Biofouling 12:205–226
Warner GF (1997) Occurrence of epifauna on the periwinkle, *Littorina littorea* (L.), and interactions with the polychaete *Polydora ciliata* (Johnston). Hydrobiologia 355:41–47
Wieczorek SK, Todd CD (1998) Inhibition and facilitation of settlement of epifaunal marine invertebrate larvae by microbial biofilm cues. Biofouling 12:81–118

Part II
Diversity Patterns and Their Causes

Coordinated by Sean D. Connell

Introduction

Sean D. Connell

Part II of this book provides an overview of some of the most conspicuous patterns of diversity (composition, structure and function of species or their traits) and the spatial processes and temporal events that shape these. The following five chapters start by recognising current patterns of local diversity as an imprint of the number of species in a regional pool (i.e. local–regional patterns; Chaps. 5 and 6) that have been shaped by evolutionary- and regional-scale events (i.e. speciation and extinction; Chap. 8). Subsequent membership of a species to local assemblages represents individual responses of populations to spatial heterogeneity (e.g. gradients; Chap. 7), which are continually modified by temporal changes (e.g. seasonal and climate) and their interaction (e.g. responses to location and timing of disturbance events). Some assemblages can respond in predictable ways to such heterogeneity, and form repeated patterns across large regions, bringing the study of local scales and regional scales onto a common stage. Interfacing local complexity into larger scales and identifying scale-dependency is a persistent and fundamental challenge to a comprehensive understanding of diversity. This requires consideration of how changes in environmental heterogeneity associated with changes in scale affect patterns of biodiversity. Hence, our need to understand appropriate methods of cross-scale observation and experimentation (Chap. 9) and be prepared for this understanding to be under constant discussion and review.

Contemporary patterns of diversity include the outcome of local processes and reflect historical events and processes operating across many timescales. Most ecological observation and experimentation, however, are done on contemporary and local assemblages. Research at these scales is intellectually attractive because biological complexity tends to be greatest at smaller scales, and their research is more tractable to punctual and unambiguous results. As a consequence, ecologists are not always mindful of the temporal context of their results. Diversity in kelp forests, for example, reflects the outcomes of photosynthetic processes that operate on the scale of fractions of a second to minutes, its conversion into biomass operates over scales of minutes to hours, and variations in weather, seasonal and long-term successional change, and their resilience to extreme events vary at the scale of decades to centuries.

M. Wahl (ed.), *Marine Hard Bottom Communities*, Ecological Studies 206,
DOI: 10.1007/978-3-540-92704-4_II, © Springer-Verlag Berlin Heidelberg 2009

The outcome of local processes can be more comprehensively understood by recognising their broader landscape and regional properties. Local environmental conditions reflect their location within mosaics or along gradients of physical and chemical variables (e.g. wave exposure and depth, oceanographic currents and upwelling, rock mineralogy and orientation, longitude and latitude). This spatial heterogeneity enhances variation in beta-diversity by providing different opportunities and constraints for organisms with alternate physical and chemical requirements. The positive relationship between species diversity and size of habitat (and reducing distances between like patches) is one of the most widely held generalities in ecology. Larger patches of habitat generally contain more species than do smaller patches of the same habitats. Larger patches contain more environmental variability, such as gradients (driving concomitant structural variation in habitat-forming species), thereby providing greater opportunities for organisms with different requirements and tolerances to find suitable sites within a patch. We recognise, therefore, that the membership of a species to a local assemblage is limited by the type and size of habitat, its location within patches (e.g. edges and centres), location among patches (e.g. within gradients and mosaics), and configuration of patches relative to alternate habitats that comprise landscapes (e.g. connectivity of interspersed habitat).

Contemporary patterns of diversity, therefore, are a product of the temporal modification of spatial opportunities and constraints for species colonisation. A common stage for this interaction centres on the events that disrupt ecological systems (i.e. disturbances). While disturbances are ubiquitous across systems, they comprise discrete events that are characterised by their type of impact (e.g. wave-induced loss), and spatial (intensity and extent) and temporal propagation (timing, frequency and duration) and damage caused. The resulting biomass loss and creation of new space represent a fundamental source of heterogeneity in sessile assemblages that often reflect mosaics of post-disturbance recovery or alternation of assemblages.

Mosaics of diversity, therefore, encompass patches of variable size and age (reflecting time since a disturbance of given intensity) populated by assemblages of species that are a product of (1) a chance arrival of propagules (lottery for space) and (2) biotic interactions at the given successional stage. The resulting patterns of diversity often represent distinctive ‘community’ types that reflect their location in spatial and temporal settings that change over time. For example, rates of key ecosystem processes (e.g. balance between primary production and loss through consumption) are shaped by variation in spatial patterns (gradients and biogeography), temporal patterns (e.g. temperatures as mediated by season and climate) and unexpected events (e.g. the coincidence of extreme events). While species-specific tolerances to abiotic factors influence the pool of species potentially able to exist within a habitat, the following chapters show how biotic interactions (i.e. strength of positive and negative interactions) determine the membership of a species to the habitat.

An unresolved difficulty of ecological textbooks is the necessary separation of knowledge into individual chapters, potentially creating the false impression that diversity can be best understood as fragments (i.e. the study of subsets of information) categorised into different scales in space (local or regional or biogeographic) and time (local or seasonal or evolutionary). It is worth recognising that the following chapters are intended as helpful abstractions that may act as a broader framework for interpretation of diversity. There are an infinitely large number of historical and environmental circumstances that can combine different species pools in a unique manner. This natural variation is the source of overwhelming detail in the literature, with an inexhaustible supply available for future generations. The following chapters, however, cut through much of the bewildering details to provide interpretive frameworks that can absorb these details in an integrated scale-dependent perspective.

Conclusion

This book part identified some useful models that account for patterns of diversity, suggesting a common basis for theoretical understanding, but it also recognises that several well-known models are incomplete. It is not surprising that no chapter procured a general model of diversity—instead, they collectively suggest an infinitely large number of historical and environmental circumstances that can combine various species pools in a unique manner. Indeed, the mere occupation of a site by a species causes changes in the conditions at that site, the sum of which determines the chances of recruitment of further species to this habitat. There was, however, some recognition of the way in which sorting of species may occur after disturbances, driven by biotic interactions and species-specific tolerances and requirements to different physical and chemical aspects of the environment (Chap. 7). Amongst the bewilderment of natural variation, therefore, we recognise the imprint of some predictable processes on patterns of diversity.

The preceding chapters of this book part considered patterns of diversity by assessing the combined effects of abiotic factors that limit the potential breadth of species distributions, and biotic interactions that determine their ecological success. Together, they recognised local patterns as a product of physical and biological events across multiple scales of time (i.e. geological through short timescales) and space (i.e. regional through local influences), signifying local knowledge to be usefully incorporated within broader-scale phenomena (e.g. biogeography, oceanography and history). On shorter timescales, succession was shown to generate common patterns through time that reflect the opportunities and constraints of alternative life history traits, as mediated by resource availability. Similarly, over smaller spatial scales, the composition of species may vary along an environmental gradient because traits that inhibit a species from one part of the gradient facilitate the same species in other parts of the gradient. Indeed, a common theme is this balance between opportunities and constraints that often switch as the successful traits change from early to late stages of succession, and from opposite ends of gradients. Across the heterogeneity of spatial gradients and temporal events, therefore, this altering membership of species and traits generates distinct patterns in diversity and composition.

The heterogeneity that temporal events and spatial gradients impose on variation in ecological success may provide the common stage on which to understand diversity on longer timescales—e.g. evolutionary implications for habitat specialization and lineage diversification. Indeed, as we seek greater integration across successively larger scales (biogeographic) and comprehensiveness (ecosystems, and their functions), there will be an increasing need to identify the common traits or functional responses to the environment and, in turn, how these modify the environment. Recognition of which community types (e.g. which subset of species and traits) occur within a particular set of environmental conditions (e.g. within a gradient or temporal event) may assist better integration of our ecological observations and concepts.

Species will continue to be critical components of study because of their linkages across scales and sources of ecological heterogeneity. Species are the units through which we observe spatial connections (e.g. population structure and dispersal as estimated by rates of gene flow) and temporal connections (e.g. species formation and extent of genetic divergence). Variation in species identity will lead to functional shifts in the community and can have fundamental effects on entire systems. These shifts apply in particular to species that have disproportionately strong effects on surrounding assemblages (variously known as strong interactors, keystone species, foundation species). Strong interactors, or sets of interacting species, have broad-scale and system-wide influences through processes of facilitation, competition, dispersal and consumption, and by modifying the habitat and the abiotic environment. Changes in the distribution and abundance of these 'key' species can have profound effects on ecosystem function (e.g. trophic dynamics) and ecosystem properties (e.g. nutrient cycling and productivity), and result in sweeping changes to species composition and diversity. We may also benefit from understanding how species diversity per se affects ecological functions, and how the diversity of ecological functions affects the maintenance and resilience of ecological systems.

In conclusion, this book part interpreted patterns of diversity within regional and historical processes through local and short-term processes. The recognition of regional processes as contributor to large-scale patterns of diversity (in addition to local processes that are repeated across regions) shows that such assessments are of legitimate concern to ecology, and that regional-scale investigations offer powerful spatial contexts for reconciling similarity and dissimilarity among local studies. Patterns of species diversity will continue to act as a focus for integrating ecology with biogeography, because of their capacity to be integrated into landscape-scale studies, as well as their susceptibility to the development of new molecular tools for interpreting population structure and evolutionary change, and ongoing statistical and analytical tools to describe ecological variation, evolutionary diversification and regional ecologies. In this regard, an essential and unstated message is that there are substantial rewards for ecologists who learn to acquire data that match the broader-scale patterns they study, and enable integration of scale-dependent information. This book part highlights the continuing importance of local studies (we cannot afford to ignore or average out heterogeneity) taken in nested sets (partitioned or observed in association with meaningful sources of heterogeneity), up to the scale of the whole region—the message being that progress will rely on good use of experimental design and statistical procedures (Chap. 9).

Finally, it would be naive to believe that the pace or direction of succession and influence of environmental gradients occur independently of human activities.

Part IV of the book demonstrates that humans have come to dominate natural systems by modifying species pools (i.e. composition of species) and the physical and chemical environment (e.g. biogeochemical cycles), and to alter the patterns of diversity explained in this book part. Human activities, therefore, need to be meaningfully absorbed into our theories about patterns of diversity, and the drivers of past and future patterns.

Chapter 5
Latitudinal Patterns of Species Richness in Hard-Bottom Communities

João Canning-Clode

5.1 Introduction

Understanding global patterns of biodiversity is a useful goal for ecologists because it has the capacity to provide insights on phenomena such as the spread of invasive species, the control of diseases and the effects of climate change (Gaston 2000). A number of hypotheses which seek to clarify spatial variation in biodiversity have been explored, including comparisons between biogeographical regions, and variation with spatial scale and along gradients across space (Gaston 2000, 2003; Mora et al. 2003; Turner 2004). However, in the marine realm there is still a relative lack of global studies as compared to terrestrial studies (Kuklinski et al. 2006). While in the last decades large-scale patterns have been acknowledged for some marine taxonomic groups (Stehli and Wells 1971; Holmes et al. 1997; Roy et al. 1998; Giangrande and Licciano 2004; Gobin and Warwick 2006), broader patterns of marine biodiversity are still poorly recognised (Clarke 1992; Clarke and Crame 1997; Gray 2001).

The almost paradigmatic '*latitudinal gradient of species richness*', which states that the tropics hold more species than do higher latitudes, is considered the oldest and most reliable concept of a large-scale ecological pattern (Rosenzweig 1995; Willig et al. 2003; Hillebrand 2004a). However, this pattern appears less consistent in the marine than in the terrestrial environment (Clarke and Crame 1997; Rivadeneira et al. 2002; Gobin and Warwick 2006). Clear gradients of decreasing diversity towards the poles have been found, e.g. for molluscs in the Western Atlantic and Eastern Pacific oceans (Roy et al. 1994, 1998, 2000), bryozoans in the North Atlantic (Clarke and Lidgard 2000), deep-sea isopods, gastropods and bivalves in the North Atlantic (Rex et al. 2000), and in intertidal sessile communities along the North-western Pacific coast of Japan (Okuda et al. 2004). In contrast, studies on polychaetes and nematodes (Boucher 1990; Mackie et al. 2005; Gobin and Warwick 2006), macroalgae (Santelices and Marquet 1998) and in marine soft sediments (Gray 2002; Ellingsen and Gray 2002) have not found evidence of a latitudinal trend in species richness.

Recently, in a meta-analytical comparison from approximately 600 published articles (Hillebrand 2004b), concluded that marine organisms generally display a

M. Wahl (ed.), *Marine Hard Bottom Communities*, Ecological Studies 206,
DOI: 10.1007/978-3-540-92704-4_5, © Springer-Verlag Berlin Heidelberg 2009

decline in species richness towards the poles but that the strength and slope of the gradient depends on regional factors, as well as habitat and organism characteristics. The same study demonstrated that the strength of the gradient is sensitive to the life form at regional and local scales. Sessile organisms (a dominant component of hard-bottom communities) and infauna revealed the weakest relationship with latitude at both spatial scales, while nekton, plankton and mobile epifauna showed stronger gradients (Hillebrand 2004b).

Several factors have been proposed as possible causes for the latitudinal diversity gradient. Aspects such as the area of the climatic zones, predation, age, competition, climate, productivity and regional species pool have been suggested as plausible explanations for the gradient (Rosenzweig 1992, 1995; Willig et al. 2003; Ricklefs 2004). In recent times, the availability of molecular, phylogenetic and paleontological data caused an increase in evolutionary and historical hypotheses to explain the pattern (Mittelbach et al. 2007).

Recently, several authors have comprehensively reviewed the current understanding of latitudinal variations in species richness (e.g. Gaston 2000; Chown and Gaston 2000; Willig et al. 2003; Hillebrand 2004a, b). However, in this chapter I specifically report on two assessments of global diversity patterns in hard-bottom assemblages which have the rare quality of representing globally replicated observations yielding data of comparable format (scale, time) and resolution. I further review a meta-analytical comparison on global patterns in benthic marine algae.

5.2 Case Studies

In a recent analysis on a global modular experiment, Canning-Clode and Wahl (2009) investigated the effects of latitude on fouling assemblages growing on artificial substratum. In 16 biogeographic regions from 42°S to 59°N, 15×15 cm polyvinylchloride (PVC) panels were submerged for colonization at approximately 0.5 m depth during a period of 5 to 8 months. They reported a significant latitudinal diversity cline at the local scale, with fouling species richness being greater in tropical areas than at higher latitudes (Fig. 5.1). Maximum local species richness occurred in Malaysia (5°N) as well as in Portugal (32°N), while Poland (54°N) and Tasmania (42°S) were the least rich regions at the local scale (Canning-Clode and Wahl 2009). Ascidians, bryozoans, barnacles and mussels were the most common sessile organisms worldwide in this type of hard-bottom community (see Table 5.1 in Canning-Clode and Wahl 2009).

With a focus on epifaunal invertebrate communities encrusting subtidal vertical rock wall habitats, Witman et al. (2004) compared 12 independent biogeographic regions from 62°S to 63°N latitude, to test the influence of latitude and regional species pools on local species richness. In their study, local richness at each site was estimated in two different ways: using the number of species observed (equivalent to the species accumulation curve) and by employing a non-parametric estimator (Chao2; Colwell and Coddington 1994). With this non-parametric technique, species

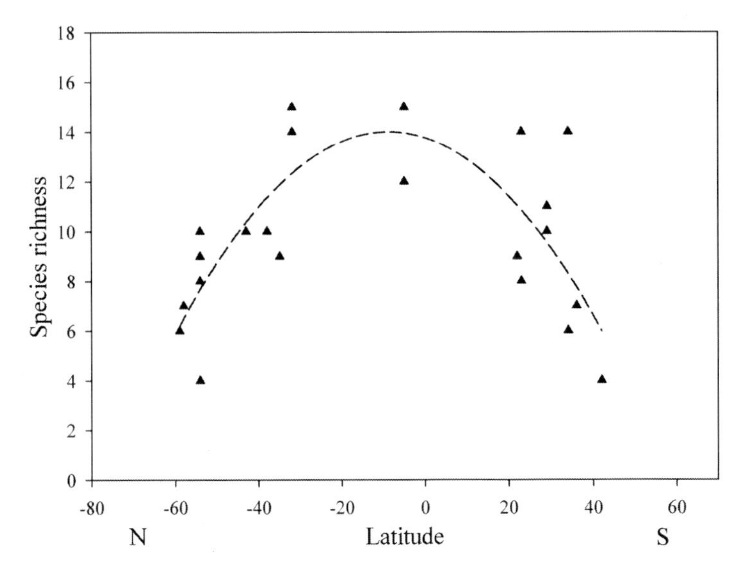

Fig. 5.1 Fouling species richness as a function of latitude. A significant polynomial relationship was found whereby fouling species richness was higher in tropical regions than at higher latitudes (modified from Canning-Clode and Wahl 2009)

richness is estimated from the prevalence of rare species in each sample. Richness is estimated, including absent species, from the proportional abundances of species contained in the total sample (Soberon and Llorente 1993). Additionally, regional diversity was supplemented from published species lists and by consulting taxonomic experts in each of the 12 different biogeographic regions. They recorded sessile communities composed mainly of ten different invertebrate phyla, with sponges, cnidarians and ascidians being the dominant groups (Witman et al. 2004). Local and regional species richness showed a significant unimodal pattern with latitude, although this trend was weaker at a local scale.

In a recent meta-analysis on studies of benthic marine algae, Kerswell (2006) compared 191 species lists from primary literature. Data were compiled at the genus level for the classes Rhodophyceae, Phaeophyceae and Chlorophyceae, and at the species level within the order Bryopsidales. This latter order was treated with higher resolution because it is well studied and taxonomically stable. Kerswell (2006) observed that algal genera display an inverse latitudinal gradient, with highest diversity in temperate regions. In contrast, a latitudinal pattern of species from the order Bryopsidales was evident with maximum species in the tropics, decreasing towards the poles (Kerswell 2006). The genus richness was highest in southern Australia and Japan, and moderate in the Indo-Australian Archipelago and the Southern Indian Ocean. Algal richness in the Atlantic Ocean was higher along the eastern coastline, with a hotspot situated at the European coast. Lowest diversity in algal genera was recorded in the polar regions (Kerswell 2006). A common explanation for a decrease in algal richness in the tropics is competition with corals and elevated herbivore pressure at low latitudes. According to the author, global patterns of algal richness can partially be explained by the *species-area hypothesis*, which proposes that larger areas hold more species, since their larger populations and more stable conditions reduce the risk of extinction. Thus, larger areas also contain more barriers and heterogeneous habitats which promote speciation (Rosenzweig 1995; Chown and Gaston 2000). In addition, levels of marine macroalgal diversity and endemism in southern Australian were reported as being the highest on the planet. This fact is associated with long isolation, as well as stable temperate conditions and oceanography which have safeguarded against mass-extinction events (Phillips 2001). These biogeographic and oceanographic breaks align well with ecological patterns and processes at these scales (Connell and Irving 2008). Bryopsidalean species richness is intimately related to corals and reef fishes diversity, implying a common regulatory mechanism. The meso-scale location of algal-richness hotspots may be determined by major ocean currents through propagule dispersal and changes in oceanic conditions (Kerswell 2006).

5.3 Discussion

The case studies I present in this overview display an evident influence of latitude on local species richness in marine hard-bottom communities. The large-scale investigations conducted by Witman et al. (2004) and by Canning-Clode and Wahl (2009) have both demonstrated that, compared to areas at higher latitudes, tropical regions hold more hard-bottom species. Moreover, in a recent meta-analysis on published studies of benthic marine algae, Kerswell (2006) also observed a distinct latitudinal pattern, with more algal species from the order Bryopsidales in the tropics than at higher latitudes. However, this pattern showed an inverse relationship with increasing latitude, when considering all algal genera (Kerswell 2006). The clear gradients found in epifaunal invertebrate communities encrusting subtidal vertical rock walls (Witman et al. 2004), in hard-bottom communities colonizing artificial substratum (Canning-Clode and Wahl 2009), and in algal species from the order Bryopsidales (Kerswell 2006) seem to be in accordance with Hillebrand's (2004b) meta-analysis of 600 published gradients, which showed a significant decrease of marine biodiversity with increasing latitude.

This latitudinal trend implies that tropical regions have higher indices of hard-bottom species diversity than do areas closer to the poles. The area of the climatic zones has been proposed as a possible explanation for the cause of the latitudinal diversity pattern. Landmasses in tropical areas contain a larger climatically similar total surface area than do landmasses at higher latitudes, which may lead to high speciation and low extinction rates in the tropics (Rosenzweig 1992). More recently, other hypotheses have been proposed: the tropical diversity maximum is geological very old; speciation rates are higher in the tropics because molecular evolution rates in ectotherms are higher in warmer climates; speciation rates are higher in the tropics because larger biome areas increase the probability for geological or ecological isolation of sub-populations, and because more biotic interactions force specialization and speciation. As a consequence, tropical diversification rates are higher due to faster speciation and slower extinction. Species originating in the tropics tend to disperse to higher latitudes but retain their presence in the tropics (Mittelbach et al. 2007).

In the analysis conducted by Canning-Clode and Wahl (2009), only one study was carried out in the tropics (Malaysia), leaving a sampling gap between 30°N and 20°S. Also, no studies were conducted at latitudes higher than 40° in the southern hemisphere. More information on hard-bottom assemblages in these areas might have yielded a different pattern (Canning-Clode and Wahl 2009). In addition, only the work conducted by Witman et al. (2004) has included a study site in the Antarctic region. Although little is known concerning hard-bottom communities in Antarctica (but see, e.g. Bowden et al. 2006; Waller 2008), for a more adequate and accurate global perspective on these communities this area should be included in future surveys (Canning-Clode and Wahl 2009).

In addition to the global-scale surveys, the effects of latitude on the diversity of hard-bottom communities have been assessed at smaller scales (e.g. within countries), particularly in rocky intertidal sessile assemblages (Gaines and Lubchenco 1982; Leonard 2000; Okuda et al. 2004; Schoch et al. 2006). However, at smaller scales the outcome is not always consistent. For example, Okuda et al. (2004) observed a clear latitudinal gradient for regional species richness of rocky intertidal sessile assemblages along the North-western Pacific coast of Japan. In contrast, in a study covering 15 degrees of latitude in the California Current region, Schoch et al. (2006) concluded that diversity declined with decreasing latitude in the low intertidal zone at different scales.

This inconsistency at lower-resolution scales emphasizes the need for large-scale assessments in hard-bottom communities to acquire a more complete understanding about global patterns and their causes in hard-bottom assemblages. Furthermore, compared to latitude, longitude has been relatively unexplored but longitudinal gradients have indeed also been recognised in both marine and terrestrial environments (Jetz and Rahbek 2001; Hughes et al. 2002; Roberts et al. 2002; Kerswell 2006). The relative ease with which hard-bottom communities can be observed and manipulated could facilitate more comprehensive comparisons at a global scale. For that, studies focusing only on latitude may be insufficient and, therefore, longitude should be incorporated in future large-scale surveys.

Acknowledgements J. Canning-Clode studies were supported by a scholarship from the German Academic Exchange Service (DAAD).

References

Boucher G (1990) Pattern of nematode species-diversity in temperate and tropical subtidal sediments. Mar Ecol PSZNI 11:133–146

Bowden DA, Clarke A, Peck LS, Barnes DKA (2006) Antarctic sessile marine benthos: colonisation and growth on artificial substrata over three years. Mar Ecol Prog Ser 316:1–16

Canning-Clode J, Wahl M (2009) Patterns of fouling on a global scale. In: Dürr S, Thomason JC (eds) Biofouling. Wiley-Blackwell, Oxford (in press)
Chown SL, Gaston KJ (2000) Areas, cradles and museums: the latitudinal gradient in species richness. Trends Ecol Evol 15:311–315
Clarke A (1992)Is there a latitudinal diversity cline in the sea. Trends Ecol Evol 7:286–287
Clarke A, Crame JA (1997) Diversity, latitude and time: patterns in the shallow sea. In: Ormond RFG, Gaje JD, Angel MV (eds) Marine biodiversity: patterns and processes. Cambridge University Press, Cambridge, pp 122–147
Clarke A, Lidgard S (2000)Spatial patterns of diversity in the sea: bryozoan species richness in the North Atlantic. J Anim Ecol 69:799–814
Colwell RK, Coddington JA (1994) Estimating terrestrial biodiversity through extrapolation. Philos Trans R Soc Lond Series B Biol Sci 345:101–118
Connell SD, Irving AD (2008) Integrating ecology with biogeography using landscape characteristics: a case study of subtidal habitat across continental Australia. J Biogeogr 35:1608–1621
Ellingsen KE, Gray JS (2002) Spatial patterns of benthic diversity: is there a latitudinal gradient along the Norwegian continental shelf? J Anim Ecol 71:373–389
Gaines SD, Lubchenco J (1982)A unified approach to marine plant-herbivore interactions.2. Biogeography. Annu Rev Ecol Syst 13:111–138
Gaston KJ (2000) Global patterns in biodiversity. Nature 405:220–227
Gaston KJ (2003) Ecology—the how and why of biodiversity. Nature 421:900–901
Giangrande A, Licciano M (2004) Factors influencing latitudinal pattern of biodiversity: an example using Sabellidae (Annelida, Polychaeta). Biodivers Conserv 13:1633–1646
Gobin JF, Warwick RM (2006) Geographical variation in species diversity: a comparison of marine polychaetes and nematodes. J Exp Mar Biol Ecol 330:234–244
Gray JS (2001) Antarctic marine benthic biodiversity in a world-wide latitudinal context. Polar Biol 24:633–641
Gray JS (2002) Species richness of marine soft sediments. Mar Ecol Prog Ser 244:285–297
Hillebrand H (2004a) On the generality of the latitudinal diversity gradient. Am Nat 163:192–211
Hillebrand H (2004b) Strength, slope and variability of marine latitudinal gradients. Mar Ecol Prog Ser 273:251–267
Holmes NJ, Harriott VJ, Banks SA (1997) Latitudinal variation in patterns of colonisation of cryptic calcareous marine organisms. Mar Ecol Prog Ser 155:103–113
Hughes TP, Bellwood DR, Connolly SR (2002) Biodiversity hotspots, centres of endemicity, and the conservation of coral reefs. Ecol Lett 5:775–784
Jetz W, Rahbek C (2001) Geometric constraints explain much of the species richness pattern in African birds. Proc Natl Acad Sci USA 98:5661–5666
Kerswell AP (2006) Global biodiversity patterns of benthic marine algae. Ecology 87:2479–2488
Kuklinski P, Barnes DKA, Taylor PD (2006) Latitudinal patterns of diversity and abundance in North Atlantic intertidal boulder-fields. Mar Biol 149:1577–1583
Leonard GH (2000) Latitudinal variation in species interactions: a test in the New England rocky intertidal zone. Ecology 81:1015–1030
Mackie ASY, Oliver PG, Darbyshire T, Mortimer K (2005) Shallow marine benthic invertebrates of the Seychelles Plateau: high diversity in a tropical oligotrophic environment. Philos Trans R Soc Lond A 363:203–227
Mittelbach GG, Schemske DW, Cornell HV, Allen AP, Brown JM, Bush MB, Harrison SP, Hurlbert AH, Knowlton N, Lessios HA, McCain CM, McCune AR, McDade LA, McPeek MA, Near TJ, Price TD, Ricklefs RE, Roy K, Sax DF, Schluter D, Sobel JM, Turelli M (2007) Evolution and the latitudinal diversity gradient: speciation, extinction and biogeography. Ecol Lett 10:315–331
Mora C, Chittaro PM, Sale PF, Kritzer JP, Ludsin SA (2003) Patterns and processes in reef fish diversity. Nature 421:933–936

Okuda T, Noda T, Yamamoto T, Ito N, Nakaoka M (2004) Latitudinal gradient of species diversity: multi-scale variability in rocky intertidal sessile assemblages along the Northwestern Pacific coast. Popul Ecol 46:159–170

Phillips JA (2001) Marine macroalgal biodiversity hotspots: why is there high species richness and endemism in southern Australian marine benthic flora? Biodivers Conserv 10:1555–1577

Rex MA, Stuart CT, Coyne G (2000) Latitudinal gradients of species richness in the deep-sea benthos of the North Atlantic. Proc Natl Acad Sci USA 97:4082–4085

Ricklefs RE (2004) A comprehensive framework for global patterns in biodiversity. Ecol Lett 7:1–15

Rivadeneira MM, Fernandez M, Navarrete SA (2002) Latitudinal trends of species diversity in rocky intertidal herbivore assemblages: spatial scale and the relationship between local and regional species richness. Mar Ecol Prog Ser 245:123–131

Roberts CM, McClean CJ, Veron JEN, Hawkins JP, Allen GR, McAllister DE, Mittermeier CG, Schueler FW, Spalding M, Wells F, Vynne C, Werner TB (2002) Marine biodiversity hotspots and conservation priorities for tropical reefs. Science 295:1280–1284

Rosenzweig ML (1992) Species-diversity gradients—we know more and less than we thought. J Mammal 73:715–730

Rosenzweig ML (1995) Species diversity in space and time. Cambridge University Press, Cambridge

Roy K, Jablonski D, Valentine JW (1994) Eastern Pacific molluscan provinces and latitudinal diversity gradient—no evidence for Rapoports Rule. Proc Natl Acad Sci USA 91:8871–8874

Roy K, Jablonski D, Valentine JW, Rosenberg G (1998) Marine latitudinal diversity gradients: tests of causal hypotheses. Proc Natl Acad Sci USA 95:3699–3702

Roy K, Jablonski D, Valentine JW (2000) Dissecting latitudinal diversity gradients: functional groups and clades of marine bivalves. Proc R Soc Lond B Biol 267:293–299

Santelices B, Marquet PA (1998) Seaweeds, latitudinal diversity patterns, and Rapoport's Rule. Divers Distrib 4:71–75

Schoch GC, Menge BA, Allison G, Kavanaugh M, Thompson SA, Wood SA (2006) Fifteen degrees of separation: latitudinal gradients of rocky intertidal biota along the California Current. Limnol Oceanogr 51:2564–2585

Soberon J, Llorente J (1993) The use of species accumulation functions for the prediction of species richness. Conserv Biol 7:480–488

Stehli FG, Wells JW (1971) Diversity and age patterns in hermatypic corals. Syst Zool 20:115–126

Turner JRG (2004) Explaining the global biodiversity gradient: energy, area, history and natural selection. Basic Appl Ecol 5:435–448

Waller CL (2008) Variability in intertidal communities along a latitudinal gradient in the Southern Ocean. Polar Biol 31:809–816

Willig MR, Kaufman DM, Stevens RD (2003) Latitudinal gradients of biodiversity: pattern, process, scale, and synthesis. Annu Rev Ecol Evol S 34:273–309

Witman JD, Etter RJ, Smith F (2004) The relationship between regional and local species diversity in marine benthic communities: a global perspective. Proc Natl Acad Sci USA 111(44): 15664–15669

Chapter 6
Regional-Scale Patterns

Jonne Kotta and Jon D. Witman

6.1 Introduction

There are broad similarities in the distribution patterns of organisms on rocky shores worldwide (Little and Kitching 1996). Species diversity and community structure differ among regions; nevertheless, macroalgae and/or sessile animals such as barnacles and mussels are the most common inhabitants of temperate intertidal rocky shores. Subtidal habitats off temperate-boreal open coasts tend to be dominated by macroalgae at shallow depths, giving way to sessile invertebrates as light levels diminish or grazing intensity increases (Witman and Dayton 2001). Subtidal hard bottom habitats in the tropics are fringed with coral reefs or dominated by other sessile invertebrates such as sponges, gorgonians, anemones, bryozoans and ascidians; macroalgae are far less abundant (Santileces et al. 2009).

The number of theories accounting for the patterns of diversity and distribution of species living in hard bottom habitats is daunting. For example, Veron (1995) listed 34 theories explaining the origin of the Indo-Pacific centre of coral diversity in an insightful review of the biogeography of stony corals. One way to simplify the search for causal processes in the face of this complexity is to recognize that the array of processes shaping the diversity of a local community act on different spatial and temporal scales (Rosensweig 1995). Diversity structuring processes range from local spatial scales of centimetres and metres to regional scales of thousands of kilometres, and from short temporal scales of ecological processes to millions of years of evolutionary history (Goldstien and Schiel, Chap. 8). The multi-factorial nature of diversity structuring processes has been represented in many conceptual models (cf. Pianka 1971; Ricklefs and Schluter 1993). Central to these frameworks is the view that local diversity represents the interplay of contemporary ecological and historical evolutionary processes. The model of Ricklefs and Schluter (1993) presents a useful way to organize this chapter on regional and local diversity patterns, and their causes in hard bottom communities. We have modified their original model (Fig. 6.1) to include other processes known to increase or decrease the diversity of marine benthic communities.

To a certain extent, the levels diversity observed in a local hard bottom community such as an intertidal mussel bed, tide pool, coral reef, kelp forest or rock wall

M. Wahl (ed.), *Marine Hard Bottom Communities*, Ecological Studies 206,
DOI: 10.1007/978-3-540-92704-4_6, © Springer-Verlag Berlin Heidelberg 2009

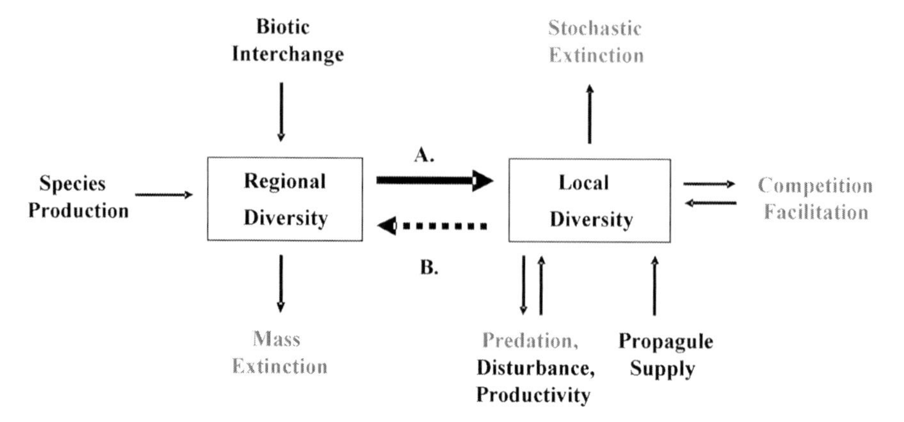

Fig. 6.1 Schematic diagram illustrating regional- and local-scale processes influencing local diversity (modified from Ricklefs and Schluter 1993). The processes covered in this chapter appear in *black*. The *A arrow* represents dispersal or habitat selection from the regional pool, the *B dashed arrow* beta diversity or between-habitat diversity. Increasing beta diversity from local-scale variability increases regional diversity

covered with epifaunal invertebrates depends on the number of species in the regional species pool, since local communities are embedded in a broader biogeographic region (Fig. 6.1). The regional species pool or metacommunity (Holyoak et al. 2005) represents the group of species potentially able to disperse to and colonize a local habitat. The number of species in the metacommunity results from evolutionary processes such as speciation (Barraclough et al. 1998), mass extinctions (Bambach et al. 2004; Cerrano and Bavestrello, Chap. 21) and the migration of species across regions (MacArthur and Wilson 1967). It is generally thought that local communities assemble from the macro- to the microscale, i.e. from the regional pool to local communities, rather than vice versa (Whittaker et al. 2001). As Keddy and Weiher (1999) put it, "*Assembly rules address a central theme in community ecology: how communities are assembled from species pools. (Evolutionary ecology by contrast, deals with the formation of the pool)*". Lawton (2000) rationalizes local community assembly from larger scales of the regional species pool by noting that the contribution of any local assemblage to the regional pool must be small, rather than the other way round. This notion is represented in Ricklefs and Schluter's (1993) model as a horizontal arrow from the regional diversity (pool) box to the local diversity box (Fig. 6.1).

6.2 Regional Diversity—Biotic Interchange and Speciation

The formation and removal of barriers has occurred repeatedly over geologic time (Vermeij 1978), with major impacts on regional diversity. Well-known Pleistocene examples include the division of the Bering land bridge, the unification of the

Indian and Pacific oceans as the Sunda Shelf was inundated, and the formation of the Panamanian isthmus cutting off the deep ocean seaway between the Pacific and Caribbean (Jackson and Budd 1996). When a barrier breaks down due to rising sea level or plate tectonics, diversity changes result from the migration of adults or larvae to the recipient region. The direction of migration is often unidirectional, with biota from species-rich regions exported to impoverished regions (Vermeij 1978). There is some evidence that molluscs dispersed from the species-rich Pacific via the Arctic Ocean to enrich the Atlantic fauna (Marincovitch 1977; also see Chap. 8 by Goldstien and Schiel). The diversity of epifaunal invertebrates is 2–3 times higher in shallow rock wall habitats in the north temperate Pacific than in the western Atlantic (Witman et al. 2004), yet the extent to which the Pacific served as a pool for Atlantic epifauna is unknown. Recent accelerated melting of the Arctic ice cap may increase inter-oceanic migration of shallow coastal marine faunas previously blocked by grounded ice, with likely effects on Atlantic and Pacific regional diversities. Barrier formation fosters vicariance speciation as the original range of a species is divided up and the isolated descendants evolve into new species. Thus, the evolutionary build up of a regional species pool depends on the interplay between speciation by isolation and leaky migration across barriers (Chap. 8).

Oceanographic currents create biotic interchanges by transporting species beyond “normal” range limits in episodic basin-wide phenomena such as the El Niño Southern Oscillation (ENSO), North Atlantic Oscillation (NAO), and Pacific Decadal Oscillation (PDO; Stenseth et al. 2003). For example, west-to-east transport of fish larvae during ENSO events has apparently enriched the Galapagos ichthyofauna (Glynn 1988). The NAO and climate warming contributed to a change in regional biodiversity of copepods by driving a northward migration of approximately ten degrees (Beaugrand et al. 2002). More frequently, eddies spinoff major currents such as the Gulf Stream as warm core rings (Mann and Lazier 1996), transporting non-indigenous species from southern to boreal-arctic biogeographic regions. Spinoff rings may create founder populations of invasive fish and invertebrates, if these survive to reproduce. Long-distance interregional dispersal of hard bottom species is related to the mode of larval development, as species with larvae capable of feeding during dispersal (planktotrophic) are generally transported farther than species which are direct developers or have non-feeding larvae (Scheltema 1989). This characterization, however, may not predict range size. Rafting on biotic and abiotic substrates is a major form of interregional dispersal for brooding or direct-developing species of corals and molluscs (Highsmith 1985).

Humans have greatly accelerated the pace of interregional migration of marine benthic species by transporting invertebrates in the ballast water or on the hulls of ships, and by releasing exotic aquarium species (Carlton and Geller 1993). Biological invasions have resulted in comparably large-scale ecological changes and economic damage worldwide. The examples of invasions in the 1980s and 1990s have shown that successful exotics may render previously stable systems unbalanced and unpredictable, and may severely affect biological diversity in the area (Mills et al. 1993; Lauringson et al. 2007; Orav-Kotta et al. 2009; Cerrano and Bavestrello, Chap. 21).

6.3 Influence of Regional Species Pools on Local Diversity

Inferences that regional effects shape local diversity are made by examining the relationship between the size of the regional species pool and the species richness of a local community. This idea has a long history in terrestrial communities, dating back to Williams' (1943) species area relations and MacArthur and Wilson's (1967) island biogeography theory (IBT). The concepts of immigration from a mainland (regional) pool and saturation of species richness on local islands in the IBT model provided theoretical background for Terborg and Faaborg's (1980) approach of testing saturation of local species richness by comparing local species richness in similar habitats subject to colonization from smaller or larger species pools. They developed the graphical dichotomy predicting regional and local influences on local species richness in terms of a linear relation between the number of species in the regional pool (regional species richness, RSR, independent variable) and local species richness (LSR, dependent variable), indicating a regional pool effect. A levelling off (or saturation) of the line would indicate limits to local species richness; localized effects of competition were invoked. Subsequent model development by Cornell (1985) and Ricklefs (1987) depicted RSR-LSR patterns as either a type I unsaturated community (linear relation), where evolutionary processes and dispersal predominate, or a type II saturated community (curvilinear function) where local ecological interactions are most important in regulating local diversity.

RSR-LSR theory has been tested for a wide variety of single taxa (e.g. Cornell 1985), with initial marine tests conducted on deep-sea gastropods (Stuart and Rex 1991) and corals (Cornell and Karlson 1996). These studies rejected the saturation hypothesis by demonstrating linear fits, suggesting that the diversity of gastropod and coral communities rises steadily with increasing regional richness, without reaching saturation. As the coral database was compiled from surveys of coral richness which varied in terms of sampling method, completeness and target habitat, the strength of the assertion of regional enrichment was unknown. In 1992, J. Witman began a collaboration with R. Etter to devise a standardized, global sampling scheme to rigorously test RSR-LSR theory in multi-phyletic marine communities in a habitat (subtidal rock walls) where the confounding influence of habitat heterogeneity on local diversity was minimal. Using multi-phyletic, rather than single taxon communities maximized the chance for local species interactions such as competition to play out and limit species richness, since competition among sponges, corals, ascidians and barnacles is multi-phyletic (Jackson 1977). In Chile, standardized sampling of subsets of rocky intertidal communities (molluscan grazers) by Rivadeneria et al. (2002) made several important advances, including the first rigorous demonstration of linear LSR-RSR relations in marine hard bottom communities, and the dependence of the LSR-RSR relation on sampling scale, with steeper slopes (greater regional enrichment) being recorded at the site scale (100m^2) than at the sampling unit scale (1.0 m^2). Although the latitudinal span was large (25°) and site replication was high (53 sites), the regional pool ranged from only 36 to 51 species; thus, it remained debatable whether local marine benthic communities embedded in a broader range of regional pool richness were saturated or unsaturated. This question was answered by the publication of convergent results from standardized field sampling, finding positive linear LSR-RSR relationships in coral (Karlson et al. 2004) and epifaunal invertebrate communities (Witman et al. 2004), indicating unsaturation. The Karlson et al. (2004) study, conducted across a gradient of regional coral biodiversity of 250 species from Indonesia to Samoa, revealed that the richness of the regional species pool explained 94–99% of the variance in local species richness, depending on the habitat (reef flat, crest, slope) sampled. The slopes, indicating the proportion of regional pool represented in local communities, ranged from 0.26 to 0.28. Global-scale (63°N–63°S, 49 sites) analysis of LSR-RSR relationships in epifaunal communities indicated that regional species pools ranging up to 1,220 species explained 73–76% of local species richness, depending on the type of local richness estimator (Witman et al. 2004; Fig. 6.2). The finding of significant regional enrichment in two types of sessile invertebrate communities, corals and multi-phyletic epifauna, where competitive interactions are strong and known to limit local richness, argues that regional processes need greater consideration as drivers of local diversity (Fig. 6.2). In the epifaunal study, the scaling of regional to local diversity was analyzed across the latitudinal gradient where both regional and local species richness displayed a humped-shaped relation with latitude, peaking at low tropical latitudes and decreasing towards high latitudes. The positive linear LSR-RSR relationships meant that the latitudinal diversity gradient was expressed at the scale of local sites, albeit more weakly than at the regional scale. Moreover, the proportion of the regional biota represented in local epifaunal communities increased from the equator to the poles, prompting five hypotheses for

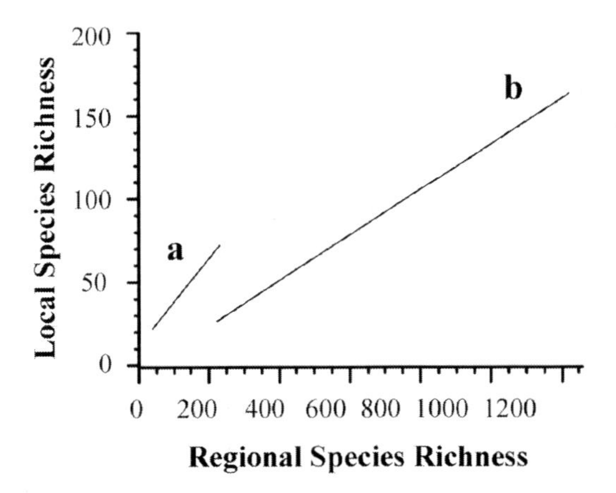

Fig. 6.2 Comparative plot of relationships between regional (metacommunity) richness and local species richness for two hard bottom community types. Regression line *a* is for corals from Karlson et al. (2004), depicting the slope habitat with an r^2 of 0.94. Line *b* represents epifaunal invertebrate communities (sponges, corals, soft corals, bryozoans, molluscs, ascidians, etc.) with an r^2 of 0.76 (Witman et al. 2004). The analyses suggest strong enrichment of local species richness from the regional species pool. Studies of regional influences on local richness in rocky intertidal habitats are based on maximum regional pool sizes of approximately 50 species in Chile (Rivadeneira et al. 2002) and 100 species along the Pacific coast of the USA (Russell et al. 2006)

further testing (Witman et. al. 2004). Using small epifaunal assemblages on pen shells, Mungia (2004) found evidence for saturation during different periods of succession. In rocky intertidal communities, the most comprehensive test of LSR-RSR relationships to date was conducted along the Pacific coast from Washington State to southern California by Russell et al. (2006), who decomposed the LSR-RSR relationships by spatial scale, species groupings and intertidal height. Analyzing the degree of saturation, they found (as did Rivadeneira et al. 2002) that the LSR-RSR relationship differs by spatial scale. Saturation was more evident at the smallest sampling scale of 0.25-m^2 quadrats (Russell et al. 2006). The effect of the regional species pool was stronger (i.e. lower saturation index) when LSR-RSR relationships were analyzed among coarse species groupings (including probable weakly interacting species) than with finer species groupings (with potentially more interactive trophic groups). Finally, the influence of the regional species pool was lower in the high intertidal zone, the most environmentally stressful habitat, suggesting a limiting effect of local environmental factors on local species richness (Russell et al. 2006).

Interpreting LSR-RSR plots is controversial, however, as it has been noted that curvilinear type II relations do not necessarily reflect a limitation by local ecological processes (Caswell and Cohen 1993). Model results suggest that the form of LSR-RSR relationships varies with different metacommunity structures (Hugueny et al. 2007). Evaluating the extent of regional enrichment, and how it varies with scale, dispersal and source-sink dynamics, is a fundamentally important question for understanding and managing the diversity of marine metacommunities. Since regionally enriched unsaturated communities are apparently prevalent in hard bottom habitats, these communities are open to invasion from exotic species migrating across regions (Lawton 2000).

6.4 Local Diversity

Our understanding of the causes of local species diversity in marine habitats originated from experimental manipulations performed at small spatial scales, which preceded the examination of regional effects. Menge and Sutherland (1987) proposed that the structure of assemblages is primarily due to four major regulating forces: environmental disturbance (temporal heterogeneity), competition, predation and recruitment. They argued that environmental stress plays a central role in shaping community structure, with complexity being inversely related to environmental stress.

Large-scale environmental stresses and disturbances—e.g. climatically driven changes in seawater temperature, sea level or the intensity of ice scouring—can synchronize population changes over wide geographical areas, if they have a direct effect on recruitment or mortality. Shifts in climatic conditions have profound ecological impacts by altering patterns of distribution, abundance and diversity of species at regional scales (Hughes 2000). As a consequence of rapid climate change, a series of large shifts in ecosystem structure have been observed. For example, barnacle populations changed from abundant *Semibalanus balanoides* in the 1930s to predominantly *Chthamalus stellatus* in the 1950s over a broad area of western Britain and Ireland (Southward and Crisp 1954). This was associated with changes in temperature (Southward et al. 1995). To date, the timing and consequences of such regime shifts are difficult to predict, as we lack the knowledge of the ecological responses involved and the tolerance thresholds of the species at different life stages.

Eutrophication occurs on regional spatial scales, and is ranked among the most serious threats to species diversity and stability of marine ecosystems worldwide. The effect of eutrophication is more pronounced in those coastal areas receiving large amounts of organic and mineral nutrients from municipal wastes, agricultural and industrial effluents (Kotta et al. 2007). Several studies have demonstrated large regional differences in the sensitivity of species to nutrient enrichment, reflecting system-specific attributes as well as direct and indirect responses which act as a filter to modulate the responses to enrichment (Cloern 2001). In general, the regions which have a high proportion of mobile and opportunistic species are more resistant to eutrophication, compared to regions characterized by perennial, long-living and sessile species. In the former regions, eutrophication has minor effects on regional-scale diversity, whereas in the latter regions eutrophication severely reduces regional-scale diversity.

Grazing, as a form of predation, varies in its effects on species diversity at both local and regional spatial scales. Since macroalgae are primary habitat providers across temperate shores worldwide (Santileces et al. 2009), grazers have key effects on both algal diversity and the diversity of associated species. The importance of grazing in the dynamics of hard bottom assemblages varies among regions. Grazing by invertebrates and competitive interactions between macroalgae are the key processes regulating the dynamics of macroalgal assemblages in many coastal areas (Lubchenco and Gaines 1981). In northern Germany, grazing periwinkles (*Littorina* spp.) determine the diversity patterns of macroalgae and invertebrates in sheltered areas (Janke 1990). Sea urchins are among the key grazers off temperate and tropical coasts (e.g. Lawrence 1975; Witman and Dayton 2001). In Australia, Connell and Irving (2008) demonstrated that biogeography has a fundamental influence on patterns of regional diversity, and that these patterns are related to regional variability in productivity and consumption. Regions of relatively high productivity (eastern Australia) are more affected by overgrazing (functional types of grazers which cause barrens of diversity) than by pollution (nutrients), and regions of relatively low productivity (west and south Australia) are less affected by grazing (no barrens) and more affected by pollution (nutrients). In the brackish Baltic Sea, biotic interactions play a minor role in controlling rocky shore communities (Herkül et al. 2006; Kotta et al. 2008) and grazing by invertebrates in general does not regulate the dynamics of macroalgal assemblages (Kotta et al. 2006). This is explained by high physical disturbance, strong seasonality, low species richness and the lack of efficient herbivores.

Connectivity, as a channel for population exchange between locations, assures the maintenance of regional-scale patterns of metapopulations and metacommunities.

With high connectivity, the dynamics of species is similar across the region (Hubbell 2001). Low or nonexistent connectivity uncouples population dynamics at the local and regional scale, promoting an increase in regional diversity and endemism. Studies of population synchrony in marine species have been rare. On Australian shores, changes in algae on the low and mid-shore appear idiosyncratic (Chapman and Underwood 1998). On the other hand, species show similar trends in South Africa and California (Roughgarden et al. 1988; Dye 1998). Wide-scale synchrony of populations in California is driven by temporal variation in coastal upwelling affecting offshore advection of larvae (Roughgarden et al. 1988). Synchrony in populations of rock lobsters (*Jasus edwardsii*) which are genetically indistinguishable between Australia and New Zealand is related to the long larval phase of this species. The dispersal of larvae from Australia to New Zealand is expected to be high enough to maintain some New Zealand populations (Chiswell et al. 2003). Spatial synchrony of population changes was studied in rocky shore communities in the British Isles (Burrows et al. 2002). The intertidal barnacle *Semibalanus balanoides* and the gastropod *Littorina saxatilis* showed synchrony among sites, while the dogwhelk *Nucella lapillus* and the macroalgae *Fucus vesiculosus*, *F. serratus* and *Mastocarpus stellatus* had very low average correlations among sites. It is interesting to note that differences in reproductive biology and ecology failed to explain patterns of synchrony among species. Species with planktonic larvae were no more likely to be synchronized than those without. This suggests that these populations are regulated by large-scale disturbances which synchronize fluctuations, rather than connectivity among populations. Spatial synchrony in population dynamics of predators (sea stars, crabs) in subtidal food webs was created on landscape-regional spatial scales in the Gulf of Maine by a massive recruitment of prey (mussels; Witman et al 2003). Since the mussels were competitive dominants, these bottom-up effects affected benthic diversity across large scales, until the mussels were eliminated by predators.

6.5 Conclusions

In this chapter, we emphasized that regional processes shape local diversity and that the realized diversity represents the interplay of local and regional processes. Local diversity is predicted by the number of species in the regional species pool. This species pool is a result of speciation, extinction and migration among regions. At smaller scales, environmental disturbance, stochastic extinctions, productivity, predation, competition, dispersal and facilitation modulate diversity patterns, limited by the size of the regional species pool. Irrespective of the significance of small-scale phenomena, large-scale abiotic processes define broad patterns of species diversity and are among the most significant factors explaining population variability in marine environments.

References

Bambach RK, Knoll AH, Wang SC (2004) Origination, extinction, and mass depletions of marine diversity. Paleobiology 30:522–542
Barraclough TG, Vogler AF, Harvey PL (1998) Revealing the factors that promote speciation. Philos Trans R Soc Lond B 353:241–249
Beaugrand G, Reid PC, Ibanez F, Lindley JA, Edwards M (2002) Reorganization of North Atlantic marine copepod biodiversity and climate. Science 296:1692–1694
Burrows MT, Moore JJ, James B (2002) Spatial synchrony of population changes in rocky shore communities in Shetland. Mar Ecol Prog Ser 240:39–48
Carlton JT, Geller JB (1993) Ecological roulette: the global transport of nonindigenous marine organisms. Science 261:78–82
Caswell H, Cohen JE (1993) Local and regional regulation of species-area relations: a patch occupancy model. In: Ricklefs RE, Schluter D (eds) Species diversity in ecological communities: historical and geographical perspectives. University of Chicago Press, Chicago, IL, pp 99–107
Chapman MG, Underwood AJ (1998) Inconsistency and variation in the development of rocky intertidal algal assemblages. J Exp Mar Biol Ecol 224:265–289
Chiswell SM, Wilkin J, Booth JD, Stanton B (2003) Trans-Tasman Sea larval transport: is Australia a source for New Zealand rock lobsters? Mar Ecol Prog Ser 247:173–182
Cloern JE (2001) Our evolving conceptual model of the coastal eutrophication problem. Mar Ecol Prog Ser 210:223–253
Connell SD, Irving AD (2008) Integrating ecology with biogeography using landscape characteristics: a case study of subtidal habitat across continental Australia. J Biogeogr 35:1608–1621
Cornell HV (1985) Local and regional richness of cynipine gall wasps on California oaks. Ecology 66:1247–1260
Cornell HV, Karlson RH (1996) Diversity of reef-building corals determined by local and regional processes. J Anim Ecol 65:233–241
Dye AH (1998) Community-level analyses of long-term changes in rocky littoral fauna from South Africa. Mar Ecol Prog Ser 164:47–57
Glynn PW (1988) El Nino–Southern Oscillation 1982-1983: nearshore population, community, and ecosystem responses. Annu Rev Ecol Syst 19:309–345
Herkül K, Kotta J, Kotta I, Orav-Kotta H (2006) Effects of physical disturbance, isolation and key macrozoobenthic species on community development, recolonisation and sedimentation processes. Oceanologia 48S:267–282
Highsmith RC (1985) Floating and algal rafting as potential dispersal mechanisms in brooding invertebrates. Mar Ecol Prog Ser 25:169–179
Holyoak M, Leibold MA, Moquet N, Holt RD, Hoopes MF (2005) Metacommunities: a framework for large-scale community ecology. In: Holyoak M, Leibold MA, Moquet N, Holt RD (eds) Metacommunities: spatial dynamics and ecological communities. University of Chicago Press, Chicago, IL, pp 1–34
Hubbell SP (2001) The unified neutral theory of biodiversity and biogeography. Princeton University Press, Princeton, NJ
Hughes L (2000) Biological consequences of global warming: is the signal already apparent? Trends Ecol Evol 15:56–61
Hugueny B, Cornell HV, Harrison S (2007) Metacommunity models predict the local-regional species richness relationship in a natural system. Ecology 88:1696–1706
Jackson JBC (1977) Competition on marine hard substrata: the adaptive significance of solitary and colonial strategies. Am Nat 111:743–767
Jackson JBC, Budd AF (1996) Evolution and environment: introduction and overview. In: Jackson JBC, Budd AF, Coates AG (eds) Evolution and environment in tropical America. University of Chicago Press, Chicago, IL, pp 1–20

Janke K (1990) Biological interactions and their role in the community structure in the rocky intertidal of Helgoland (German Bight, North Sea). Helgol Wiss Meeresunters 44:219–263
Karlson RH, Cornell HV, Hughes TP (2004) Coral communities are regionally enriched along an oceanic biodiversity gradient. Nature 429:867–870
Keddy P, Weiher E (1999) Introduction: the scope and goals of research on assembly rules. In: Weiher E, Keddy P (eds) Ecological assembly rules. Cambridge University Press, Cambridge, pp 1–22
Kotta J, Orav-Kotta H, Paalme T, Kotta I, Kukk H (2006) Seasonal changes in situ grazing of the mesoherbivores *Idotea baltica* and *Gammarus oceanicus* on the brown algae *Fucus vesiculosus* and *Pylaiella littoralis* in the central Gulf of Finland, Baltic Sea. Hydrobiologia 554:117–125
Kotta J, Lauringson V, Kotta I (2007) Role of functional diversity and physical environment on the response of zoobenthos communities to changing eutrophication. Hydrobiologia 580:97–108
Kotta J, Lauringson V, Martin G, Simm M, Kotta I, Herkül K, Ojaveer H (2008) Gulf of Riga and Pärnu Bay. In: Schiewer U (ed) Ecology of Baltic coastal waters. Ecological Studies 197. Springer, Berlin Heidelberg New York, pp 217–243
Lauringson V, Mälton E, Kotta J, Kangur K, Orav-Kotta H, Kotta I (2007) Environmental factors influencing the biodeposition of the suspension feeding bivalve *Dreissena polymorpha* (Pallas): comparison of brackish and fresh water populations in the Northern Baltic Sea and Lake Peipsi. Estuarine Coast Shelf Sci 75:459–467
Lawrence JM (1975) On the relationship between marine plants and sea urchins. Oceanogr Mar Biol Annu Rev 13:213–286
Lawton JH (2000) Community ecology in a changing world. Ecology Institute, Oldendorf Little C, Kitching JA (1996) The biology of the rocky shores. Oxford University Press, New York
Little C, Kitching JA (1996) The biology of the rocky shores. Oxford University Press, New York
Lubchenco J, Gaines SD (1981) A unified approach to marine-plant herbivore interactions. I. Populations and communities. Annu Rev Ecol Syst 12:405–437
MacArthur RH, Wilson EO (1967) The theory of island biogeography. Princeton University Press, Princeton, NJ
Mann KH, Lazier JRN (1996) Dynamics of marine ecosystems: biological-physical interactions in the oceans. Blackwell, Cambridge, MA
Marincovitch L (1977) Cenozoic Naticidae (Mollusca Gastropoda) of the northeastern Pacific. Bull Am Paleontol 70:169–494
Menge BA, Sutherland JP (1987) Community regulation: variation in disturbance, competition and predation in relation to environmental stress and recruitment. Am Nat 130:730–757
Mills EL, Leach JH, Carlton JT, Secor CL (1993) Exotic species in the Great Lakes: a history of biotic crises and anthropogenic introductions. J Great Lakes Res 19:1–54
Mungia P (2004) Successional patterns on pen shell communities at local and regional scales. J Anim Ecol 73:64–74
Orav-Kotta H, Kotta J, Herkül K, Kotta I, Paalme T (2009) Seasonal variability in the grazing potential of the invasive amphipod *Gammarus tigrinus* and the native amphipod *Gammarus salinus* in the northern Baltic Sea. Biol Invasions 11:597–608
Pianka ER (1971) Species diversity. In: Kramer A (ed) Topics in the study of life: The Bio Source Book. Harper & Row, New York, pp 401–406
Ricklefs RE (1987) Community diversity: relative roles of local and regional processes. Science 235:167–171
Ricklefs RE, Schluter D (1993) Species diversity: regional and historical influences In: Ricklefs RE, Schluter D (eds) Species diversity in ecological communities: historical and geographical perspectives. University of Chicago Press, Chicago, IL, pp 350–364
Rivadeneira MM, Fernández M, Navarrete SA (2002) Latitudinal trends of species diversity in rocky intertidal herbivore assemblages: spatial scale and the relationship between local and regional species richness. Mar Ecol Prog Ser 245:123–131
Rosensweig ML (1995) Species diversity in space and time. Cambridge University Press, Cambridge

Roughgarden J, Gaines S, Possingham HP (1988) Recruitment dynamics in complex life cycles. Science 241:1460–1466

Russell R, Wood SA, Allison G, Menge BA (2006) Scale, environment and trophic status: the context dependency of community saturation in rocky intertidal communities. Am Nat 167:E158–E170

Santileces B, Bolton JJ, Meneses I (2009) Marine algal communities. In: Witman JD, Roy K (eds) Marine macroecology. University of Chicago Press, Chicago, IL (in press)

Scheltema RS (1989) Planktonic and non-planktonic development among prosobranch gastropods and its relationships to the geographic ranges of species. In: Ryland JS, Tyler PA (eds) Reproduction, genetics, and distributions of marine organisms. Olsen and Olsen, Fredensborg, pp 183–188

Southward AJ, Crisp DJ (1954) Recent changes in the distribution of the intertidal barnacles *Chthamalus stellatus* Poli and *Balanus balanoides* L. in the British Isles. J Anim Ecol 23:163–177

Southward AJ, Hawkins SJ, Burrows MT (1995) Seventy years' observations of changes in distribution and abundance of zooplankton and intertidal organisms in the western English Channel in relation to rising sea temperature. J Therm Biol 20:127–155

Stenseth NC, Ottersen G, Hurrell JW, Mysterud A, Lima M, Chan KS, Yoccoz NG, Adlandsvik B (2003) Studying climate effects on ecology through the use of climate indices: the North Atlantic Oscillation, El Nino Southern Oscillation and beyond. Proc R Soc Lond B Biol Sci 270:2087–2096

Stuart CT, Rex MA (1991) Larval development and species diversity. The relationship between developmental pattern and species diversity in deep-sea prosobranch snails. In: Young CM, Eckelbarger KJ (eds) Reproduction, larval biology, and recruitment of the deep-sea benthos. Columbia University Press, New York, pp 118–136

Terborg JW, Faaborg J (1980) Saturation of bird communities in the West Indies. Am Nat 116:178–195

Vermeij GJ (1978) Biogeography and adaptation: patterns of marine life. Harvard University Press, Cambridge

Veron JEN (1995) Corals in space and time. UNSW Press, Sydney

Whittaker RJ, Willis KJ, Field R (2001) Scale and species richness: towards a general, hierarchical theory of species diversity. J Biogeogr 28:453–470

Williams CB (1943) Area and the number of species. Nature 152:264–267

Witman JD, Dayton PK (2001) Rocky subtidal communities. In: Bertness MD, Gaines SD, Hay M (eds) Marine community ecology. Sinauer Press, Sunderland, MA, pp 339–366

Witman JD, Genovese SJ, Bruno JF, McLaughlin JW, Pavlin BI (2003) Massive prey recruitment and the control of rocky subtidal communities on large spatial scales. Ecol Monogr 73:441–462

Witman JD, Etter RJ, Smith F (2004) The relationship between regional and local species diversity in marine benthic communities: a global perspective. Proc Natl Acad Sci USA 101:15664–15669

Chapter 7
Patterns Along Environmental Gradients

Antonio Terlizzi and David R. Schiel

7.1 Introduction

Environmental gradients both limit and characterise marine communities, from light-limited abyssal depths to the often-exposed, high intertidal zone. Within geographical areas, gradients in light, temperature, salinity, wave forces and habitat types influence the species present, trophic relationships and community structure (Schiel and Foster 1986; Menge and Olson 1990), while broad gradients, from polar regions to the tropics, occur in species assemblages and diversity (Pianka 1966; Gaston 2000). Early studies on rocky shores stressed the role of environmental gradients as fundamental to the processes structuring communities, and there was considerable debate about the nuances and exceptions that defined gradients in different conditions and locations (Stephenson and Stephenson 1949; Lewis 1964), and even the worth of such models for understanding structure (Underwood 1978). Early experimentalists pointed out that variation in communities involved ecological processes such as competition and predation interacting with individual species, and that life history features, tolerances and interactions varied over tidal gradients (e.g. Connell 1961). This was often polarized into basically two models (May 1984) involving either environmental or biotic control, although both types of processes clearly interact in community structure (Quinn and Dunham 1983; Underwood and Denley 1984). Later models incorporated gradients of many important biotic and abiotic processes (e.g. Menge and Sutherland 1987), as well as disturbances (Connell 1978), to account for the complex mosaic of patches of rocky shore communities (but see Chap. 15 by Noël et al. for details). Other studies clarified facilitative processes, and emphasized the positive interactions of species that can enhance diversity and ecological functioning of communities (Crain and Bertness 2006; Lilley and Schiel 2006; see Chap. 15 for an example in rockpool communities). These and other studies have contributed substantially to the way patterns and underlying processes in rocky shore communities are perceived, and offer challenges for advancing innovative experiments and statistical tools to deal with the complexity of diversity patterns (Benedetti-Cecchi 2004).

Patchiness is generated by complex interactions of physical disturbances, climate, substratum characteristics, species relationships and recruitment processes

M. Wahl (ed.), *Marine Hard Bottom Communities*, Ecological Studies 206,
DOI: 10.1007/978-3-540-92704-4_7,

(Sousa 1984; Picket and White 1985), and is intrinsically related to measurement scale (e.g. Levin 1992; Underwood and Petraitis 1993). There are important ecological considerations in discerning and understanding gradients in processes and structure (Constable 1999; Underwood et al. 2000). First, because different processes are likely to act differently in space and time, it is necessary to identify relevant scales of variation before causal processes underlying patterns can be properly framed and tested (Andrew and Mapstone 1987; Wu and Loucks 1995). Second, investigating patterns across scales helps in determining the level at which small-scale processes can be generalised (Lawton 1996), and how local processes may scale up to generate patterns at regional scales (Wootton 2001). Third, identification of relevant scales of variation in structure and processes is required both for understanding communities and for environmental management because this can lead to reliable predictions about changes caused by multiple stressors (Bishop et al. 2002; Schiel, Chap. 20).

7.1.1 Definitions of Diversity

"Biological diversity" is often defined ambiguously or formulated imprecisely, and used to mean different things (Magurran 2004). Diversity is usually designated as being α-diversity (the diversity within a given habitat), β-diversity (the degree to which communities show spatial variability in species composition from place to place) and γ-diversity (the overall diversity in a whole region; Whittaker 1975). At the species level in a given assemblage, α-diversity can be regarded as either the number of species present ("species richness"), the proportional abundance or homogeneity of individual species ("evenness" or "equitability") or, more commonly, a combination of both. Sensu stricto, species diversity has both richness and evenness components that consider the number of species present and their relative frequencies.

Structurally complex habitats influence diversity (McCoy and Bell 1991). On rocky substrates, biogenic habitat complexity is generated mostly by macroalgal and sessile invertebrate assemblages. Therefore, habitat complexity varies in relation to patchiness (habitat heterogeneity) and its changes along environmental gradients. "Diversity" is used here in a broad sense to refer to habitat heterogeneity, including changes in the mean number of species, species identities, evenness, taxonomic relatedness among species, and multivariate structure of assemblages. Our goal is to review briefly some of the ways in which diversity patterns have been described and experimentally clarified in relation to environmental gradients, and how they have contributed to understanding the distribution of biological diversity in marine rocky habitats. We draw attention to the knowledge gaps concerning spatial heterogeneity in rocky assemblages, to potential risks in interpreting patterns only in relation to obvious environmental gradients, and to approaches needed to deal with this issue in temperate rocky habitats. We limit the use of environmental gradients to gradual and continuous changes in abiotic variables. Given the important role that consumer pressure can play in diversity patterns, we also consider how changes in environmental conditions can influence biotic gradients by modulating the physiological performances of consumers.

We cite recent case studies for methodologies, and to illustrate how understanding changes along environmental gradients must necessarily underpin conservation policies concerning marine diversity.

7.2 Zonation

Conspicuous zonational patterns and strong environmental gradients have long attracted ecologists and physiologists to rocky shores (Benson 2002). "Zonation" is firmly entrenched in the jargon of marine ecologists but the term is now used more in a loose sense relating to broad distributional patterns, rather than in the mechanistic sense of ecologists in the 1940s–1960s. The topic of zonation occupied much of early marine ecology, and essentially involved the description of pattern and species' occurrences along perceived gradients.

Temperature, light, moisture, wave exposure and pressure are usually considered as major factors determining zonational patterns. Pérès (1982) described vertical zones (or *étage*) as the "depth interval of the benthic domain where the ecological conditions related to the main environmental factors are homogeneous … where the boundary between two adjacent vertical zones corresponds to a sharp change in the living assemblage composition". The occurrence of seawater spray during storms, tidal range, the lower limit of distribution of seagrasses or photophilic algae, and the depth compatible with the life of photosynthetic organisms were criteria for identifying zones (namely supralittoral, mediolittoral, infralittoral and circalittoral). Changes in water movement across the tidal gradient have also been proposed as a determinant of zonation of sessile organisms. Riedl (1966), for example, distinguished five zones characterised by homogeneous conditions of water movement and three critical depths.

The classical schemes of zonation have influenced generations of marine biologists. Lacking, however, were experiments to determine how bounded species were by these gradients, and the relative roles of ecological processes, especially competition and predation, and early life history processes, particularly settlement and recruitment, that determined whether species actually arrived to different parts of shores. A more recent context involves positive interactions, ecological "function" and the roles of foundation or key species as ecological determinants of diversity. In this sense, there is renewed interest in dominant species, perhaps especially large perennial autotrophs, their distributional patterns and interactions with biotic and physical gradients, and the relevant spatial and temporal scales at which they operate.

7.2.1 Intertidal Zone

The intertidal zone comprises a relatively thin band and only a small portion of nearshore benthic habitat on most shores, yet this zone has been the focus of most studies on environmental gradients. Not only is there a huge range of conditions over small spatial scales but this zone is accessible, visible at low tide, amenable to experimentation, and often the first responder to many anthropogenic stressors. In the early days of marine ecology, Stephenson and Stephenson (1949) described "universal" patterns of zonation with respect to two dominant gradients of physical conditions, wave exposure and critical tidal heights. Underlying mechanisms involved physiological limits and tolerances of species, and the schema for zonational patterns became rather elaborate. Although recognising a relationship between species tolerances and their occurrences, the plug was pulled on the mechanistic underpinnings by the loose or nonexistent correlation of species occurrences with tidal height (Underwood 1978), patchiness within zones, the many exceptions, and increasing experimental evidence that the gradient space of species (their fundamental niche) was usually much broader than where they occurred. Connell (1961), for example, demonstrated the role of differential predation, settlement, competition and growth along environmental gradients in the intertidal zone. His model of increased influence of physical factors high on the shore and of biotic factors low on the shore still underpins much of experimental ecology. Southward (1964) did a very simple removal experiment with limpets and showed that algal communities could occur widely across tidal gradients, if they were not continually grazed. Dayton (1975) then showed that predation, competition and species' life histories interacted over wave force gradients to account for patchiness within and between sites. Paine (1966) demonstrated that strong predation on exposed shores could prevent competitive dominance by a single species, thereby enabling higher diversity to persist regionally. Underwood (1981) showed that, in the absence of strong predation, both intra- and inter-specific competition could account for patchiness within zones and differences between zones.

Many other excellent studies worldwide have helped clarify the changing roles of interactions across gradients, including those relating to temperature, desiccation, wave exposure, shore aspect and morphology. These have been synthesized into several models (e.g. Connell 1978; Menge and Sutherland 1987). If there are generalisations, they are that we know the sorts of things to look for when analysing patterns of diversity across physical gradients but that almost everything will be context-dependent to some extent because of unique combinations and abundances of species and conditions present on different shores. This has several implications, not least of which is the continued need for innovative ways of detecting and measuring patterns and gradients for more communities and in more places. Perhaps the greatest need is to work out meaningful ways of integrating processes across spatial and temporal scales. For example, settlement processes for most marine organisms occur in boundary layer conditions along hard substrata but larval transport often occurs over scales of tens to thousands of meters (Kinlan and Gaines 2003; Fraschetti et al. 2003). Successful replenishment and maintenance of populations are the results of processes acting at several spatial scales (Schiel 2004; Fraschetti et al. 2005). Similarly, waves occur over a matter of seconds, El Niño Southern Oscillation (ENSO) events over years, and other events such as the Pacific Decadal Oscillation (PDO) over decades, all of which can affect marine communities.

How much of what we know from small-scale experiments in the intertidal zone will scale up to larger areas across these multi-scale influences? Mostly, we do not know the answer to this.

The intertidal zone still serves as a crucible for understanding impacts and changes. In situ data loggers recording the temperatures experienced by mussels, for example, help in understanding physiological tolerances, constraints and ecological contexts under which organisms can operate in often stressful circumstances (Helmuth 1999). Upwelling episodes that bring nutrients to shore, and relaxation events that allow larvae to come inshore to settle are now being clarified in their roles in community structure (Connolly and Roughgarden 1999; Menge et al. 1999). Highly localised issues, such as effects of people pressure from walking and foraging in the intertidal zone, have increasing importance worldwide. Trampling gradients affect dominant species, diversity and trophic relationships, with short-term damage often leading to long-term changes (Keough and Quinn 1998; Schiel and Taylor 1999). Other long-term influences are the arrival of non-indigenous species, which can alter the character of local assemblages (Ruiz et al. 2000). Detecting and understanding the interaction of all these factors continues to challenge ecologists.

7.2.2 Subtidal Zone

Numerous subtidal studies have related vegetational patterns, as well as population and community processes to gradients in physical and biotic conditions (e.g. Dayton 1985). Major physical gradients nearshore involve light and wave forces on localised scales, and temperature, nutrients and oceanographic conditions on broader scales. Most open rocky reefs are dominated by kelps and fucoids, individual species of which respond quite differently physiologically and demographically across dynamic nearshore waters (Schiel and Foster 2006). These large brown algae occur over a wide temperature range but are mostly restricted to waters with average monthly temperatures <25°C. Vegetational layering greatly reduces light to areas below, so that up to 2% of surface light reaches the substratum at around 20 m depth below kelp canopies (Reed and Foster 1984). Few kelps extend into the intertidal zone because they are intolerant of desiccation (e.g. Dayton 1975). Fucoid algal beds are mostly confined to waters less than 10 m deep because their light requirements tend to be higher than for kelps. Fucoids are generally more tolerant of desiccation and wave exposure.

The ecological literature is dominated by trophic interactions affecting kelp forests and community structure, rather than by kelp dynamics and life histories (Bartsch et al. 2008). Recent reviews have highlighted the diminishing role of overfished large predators (Steneck et al. 2002), with purportedly great changes in coastal communities, especially where sea otters (California, Alaska) or predacious fishes or invertebrates have been affected, such as in the north-eastern US (Jackson et al. 2001). The extent to which this applies to other areas is unknown.

Temperature, wave force, nutrients, light and coastal sediments interact at various spatial and temporal scales with each other and with biotic effects. For example, thermally stratified nearshore waters can lead to nutrient depletion and deterioration of surface kelp canopies (Tegner and Dayton 1991). Foraging patterns of echinoderms may be temperature-related (Sanford 2002), as can vectoring of diseases. For example, mass mortalities of sea urchins (Scheibling 1986; Pearse and Lockhart 2004) and abalone (Tissot 1995), and sessile epibenthic invertebrates, particularly sponges, anthozoans, bivalves and ascidians, have been linked to increased temperature and related effects (Cerrano et al. 2000).

The effects of sedimentation on algae (Airoldi and Virgilio 1998; Irving and Connell 2002) and sessile invertebrates (Naranjo et al. 1996) are widespread, and there is an increasing concern about the consequences for diversity of rocky assemblages (Airoldi 2003; Balata et al. 2007). Light penetration can be severely reduced by suspended material, exacerbating the effects of the sediments *per se* on recruitment and growth processes (Connell 2005).

Gradients of stressors can have long-term effects, such as in the Baltic Sea where increased sedimentation and nutrients have led to restricted distributions of many species (e.g. Kautsky et al. 1986). Overall, we have little direct knowledge of how gradients of new stressors act with those already present and, for many important habitats, how these affect diversity, species identities, demographics, variation and trophic relationships. Understudied habitats include caves and rock walls dominated by sessile invertebrates (Terlizzi et al. 2003a; Bussotti et al. 2006), canyons, nearshore trenches, and coastal plains out to the continental shelf that increasingly are being impacted by trawling and extractions (Thrush and Dayton 2002).

7.3 Gaps in Knowledge

Obvious gradients account for only a portion of the variability in species distributions and abundances, and can bias interpretations of patterns unless multiple factors and underlying ecological processes are examined within and across gradients (Underwood 2000). One useful type of analysis uses fixed sources of variation (i.e. the differences among positions along a given gradient) at a hierarchy of spatial scale. This enables the inclusion of a wide range of ecological processes that may influence patchiness of assemblages, regardless of whether or not specific models about relevant processes can be proposed. The analysis of spatial patterns along environmental gradients at different scales is also a logical requirement to deal with spatial and temporal confounding (Hurlbert 1984), and provides tests for generality of models for the distribution and abundance of organisms. Despite the general utility of this approach, pragmatic considerations of time, resource and logistics have limited its adoption and there are few analyses that have tested the consistency of patterns along sharp environmental gradients at hierarchies of spatial scales (see Benedetti-Cecchi 2001 for an example).

Quantifying variability at hierarchies of spatial scales across environmental gradients can help discern whether ecological factors underlying multi-scale patterns of horizontal variation can vary across vertical gradients of physical variables. For example, there is relatively little information on recruitment along depths in subtidal assemblages. The speed of wind-driven and wave-induced streams decreases with depth (Denny and Wethey 2001). Witman and Dayton (2001) have proposed that, in temperate environments, this pattern can have several implications for the scale of dispersion of larvae, spores and asexual propagules, influencing the connectivity between sites. At the scale of hundreds of metres, this implies that greater connectivity could occur between sites in the shallow than in the deep subtidal (below the thermocline), where assemblages are less linked by advective transport of larvae and/or propagules. This hypothesis represents a challenging issue for marine ecologists, and provides an additional perspective on connectedness in rocky assemblages across environmental gradients (Terlizzi et al. 2007).

Patterns of diversity along environmental gradients in peculiar or iconic habitats generate considerable interest, especially for conservation. The novelty of particular systems is often a major impetus for establishing marine protected areas. For example, marine submerged caves are characterised by unique faunistic and ecological features (Harmelin et al. 1985). Bussotti et al. (2006) showed that the number and identity of taxa within Mediterranean marine caves changed sharply along the exterior-interior axis of caves, that several taxa were exclusive to only one cave and, often, different taxa characterised assemblages at equivalent distances from cave entrances. This is a further demonstration that multi-scale mensurative experiments (cf. Hurlbert 1984) provide useful information about the consistency of biodiversity patterns.

7.4 Concluding Remarks

The integration of different scales in tests of hypotheses about spatial and temporal patterns of biodiversity distribution across environmental gradients is methodologically complex but is continually challenged by the development of innovative statistical procedures (e.g. Rossi et al. 1992; Anderson et al. 2006). There is, therefore, a potential for much better sampling designs, coupled with new statistical tools, to provide a powerful basis for future analyses of biodiversity responses to environmental factors (Anderson and Thompson 2004). Many pressing problems call for sound experimental quantification of biodiversity patterns in relation to environmental gradients. A better understanding of the role played by abiotic variables is a key prerequisite for forecasting the effects of shifts in environmental conditions on biodiversity as a result of human pressure, and for setting up adequate policies of marine conservation and management. For example, global warming and variations of extreme weather result in shifts in mean intensity and temporal variance of climatic variables, affecting biological systems at different hierarchies of organisation. Changes in physiological responses to environmental stress, fragmentation of populations, modification of distributional patterns of key species, and direct or indirect alteration of interactions among population are examples of likely impacts of climate change on marine systems (Walther et al. 2002). Despite increasing concern about societal and ecological implications of climate change, its possible effects on marine biodiversity are poorly addressed quantitatively, and we do not have a clear understanding of the roles of short- vs. long-term environmental stochasticity and population-intrinsic processes in controlling community dynamics. Reconstructing past changes in biodiversity patterns and modelling potential future variation under climate change scenarios will necessarily involve correlations between long-term data on biodiversity distribution vs. long-term climate data, and complex manipulative experiments to test and measure how simulated changes in intensity and temporal variance of environmental variables affect diversity in natural assemblages (Benedetti-Cecchi et al. 2006).

Biodiversity is usually treated as a response variable. Of considerable importance, however, is using it as a predictor variable to delineate how biodiversity is causally linked to ecosystem properties and how it affects ecosystem functions (Naeem 2002; Boero and Bonsdorff 2007). There is increasing evidence that ecosystem properties are strongly influenced by biodiversity in terms of the functional characteristics of the species present, and their distribution and abundance over space and time. Biodiversity interacts with climate, disturbance regimes and resource availability, determining changes in ecosystem properties (Hooper et al. 2005). Given the worldwide impact of human activities on biodiversity, there is broad concern about many aspects of the relationships between biodiversity and ecosystem functioning (BEF), including several points relevant to management of ecosystems. The BEF approach requires increasing accuracy of causal inference in experimental analyses of biodiversity, and detailed studies and integration of knowledge about biotic and abiotic controls on ecosystem properties, how ecological communities are structured, and the forces driving species extinctions and invasions.

Whether biodiversity is treated as a response or predictor variable, a key issue is the role of taxonomy and how the perception of changes of what we define as "biodiversity" is affected by the way it is described (Terlizzi et al. 2003b). When based on a solid taxonomic basis, well-designed long-term experiments can determine if abrupt changes have been occurring (a realistic model under major shifts in environmental variables), and which taxonomic level can be used as an effective surrogate for understanding how shifts in environmental variables affect diversity. Multi-scale analyses of biodiversity using novel analytical tools, well-designed long-term experimental studies at large spatial scales, and experimental tests of explanatory models will yield a better understanding of the relationships between biodiversity and ecosystem functioning. This highlights the need for multidisciplinary studies that include a coalescence of taxonomic, ecological, statistical, physiological and oceanographic skills.

References

Airoldi L (2003) The effects of sedimentation on rocky coast assemblages. Oceanogr Mar Biol Annu Rev 41:161–236
Airoldi L, Virgilio M (1998) Responses of turf-forming algae to spatial variations in the deposition of sediments. Mar Ecol Prog Ser 165:271–282
Anderson MJ, Thompson AA (2004) Multivariate control charts for ecological and environmental monitoring. Ecol Appl 14:1921–1935
Anderson MJ, Ellingsen KE, McArdle BH (2006) Multivariate dispersion as a measure of beta diversity. Ecol Lett 9:683–693
Andrew NL, Mapstone BD (1987) Sampling and the description of spatial pattern in marine ecology. Oceanogr Mar Biol Annu Rev 25:39–90
Balata D, Piazzi L, Benedetti-Cecchi L (2007) Sediment disturbance and loss of beta diversity on subtidal rocky reefs. Ecology 88:2455–2461
Bartsch I, Wiencke C, Bischof K, Buchholz CM, Buck BH, Eggert A, Feuerpfeil P, Hanelt D, Jacobsen S, Karez R, Karsten U, Molis M, Roleda MY, Schubert H, Schumann R, Valentin K, Weinberger F, Wiese J (2008) The genus Laminaria sensu lato: recent insights and developments. Eur J Phycol 43:1–86
Benedetti-Cecchi L (2001) Variability in abundance of algae and invertebrates at different spatial scales on rocky sea shores. Mar Ecol Prog Ser 215:79–92
Benedetti-Cecchi L (2004) Increasing accuracy of causal inference in experimental analyses of biodiversity. Funct Ecol 18:761–768
Benedetti-Cecchi L, Bertocci I, Vaselli S, Maggi E (2006) Temporal variance reverses the impact of high mean intensity of stress in climate change experiments. Ecology 87:2489–2499
Benson KR (2002) The study of vertical zonation on rocky intertidal shores. A historical perspective. Integ Comp Biol 42:776–779
Bishop M, Underwood AJ, Archambault P (2002) Sewage and environmental impacts on rocky shores: necessity of identifying relevant spatial scales. Mar Ecol Prog Ser 236:121–128
Boero F, Bonsdorff E (2007) A conceptual framework for marine biodiversity and ecosystem functioning. Mar Ecol 28 suppl 1:134–145
Bussotti S, Terlizzi A, Fraschetti S, Belmonte G, Boero F (2006) Spatial and temporal variability of sessile benthos in shallow Mediterranean marine caves. Mar Ecol Prog Ser 325:109–119
Cerrano C, Bavestrello G, Bianchi CN, Cattaneo-Vietti R, Bava S, Moranti C, Morri C, Picco P, Sara G, Schiaparelli S, Siccardi A, Sponga F (2000) A catastrophic mass-mortality episode of gorgonians and other organisms in the Ligurian Sea (Northwestern Mediterranean), summer 1999. Ecol Lett 3:284–293
Connell JH (1961) Effects of competition, predation by *Thais lapillus* and other factors on natural populations of the barnacle *Balanus balanoides*. Ecol Monogr 31:61–104
Connell JH (1978) Diversity in tropical rain forests and coral reefs: high diversity of trees and corals is maintained only in a nonequilibrium state. Science 199:1302–1310
Connell SD (2005) Assembly and maintenance of subtidal habitat heterogeneity: synergistic effects of light penetration and sedimentation. Mar Ecol Prog Ser 289:53–61
Connolly SR, Roughgarden J (1999) Theory of marine communities: competition, predation, and recruitment strength. Ecol Monogr 69:277–296
Constable AJ (1999) Ecology of benthic macro-invertebrates in soft-sediment environments: a review of progress towards quantitative models and predictions. Aust J Ecol 24:452–476
Crain CM, Bertness MD (2006) Ecosystem engineering across environmental gradients: implications for conservation and management. Bioscience 56:211–218
Dayton PK (1975) Experimental evaluation of ecological dominance in a rocky intertidal algal community. Ecol Monogr 45:137–159
Dayton PK (1985) Ecology of kelp communities. Annu Rev Ecol Syst 16:215–245
Denny M, Wethey D (2001) Physical processes that generate patterns in marine communities. In: Bertness MD, Gaines SD, Hay ME (eds) Marine community ecology. Sinauer, Sunderland, MA, pp 3–37

Fraschetti S, Giangrande A, Terlizzi A, Boero F (2003) Pre- and post-settlement events in benthic community dynamics. Oceanol Acta 25:285–295
Fraschetti S, Terlizzi A, Benedetti-Cecchi L (2005) Patterns of distribution of rocky marine assemblages: evidence of relevant scales of variation. Mar Ecol Prog Ser 296:13–29
Gaston KJ (2000) Global patterns in biodiversity. Nature 405:220–227
Harmelin JG (1985) Organisation spatiale des communautés sessiles des grottes sous-marines de Méditerranée. Rapp Comm Int Mer Médit 5:149–153
Helmuth B (1999) Thermal biology of rocky intertidal mussels: quantifying body temperatures using climatological data. Ecology 80:15–34
Hooper DU, Chapin FS, Ewel JJ, Hector A, Inchausti P, Lavorel S, Lawton JH, Lodge DM, Loreau M, Naeem S, Schmid B, Setälä H, Symstad AJ, Vandermeer J, Wardle DA (2005) Effects of biodiversity on ecosystem functioning: a consensus of current knowledge. Ecol Monogr 75:3–35
Hurlbert SH (1984) Pseudoreplication and the design of ecological experiments. Ecol Monogr 54:187–211
Irving AD, Connell SD (2002) Interactive effects of sedimentation and microtopography on the abundance of subtidal turfing algae. Phycologia 41:517–522
Jackson JBC, Kirby MX, Berger WH, Bjorndal KA, Botsford LW, Bourque BJ, Bradbury RH, Cooke R, Erlandson J, Estes JA, Hughes TP, Kidwell S, Lange CB, Lenihan HS, Pandolfi JM, Peterson CH, Steneck RS, Tegner MJ, Warner RR (2001) Historical overfishing and the recent collapse of coastal ecosystems. Science 293:629–638
Kautsky N, Kautsky H, Kautsky U, Waern M (1986) Decreased depth penetration of *Fucus vesiculosus* (L.) since the 1940's indicates eutrophication of the Baltic Sea. Mar Ecol Prog Ser 28:1–8
Keough MJ, Quinn GP (1998) Effects of periodic disturbances from trampling on rocky intertidal algal beds. Ecol Appl 8:141–161
Kinlan BP, Gaines SD (2003) Propagule dispersal in marine and terrestrial environments: a community perspective. Ecology 84:2007–2020
Lawton JH (1996) Patterns in ecology. Oikos 75:145–147
Levin SA (1992) The problem of pattern and scale in ecology. Ecology 73:1943–1967
Lewis JR (1964) The ecology of rocky shores. English University Press, London
Lilley SA, Schiel DR (2006) Community effects following the deletion of a habitat-forming alga from rocky marine shores. Oecologia 148:672–681
Magurran AE (2004) Measuring biological diversity. Blackwell, Oxford
May RM (1984) An overview: real and apparent patterns in community structure. In: Strong DR, Simberloff D, Abeler LG, Thistlef AB (eds) Ecological communities: conceptual issues and the evidence. Princeton University Press, Princeton, NJ, pp 3–16
McCoy ED, Bell SS (1991) Habitat structure: the evolution and diversification of a complex topic. In: Bell SS, McCoy ED, Mushinsky HR (eds) Habitat structure: the physical arrangement of objects in space. Chapman & Hall, London, pp 3–27
Menge BA, Olson AM (1990) Role of scale and environmental factors in regulation of community structure. Trends Ecol Evol 5:52–57
Menge BA, Sutherland JP (1987) Community regulation: variation disturbance, competition, and predation in relation to environmental stress and recruitment. Am Nat 130:730–757
Menge BA, Daley BA, Lubchenco J, Sanford E, Dahlhoff E, Halpin PM, Hudson G, Burnaford JL (1999) Top-down and bottom-up regulation of New Zealand rocky intertidal communities. Ecol Monogr 69:297–330
Naeem S (2002) Ecosystem consequences of biodiversity loss: the evolution of a paradigm. Ecology 83:1537–1552
Naranjo SA, Carballo JL, García-Gomez (1996) Effects of environmental stress on ascidian populations in Algeciras Bay (southern Spain): possible marine bioindicators? Mar Ecol Prog Ser 144:45–71
Paine RT (1966) Food web complexity and species diversity. Am Nat 100:65–75
Pearse JS, Lockhart SJ (2004) Reproduction in cold water: paradigm changes in the 20th century and a role for cidaroid sea urchins. Deep-Sea Res II 51:1533–1549
Pérès JM (1982) Major benthic assemblages. In: Kinne OJ (ed) Marine Ecology, vol 5. Wiley, London, pp 373–508

Pianka ER (1966) Latitudinal gradients in species diversity: a review of concepts. Am Nat 100:33–46
Pickett STA, White PS (1985) The ecology of natural disturbance and patch dynamics. Academic Press, Orlando, FL
Quinn JF, Dunham AE (1983) On hypothesis testing in ecology and evolution. Am Nat 122:602–617
Reed DC, Foster MS (1984) The effects of canopy shading on algal recruitment and growth in a giant kelp forest. Ecology 5:937–948
Riedl R (1966) Biologie der Meereshöhlen. Paul Parey, Hamburg
Rossi RE, Mulla DJ, Journel AG, Franz EH (1992) Geostatistical tools for modeling and interpreting ecological spatial dependence. Ecol Monogr 62:277–314
Ruiz GM, Fofonoff PW, Carlton JT, Wonham MJ, Hines AH (2000) Invasion of coastal marine communities in North America: apparent patterns, processes, and biases. Annu Rev Ecol Syst 31:481–531
Sanford E (2002) Water temperature, predation, and the neglected role of physiological rate effects in rocky intertidal communities. Integ Comp Biol 42:881–891
Scheibling RE (1986) Increased macroalgal abundance following mass mortalities of sea urchins (*Strongylocentrotus droebachiensis*) along the Atlantic coast of Nova Scotia. Oecologia 68:186–198
Schiel DR (2004) The structure and replenishment of rocky shore intertidal communities and biogeographic comparisons. J Exp Mar Biol Ecol 300:309–342
Schiel DR, Foster MS (1986) The structure of subtidal algal stands in temperature waters. Oceanogr Mar Biol Annu Rev 24:265–307
Schiel DR, Foster MS (2006) The population biology of large brown seaweeds: ecological consequences of multiphase life histories in dynamic coastal environments. Annu Rev Ecol Evol Syst 37:343–372
Schiel DR, Taylor DI (1999) Effects of trampling on a rocky intertidal algal assemblage in southern New Zealand. J Exp Mar Biol Ecol 235:213–235
Sousa WP (1984) Intertidal mosaics: patch size, propagule availability, and spatially variable patterns of succession. Ecology 65:1918–1935
Southward AJ (1964) Limpet grazing and the control of vegetation on rocky shores. In: Crisp DJ (ed) Grazing in terrestrial and marine environments. Blackwell, Oxford, pp 265–273
Steneck RS, Graham MH, Bourque BJ, Corbett D, Erlandson JM, Estes JA, Tegner MJ (2002) Kelp forest ecosystem: biodiversity, stability, resilience and their future. Environ Conserv 29:436–459
Stephenson TA, Stephenson A (1949) The universal features of zonation between the tidemarks on rocky coasts. J Ecol 38:289–305
Tegner MJ, Dayton PK (1991) Sea urchins, El Ninos, and the long term stability of Southern California kelp forest communities. Mar Ecol Progr Ser 77:49–63
Terlizzi A, Scuderi D, Fraschetti S, Guidetti P, Boero F (2003a) Molluscs on subtidal cliffs: patterns of spatial distribution. J Mar Biol Assoc UK 83:165–172
Terlizzi A, Bevilacqua S, Fraschetti S, Boero F (2003b) Taxonomic sufficiency and the increasing insufficiency of taxonomic expertise. Mar Pollut Bull 46:544–560
Terlizzi A, Anderson MJ, Fraschetti S, Benedetti-Cecchi L (2007) Scales of spatial variation in Mediterranean subtidal sessile assemblages at different depths. Mar Ecol Prog Ser 332:25–39
Thrush SF, Dayton PK (2002) Disturbance to marine benthic habitats by trawling and dredging: implications for marine biodiversity Annu Rev Ecol Syst 33:449–473
Tissot BN (1995) Recruitment, growth, and survivorship of black abalone on Santa Cruz Island following mass mortality. Bull South Calif Acad Sci 94:179–189
Underwood AJ (1978) The detection of non-random patterns of distribution of species along a gradient. Oecologia 36:317–326
Underwood AJ (1981) An experimental evaluation of competition between three species of intertidal prosobranch gastropods. Oecologia 33:185–202

Underwood AJ (2000) Experimental ecology on rocky intertidal habitats: what are we learning? J Exp Mar Biol Ecol 250:51–76

Underwood AJ, Denley EJ (1984) Paradigms, explanations and generalizations in models for the structure of intertidal communities on rocky shores. In: Strong DR, Simberloff D, Abele LG, Thistle A (eds) Ecological communities: conceptual issues and the evidence. Princeton University Press, Princeton, NJ, pp 151–180

Underwood AJ, Petraitis PS (1993) Structure of intertidal assemblages in different locations: how can local processes be compared? In: Ricklefs R, Schutler D (eds) Species diversity in ecological communities. University of Chicago Press, Chicago, IL, pp 38–51

Underwood AJ, Chapman MG, Connell SD (2000) Observation in ecology: you can't make progress on processes without understanding the patterns. J Exp Mar Biol Ecol 250:97–115

Walther GR, Post E, Convey P, Menzel A, Parmesank C, Beebee TJC, Fromentin JM, Hoegh-Guldberg O, Bairlein F (2002) Ecological responses to recent climate change. Nature 416:389–395

Whittaker RH (1975) Communities and ecosystems. Macmillan, New York

Witman JD, Dayton PK (2001) Rocky subtidal communities. In: Bertness MD, Gaines SD, Hay ME (eds) Marine community ecology. Sinauer, Sunderland, MA, pp 339–366

Wootton JT (2001) Local interactions predict large-scale pattern in empirically derived cellular automata. Nature 413:841–843

Wu J, Loucks OL (1995) From balance of nature to hierarchical patch dynamics: a paradigm shift in ecology. Q Rev Biol 70:439–466

Chapter 8
Evolutionary Patterns of Diversity and Their Causes

Sharyn J. Goldstien and David R. Schiel

8.1 Introduction

There is considerable debate and some contention relating to evolutionary processes, marine speciation and biodiversity. The uncertainties involve the high dispersal potential and often large population sizes of many marine species, and the degree to which they result from and influence evolutionary processes (Cowie and Holland 2006). Phenotypic and genetic diversity among temperate reef communities are often greater than would be expected from marine species in general (Palumbi 1994). This disparity results from difficulties in assessing dispersal potential, historical associations of speciation, diversity and evolution, and putative dispersal barriers. Additionally, many conspecific assemblages appear to coexist and exhibit similar ecologies, yet differ in biogeographical range, albeit with some degree of overlap (Kelly and Eernisse 2007), suggesting that intrinsic factors may be important evolutionary determinants or that historical associations are obscured by contemporary patterns.

Here, we discuss the evolutionary processes which may account for biodiversity observed in temperate reef communities, focusing on three biogeographic regions. We also address the questions: is there concordance among communities? Can we generalise evolutionary causes of marine diversity? What can evolutionary patterns predict for future diversity of temperate reef communities in a human-dominated world?

8.2 Evolutionary Process

The evolution of species results from a host of mechanisms acting over long periods to change allele frequencies within populations (Freeman and Herron 2007). Subsequently, speciation or extinction (Fig. 8.1) will occur if change is associated with reproductive isolation and the reduction of genetic exchange within or among populations (i.e. gene flow). Several mechanisms may be responsible for the speciation of marine organisms (see excellent reviews by Avise 1998; Landry et al. 2003; Briggs 2007b) but the two main ones are allopatric and sympatric speciation

M. Wahl (ed.), *Marine Hard Bottom Communities*, Ecological Studies 206,
DOI: 10.1007/978-3-540-92704-4_8, © Springer-Verlag Berlin Heidelberg 2009

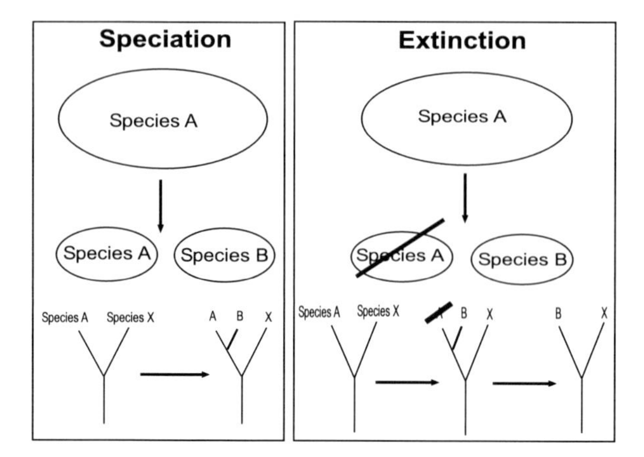

Fig. 8.1 Diagrammatic representation of speciation and extinction, including the phylogenetic consequence of each process. Speciation results in increased biodiversity due to the creation of sister species. Extinction results in the reduction of biodiversity due to the removal of lineages

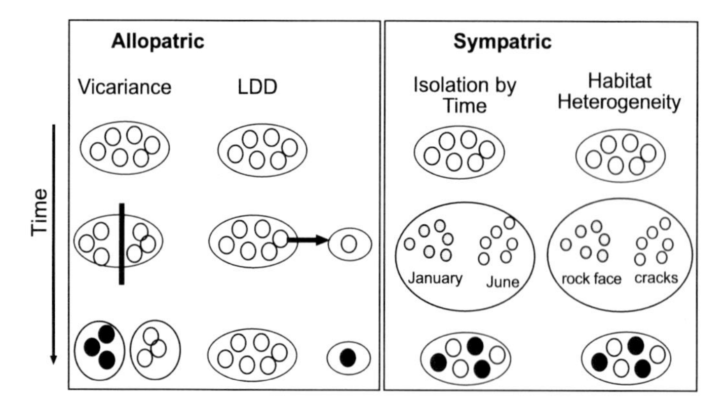

Fig. 8.2 Diagrammatic representation of allopatric and sympatric speciation through time. Allopatric speciation mechanisms include *vicariance*—the separation of populations by a physical barrier—and *long-distance dispersal* (LDD)—isolation of a subset of individuals from a population due to chance dispersal. Two sympatric speciation mechanisms include *isolation by time*—the isolation of individuals due to asynchronous spawning periods—and *habitat heterogeneity*—the adaptation to specific niches within a habitat

(Fig. 8.2). *Allopatric speciation* (splitting of contiguous populations into isolated demes) is readily accepted as the main process by which speciation occurs because there is clear association of geographic and reproductive isolation. Allopatry also fits the early idea that speciation in marine populations is slow, because of their high dispersal and connectivity, and results mostly from physical dispersal barriers.

Sympatric speciation (ecological niche capture and competition in the absence of geographic separation) is a difficult process to test because observed biogeographic patterns often obscure historical isolation (Mayr 1954; Losos and Glor 2003). Furthermore, an association is required between traits under selection and traits resulting in reproductive isolation (Bolnick and Fitzpatrick 2007). The ambiguous nature of sympatry makes it a more contentious process in evolutionary theory (Bolnick and Fitzpatrick 2007), although there are many examples of sympatry in terrestrial and aquatic systems (cichlids, Genner et al. 2007; polychaetes, Kruse et al. 2004; *Echinometra* spp., Landry et al. 2003).

There are two schools of thought relating to the mechanisms of allopatric speciation (McGlone 2005):

1. *Vicariance* is a means of isolation through extrinsic barriers such as mountain building, glacial habitat fragmentation and oceanic barriers. Vicariant events create and maintain global biodiversity of species through the splitting of contiguous habitat, resulting in the isolation of populations, leading to the creation of endemic sister species (Briggs 2006).
2. *Long-distance dispersal* (LDD) involves isolation of populations resulting from stochastic dispersal events (Trakhtenbrot et al. 2005). These take individuals into habitats beyond their normal dispersal range and may lead to range expansion or speciation.

LDD is often invoked to counteract vicariant theory (Trewick et al. 2007) and explain extraordinary patterns of species' distributions and diversity. Many biogeographers doubt that such stochastic events can account for the development of communities (Carlquist 1981), while others claim LDD to be a mechanism crucial to population spread and genetic connectivity (Trakhtenbrot et al. 2005). Rafting, primarily on large kelps, is now recognised as a major mechanism by which marine communities can be transported worldwide (Thiel and Haye 2006). Phylogenetic reconstructions of marine and terrestrial organisms show that speciation events and intraspecific genetic structure do not always correspond to geological dates of previously invoked vicariant events, but are more concordant with recent colonisation by LDD (Waters and Craw 2006).

Finally, extinction is an evolutionary process which affects diversity by counteracting speciation and, therefore, is considered an evolutionary mechanism of community diversity. Extinction is related to individual fitness and the preservation or decline of a species (Levin 1994). Biogeographic patterns such as the East Indies centre of origin (Briggs 1999), anti-tropical distributions (Briggs 1987), and tropical assemblage turnover (Jackson et al. 1993) have been attributed, at least partially, to interchange or extinction and replacement of species. Mechanisms of extinction include inherent traits such as reproductive strategy and timing (Calabrese and Fagan 2004), or species tolerance to environmental conditions. Reproductive strategy, such as brooding live juveniles, free-spawning of gametes and internal fertilisation, can influence population dispersion (Levin and Bridges 1995) and, in turn, determine its range and potential to survive localised extinction events (Morgan 1996). Asynchronous spawning events within a reproductive season can reduce the temporal overlap of male and female gametes and the probability of successful mating (Calabrese and Fagan 2004). Similarly, the tolerance range or plasticity of species traits will determine individual and population responses to environmental conditions (Shiota and Kimura 2007), habitat heterogeneity (Levin 1994) and resource competition, which are all major evolutionary determinants of species diversity. The complexity of the marine environment and the interaction with both the nearshore waters and offshore environment make understanding marine evolution challenging.

8.3 Regional Biogeographic Patterns

A global overview of the marine environment (Fig. 8.3) shows latitudinal trends in sea-surface temperature (SST) and hydrographical features which have a bearing on the transport of species and the genetic structure of populations. At a regional scale, heterogeneity of habitat and species' ecology is expected to have a dominant role in the persistence and range expansion of species. Here, we use three biogeographic areas to highlight both the problems and the current state of understanding of evolutionary patterns of diversity and their causes.

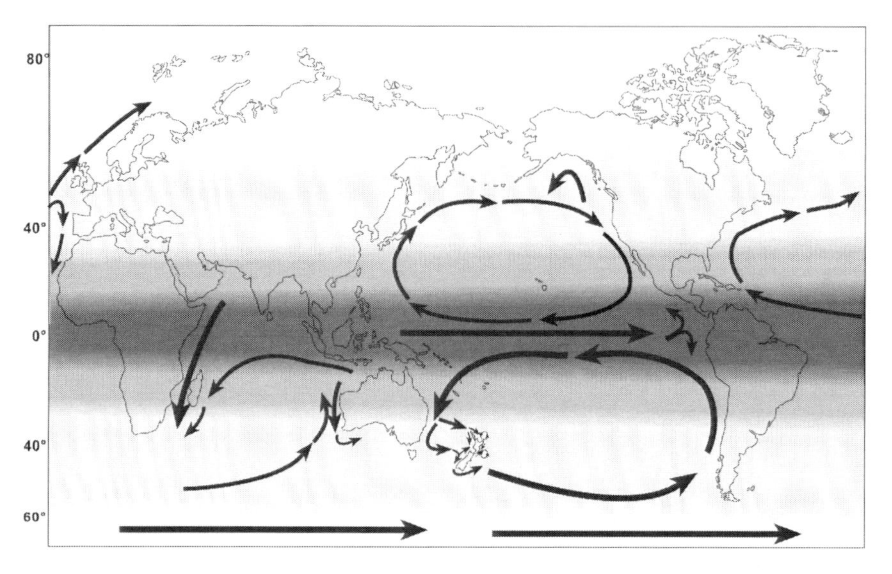

Fig. 8.3 World map illustrating the latitudinal trends in sea-surface temperature (*dark grey* warm water, *light grey* cooler water) and the major sea-surface currents (*black arrows*) influencing the three biogeographic zones discussed in the text (North America, Australasia and Europe)

8.3.1 North America

The long expanse of contiguous coastline of the Americas gives a high potential for connectivity and stability of coastal populations. However, assemblage turnover, biogeographic disjunctions and intraspecific genetic structure are readily observed along the Pacific and Atlantic coasts of North America.

The coastal molluscan assemblage is a good example of long-term changes on the Pacific coast. Fossil samples from the basal part of the marine Tejon formation, southern California have revealed that at least four *Patelloida* limpet species inhabited the Californian coast in the middle Eocene period (Lindberg and Squires 1990). In contrast, extant *Patelloida* are restricted to tropical and subtropical waters, and the present limpet guild of western North America is dominated by acmaeid limpets (Lindberg and Hickman 1986). These findings suggest that convergent morphology may have occurred within the *Patelloida* lineage and that the molluscan assemblage of this region has changed since the Eocene (Lindberg and Squires 1990). Additional limpet fossils of the late Eocene, collected from the Oregon coast, include three dominant limpet taxa of the modern Australian limpet guild including nacellid, siphonarid and patelloida limpets which no longer exist on this coastline (Lindberg and Hickman 1986), again supporting a turnover of molluscan species throughout this period.

In addition to species turnover, present biogeographic disjunctions delineate the ranges of extant species into areas of peculiar species assemblages, which occur mostly as gradients along adjacent shores (Dawson 2001). Several locations along the Pacific coast of America show demarcation of biogeographic provinces corresponding to SST gradients, upwelling and strong directional currents. For instance, Point Conception is located between 34 and 35°N, and species diversity is considerably less than in surrounding regions. North and south of Point Conception, transition zones, peaks in species range termini and species diversity suggest that processes are acting across different taxa to impart species limitations along the coastal reefs (Dawson 2001).

For example, two sister species of the prosobranch mollusc *Nucella* have overlapping distributions along the Pacific coast from Oregon to north of Point Conception (Marko 1998). Genetic analyses suggest that *Nucella emarginata* experienced post-glacial northward range expansion across Point Conception, with peripheral isolation potentially leading to speciation and the creation of the younger species *N. ostrina* (Marko 1998). Peripheral isolation may have been caused by retention within upwelling regions around Point Conception, leading to speciation, with secondary contact resulting in the overlapping range now observed. In this case, what would appear to be sympatric speciation could be more readily explained by vicariance and dispersal.

In addition to speciation, coastal populations continually experience evolutionary change at more contemporary timescales. For instance, a comparative study of 28 intertidal chiton species along the Pacific coast (Kelly and Eernisse 2007) shows a latitudinal trend in gene flow. Chiton species in southern California and on both sides of the Baja California peninsula showed increasing genetic structure and isolation of populations towards the equator. Populations of species restricted to central California and the Pacific Northwest appeared to be panmictic, with no significant intraspecific genetic structure observed (Kelly and Eernisse 2007). There was high concordance among species related to geographic location. The authors proposed many possible explanations for the observed species similarities, including temperature gradients acting on larval longevity, reducing the dispersal distance of southern species. Reduced dispersal could have created a stepping stone or isolation-by-distance effect, especially when currents in the south are weaker and less continuous than those in the north. It appears that chiton species of similar ecologies and distribution are at least responding to their environment in the same way and, therefore, may have phylogenetic constraints on their response to evolutionary causes of species diversity.

The North American coastal environment is one of the most intensively studied biogeographic zones in the world. Here, we have shown only a small subset of available data which highlights the variety of evolutionary processes occurring within a single region. Long-term processes (millions of years) have altered whole assemblages of molluscs, while more recent (thousands of years) interactions of larval transport and tolerances to climatic shifts and oceanic processes are maintaining population isolation and encouraging species diversity (Sotka et al. 2004). The "leaky border" around Point Conception shows that some species are more sensitive than others to biogeographic processes, implicating the interaction of ecology and evolution as being quite important.

8.3.2 Australasia

Australasia has been a strong focal point for the vicariance versus LDD debate (McGlone 2005), with much of the extant taxa originally thought to be derived from Gondwanan relicts. This focus stems from the taxonomic similarities of Antarctica, Australia, Africa, New Caledonia, New Zealand and South America, which were once joined in the southern landmass Gondwanaland (Yoder and Nowak 2006). The continents are now separated by vast oceans and strong currents (e.g. the Circumpolar Current). The isolation of a once contiguous land logically leads to the belief that vicariance enhanced biodiversity by creating endemic sister taxa. New Zealand's island status and long (50–80 Ma) isolation by 800–2,000 km presents particularly strong support for the probability that these assemblages are Gondwanan relicts. Geological evidence, however, suggests that most, if not all, of the New Zealand landmass was underwater during the Oligocene (King 2000). In this instance, most extant taxa would have been derived from post-Oligocene re-colonisation either from trans-Tasman dispersal (Waters and Craw 2006), northern Pacific waters (Knox 1980) or southern polar waters (Crame 1986). There is phylogenetic support for trans-Tasman dispersal in several taxa, such as lobsters (*Jasus edwardsii*; Ward et al. 2006) and molluscs (*Nerita*; Waters et al. 2005). Other studies give equal weight to the probability of Australian or Pacific links for ancient taxa such as *Haliotis* abalone (Degnan et al. 2006) and *Cellana* limpets (Goldstien et al. 2006b).

There is evidence from the Australian continent of more recent vicariant events, such as the faunal disjunction noted around Wilson's Promontory (Waters and Roy 2003). Similar to Point Conception, Wilson's Promontory within Bass Strait (south-eastern Australia, separating Tasmania from the mainland) is designated as the point at which western and eastern Australian assemblages become distinct due to an historical land bridge which connected Tasmania to the mainland during glacial maxima of the Pleistocene (Bennett and Pope 1960).

Many intertidal and subtidal organisms exhibit a strong genetic disjunction across this region of historical isolation, including the gastropods *Austrolittorina* (Waters et al. 2007) and *Nerita* (Waters et al. 2005). Furthermore, the coast of Australia is largely influenced by two counter-currents. The Leeuwin Current, flowing southwards on the western coast, can reach as far as Tasmania, with variation in strength noticeable throughout the year, and the cooler Capes Current flows northwards on the south-western coast (Cresswell and Griffin 2004). The Leeuwin Current in particular is thought to affect the dispersal of coastal species and create temporal fluctuations in propagule transport from the Indian Ocean along the western coast of Australia, resulting in further incongruence between western and eastern coastal taxa (Waters and Roy 2003).

Similar to Australia, New Zealand has several identified regions of species disjunctions (Pawson 1961; Nelson 1994). In particular, Cook Strait now separates the North and South islands of New Zealand but has been a transient feature over the last 500,000 years. This area was once continuous coastline until sea-level changes scoured out a channel. In addition to demarcation of faunal and floral assemblages in this region, there are many coastal invertebrate species showing strong genetic structure associated with the isolation of once continuous populations north and south of the Strait (Goldstien et al. 2006a).

The Australasian biogeographic area has been subjected to geological instability with the break-up of Gondwanaland (80–100 Ma), which has physically separated once continuous populations, as well as initiating many directional currents leading to climatic change and new transport pathways (Sanmartin et al. 2007). These long-term processes influence many life-history phases, making it difficult to tease apart the mechanisms influencing the evolution of species in this region. Some species are able to disperse across the 2,000-km oceanic expanse of the Tasman Sea, while others are isolated by a 26-km stretch of water which has been historically unstable, making this region a challenging system for biogeographers.

8.3.3 Europe

The Atlantic coast of Europe encompasses the British Isles, and the continental countries north to Norway and south to Spain, including the Mediterranean Sea. Many physical barriers along the European coast suggest reduced connectivity among coastal populations (Jolly et al. 2005). Examples include the Skagerrak-Kattegat-Baltic (SKB) Seas region of reduced salinity, oceanic fronts separating the Mediterranean Sea from the Atlantic Ocean, and contemporary oceanographic features such as directional currents and large annual fluctuations in SST (Peacock 1989). However, Europe has also experienced geological change during Pleistocene ice ages, whereby habitat fragmentation and temperature shifts associated with the southward shift of the Polar Front have had a strong influence on the persistence and ranges of many nearshore species over the last 15,000 years or so (Peacock 1989).

There are numerous examples of glacial processes affecting the population genetic structure of species (Coyer et al. 2003) but little to suggest that these cycles affect species diversity. For instance, the red seaweed *Palmaria palmata* ranges from Spitsbergen (80°N) to Portugal (40°N) and is endemic to the North Atlantic. The most parsimonious phylogenetic reconstruction of the Palmariaceae suggests this taxon originates from the Pacific Ocean and that *P. palmata* colonised the North Atlantic around 3.5 Ma through LDD associated with the opening of the Bering Strait, with subsequent extinction of its ancestral species in the Pacific (Lindstrom et al. 1996). *P. palmata* is now widely distributed throughout the North Atlantic and, while glaciations, sea-level and SST changes have provided opportunity for rapid speciation of Palmariaceae in the Pacific, similar events have not led to further radiation of this taxon in the North Atlantic. This lack of speciation persists despite genetic evidence that *P. palmata* has re-colonised much of its present distribution over the last 20,000 years from two separate glacial refugia, and has low dispersal potential (Provan et al. 2005). Asymmetry in species diversity between ocean basins may suggest fundamental differences between the genetically distinct *P. palmata* and the other Palmariaceae, rather than environmental differences between the Pacific and North Atlantic oceans (Lindstrom et al. 1996).

Contrasting patterns of speciation occur in the brown seaweed genus *Fucus* (Coyer et al. 2006) which, based on phylogenetic relationships with sister taxa, also appears to originate from the North Pacific, is widely distributed along the coast of Europe, and has a low dispersal potential. Unlike Palmariaceae, the North Atlantic assemblage of *Fucus* is species-rich, although hybridisation and morphological plasticity make it difficult to assign species accurately (Coyer et al. 2006). The shallow and mostly unresolved phylogenetic reconstruction of *Fucus* suggests that colonisation and speciation within the North Atlantic was rapid (Coyer et al. 2006). It is thought that the ancestral *Fucus* sp. was hermaphroditic and entered the North Atlantic through the Bering Strait 3–7 Ma ago, with extensive changes and local adaptation in morphology and reproductive strategy (dioecy-derived), and the onset of asynchronous reproductive seasons within species. Furthermore, significant genetic structure observed among *F. serratus* populations (Coyer et al. 2003) indicates glacial refugia, as well as repeated bottlenecks from thermally induced cycles of re-colonisation and extinction.

In marked contrast to the genetic structure and species diversity in seaweed taxa of the North Atlantic is the intertidal mollusc *Littorina*, one of the most widely studied taxa of temperate reefs. Two European species have been pivotal in understanding sympatric speciation and phenotypic diversity. *Littorina saxatillis* and

Littorina striata have different shell morphologies, and the variants are heritable ecotypes. There are two distinct ecotypes of *L. saxatillis* in each of Britain and Spain. In both places, the ecotypes exhibit assortative mating based on shell morphology (Absher et al. 2003; Hollander 2005). Similarly, in Sweden *L. saxatillis* showed genetic structure partitioned more by habitat than by islands of the Swedish archipelago, providing support for local adaptation (Johanesson et al. 2004). In contrast, the different ecotypes of *L. striata* showed no genetic structure, with high levels of gene flow and mating among the ecotypes, despite the persistence of shell morphologies (De Wolf et al. 2000).

The European coast includes a vast, heterogeneous system of embayments and archipelagos, as well as latitudinal and localised gradients of SST and salinity. This region has also been strongly influenced by recent (thousands of years) glaciations and sea-level changes. This makes for a dynamic evolutionary history for the coastal biota, which is evident in the small subset of examples presented here. Taxa such as *Palmeria palmata* and *Fucus* have taken the opportunity of new transport pathways to colonise the North Atlantic, where environmental change has caused population-level changes in some taxa and more pronounced speciation in others. *Fucus* has been particularly successful in colonising and adapting to a wide variety of habitats, with a subsequent increase in species, morphological and reproductive diversity. In contrast, some littorinid species are undergoing sympatric speciation due to morphological adaptation in line with the heterogeneous habitat along the European coastline.

8.4 Discussion

Can we get a clear view of evolution? The real question is what do we get from understanding the complexity and intricate linkages between ecology and evolution. General evolutionary patterns borne from extensive regional and cross-taxa studies are invaluable in understanding the origin of species and the building of biodiversity, and also in gauging how communities may be affected by future climate change, habitat fragmentation and non-indigenous species (NIS). However, it is important to recognise the limitation of evolutionary research and to interpret evolutionary processes with caution.

The Australasian region in particular highlights the intricacy of evolutionary mechanisms. Vicariant events such as the split-up of Gondwanaland have been pivotal in interpreting biodiversity of the southern hemisphere. However, there is potential for circularity to occur when correlating geological events with species biogeography. While many species patterns may be explained by vicariance, dispersal and vicariance are not mutually exclusive, and the lack of independence between geological and biogeographic studies could mislead the interpretation of data in these fields (Waters and Craw 2006).

Phylogenetics and population genetic studies are essential to evolutionary studies, and are often used to support or dismiss dates of speciation and mechanisms of biodiversity maintenance. However, molecular data do not stand alone and should be used with caution. Similar to the inaccuracies of fossil dates and geological events, molecular dating is still in its infancy. The rate of genetic mutation, substitution and divergence is based primarily on the correlation of genetic substitutions, fossil data and geological events which together produce an estimate of evolutionary change. The most readily accepted event used to estimate rates of change is the emergence of the Isthmus of Panama (Knowlton and Weigt 1998) and, yet, the timing and processes involved in isolating populations in this region are still under debate. While the correlation of data from three fields supports time estimates, there is room for misinterpretation, as the standard error on most dates is in the order of thousands to millions of years.

Despite the shortcomings of "accurate" dating tools, the concordance of evolutionary patterns observed within and among biogeographic regions highlights the integrative nature of ecology and evolution. It is now clear that larval transport plays only one part in the evolution of species. Transport is affected by reproductive modes and strategies, timing of larval release, larval longevity, and the response of larvae to surrounding environmental conditions such as temperature gradients, habitats, competition and predation. Subsequent recruitment into a habitat, whether it is a new one reached through LDD or an adjacent one common to the larval pool, also plays a key role in the survival and expansion of a species. However, the main process of speciation leading to biodiversity is the reproductive isolation of populations which have successfully dispersed, recruited and effectively reproduced. The mechanisms involved in these processes are clearly dynamic and strongly influenced by species' ecologies, tolerances and reproductive success, all of which influence the success of reproductive strategies and recruitment in a changing environment.

The evolutionary patterns discussed above have increased importance as more species are transported worldwide through shipping, the aquarium trade and other human activities. The prevalence of "leaky borders" therefore applies not only to biogeographic transition zones but also to entire countries like New Zealand, where the advent of new species is recognised as a major threat to the integrity of coastal ecosystems (Grosholz 2002). What were once extreme chance events now occur repeatedly in ecological time frames (Mack et al. 2000). The spread and traction of NIS into native communities is exacerbated by multiple incursions, affecting both the genetic structure of populations and the ability of species to spread from multiple source populations along coastlines. Rather than extinction of benthic species, the current synopsis is that diversity is increasing (Briggs 2007a), with unknown consequences which will be a product of which species arrive, and the habitats and ecosystems they manage to get traction within.

References

Absher TM, Boehs G, Feijó AR, da Cruz AC (2003) Pelagic larvae of benthic gastropods from shallow Antarctic waters of Admiralty Bay, King George Island. Polar Biol 26:359–364

Avise JC (1998) The history and purvue of phylogeography: a personal reflection. Mol Ecol 7:371–379

Bennett I, Pope EC (1960) Intertidal zonation of exposed rocky shores of Tasmania and its relationship with the rest of Australia. Aust J Mar Freshw Res 11:182–221

Bolnick DI, Fitzpatrick BM (2007) Sympatric speciation: models and empirical evidence. Annu Rev Ecol Evol Syst 38:459–487

Briggs JC (1987) Antitropical distribution and evolution in the Indo-West Pacific Ocean. Syst Zool 36:237–247

Briggs JC (1999) Coincident biogeographic patterns: Indo-West Pacific Ocean. Evolution 53:326–335

Briggs JC (2006) Proximate sources of marine biodiversity. J Biogeogr 33:1–10

Briggs JC (2007a) Marine biogeography and ecology: invasions and introductions. J Biogeogr 34:193–198

Briggs JC (2007b) Marine longitudinal biodiversity: causes and conservation. Divers Distrib 13:544–555 doi:10.1111/j.1472-4642.2007.00362.x

Calabrese JM, Fagan WF (2004) Lost in time, lonely, and single: reproductive asynchrony and the Allee effect. Am Nat 164:25–37

Carlquist S (1981) Chance dispersal. Am Sci 69:509–516

Cowie RH, Holland BS (2006) Dispersal is fundamental to biogeography and the evolution of biodiversity on oceanic islands. J Biogeogr 33:193–198

Coyer JA, Peters A, Stam W, Olsen J (2003) Post-ice age recolonization and differentiation of Fucus serratus L (Phaeophyceae: Fucaceae) populations in Northern Europe. Mol Ecol 12:1817–1829

Coyer JA, Hoarua G, Oudot-Le Secq M, Stam W, Olsen J (2006) A mtDNA-based phylogeny of the brown algal genus Fucus (Heterokontophyta: Phaeophyta). Mol Phylog Evol 39:209–222

Crame JA (1986) Polar origins of marine invertebrate faunas. Palaios 1:616–617

Cresswell GR, Griffin DA (2004) The Leeuwin Current, eddies and sub-Antarctic waters off south-western Australia. Mar Freshw Res 55:267–276

Dawson MN (2001) Phylogeography in coastal marine animals: a solution from California? J Biogeogr 28:723–736

Degnan S, Geiger DL, Degnan BM (2006) Evolution in temperate and tropical seas: disparate patterns in southern hemisphere abalone (Mollusca: Vetigastropoda: Haliotidae). Mol Phylog Evol 41:249–256

De Wolf H, Verhagen R, Backeljau T (2000) Large scale population structure and gene flow in the planktonic developing periwinkle, *Littorina striata*, in Macronesia (Mollusca: Gastropoda). J Exp Mar Biol Ecol 246:69–83

Freeman S, Herron JC (2007) Evolutionary analysis. Pearson Education, Upper Saddle River, NJ

Genner MJ, Nichols P, Carvalho GR, Robinson RL, Shaw PW, Turner GF (2007) Reproductive isolation among deep-water cichlid fishes of Lake Malawi differing in monochromatic male breeding dress. Mol Ecol 16:651–662

Goldstien SJ, Gemmell NJ, Schiel DR (2006a) Comparative phylogeography of coastal limpets across a marine disjunction in New Zealand. Mol Ecol 15:3259–3268

Goldstien SJ, Gemmell NJ, Schiel DR (2006b) Molecular phylogenetics and biogeography of the nacellid limpets of New Zealand (Mollusca: Patellogastropoda). Mol Phylog Evol 38:261–265

Grosholz E (2002) Ecological and evolutionary consequences of coastal invasions. Trends Ecol Evol 17:22–27

Hollander J (2005) Local adaptation but not geographical separation promotes assortative mating in a snail. Animal Behav 70:1209–1219

Jackson BCC, Jung P, Coates A, Collins LS (1993) Diversity and extinction of tropical American mollusks and emergence of the Isthmus of Panama. Science 260:1624–1626

Johanesson K, Lundberg J, André C, Nilsson PG (2004) Island isolation and habitat heterogeneity correlate with DNA variation in a marine snail (*Littorina saxatilis*). Biol J Linn Soc 82:377–384

Jolly M, Jollivet D, Gentil F, Thiébaut E, Viard F (2005) Sharp genetic break between Atlantic and English Channel populations of the polychaete *Pectinaria koreni*, along the North coast of France. Heredity 94:23–32

Kelly RP, Eernisse D (2007) Southern hospitality: a latitudinal gradient in gene flow in the marine environment. Evolution 61:700–707
King PR (2000) Tectonic reconstructions of New Zealand: 40 Ma to Present. N Z J Geol Geophys 43:611–638
Knowlton N, Weigt LA (1998) New dates and new rates for divergence across the Isthmus of Panama. Proc R Soc Lond 265:2257–2263
Knox GA (1980) Plate tectonics and the evolution of intertidal and shallow-water benthic biotic distribution patterns of the southwest Pacific. Palaeogeogr Palaeoclimatol Palaeoecol 31:267–297
Kruse I, Strasser M, Thiermann F (2004) The role of ecological divergence in speciation between intertidal and subtidal *Scoloplos armiger* (Polychaeta, Orbiniidae). J Sea Res 51:53–62
Landry C, Geyer LB, Arakaki Y, Uehara T, Palumbi SR (2003) Recent speciation in the Indo-West Pacific: rapid evolution of gamete recognition and sperm morphology in cryptic species of sea urchin. Proc R Soc Lond 270:1839–1847
Levin SA (1994) Patchiness in marine and terrestrial systems: from individuals to populations. Philos Trans R Soc Lond 343:99–103
Levin L, Bridges T (1995) Pattern and diversity in reproduction and development. In: McEdward L (ed) Ecology of marine invertebrate larvae. CRC Press LLC, Boca Raton, FL, pp 1–48
Lindberg DR, Hickman CS (1986) A new anomalous giant limpet from the Oregon Eocene (Mollusca: Patellidae). J Paleontol 60:661–668
Lindberg DR, Squires RL (1990) Patellogastropods (Mollusca) from the Eocene Tejon formation of southern California. J Paleontol 64:578–587
Lindstrom SC, Olsen JL, Stam WT (1996) Recent radiation of the Palmariaceae (Rhodophyta). J Phycol 32:457–468
Losos JB, Glor RE (2003) Phylogenetic comparative methods and the geography of speciation. Trends Ecol Evol 18:220–227
Mack RN, Simberloff D, Lonsdale WM, Evans H, Clout M, Bazzaz FA (2000) Biotic invasions: causes, epidemiology, global consequences, and control. Ecol Appl 10:689–710
Marko PB (1998) Historical allopatry and the biogeography of speciation in the prosobranch snail genus *Nucella*. Evolution 52:757–774
Mayr E (1954) Geographic speciation in tropical echinoids. Evolution 8:1–18
McGlone M (2005) Goodbye Gondwana. J Biogeogr 32:739–740
Morgan SG (1996) Influence of tidal variation on reproductive timing. J Exp Mar Biol Ecol 206:237–251
Nelson WA (1994) Distribution of macroalgae in New Zealand: an archipelago in space and time. Bot Mar 37:221–233
Palumbi SR (1994) Genetic divergence, reproductive isolation, and marine speciation. Annu Rev Ecol Syst 25:547–572
Pawson DL (1961) Distribution patterns of New Zealand echinoderms. Tuatara 9:9–18
Peacock JD (1989) Marine molluscs and late quaternary environmental studies with particular reference to the late-glacial period in northwest Europe: a review. Quat Sci Rev 8:179–192
Provan J, Wattier RA, Maggs CA (2005) Phylogeographic analysis of the red seaweed *Palmaria palmata* reveals a Pleistocene marine glacial refugium in the English Channel. Mol Ecol 14:793–803
Sanmartin I, Wanntorp L, Winkworth RC (2007) West Wind Drift revisited: testing for directional dispersal in the Southern Hemisphere using event-based tree fitting. J Biogeogr 34:398–416
Shiota H, Kimura MT (2007) Evolutionary trade-offs between thermal tolerance and locomotor and developmental performance in drosophilid flies. Biol J Linn Soc 90:375–380
Sotka E, Wares J, Barth J, Grosberg R, Palumbi S (2004) Strong genetic clines and geographical variation in gene flow in the rocky intertidal barnacel *Balanus glandula*. Mol Ecol 13:2143–2156
Thiel M, Haye P (2006) The ecology of rafting in the marine environment. III. Biogeographical and evolutionary consequences. Oceanogr Mar Biol 44:323–329
Trakhtenbrot A, Nathan R, Perry G, Richardson DM (2005) The importance of long-distance dispersal in biodiversity conservation. Divers Distrib 11:173–181
Trewick SA, Paterson AM, Campbell HJ (2007) Hello New Zealand. J Biogeogr 34:1–6

Ward RD, Ovenden JR, Meadows JRS, Grewe PM, Lehnert SA (2006) Population genetic structure of the brown tiger prawn, Penaeus esculentus, in tropical northern Australia. Mar Biol 148: 599–607

Waters JM, Craw D (2006) Goodbye Gondwana? New Zealand biogeography, geology, and the problem of circularity. Syst Biol 55:351–356

Waters JM, Roy MS (2003) Global phylogeography of the fissiparous sea-star genus *Coscinasterias*. Mar Biol 142:185–191

Waters JM, King TM, O'Loughlin PM, Spencer HG (2005) Phylogeographical disjunction in abundant high-dispersal littoral gastropods. Mol Ecol 14:2789–2802

Waters JM, McCulloch GA, Eason JA (2007) Marine biogeographical structure in two highly dispersive gastropods: implications for trans-Tasman dispersal. J Biogeogr 34:678–687

Yoder AD, Nowak MD (2006) Has vicariance or dispersal been the predominant biogeographic force in Madagascar? Only time will tell. Annu Rev Ecol Evol Syst 37:405–431

Chapter 9
Environmental Variability: Analysis and Ecological Implications

Lisandro Benedetti-Cecchi

> *"Man I'm losing sound and sight of all those who can tell me wrong from right when all things beautiful and bright sink in the night yet there's still something in my heart that can find a way to make a start to turn up the signal wipe out the noise."*
>
> Signal to Noise, Peter Gabriel

9.1 Introduction

The concept of variability is pervasive in marine ecology, although its meaning may vary among studies and researchers. Explicit reference to ecological variability is usually made with respect to spatial and temporal change in the distribution, abundance and richness of species and in the structure of assemblages, to the processes causing these changes and to unexplained variation in observational or manipulative studies (Andrewartha and Birch 1984). These aspects of variability are obviously interrelated: progress in understanding requires the explanation of pattern in terms of the underlying processes with a consequent reduction in unexplained variation.

Why understanding the causes and consequences of ecological variability is important? Species and assemblages fluctuate at various scales in space and time in response to changes in physical and biological processes, and these fluctuations have important ecological consequences. Temporal fluctuations may mitigate competitive interactions and consumer-resource interactions, facilitating species coexistence and ultimately enhancing species diversity (Chesson and Warner 1981; Chesson 2000). Spatial heterogeneity in species distribution may enhance productivity, reduce the spread of diseases and increase resistance to disturbance (Hutchinson et al. 2003). Environmental changes promoting fluctuations in species and assemblages are therefore important ecological forces. Large fluctuations may, however, be detrimental to species, reducing the persistence of viable populations and ultimately increasing the risk of extinction (Lawton 1988; Pimm 1991; Lande 1993; Drake 2005).

Anthropogenic influences on the biosphere are enhancing spatial and temporal heterogeneity in environmental conditions (Easterling et al. 2000). Models of climate change predict an increase in the variance of climate events such as storms

M. Wahl (ed.), *Marine Hard Bottom Communities*, Ecological Studies 206,
DOI: 10.1007/978-3-540-92704-4_9, © Springer-Verlag Berlin Heidelberg 2009

and droughts due to increased likelihood in the occurrence of extreme weather conditions (Wilson and Mitchell 1993; Houghton 1997; Michener et al. 1997; Muller and Stone 2001). These fluctuations are superimposed onto long-term trends in mean values of climate variables that in combination may expose organisms to unfamiliar environmental settings. Loss and fragmentation of habitats, the introduction of exotic species and a variety of local anthropogenic disturbances (e.g. pollution and urban development) contribute to generate variability in populations and assemblages (Fahrig 2003; Sax and Gaines 2003; Bulleri 2006).

Understanding and predicting the consequences of enhanced environmental fluctuations on the abundance, distribution and richness of species (for the purpose of this paper, I assume that any of these ecological responses may contribute to generate a pattern in diversity) requires more research on the linkages between environmental change and ecological variation. In this paper, I first distinguish between deterministic and stochastic approaches to the analysis of ecological variability. Then, I discuss the most common strategies used to describe variability in marine hard bottom assemblages, and illustrate experiments that have examined the effects of changing the mean intensity and spatial or temporal variances of ecological drivers simultaneously, drawing from studies undertaken in my laboratory in the last 3 years. Finally, I relate this approach to others that also incorporate random variation in ecological drivers, discuss the utility of these stochastic approaches to address scaling problems and highlight future directions for experimental research on these topics. I will not consider well-established relationships between environmental change and diversity, like the intermediate disturbance/predation hypothesis, nor I will delve into macroecological patterns of diversity, both of which are covered extensively elsewhere (Bertness et al. 2001).

9.2 A Framework for Investigating Ecological Variability

When addressing issues of environmental variability, it is useful to distinguish between variability in the processes driving change in assemblages, the ecological drivers, and the variables that are influenced by these forces, the ecological responses (Fig. 9.1). In principle, both drivers and responses can be treated as random variates with frequency distributions defined by parameters such as the mean and the variance. Hence, spatial and temporal variability in assemblages may be generated by a shift in the location parameter of a driving force (referred to as the mean intensity of the ecological driver), by changes in its variance or by a combination of these events. In turn, changes in ecological drivers can affect the mean, the variance or both parameters of the distributions of targeted ecological responses (Benedetti-Cecchi 2003; Fig. 9.1).

This framework recognizes the importance of both trends and stochastic variation in ecological drivers and responses. Stochastic approaches to ecological problems have a long history in ecological research (Lande et al. 2003). Variances have been used to characterize random fluctuations in the size of populations (den Boer 1968; Andrewartha and Birch 1984), and variance-mean relationships have been

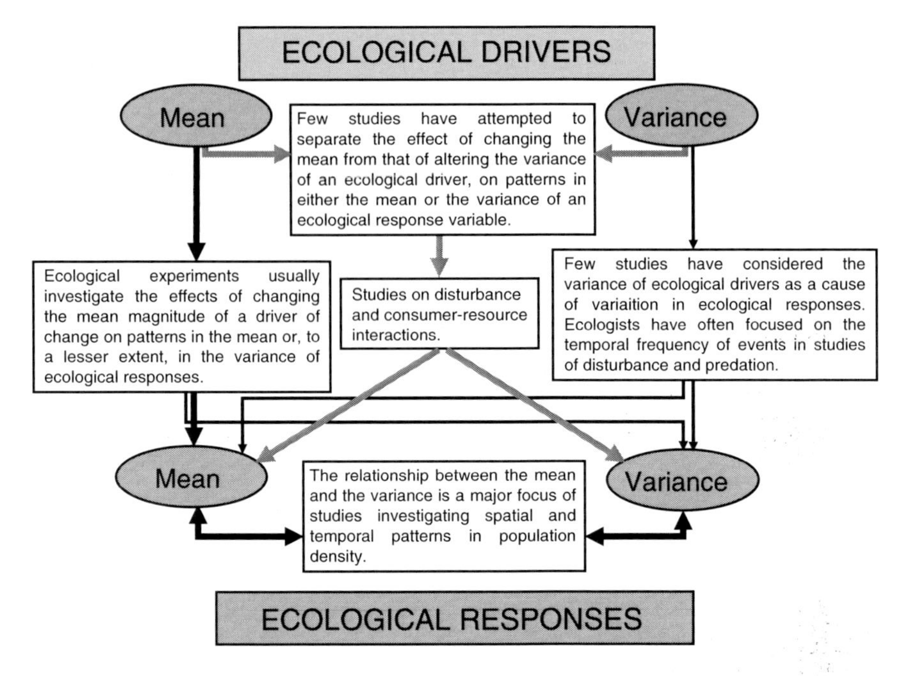

Fig. 9.1 Distinction between mean and variance in ecological drivers of change and ecological responses, and how they are addressed in ecological studies. Thickness of *black arrows* is proportional to the research efforts dedicated to a particular relationship. *Gray arrows* identify gaps in ecological research that are addressed in the present paper. See text for further details and for references (modified with permission from Benedetti-Cecchi 2003)

investigated extensively to uncover the mechanisms underlying these fluctuations (Taylor 1961; Pielou 1969; McArdle and Gaston 1992; Gaston and McArdle 1993). The importance of random variation in ecology has been revitalized by recent theoretical advances, like the neutral theory of biodiversity (Hubbell 2001), and renewed emphasis on demographic and environmental stochasticity (Engen et al. 1998; Boyce et al. 2006). Unfortunately, these approaches are used mostly in the context of descriptive studies of spatial or temporal variation in populations and assemblages, whilst experimental manipulation of spatial or temporal variances of ecological drivers are much less common (Benedetti-Cecchi 2003; Fig. 9.1).

These tendencies are also evident in the marine literature. Common approaches to characterize variation in hard bottom systems have involved comparisons of the structure of assemblages in contrasting environmental scenarios (Colman 1933; Stephenson and Stephenson 1949; Underwood 1978), and analyses of variability at multiple spatial or temporal scales (Underwood and Chapman 1996; Benedetti-Cecchi 2001; Fraschetti et al. 2005; Terlizzi et al. 2007). In the first approach, changes in mean values of response variables have been related to changes in average environmental conditions. The goal has been that of documenting consistent structures of assemblages in relatively homogenous environments (patterns) and trends in ecological responses across ranges of environmental conditions (patterns of change). This research has been influenced by deterministic views of the structure and dynamics of assemblages. The second approach, in contrast, has its foundation in the notion that populations and assemblages vary randomly in time and space, and has been designed to examine how variances in ecological responses change with the scale of observation (Kotliar and Wiens 1990; Menge and Olsen 1990; Holling 1992; Levin 1992).

As in other systems, experimental studies addressing the causes of variation in marine benthic assemblages have largely focused on the consequences of changing the mean intensity of ecological drivers on the means or, to a lesser extent, on the variances of ecological responses (Benedetti-Cecchi 2003; Fig. 9.1). Consistency and variation in these effects have been investigated by repeating experiments in contrasting environmental conditions (e.g. Menge et al. 2003), offering insights into deterministic changes of effects sizes, or across random samples of localities (e.g. Coleman et al. 2006), and providing information on spatial or temporal variation in effects sizes. In contrast, experimental manipulation of the variability of ecological drivers, either in terms of frequency of events (e.g. Navarrete 1996; Lenz et al. 2004) or as changes in spatial or temporal variance in association with changes in mean intensity (e.g. Bertocci et al. 2005), have been undertaken only occasionally (Fig. 9.1). Only by manipulating the parameters of the frequency distribution of ecological drivers explicitly (e.g. means and variances) can insights be obtained into the role of stochastic variation as a force shaping the structure of assemblages.

It is clear that deterministic and stochastic approaches provide distinct frameworks to investigate ecological change, the former targeting well-defined causes of variation, the latter offering a probabilistic context to deal with random variation. These approaches are, however, not mutually exclusive and marine ecologists are not limited in the use of one or the other. Rather, an integration of the two approaches, when appropriate, seems a profitable strategy to improve ecological understanding.

9.3 Observational Approaches: Variability in Ecological Responses

Early descriptions of the structure of hard bottom assemblages documented pronounced changes in numbers, identities and abundances of species in relation to wave exposure, depth and tidal height (Stephenson and Stephenson 1949; Southward 1958; Lewis 1964). Temporal variation has been examined mostly with respect to seasonality or to periodic disturbances like El Niño (e.g. Dayton et al. 1984). While relating variation in assemblages to specific ecological drivers, these studies have rarely quantified variability across scales in space and time. It was only after the recognition of the importance of incorporating multiple scales of observation into descriptive ecological investigations that this approach has become popular among marine ecologists (reviewed by Fraschetti et al. 2005).

Studies examining variation at multiple scales usually do not quantify ecological change in relation to explicit gradients in physical factors. Instead, variability is partitioned among a set of nested spatial or, less frequently, temporal scales selected within an apparently homogeneous set of environmental circumstances. For example, a typical study of spatial variation at three spatial scales includes a set of shores 100s m apart, replicate plots within each shore 10s m apart and replicate observations within each plot 10s to 100s cm apart (Fraschetti et al. 2005). The levels of each factor in this hierarchical design are selected randomly from a population of possible levels and, being random, do not include systematic differences in physical or biological conditions. This design enables unbiased estimates of variability to be obtained at each spatial scale but not the identification of specific ecological drivers of change.

Although the nature of the processes causing variation cannot be ascertained precisely from any observational study, identifying one or more characteristic scales of variation in hierarchical analyses of variability can hint at the relevant processes to which species and assemblages are most sensitive. This correlation is informative because many of the factors causing variation in assemblages operate at specific scales in space and time. For example, in marine benthic environments behavioural processes, biotic interactions and substratum topography often vary at scales of 10s to 100s cm (Underwood and Chapman 1996), recruitment variation is often in the range of 10s to 1,000s m (Kinlan and Gaines 2003) and climatic conditions vary at scales of 100s to 1,000s km (Leonard 2000). The synthesis of the available evidence from hierarchical analyses conducted by Fraschetti et al. (2005) has revealed that small-scale variability is an ubiquitous pattern in the marine benthos, regardless of habitat (rocky or soft bottoms) and taxa.

Denny et al. (2004) have discussed the importance of integrating variation associated with changes in well-defined environmental conditions (e.g. wave exposure) when examining spatial or temporal variation in biological variables. These authors used different definitions of scale to characterize biological variation in a rocky intertidal system in California and to relate this to changes in the physical environment. Although characteristic spatial and temporal scales of variation were identified for several of the physical and biological variables analysed, results were not consistent among methods, so that well-defined scales of variability could be identified only occasionally. In contrast, the data were better interpreted in terms of $1/f$ noise processes, with variance changing continuously with the scale of measurement as $V = 1/f^{\beta}$ (Halley 1996). In this formulation, f is the frequency of the signal (the spatial or temporal scale of measurement) and β is the spectral exponent that can be estimated with standard spectral techniques (e.g. Platt and Denman 1975; see also Appendix A in Denny et al. 2004 for a brief introduction). Denny et al. (2004) concluded that the concept of scale may not always be appropriate to describe variation in ecological variables and emphasized the alternative approach of considering the entire spectrum of variability of physical or biological measures.

9.4 Experimental Approaches: Manipulation of Intensity and Variance of Ecological Drivers

Manipulative experiments are necessary in order to understand and predict the ecological consequences of environmental change. Consider a simple experiment in which a consumer is removed from a set of exclosure plots, with another set of plots left undisturbed as controls (Fig. 9.2). For simplicity, assume there are no artefacts associated with the cages used to exclude the consumer. Thus, the experiment consists of an equal number of exclosure and open plots interspersed in the study area. Now, suppose that the cages are effective in excluding the consumers, so their average density in the exclosure plots is zero over the course of the experiment (the zeros above the +consumer boxes in Fig. 9.2). In contrast, consumers have free access to the control plots and their density may vary naturally from plot to plot (the numbers above the -consumer boxes in Fig. 9.2 indicate the average density of consumers in these plots during the course of the experiment). The density of prey is recorded at the end of the experiment in each plot (the numbers within boxes in Fig. 9.2) and compared between treatments using either a *t*-test or analysis of variance. Whatever the technique employed, the comparison is a test of the effect of changing the mean number of consumers from 4 to 0 on the mean density of prey (4 and 8 preys in control and exposure plots respectively; Fig. 9.2). Only the effect due to the difference in the average number of consumers between exclosure and control plots is tested here. Any effect due to spatial variation in the distribution of consumers in the study area, as revealed by differences in numbers of consumers in control plots, is subsumed in the within-treatment (residual) variation and cannot be explained by the experiment.

Of course, one might run the experiment in such a way to ensure that consumer pressure is kept constant among the +consumer plots (e.g. by using enclosures).

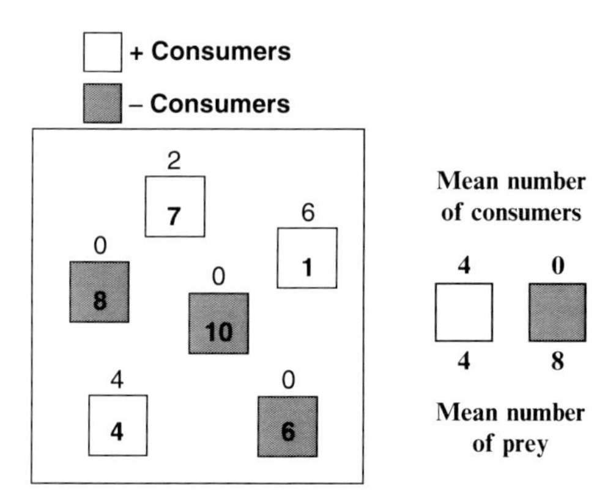

Fig. 9.2 Schematic representation of a hypothetical experiment in which a consumer is excluded from three replicate plots and left undisturbed in three control plots. *Numbers above* experimental plots indicate the mean number of consumers found in the plots during the experiment. *Numbers within* plots indicate the density of prey at the end of the experiment. See text for explanation

This solution would not alter the essence of the experiment, which would still examine the effect of a difference in the mean number of consumers (with variation around means eliminated experimentally in this case) on the mean density of prey. Both examples provide an appropriate test of the effect of changing the mean intensity of foraging but both ignore the consequences of spatial variation in foraging. With or without concomitant changes in mean density, consumers may become more or less aggregated in a given area in response to some environmental cues. The various scenarios defined by the different levels of mean intensity and spatial variance of foraging may have different impacts on the abundance and spatial distribution of prey. These effects have been discussed in detail elsewhere (Benedetti-Cecchi 2000, 2003) and will not be reiterated here. The point is that a simple removal experiment is not sufficient to examine these complexities. The same argument applies to the majority of ecological experiments, which are designed to assess responses to changes in mean intensity of ecological drivers, not to changes in their spatial or temporal variance (Benedetti-Cecchi 2000, 2003).

A suitable experimental design to assess the effects of changing spatial variation and mean intensity of foraging is presented in Benedetti-Cecchi (2003) and a real example is provided in Benedetti-Cecchi et al. (2005). These authors examined the effects of grazing limpets on algal cover in rock pools of the northwest Mediterranean. Individual levels of experimental treatments were generated from groups of four plots, two from pools with limpets and two from pools where limpets had been removed. Effects were quantified as response ratios (Osenberg et al. 1997) calculated from values of algal cover in plots with and without limpets. Limpets were maintained at the average density of one or two individuals per plot, so that the response ratio comparing these plots with those without limpets resulted in low and high levels of intensity of grazing respectively. Each level of density was crossed with two levels of spatial variation of grazing, obtained by maintaining either equal or unequal numbers of limpets in the two plots of a given pair, reflecting homogeneous or heterogeneous patterns of grazing respectively (see also Fig. 1 in Benedetti-Cecchi et al. 2005). As expected, high intensity of grazing had a strong negative impact on mean algal cover, independently of changes in spatial variance. A heterogeneous pattern of grazing, however, affected mean algal cover relative to ungrazed plots significantly less than did the homogeneous pattern of grazing. This indicated that limpets had lower effects on mean algal cover when spatially aggregated than when distributed homogeneously among pools.

The experiment also enabled an analysis of the effects of limpets on spatial variance in algal cover. In this case, the response ratio was calculated using the variance in algal cover between plots with limpets and the variance between plots without limpets. Results indicated a positive effect of spatial variance of grazing on spatial variance in algal cover, although this effect occurred under low intensity of grazing in a first run of the experiment and under high intensity of grazing in a second run. This experiment illustrated the importance of spatial variation of grazing in maintaining spatial variation in algal cover among pools, and the interactive nature of intensity and variance of foraging. These effects could be detected because spatial variation in density of limpets was manipulated explicitly in the experiment, whilst they would have been classified as unexplained variation in a classical removal experiment.

The problem of separating the mean intensity from the variance of ecological drivers is also present in studies of temporal variation. One way of examining the consequences of temporal variation in ecological drivers is that of manipulating the frequency of events. Frequency—the number of events per unit of time—is an appropriate descriptor of variability in studies aimed primarily at examining variation in the occurrence of events, like fires in forests or other disturbances (Collins 1992, 2000; McCabe and Gotelli 2000). Frequency, in contrast, is not appropriate when the objective of the study is that of examining effects of temporal variance per se, or interactions between mean intensity and variance of an ecological driver. This is because a change in frequency of an event inevitably modifies both the intensity and the temporal variance of the process that has generated the event, over the course of the study (Fig. 9.3a).

In order to separate the effect of variance from that of mean intensity, it is necessary to break the correlation between these parameters experimentally (Fig. 9.3b). This can be done by distinguishing between the levels of each experimental factor, intensity and variance, on the basis of the magnitude of each single event and the temporal patterning of series of events respectively, while ensuring that the same number of events is applied to all treatments during the course of the study. For example, in a study of disturbance, 12 events can be distributed over 2 years either regularly (one event every 2 months) to produce a treatment of zero temporal variance (defined as the variance in the interval of time between successive disturbances) or clustered in time to produce a temporally variable regime of disturbance (with temporal variance determined by the degree of clustering). Each of these conditions can be repeated by using series of mild events of disturbance (low intensity) and series of strong events (high intensity) in factorial combinations.

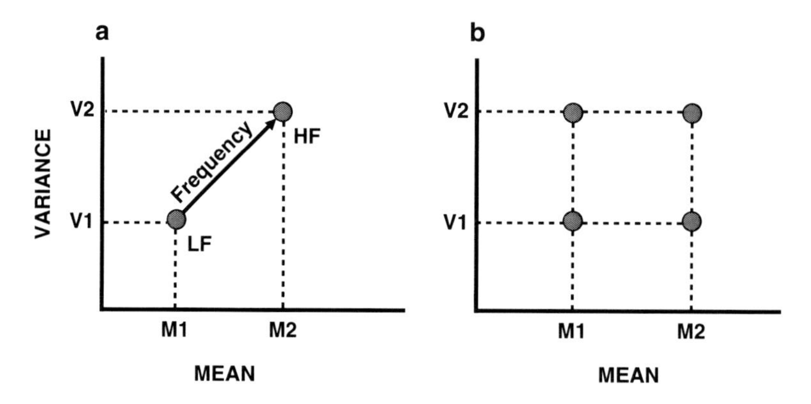

Fig. 9.3 Relationships among mean intensity, variance (spatial or temporal) and frequency in experimental manipulations of ecological drivers of change: **a** an increase in frequency inevitably results in enhanced levels of mean intensity and variance of the ecological driver; **b** diagrammatic illustration of a factorial experiment to separate the effects of changing the mean intensity and the variance of an ecological driver. *M1* and *M2* refer to two hypothetical levels of mean intensity of the driver, *V1* and *V2* to two hypothetical levels of the variance of the driver, and *LF* and *HF* to low frequency and high frequency of events respectively

One might argue that intensity and temporal variance of ecological drivers may be correlated naturally, so any attempt to break this correlation may be inappropriate. I argue that this is not a real problem. Factorial designs break the covariance among manipulated factors, making these independent (Petraitis 1998), which is a requisite for, not an impediment to a correct analysis of main effects and interactions. In addition, there are circumstances in which means and variances of natural phenomena are uncorrelated, at least within certain ranges of values. These issues are discussed at length in Benedetti-Cecchi (2003) and will not be detailed further here.

The rationale and design for experimentally manipulating the mean intensity and temporal variance of ecological drivers of change are illustrated in detail in Benedetti-Cecchi (2003). Real examples include experimental analyses of effects of variation in the regime of mechanical disturbance (Bertocci et al. 2005) and aerial exposure (Benedetti-Cecchi et al. 2006; Bertocci et al. 2007) on assemblages of algae and invertebrates of rocky shores in the northwest Mediterranean. The hypothesis investigated in these studies was that intensity and temporal variance of the ecological drivers (disturbance and aerial exposure) would have affected means and variances of ecological responses (abundance of individual taxa and similarity of assemblages) interactively, with the largest effects occurring under a regime of high intensity and high variance of events. Under these conditions, organisms experienced a highly fluctuating environment with strong events occurring over short periods, alternating with relatively long periods of benign conditions. Ecological effects were expected to range from the extirpation of species during periods of severe perturbations to the recovery of fast-colonizing organisms during less perturbed periods. Thus, both positive and negative effects were expected, depending on the ability of organisms to resist to and to recover from intense perturbations.

Bertocci et al. (2005) exposed assemblages to various regimes of mechanical disturbance to mimic variation in intensity and temporal variance of wave shock associated with large storms. These authors found a positive effect of intensity of disturbance on multivariate temporal variability in assemblages, whilst temporal variance in disturbance decreased multivariate variability independently of effects of intensity. Benedetti-Cecchi et al. (2006) manipulated intensity and temporal variance of aerial exposure by transplanting low-shore assemblages upwards at either regular or variable intervals. Results showed interactive effects of intensity and temporal variance of aerial exposure on diversity and abundance of common taxa. Organisms that were common in the low-shore habitat (e.g. filamentous and coarsely branched algae) displayed a strong reduction in percentage cover when transplanted upshore at regular but not at variable intervals. A similar effect was observed on mean number of taxa, with high temporal variance of aerial exposure mitigating the negative effect on diversity observed when assemblages were manipulated at regular intervals. Organisms that were more common in the high-shore habitat (e.g. barnacles), in contrast, increased in abundance when assemblages were transplanted higher on the shore but, again, this effect was reversed under a variable regime of aerial exposure. Intensity and temporal variance of aerial exposure also affected temporal variation in ecological responses, with effects mimicking those observed on the means (Bertocci et al. 2007).

9.5 Future Directions

The limited evidence available indicates that environmental variance can be an important driver of change in marine benthic assemblages, either in isolation or in interaction with mean intensity. In perturbation experiments, effects of variance are generally in directions opposite to those elicited by changes in the mean. This is observed with mechanical disturbance on multivariate temporal variability of assemblages, and with aerial exposure on the diversity and mean abundance of algae and invertebrates. That an increase in variance of an ecological driver may mitigate the impact of an increase in mean intensity suggests that rocky shore assemblages have a greater capacity to withstand environmental change than currently recognized. Whether these effects will be efficient to buffer marine benthic diversity against the simultaneous increases in means and variances predicted for several climate variables is currently unknown. Answering this question requires much more experimentation on the effects of altering the parameters that describe the frequency distributions of ecological drivers of change in a wide range of marine habitats. Future research should also identify the mechanisms that drive ecological responses to environmental change. This requires comparisons of responses among species with different life histories, patterns of dispersal and colonizing capabilities, in addition to more attention to physiological responses.

In general, sensitivity to environmental variance underscores nonlinear relationships between ecological drivers and responses. For functional relationships that curve in a consistent direction (concave up or concave down), a change in variance around a mean value of the variable on the abscissa (the ecological driver) affects the mean of the variable on the ordinate (the ecological response) as a consequence of Jensen's inequality (Jensen 1906; see Chesson 1991; Ruel and Ayers 1999 and Benedetti-Cecchi 2005 for further details in ecological contexts).

Chesson and co-workers have used the properties of nonlinear functions to develop the scale transition theory, which enables the scaling of local nonlinear population dynamics to regional spatial scales (Chesson 1998; Chesson et al. 2005; Melbourne and Chesson 2006). This theory uses nonlinear averaging to make predictions of population dynamics at large scales from density-dependent models describing changes in population size at smaller scales. If dynamics were linear, then simple averages of local measurements would predict large-scale patterns. In more realistic nonlinear scenarios, in contrast, scale transition terms need to be estimated to correct the bias introduced by taking averages of nonlinear functions—i.e. to account for the effect of Jensen's inequality. This requires quantifying the change in spatial variance of population densities from small to large scales, and the degree of nonlinearity in local population dynamics. If descriptive measures of variation in ecological responses across spatial (or, by extension, temporal) scales are sufficient for the purpose of applying the scale transition theory, experiments are needed to understand which ecological drivers contribute most to the measured variability. Experiments like those described by Schiel in Chapter 20 would complement the measurements necessary to derive the scale transition terms, contributing to a more comprehensive view of the ecological drivers of change in populations dynamics across scales.

Altering the variance of ecological drivers of change is not the only way to generate fluctuating environments and to test hypotheses about ecological responses to these fluctuations. Autoregressive and $1/f$ models provide two alternative approaches to manipulate and analyse environmental heterogeneity. Both approaches model the temporal or spatial correlation among values of an ecological variable (either a driver or a response), although they differ in the way variance changes with scale, levelling out at low frequencies (i.e. large scales; see Chap. 7 by Terlizzi and Schiel) in autoregressive models, whilst increasing indefinitely with scale in $1/f$ models. There is a rich body of literature addressing how temporal and, less often, spatial changes in environmental autocorrelation affect fluctuations in population densities and the risk of extinction (Ripa and Lundberg 1996; Petchey et al. 1997; Heino 1998; Miramontes and Rohani 1998; Benton et al. 2001; Pike et al. 2004; Vasseur and McCann 2007). Autocorrelation in environmental variables is considered an important ecological driver of change, implying long runs of either favourable or unfavourable conditions for population growth and persistence. Experiments that involve the manipulation of the variance of an ecological driver, like those discussed in this paper, also modify the spatial or temporal pattern of autocorrelation of the driver. For example, the clustering of perturbations over short periods of time to increase temporal variance also results in an increase of autocorrelation, compared to the uniform distribution of events in time. It is unclear whether autocorrelation and variance are different ways to characterize the same aspects of environmental change or, rather, if they impinge on substantially different demographic and life history traits of species. To date, the ecological effects of correlated environmental fluctuations have been examined with mathematical models, simulations and laboratory experiments. To the best of my knowledge, there has been no attempt to manipulate the autocorrelation or the spectral characteristics of ecological drivers in the field. Marine ecologists have the unique opportunity to contribute to this fervid area of research by taking advantage of the tractability of benthic systems as natural laboratories for field experiments.

There is increasing evidence that power laws like $1/f$ models provide appropriate descriptions of patterns of variability in ecological drivers and responses (Steele 1985; Vasseur and Yodzis 2004; Denny et al. 2004). The extent to which these patterns are causally related remains, however, elusive. Since the seminal contribution to the subject by Steele (1985), there has been a tendency to explain variability in ecological responses in terms of the power spectra of individual ecological drivers, as if each were operating in isolation from the others. Species and assemblages are, however, exposed to the simultaneous influence of many drivers and these may be characterized by different spectra of variability. It can be proposed that the variance observed at various spatial and temporal scales in ecological responses reflects additive or interactive effects amongst drivers with distinct spectra of variability. Nearly independent or negatively autocorrelated events contributing to small-scale variability of ecological responses would be embedded within positively autocorrelated ecological drivers dominated by low-frequency events, which are responsible for large-scale changes. This proposition leads to the prediction that the scales of spatial or temporal heterogeneity in ecological responses can be altered experimentally by manipulating the power spectra of ecological drivers. Testing this prediction would require the manipulation of ecological drivers over long periods or large spatial extents (e.g. entire shores), to generate the desired spectra of variation. Different variance spectra could be manipulated in factorial combinations to examine how they combine to generate variation in ecological responses. These experiments could also be used to clarify the mechanisms that prevent the extrapolation of small-scale measurements to large spatial and temporal extents. Interactions among ecological drivers with different power spectra may be at the origin of the nonlinearities and environmental variance that make nonlinear averaging a poor predictor of large-scale patterns in scale transition theory. These effects could be investigated explicitly with large-scale manipulations of the spatial and temporal autocorrelation of ecological drivers.

9.6 Conclusions

Spatial and temporal variability in species abundance, distribution and richness is ubiquitous in marine hard bottom systems. This variation appears as systematic change along environmental gradients and as stochastic fluctuations as a function of the scale of observation. Studies adopting hierarchical sampling designs indicate that small-scale spatial and temporal patchiness is a main feature of marine benthic assemblages, whilst variation at larger scales may or may not be present, depending on the specific ecological response considered (Fraschetti et al. 2005). Studies examining variation continuously in space or time may fail to detect a characteristic scale of variation, with variability occurring at every scale (e.g. Denny et al. 2004). The extent to which this discrepancy reflects true differences among studied assemblages, or is also affected by differences among analytical methods, remains to be determined. Identifying the causes of variation in assemblages rests, however, on experimentation. Manipulating the mean intensity and variance of ecological drivers over explicit spatial and temporal scales enables one to examine the consequences of systematic and random variation of ecological drivers simultaneously (Benedetti-Cecchi 2000, 2003). This is an area of experimental ecology that deserves further attention (Fig. 9.1), with important implications for understanding the ecological consequences of climate change and in addressing scaling issues. Experiments examining the effects of environmental fluctuations on ecological responses can be extended to autoregressive and $1/f$ models of variation. There is a rich body of theories based on these models that have been developed in other areas of ecology (Vasseur and McCann 2007) and that can explain variation in benthic assemblages. Marine ecologists have the almost unique opportunity to contribute to the growth of these theories by testing their predictions with field experiments.

Acknowledgements This study was supported by grants from the University of Pisa, the Census of Marine Life Programme NaGISA and the MARBEF (Marine Biodiversity and Ecosystem Functioning) Network of Excellence funded in the European Community's Sixth Framework Programme. This is contribution number 8042 of MARBEF. I wish to thank Fabio Bulleri, Elena Maggi and Laura Tamburello for comments on earlier versions of the manuscript. Some of the ideas expressed in this paper were stimulated during discussions at the EMBED (Environmental Modulation and Ecosystem Dynamics) workshop organized in Pisa (27–29 April 2006) and supported by the European Census of Marine Life.

References

Andrewartha HG, Birch LC (1984) The ecological web. University of Chicago Press, Chicago, IL

Benedetti-Cecchi L (2000) Variance in ecological consumer-resource interactions. Nature 407:370–374

Benedetti-Cecchi L (2001) Variability in abundance of algae and invertebrates at different spatial scales on rocky sea shores. Mar Ecol Prog Ser 215:79–92

Benedetti-Cecchi L (2003) The importance of the variance around the mean effect size of ecological processes. Ecology 84:2335–2346

Benedetti-Cecchi L (2005) Unanticipated impacts of spatial variance of biodiversity on plant productivity. Ecol Lett 8:791–799

Benedetti-Cecchi L, Vaselli S, Maggi E, Bertocci I (2005) Interactive effects of spatial variance and mean intensity of grazing on algal cover in rock pools. Ecology 86:2212–2222

Benedetti-Cecchi L, Bertocci I, Vaselli S, Maggi E (2006) Temporal variance reverses the impact of high mean intensity of stress in climate change experiments. Ecology 87:2489–2499

Benton TG, Lapsley CT, Beckerman AP (2001) Population synchrony and environmental variation: an experimental demonstration. Ecol Lett 4:236–243

Bertness MD, Gaines SD, Hay ME (2001) Marine community ecology. Sinauer, Sunderland, MA

Bertocci I, Maggi E, Vaselli S, Benedetti-Cecchi L (2005) Contrasting effects of mean intensity and temporal variation of disturbance on a rocky seashore. Ecology 86:2061–2067

Bertocci I, Vaselli S, Maggi E, Benedetti-Cecchi L (2007) Changes in temporal variance of rocky shore organism abundances in response to manipulation of mean intensity and temporal variability of aerial exposure. Mar Ecol Prog Ser 338:11–20

Boyce MS, Haridas CV, Lee CT, NCEAS Stochastic Demography Working Group (2006) Demography in an increasingly variable world. Trends Ecol Evol 21:141–148

Bulleri F (2006) Is it time for urban ecology to include the marine realm? Trends Ecol Evol 21:658–659

Chesson P (1991) Stochastic population models. In: Kolasa J, Pickett STA (eds) Ecological heterogeneity. Springer, Berlin Heidelberg New York, pp 123–143

Chesson P (1998) Making sense of spatial models in ecology. In: Bascompte J, Solè RV (eds) Modeling spatiotemporal dynamics in ecology. Springer, Berlin Heidelberg New York, pp 151–166

Chesson P (2000) Mechanisms of maintenances of species diversity. Annu Rev Ecol Syst 31:343–366

Chesson PL, Warner RR (1981) Environmental variability promotes coexistence in lottery competitive systems. Am Nat 117:923–943

Chesson P, Donahue MJ, Melbourne BA, Sears AL (2005) Scale transition theory for understanding mechanisms in metacommunities. In: Holyoak M, Leibold MA, Holt RD (eds) Metacommunities: spatial dynamics and ecological communities. University of Chicago Press, Chicago, IL, pp 279–306

Coleman RA, Underwood AJ, Benedetti-Cecchi L, berg P, Arenás F, Arrontes J, Castro J, Hartnoll RG, Jenkins SR, Paula J, Della Santina P, Hawkins SJ (2006) A continental scale evaluation of the role of limpet grazing on rocky shores. Oecologia 147:556–564

Collins SL (1992) Fire frequency and community heterogeneity in tallgrass prairie vegetation. Ecology 73:2001–2006

Collins SL (2000) Disturbance frequency and community stability in native tallgrass prairie. Am Nat 155:311–325

Colman J (1933) The nature of the intertidal zonation of plants and animals. J Mar Biol Assoc UK 18:435–476
Dayton PK, Currie V, Gerrodette T, Keller BD, Rosenthal R, Ven Tresca D (1984) Patch dynamics and stability of some California kelp communities. Ecol Monogr 54:253–289
den Boer PJ (1968) Spreading of risk and stabilization of animal numbers. Acta Biotheoretica 18:165–194
Denny MW, Helmuth B, Leonard GH, Harley CDG, Hunt LJH, Nelson EK (2004) Quantifying scale in ecology: lessons from a wave-swept shore. Ecol Monogr 74:513–532
Drake JM (2005) Population effects of increased climate variation. Proc R Soc Lond B 272:1823–1827
Easterling DR, Meehl GA, Parmesan C, Changnon SA, Karl TR, Mearns LO (2000) Climate extremes: observations, modeling, and impacts. Science 289:2068–2074
Engen S, Bakke O, Islam A (1998) Demographic and environmental stochasticity-concepts and definitions. Biometrics 54:840–846
Fahrig L (2003) Effects of habitat fragmentation on biodiversity. Annu Rev Ecol Evol Syst 34: 487–515
Fraschetti S, Terlizzi A, Benedetti-Cecchi L (2005) Patterns of distribution of marine assemblages from rocky shores: evidence of relevant scales of variation. Mar Ecol Prog Ser 296:13–29
Gaston KJ, McArdle BH (1993) Measurement of variation in the size of populations in space and time: some points of clarification. Oikos 68:357–360
Halley M (1996) Ecology, evolution and 1/f noise. Trends Ecol Evol 11:33–37
Heino M (1998) Noise colour, synchrony and extinctions in spatially structured populations. Oikos 83:368–375
Holling CS (1992) Cross-scale morphology, geometry and dynamics of ecosystems. Ecol Monogr 62:447–502
Houghton J (1997) Global warming: the complete briefing. Cambridge University Press, Cambridge
Hubbell SP (2001) The unified neutral theory of biodiversity and biogeography. Princeton University Press, Princeton, NJ
Hutchinson MJ, John EA, Wijesinghe DK (2003) Toward understanding the consequences of soil heterogeneity for plant populations and communities. Ecology 84:2322–2334
Jensen JL (1906) Sur les fonctions convexes et les inégalités entre les valeurs moyennes. Acta Math 30:175–193
Kinlan BP, Gaines SD (2003) Propagule dispersal in marine and terrestrial environments: a community perspective. Ecology 84:2007–2020
Kotliar NB, Wiens JA (1990) Multiple scale of patchiness and patch structure-a hierarchical framework for the study of heterogeneity. Oikos 59:253–260
Lande R (1993) Risk of population extinction from demographic and environmental stochasticity and random catastrophes. Am Nat 142:911–927
Lande R, Engen S, Saether BE (2003) Stochastic population dynamics in ecology and conservation. Oxford University Press, Oxford
Lawton JH (1988) More time means more variation. Nature 334:563
Lenz M, Molis M, Wahl M (2004) Experimental test of the intermediate disturbance hypothesis: frequency effects of emersion on fouling communities. J Exp Mar Biol Ecol 305:247–266
Leonard GH (2000) Latitudinal variation in species interactions: a test in the New England rocky intertidal zone. Ecology 81:1015–1030
Levin SA (1992) The problem of pattern and scale in ecology. Ecology 73:1943–1967
Lewis JR (1964) The ecology of rocky shores. English University Press, London
McArdle BH, Gaston KJ (1992) Comparing population variabilities. Oikos 64:610–612
McCabe DJ, Gotelli NJ (2000) Effects of disturbance frequency, intensity, and area on assemblages of stream invertebrates. Oecologia 124:270–279
Melbourne BA, Chesson P (2006) The scale transition: scaling up population dynamics with field data. Ecology 87:1478–1488
Menge BA, Olson AM (1990) Role of scale and environmental-factors in regulation of community structure. Trends Ecol Evol 5:52–57

Menge BA, Lubchenco J, Bracken MES, Chan F, Foley MM, Freldenburg TL, Galnes SD, Hudson G, Krenz C, Leslle H, Menge DNL, Russell R, Webster MS (2003) Coastal oceanography sets the pace of rocky intertidal community dynamics. PNAS 100:12229–12234

Michener WK, Blood ER, Bildstein KL, Brinson MM (1997) Climate change, hurricanes and tropical storms, and rising sea level in coastal wetlands. Ecol Appl 7:770–801

Miramontes O, Rohani P (1998) Intrinsically generated coloured noise in laboratory insect populations. Proc R Soc Lond B 265:785–792

Muller RA, Stone GW (2001) A climatology of tropical storm and hurricane strikes to enhance vulnerability prediction for the southeast US coast. J Coast Res 17:949–956

Navarrete SA (1996) Variable predation: effects of whelks on a mid-intertidal successional community. Ecol Monogr 66:301–321

Osenberg CW, Sernelle O, Cooper SD (1997) Effect size in ecological experiments: the application of biological models in meta-analysis. Am Nat 150:798–812

Petchey OL, Gonzalez A, Wilson HB (1997) Effects on population persistence: the interaction between environmental noise colour, intraspecific competition and space. Proc R Soc Lond B 264:1841–1847

Petraitis PS (1998) How can we compare the importance of ecological processes if we never ask, "compared to what"? In: Resetaris WJ, Bernardo J (eds) Experimental ecology. Issues and perspectives. Oxford University Press, Oxford, pp 183–201

Pielou EC (1969) An introduction to mathematical ecology. Wiley, New York

Pike N, Tully T, Haccou P, Ferrière R (2004) The effect of autocorrelation in environmental variability on the persistence of populations: an experimental test. Proc R Soc Lond B 271:2143–2148

Pimm SL (1991) The balance of nature? University of Chicago Press, Chicago, IL

Platt T, Denman KL (1975) Spectral analysis in ecology. Annu Rev Ecol Syst 6:189–210

Ripa J, Lundberg P (1996) Noise colour and the risk of population extinctions. Proc R Soc Lond B 263:1751–1753

Ruel JJ, Ayers MP (1999) Jensen's inequality predicts effects of environmental variation. Trends Ecol Evol 14:361–366

Sax DF, Gaines SD (2003) Species diversity: from global decreases to local increases. Trends Ecol Evol 18:561–566

Southward AJ (1958) The zonation of plants and animals on rocky sea shores. Biol Rev 33:137–177

Steele JH (1985) A comparison of terrestrial and marine ecological systems. Nature 313:355–358

Stephenson TA, Stephenson A (1949) The universal features of zonation between tide marks on rocky coasts. J Ecol 37:289–305

Taylor RL (1961) Aggregation, variance and the mean. Nature 189:732–735

Terlizzi A, Anderson MJ, Fraschetti S, Benedetti-Cecchi L (2007) Scales of spatial variation in Mediterranean subtidal sessile assemblages at different depths. Mar Ecol Prog Ser 332:25–39

Underwood AJ (1978) A refutation of critical tidal levels as determinants of the structure of intertidal communities on British shores. J Exp Mar Biol Ecol 33:261–276

Underwood AJ, Chapman MG (1996) Scales of spatial patterns of distribution of intertidal invertebrates. Oecologia 107:212–224

Vasseur DA, McCann (2007) The impact of environmental variability on ecological systems. Springer, Dordrecht

Vasseur DA, Yodzis P (2004) The color of environmental noise. Ecology 85:1146–1152

Wilson CA, Mitchell JFB (1993) Simulation of climate and CO2 induced climate changes over Western Europe. Climatic Change 10:11–42

Part III
Community Dynamics

Coordinated by Christopher D. McQuaid

Introduction

Christopher D. McQuaid

Part III of this book deals with community dynamics, not so much in a descriptive as in an analytical way. Perhaps most striking is the shift from the emphasis on species interactions, such as predation and competition, which dominated so much of ecology over the last several decades. Instead, there is a strong emphasis on "supply-side ecology" in a broad sense. This widens us from a focus on interactions among individuals that have successfully joined the adult population, to include the control of recruitment. This may be taken as an example of the discipline maturing to the level where a single chapter (Chap. 16) can effectively synthesise simple and complex interactions in a meaningful way, identifying the broad factors that determine the level of complexity in an interaction web. In contrast, three chapters (Chaps. 10–12) are required to mark out the progress from gamete production to successful recruitment into the adult population. This involves an odyssey of critical steps from gamete production to fertilisation to larval dispersal and supply to settlement and post-settlement mortality. Failure at any of these steps has profound consequences for fitness. Subsequently imposed on the resultant adult populations are the *relatively* predictable effects of seasonality (Chap. 13) and succession (Chap. 15), and the consequences of stochastic disturbances, which are almost definitively unpredictable, at least in the short term (Chap. 14).

Because of space constraints, none of the chapters in this book part attempts a comprehensive review of its topic but, rather, each emphasises recent advances and changes in our understanding, while highlighting present controversies. In doing so, they attempt to identify future directions that are likely to be critical. Fertilisation success is problematic in different ways for internal and external fertilisers and is, of course, subject to enormous selective pressure. Many species show convergence in their adaptations to enhance fertilisation success, but it remains unclear whether sperm limitation is critical. Chapter 10 examines the leading hypotheses and controversies surrounding this topic. Chapter 11 covers the wide range of factors influencing larval dispersal and supply, including variation in individual fecundity/reproductive success, the fate of planktonic larvae and the processes influencing their transport, including both oceanography and larval behaviour. Chapter 12 takes up the story at the point of settlement, identifying the patterns of settlement that occur at different spatial scales. Settlement involves a number of interactions influenced

M. Wahl (ed.), *Marine Hard Bottom Communities*, Ecological Studies 206,
DOI: 10.1007/978-3-540-92704-4_III, © Springer-Verlag Berlin Heidelberg 2009

by both physical and biological factors and, again, the importance of larval behaviour is emphasised. Propagules need not be passive particles, and even simple behaviour patterns can powerfully influence both dispersal and settlement. These chapters lay the ground work by discussing how new recruits enter an adult population. The following chapters broaden the focus from the individual/species level to include community-level issues. Successful recruits must cope with environmental changes that may be predictable or stochastic. Seasonality is pervasive in almost all ecosystems, and organisms have many ways of detecting and reacting to it. Chapter 13 makes the important distinction that seasonality can affect ecosystems not only through the physiological or behavioural response of the individual. Seasonality also leads to changes in mass rates of primary production and mortality that may not involve individual responses. This chapter describes some of the many seasonal patterns in the marine benthos and identifies overlooked aspects. Disturbances can determine patterns of species distribution and are especially obvious in the marine benthos. Despite an intuitive understanding of the concept, it can be difficult even to define a disturbance, and Chapter 14 adopts a definition with clear applicability to systems in which primary space is often a limiting resource: disturbances cause the removal of living biomass, although in this chapter consumers are explicitly excluded. It is also possible to distinguish between human and natural disturbances, and this chapter shows how an understanding of the effects of disturbance is important in deciding how to respond to the effects of people on the environment. Chapter 15 deals with succession, beginning at the very early stages when initial colonisation of free space is by microorganisms and occurs over microscales. This chapter emphasises the multiple factors that can influence trajectories of succession and highlights the need to explore the consequences of successional changes to ecosystem functioning. Finally, organisms not only react to (or sometimes shape) their environment, they also interact among themselves in an enormous variety of ways. Chapter 16 describes some of the many types of interactions, considering a range from simple to very complex interactions. To a degree, the level of complexity is set by the number of interacting species, from intraspecific to community level interactions, but this is a very simplistic view as density-dependence and indirect effects also determine the level of complexity. For example, species traits may interact with density, and density-dependence can simultaneously alter both inter- and intraspecific interactions. This closing chapter focuses on how levels of complexity are shaped.

Conclusion

In the broadest sense, ecology seeks to explain the distributions and abundances of species. A very clear common theme emerges from an overview of the chapters in this book part, and this theme goes to the root of ecology. In all seven chapters, it is clear that the phenomenon studied represents the resolution of competing forces and, in most cases, is highly context-dependent. Present patterns of life histories and fertilisation strategies are seen to represent a balance between sperm limitation and sperm competition. It is suggested that the quest for identifying the primary drivers of larval supply may be doomed because each settlement event is the result of a unique combination of factors. Settlement is complicated by the interaction of larval behaviour with both biotic and abiotic factors and, subsequently, settlement patterns can be hugely modified by post-settlement mortality in many ways to shape adult distribution … and so on. Essentially, this reflects the fact, highlighted in Chapter 13, that organisms do not respond to single factors but to a set of environmental and biological conditions. These reactions in turn are subject to possible feedbacks, resulting in enormous plasticity of outcomes. Likewise at the community level, patterns of succession are nonlinear, and can be shaped by both abiotic and biotic effects. Yet, despite context dependency, broad patterns emerge and this brings out the importance of scales, emphasised so strongly in Chapter 14. There is no doubt that our perception of patterns and of the relative importance of different factors depends on the spatial scale of study. At microscales, nutrient uptake by phytoplankton is limited by diffusion and boundary layer effects, while at large scales it may be limited by nutrient supply through upwelling or other oceanographic processes. These issues are perhaps particularly clear in marine ecosystems because of the nature of water as a medium. Chemical cues are more persistent and are carried farther in water than in air (Chap. 16) and, through its properties as a transport medium, water promotes long-range ecological connections, often over very large distances—for example, in the supply of propagules (Chap. 11). It is also clear that temporal scale in relation to organism longevity influences whether we perceive a disturbance, for example, as being chronic or stochastic (Chap. 14).

It is difficult to summarise such disparate chapters in a few words without repeating their own conclusions. Nevertheless, it is worth concluding that:

- the ecology and evolution of fertilisation appear to have been driven by sperm limitation and sperm competition;
- despite the absence of clear common drivers of larval supply, there remains a level of predictability in terms of site rankings, biogeographic trends, etc.;
- settlement patterns alone are not enough to predict adult distribution;
- we need to separate the effects of endogenous circannual rhythms reset by temperature, from direct responses to annual cues;
- the absence of direct links between life histories and patterns of succession in the sea may reflect the uncertainty of recruitment;
- if we are to understand the role of diversity in ecosystem functioning, then we must clarify how it is affected by species identity and abundances; and
- single experiments on interactions sample only a part of possible species combinations and so have limited generality.

These conclusions advance our understanding of the ecology of marine benthic ecosystems by helping us to understand how the distributions and abundances of organisms are determined and how they drive ecosystem functions. Organism recruitment and survival depend on complex and context-dependent interaction of a huge array of factors—yet, viewed at the correct scales, clear patterns emerge from these bewildering effects. Several authors also make the point that in order to manage the effects of man's activities on the natural world, we need to understand how responses to natural and man-induced effects differ.

10.4 Gamete Traits that Influence Fertilization Success

10.4.1 In Broadcast Spawners

A variety of both male and female gamete traits have been reported to influence the success of fertilization in laboratory experiments and in the field, particularly traits that enhance sperm:egg encounters. For eggs, these traits include cell size, the size of their accessory structures, and sperm chemoattractants. The importance of physical egg size was first highlighted by Levitan (1993), a hitherto largely unnoticed consequence of Vogel et al.'s (1982) model: larger eggs provide bigger targets for sperm and, therefore, should be fertilized more easily. Subsequent studies supported the assertion that physical size of the egg cell influenced fertilization success (e.g. Levitan and Irvine 2001), whereas others—sometimes working on the very same species—argued for an equally important role of egg accessory structures such as egg jelly or follicle cells (e.g. Podolsky 2001). Sperm availability appeared to influence selection on egg cell size, raising the debate of whether accessory structures should be selected for in preference to increasing egg size with energetically costly cytoplasm (Podolsky 2004). Sperm chemoattractants are another, probably energetically inexpensive means to increase the "effective" egg size. Under sperm-limiting conditions, chemoattractant halos around freely spawned tunicate and abalone eggs could more than double the egg target area and fertilization rates (Jantzen et al. 2001; Riffell et al. 2004). Given the prevalence of sperm chemoattractants in free-spawning marine species (Miller 1985; Maier and Müller 1986; although interestingly not in the echinoids that have been the focus of much fertilization research), this could be an important mechanism decoupling physical egg size from fertilization success. If chemoattractant production is directly linked to egg size, then this same mechanism would still result in selection for larger eggs, as modelled by Dusenbery (2000). Additionally, "fertilization efficiency" (the number of sperm required to obtain a fertilization; β/β_0 in Vogel et al.'s (1982) model; F_e in Styan 1998) of eggs may vary independently of egg size: in two congeneric species of echinoderms, large eggs required significantly *more* sperm to fertilize than small eggs (Styan et al. 2005). Thus, although physical egg size may be important in some species, other traits such as chemoattractants, accessory structures and fertilization efficiency all play vital roles when sperm are limiting. As yet, there are too few data to form generalizations about which of these is most important under which circumstances, and much remains to be done in this field.

Concerning the role of sperm traits, it has long been suggested that higher sperm velocity and/or longevity increases fertilization success (Rothschild and Swann 1951; Vogel et al. 1982) but that these trade-off, likely a consequence of optimization of energetic reserves (Levitan 2000). In general, sperm start to swim following dilution upon spawning, with increasing dilution leading to increasing activity—the so-called respiratory dilution effect. In brown algae, the variable amount of mucilage released with sperm might delay activation by seawater, yet fertilization has also been reported to occur within the emersed mucilage at low tide (reviewed by Brawley et al. 1999). Whether mucilage influences longevity is not known but, following dilution, motility of male gametes from brown and green algae has been reported for a variable (1–7 h) range of durations (e.g. Togashi et al. 1998; Clifton and Clifton 1999; also reviewed in Brawley and Johnson 1992) and to be affected by salinity when marine taxa colonize brackish habitats (Serrão et al. 1996a). For marine invertebrates, sperm longevity varies widely. Echinoderm sperm appear to be short-lived (typically, minutes to perhaps a few hours; e.g. Yund 1990; Benzie and Dixon 1994), whereas in many other taxa sperm may remain viable for many hours and even as long as a day (e.g. Havenhand 1991; Johnson and Yund 2004). This may arise due to slow sperm dilution during spawning (e.g. Meidel and Yund 2001; Marshall 2002) and/or the requirement for sperm activation by chemoattractants from the egg (e.g. tunicates, Miller 1982; Bolton and Havenhand 1996), yet extended sperm longevity has also been recorded in the absence of such factors (e.g. Havenhand 1991; Williams and Bentley 2002). Sperm swimming velocities (e.g. Serrão et al. 1996a; Kupriyanova and Havenhand 2002) are negligible relative to the magnitude of sea currents; however, sperm swimming is not irrelevant. The length scales at which sperm swim are typically close to or below the length scales of turbulent eddies and, consequently, sperm and eggs are not advected independently. Moreover, within approximately 1 mm of an egg, turbulence is negligible (Mitchell et al. 1985) and sperm swimming can play a major role. Phototactic responses of algal gametes, positive or negative, by concentrating the gametes at the surface or bottom, may lower gamete dilution (suggested in Brawley and Johnson 1992).

Egg longevity is generally longer than that of active sperm, up to several days (e.g. Havenhand 1991; Meidel and Yund 2001; Williams and Bentley 2002; also motile female gametes of green algae, Togashi et al. 1998). Again, the selective benefit of this trait seems questionable, since turbulence and advection might dilute gametes beyond the possibility of fertilization long before they become unviable (Levitan and Petersen 1995). In many cases, however, eggs may be adhesive, released under minimal turbulence (see below) and/or released in adhesive media such that they remain within a given location, thereby integrating the fluctuating sperm concentration over time as in spermcasters (e.g. tunicates, Svane and Havenhand 1993; Marshall 2002; echinoids, Thomas 1994; Meidel and Yund 2001; some fucoid algae, Engelen et al. 2008). Eggs retained in this manner can obtain far higher fertilization rates than if advected freely (Svane and Havenhand 1993; Yund and Meidel 2003). Demonstrating that such behaviours are ecologically relevant requires substantial investment in fieldwork and, consequently, the extent to which similar behaviours occur in other taxa is not well understood.

10.4.2 In Spermcasters

Unlike broadcast spawners, the influence of gamete traits on fertilization success of spermcasting species is limited largely to the sperm, with the possible exception of taxa where the egg is not enclosed internally (e.g. kelps). In all other cases, spermcasters retain unfertilized eggs, and sequester sperm from the water column prior to using these for fertilizations. Perhaps unsurprisingly, therefore, extended sperm longevity has been reported in a number of spermcasters such as ascidians (e.g. Bishop 1998; Johnson and Yund 2004) and bryozoans (e.g. Manríquez et al. 2001). Clearly, this is an important adaptation that permits sperm dispersal and dilution over considerable distances without compromising fertilization success. This contrasts with broadcasters that typically have shorter-lived and more active sperm (see above) that may require rapid fertilization before gamete dilution eradicates the possibility of further gamete encounters.

Many spermcasters are filter-feeding species with mechanisms enabling efficient utilization of dilute sperm (Bishop 1998 and references therein), thereby overcoming many issues that limit fertilization in broadcasters. Mechanisms of sperm capture are unknown (Bishop and Pemberton 2006) but sperm can be captured from extremely dilute suspensions (as low as 10 ml^{-1}; Bishop 1998) and stored and used several weeks thereafter (Bishop and Ryland 1991). Presumably, fewer sperm are required for internal, as opposed to external, fertilization (Bishop 1998). Sperm limitation is thought to have been the selective pressure for mitotic cloning of the zygote into carpospores in red algae (reviewed in Santelices 2002). This is contradicted by the high fertilization success recorded in red algae (Kaczmarska and Dowe 1997; Engel et al. 1999), which is puzzling given that red algae are spermcasters with non-motile male gametes (spermatia), and have no known mechanisms to concentrate spermatia (although several competing spermatia have been found around trichogynes; Kaczmarska and Dowe 1997). Gamete encounters may be facilitated by extracellular projections in spermatia and release of mucilage with the gametes (reviewed by Brawley and Johnson 1992); however, more work is required to clarify the intriguing fertilization success of red algae.

10.4.3 In Copulatory Fertilizers

In comparison with broadcasters and spermcasters, processes of fertilization in copulating hard substratum taxa are poorly known. This type of internal fertilization is found primarily in crustaceans and most gastropods (limpets being an obvious exception), as well as in several lesser phyla such as flatworms. Mating behaviour and potential for selfing have been investigated in such taxa (e.g. Furman and Yule 1990), although we found no literature detailing the influence of gamete traits on fertilization success in copulatory fertilizers from hard substrata. This almost certainly arises because of the difficulties in demonstrating this (cf. fertilization is internal and, therefore, difficult to study without impacting the process itself) but also because the potential for post-copulatory manipulation of gametes by the female is so great. However, we also found no literature investigating the extent of sperm competition and sexual selection in these taxa. Given the ease with which snails, for example, can be manipulated, this is surprising. This is clearly an area where considerable progress could be made.

10.5 Gamete Mixing

10.5.1 Role of Hydrodynamics

The importance of water flow for the fertilization of broadcasters was first documented by Pennington (1985) who found reduced fertilization success of urchins induced to spawn in high flows (>0.2 m s^{-1} versus <0.2 m s^{-1}), and attributed this to increased sperm dispersion and dilution. Subsequent modelling showed how advection and turbulence can result in low fertilization success (Denny 1988; Denny and Shibata 1989). This pattern of flow-mediated gamete dilution, and consequent sperm limitation, became a well-established paradigm (e.g. Levitan and Petersen 1995; Mead and Denny 1995; cf. Denny et al. 2002).

This body of literature did not, however, consider the importance of (1) variation in the probability of fertilization across time and space in a population and (2) variation in hydrodynamically cued gamete-release behaviours within a population. More recent research has focused on these issues, not least on the caveat raised by several authors (including Denny and Shibata 1989) that these models provide time-integrated, rather than instantaneous estimates of gamete concentrations. Downstream of a spawning adult, gametes have highly heterogeneous distributions in much the same way that smoke from a campfire travels in concentrated filaments. Spermcasters and broadcasting taxa that retain eggs at a given location will experience extreme variations in gamete concentration, encountering periods of no sperm, and periodically exposed to quite high sperm concentration, therefore integrating the fluctuating sperm concentrations over their retention time (e.g. Wahle and Gilbert 2002). In contrast, broadcasters will experience high spatiotemporal variation in fertilization success due to varying coincidence of sperm and egg filaments in the water column (e.g. Coma and Lasker 1997). Recent modelling has shown that turbulent structures in the flow field can cause coalescence between high-concentration filaments of freely spawned egg and sperm, enhancing fertilization rates (Crimaldi and Browning 2004). Release of gametes in viscous media (e.g. Svane and Havenhand 1993; Thomas 1994; Yund and Meidel 2003) will enhance this process and substantially raise fertilization rates above those predicted for freely spawned gametes (Crimaldi and Browning 2004), even for mild viscosity increases.

The influential paradigm of a decade ago—that flow-induced gamete dilution restricts fertilization rates in broadcast spawners—is not as pervasive as once thought, and focus is now turning to the importance of small-scale spatiotemporal variation, not least in turbulent structures, leading to significant variation in fertilization rates. This exciting development is ripe for testing with empirical data.

Flow-mediated spawning behaviour has been studied primarily in fucoid algae (reviewed by Pearson and Serrão 2006; see also Gordon and Brawley 2004 for kelp and green algae). Avoidance of spawning during high water motion periods is a conserved response superimposed on their common semilunar release patterns, resulting in high natural fertilization success (e.g. Serrão et al. 1996b). Diffusion-limited supply of dissolved inorganic carbon (DIC) for photosynthesis has been shown to act as a signal for gamete release under favourable (i.e. calm) conditions (Pearson et al. 1998). Photoreceptors to specific blue and green light wavelengths also coordinate gamete release by some fucoid algae, and may contribute to restrict spawning to calmer intervals (Pearson et al. 2004).

Comparable information on flow-mediated spawning behaviours of marine invertebrates is limited. Several intertidal species are known to restrict spawning to periods of low tide (e.g. gastropods, Counihan et al. 2001; tunicates, Marshall 2002; anthozoans, Marshall et al. 2004). Again, this topic is in need of detailed investigations to determine the extent to which spawning behaviours have been selected for different flow regimes.

Gamete mixing of copulatory species is not influenced directly by flow environment, although by reducing densities, dislodgement due to wave action can have major impacts on fertilization success. As dislodgement risk is a product of drag and adhesion (Denny 1988), species in which the males mount the females in order to copulate face increased risk of dislodgement (Johannesson et al. 2008).

10.5.2 Role of Density/Aggregation Spawning

Gregarious and/or aggregative settlement behaviour in sessile marine species is now well documented (e.g. Burke 1986). Population density is one of the primary determinants of gamete concentrations in the water column during spawning and, hence, a key component of fertilization success in broadcast spawners. In sea urchins, fertilization success varies inversely with distance between spawning pairs (Pennington 1985), and is higher at central (rather than marginal) locations in an aggregation and in greater-density aggregations (Levitan et al. 1992). Similar results have been reported by other workers (e.g. Yund 1995; Levitan 2002a; Gaudette et al. 2006), and this general pattern has become broadly accepted. Nonetheless, there are studies that have found either no such relationship (e.g. Phillipi et al. 2004; although, interestingly, this result is for a spermcasting tunicate) or only a weak inverse relationship, with fertilization rates still comparatively high at great distances (e.g. Babcock and Mundy 1992). Moreover, aggregation size interacts with spawning synchrony to influence overall fertilization success (Gaudette et al. 2006).

10.5.3 Role of Spawning Synchrony

Spawning synchrony is key to maximizing the probability of gamete encounters in the sea, and has been shown for a wide variety of rocky shore organisms (see references in Table 10.1). Although spawning synchrony in broadcasters is expected for reproductive assurance, intriguingly it may also occur in selfing hermaphrodites (discussed in Pearson and Serrão 2006). Spawning synchrony induced by direct communication occurs in several species including soft corals (Slattery et al. 1999), polychaetes (Watson et al. 2003) and kelps, where sperm release is induced by a pheromone secreted by the egg (Luning and Müller 1978) during calm periods, which reduces pheromone dilution (Gordon and Brawley 2004). Synchrony on broader scale may occur as a response to single or combinations of environmental signals such as lunar or tidal phases (e.g. Counihan et al. 2001; Levitan et al. 2004), hydrodynamics (e.g. Serrão et al. 1996b), temperature (e.g. Olive 1995) or availability of food such as phytoplankton (e.g. Starr et al. 1990). On a circadian scale, synchronous gamete release in many species occurs in response to light–dark or dark–light shifts (e.g. Dybern 1965; Brawley and Johnson 1992), as in the remarkable synchronous events of early morning multi-species spawning by green algae on coral reefs (Clifton 1997; Clifton and Clifton 1999), the fate of which remains totally unexplored in terms of fertilization success. Multi-species spawning synchrony (and resultant fertilization success) is, however, perhaps best known for corals (e.g. Levitan et al. 2004; Guest et al. 2005). Despite widespread correlations between spawning and various possible environmental cues, particularly with lunar cycles (for many taxa throughout the world), the physiological mechanisms behind such patterns are poorly understood (but see Pearson and Brawley 1998; Pearson et al. 2004; Guest 2008 and references therein).

10.6 Risk of Polyspermy and the Role of Polyspermy Blocks

Polyspermy, the fertilization of an egg by more than one sperm, is usually a lethal condition but—how often do free spawners encounter the circumstance where there are too many sperm? Although the cellular mechanisms involved in polyspermy have been studied extensively (reviewed by Gould and Stephano 2003), this question has been asked rarely and, consequently, this aspect remains unclear. An indicative answer can be inferred from the fact that polyspermy-avoidance mechanisms have been found in all free-spawning marine invertebrates and marine algae investigated to date (Brawley 1991; Gould and Stephano 2003), suggesting that polyspermy imposes a strong negative selective pressure. In situ rates of polyspermy in induced sea urchins reached 42% at high subtidal population densities (Levitan 2004) and up to 63% in tide pools (Franke et al.2002), which may reflect the physical constraint to gamete dispersion in a tide-pool habitat. Yet, modelling also shows that at sperm concentrations where as few as 60% of eggs are fertilized, up to 11% of zygotes can still be polyspermic (C.A. Styan, unpublished simulations for urchins). So, the scant available data for invertebrates support the hypothesis that polyspermy is common, and may well be prevalent at sperm concentrations far lower than previously supposed. For broadcasting marine algae, however, natural polyspermy levels are reported to be comparatively low (usually <10%) in intertidal (Brawley 1992), tide-pool (Pearson and Brawley 1996) and subtidal (Serrão et al. 1999) habitats. Interestingly, the sodium-ion dependence of the fast polyspermy block may be compromised when marine taxa colonize ionically dilute brackish seas such as the Baltic and, in such environments, relatively high natural polyspermy levels (10–30%) have been recorded (Serrão et al. 1999). We could find no comparable data for marine invertebrates.

Polyspermy risk may result in sexual conflict, in which traits that increase male reproductive success will simultaneously decrease female reproductive success. The scenario through which this arises is simple: in sperm-limiting conditions, adaptations that maximize fertilizations are selectively beneficial for both males and females. Once sperm are no longer limiting, however, these adaptations have a high cost for females (high sperm concentrations increase polyspermy) and, therefore, mechanisms that reduce fertilizations are selected for, whereas in males sperm are selected for even faster swimming, binding, and egg penetration. This conflict between male and female function has been suggested as a driver of rapid evolution of gamete recognition proteins in sea urchins (e.g. Vacquier 1998) and, indeed, sexual conflict theory predicts this form of variation in sperm:egg binding compatibility in broadcasting marine species (e.g. Gavrilets 2000). Available empirical data support this hypothesis (Levitan and Ferrell 2006) but much remains to be done in this field.

10.7 Fertilization Compatibility

Intraspecific variability in fertilization success appears to be common in hard substratum taxa (e.g. urchins, Palumbi 1999; Levitan and Ferrell 2006; Evans et al. 2007; mussels, Beaumont et al. 2004; oysters, Gaffney et al. 1993; polychaetes, Kupriyanova and Havenhand 2002; abalone, Havenhand and others, unpublished data). As outlined above, the theoretical basis for this is well established; however, the consequences of variable fertilization probability *after* sperm:egg encounter have been largely overlooked in rocky shores taxa (but see Levitan and Ferrell 2006). Variable compatibility will reduce the effective population size, restricting reproductive success and intra-population gene flow, with cascading consequences for population dynamics. Again, the extent, nature and causes of variability in fertilization compatibility have yet to be explored in the vast majority of hard substratum taxa.

A second consequence of variable gamete compatibility is mediating hybridization in synchronous broadcast spawning of closely related species. This may be aggravated by chemical interference due to the often non-specific nature of pheromones (e.g. Maier and Müller 1986; Bolton and Havenhand 1996). Yet, subtle (i.e. minutes) species spawning asynchronies have been reported (Clifton and Clifton 1999) and at least a 2-h shift in spawning peak time appears sufficient to avoid hybridization (Levitan et al. 2004). Several recently diverged rocky shore species appear to have undergone incomplete reproductive isolation and still hybridize occasionally (e.g. brown algae, Coyer et al. 2002; Engel et al. 2005; sea urchins, Levitan 2002b; bivalves, Bierne et al. 2002; Beaumont et al. 2004; corals, Vollmer and Palumbi 2002; Levitan et al. 2004). Speciation/hybridization processes trade-off under sperm-limiting conditions: specialization of gamete recognition proteins reduces risk of hybridization but may simultaneously reduce success of intraspecific fertilizations. Consequently, pre-zygotic hybridization barriers may be more likely to evolve under sperm competition (e.g. in high-density populations; Levitan 2002b).

10.8 Conclusions

Rocky shore taxa show many convergent adaptations to increase fertilization probabilities, yet the debate as to whether sperm limit, or compete for, external fertilizations (e.g. Levitan and Petersen 1995; Yund 2000) is still active. We conclude that both sperm limitation and sperm competition play essential roles in driving the ecology and evolution of fertilization in rocky shore species. We have described multiple reported adaptations to avoid sperm limitation, thereby highlighting that this is an important selective factor. Nonetheless, it is difficult to demonstrate naturally occurring sperm limitation: populations that have not evolved adaptations to overcome sperm limitation are less likely to persist and be observed. Conversely, it is equally apparent that adaptations to avoid polyspermy are diverse and common, suggesting that hydrodynamic conditions, perhaps in combination with adaptations to avoid sperm limitation, can often lead to sperm competition. It is the balance between these competing selective pressures that has created the patterns we see today in rocky substratum taxa. Understanding how these patterns evolve will require a detailed understanding of the micro- to meso-scale variation in fertilization success, both in space and time. This presents exciting prospects for the coming decade.

Acknowledgements This chapter was improved by helpful suggestions (on a previous longer version) from Susan Brawley, Don Levitan and Phil Yund—thank you to all. The author order for the chapter is based on a head-or-tail decision (because of equal contribution) ... we thank Kerstin Johannesson for tossing the coin.

References

Babcock RC, Mundy CN (1992) Reproductive biology, spawning and field fertilization rates of *Acanthaster planci*. Aust J Mar Freshw Res 43:525–534

Beaumont AR, Turner G, Wood AR, Skibinski DOF (2004) Hybridisations between *Mytilus edulis* and *Mytilus galloprovincialis* and performance of pure species and hybrid veliger larvae at different temperatures. J Exp Mar Biol Ecol 302:177–188

Benzie JAH, Dixon P (1994) The effects of sperm concentration, sperm-egg ratio, and gamete age on fertilization success in crown-of-thorns starfish (*Acanthaster planci*) in the laboratory. Biol Bull 186:139–152

Berndt ML, Callow JA, Brawley SH (2002) Gamete concentrations and timing and success of fertilization in a rocky shore seaweed. Mar Ecol Prog Ser 226:273–285

Bierne N, David P, Boudry P, Bonhomme F (2002) Assortative fertilization and selection at larval stage in the mussels *Mytilus edulis* and *M galloprovincialis*. Evolution 56:292–298

Bishop JDD (1998) Fertilization in the sea: are the hazards of broadcast spawning avoided when free-spawned sperm fertilize retained eggs? Proc R Soc Lond B 265:725–731

Bishop JDD, Pemberton AJ (2006) The third way: spermcast mating in sessile marine invertebrates. Integr Comp Biol 46:398–406
Bishop JDD, Ryland JS (1991) Storage of exogenous sperm by the compound ascidian *Diplosoma listerianum*. Mar Biol 108:111–118
Bolton TF, Havenhand JN (1996) Chemical mediation of sperm activity and longevity in the solitary ascidians *Ciona intestinalis* and *Ascidiella aspersa*. Biol Bull 190:329–335
Brawley SH (1991) The fast block against polyspermy in fucoid algae is an electrical block. Dev Biol 144:94–106
Brawley SH (1992) Fertilization in natural populations of the dioecious brown alga *Fucus ceranoides* L. and the importance of the polyspermy block. Mar Biol 113:145–157
Brawley SH, Johnson LE (1992) Gametogenesis, gametes and zygotes: an ecological perspective on sexual reproduction in the algae. Eur J Phycol 27:233–252
Brawley SH, Johnson LE, Pearson GA, Speransky V, Li R, Serrao E (1999) Gamete release at low tide in fucoid algae: maladaptive or advantageous? Am Zool 39:218–229
Burke RD (1986) Pheromones and the gregarious settlement of marine invertebrate larvae. Bull Mar Sci 39:323–331
Clifton KE (1997) Mass spawning by green algae on coral reefs. Science 275:1116–1118
Clifton KE, Clifton LM (1999) The phenology of sexual reproduction by green algae (Bryopsidales) on Caribbean coral reefs. J Phycol 35:24–34
Coma R, Lasker HR (1997) Small-scale heterogeneity of fertilization success in a broadcast spawning octocoral. J Exp Mar Biol Ecol 214:107–120
Counihan RT, McNamara DC, Souter DC, Jebreen EJ, Preston NP, Johnson CR, Degnan BM (2001) Pattern, synchrony and predictability of spawning of the tropical abalone *Haliotis asinina* from Heron Reef, Australia. Mar Ecol Prog Ser 213:193–202
Coyer JA, Peters AF, Hoarau G, Stam WT, Olsen JL (2002) Hybridisation of the marine seaweeds, *Fucus serratus* and *F evanescens* in a century-old zone of secondary contact. Proc R Soc Lond B 269:1829–1834
Crimaldi JP, Browning HS (2004) A proposed mechanism for turbulent enhancement of broadcast spawning efficiency. J Mar Syst 49:3–18
Denny MW (1988) Biology and the mechanics of the wave-swept environment. Princeton University Press, Princeton, NJ
Denny MW, Shibata MF (1989) Consequences of surf-zone turbulence for settlement and external fertilization. Am Nat 134:859–889
Denny MW, Nelson EK, Mead KS (2002) Revised estimates of the effects of turbulence on fertilization in the purple sea urchin, *Strongylocentrotus purpuratus*. Biol Bull 203:275–277
Dusenbery DA (2000) Selection for high gamete encounter rates explains the success of male and female mating types. J Theor Biol 202:1–10
Dybern BI (1965) The life cycle of *Ciona intestinalis* (L.) F. typica in relation to the environmental temperature. Oikos 16:109–131
Engel CR, Wattier R, Destombe C, Valero M (1999) Performance of non-motile male gametes in the sea: analysis of paternity and fertilization success in a natural population of a red seaweed, *Gracilaria gracilis*. Proc R Soc Lond B 266:1879–1886
Engel CR, Daguin C, Serrão EA (2005) Genetic entities and mating system in hermaphroditic *Fucus spiralis* and its close dioecious relative *F. vesiculosus*. Mol Ecol 14:2033–2046
Engelen A, Espirito-Santo C, Simões T, Monteiro C, Serrão EA, Pearson GA, Santos R (2008) Periodicity of propagule expulsion and settlement in the competing native and invasive brown seaweeds, *Cystoseira humilis* and *Sargassum muticum*. Eur J Phycol (in press)
Evans JP, García-González F, Marshall DJ (2007) Sources of genetic and phenotypic variance in fertilization rates and larval traits in a sea urchin. Evolution 61:2832–2838
Franke ES, Babcock RC, Styan CA (2002) Sexual conflict and polyspermy under sperm-limited conditions: in situ evidence from field simulations with the free-spawning marine echinoid *Evechinus chloroticus*. Am Nat 160:485–496
Furman ER, Yule AB (1990) Self-fertilization in *Balanus improvisus* Darwin. J Exp Mar Biol Ecol 144:235–239

Gaffney PM, Bernat CM, Allen SK (1993) Gametic incompatibility in wild and cultured populations of the eastern oyster *Crassostrea virginica* (Gmelin). Aquaculture 115:273–284

Gaudette J, Wahle RA, Himmelman JH (2006) Spawning events in small and large populations of the green sea urchin *Strongylocentrotus droebachiensis* as recorded using fertilization assays. Limnol Oceanogr 51:1485–1496

Gavrilets S (2000) Rapid evolution of reproductive barriers driven by sexual conflict. Nature 403:886–889

Gordon R, Brawley SH (2004) Effects of water motion on propagule release from algae with complex life histories. Mar Biol 145:21–29

Gould MC, Stephano JL (2003) Polyspermy prevention in marine invertebrates. Microsc Res Technol 61:379–388

Guest J (2008) How reefs respond to mass coral spawning. Science 320:621–623

Guest JR, Baird AH, Goh BPL, Chou LM (2005) Seasonal reproduction in equatorial reef corals. Invert Reprod Dev 48:207–218

Harrison PL, Babcock RC, Bull GD, Oliver JK, Wallace CC, Willis BL (1984) Mass spawning in tropical reef corals. Science 223:1186–1189

Havenhand JN (1991) Fertilisation and the potential for dispersal of gametes and larvae in the solitary ascidian *Ascidia mentula*. Ophelia 33:1–15

Jantzen TM, de Nys R, Havenhand JN (2001) Fertilization success and the effects of sperm chemoattractants on effective egg size in marine invertebrates. Mar Biol 138:1153–1161

Johannesson K, Havenhand JN, Jonsson PR, Lindegarth M, Sundin A, Hollander J (2008) Male discrimination of female mucous trails permits assortative mating in a marine snail species. Evolution (in press)

Johnson SL, Yund PO (2004) Remarkable longevity of dilute sperm in a free-spawning colonial ascidian. Biol Bull 206:144–151

Kaczmarska I, Dowe LL (1997) Reproductive biology of the red alga *Polysiphonia lanosa* (Ceramiales) in the Bay of Fundy, Canada. Mar Biol 128:695–703

Kupriyanova E, Havenhand JN (2002) Variation in sperm swimming behaviour and its effect on fertilization success in the serpulid polychaete *Galeolaria caespitosa*. Invert Reprod Dev 41:21–26

Levitan D (1993) The importance of sperm limitation to the evolution of egg-size in marine invertebrates. Am Nat 141:517–536

Levitan DR (1998) Sperm limitation, gamete competition, and sexual selection in external fertilizers. In: Birkhead TR, Moller AP (eds) Sperm competition and sexual selection. Academic Press, London, pp 175–217

Levitan DR (2000) Sperm velocity and longevity trade off each other and influence fertilization in the sea urchin *Lytechinus variegatus*. Proc R Soc Lond B 267:531–534

Levitan DR (2002a) Density-dependent selection on gamete traits in three congeneric sea urchins. Ecology 83:464–479

Levitan DR (2002b) The relationship between conspecific fertilization success and reproductive isolation among three congeneric sea urchins. Evolution 56:1599–1609

Levitan DR (2004) Density-dependent sexual selection in external fertilizers: variances in male and female fertilization success along the continuum from sperm limitation to sexual conflict in the sea urchin *Strongylocentrotus franciscanus*. Am Nat 164:298–309

Levitan DR, Ferrell DL (2006) Selection on gamete recognition proteins depends on sex, density, and genotype frequency. Science 312:267–269

Levitan DR, Irvine SD (2001) Fertilization selection on egg and jelly-coat size in the sand dollar *Dendraster excentricus*. Evolution 55:2479–2483

Levitan DR, Petersen C (1995) Sperm limitation in the sea. Trends Ecol Evol 10:228–231

Levitan DR, Sewell MA, Chia F-S (1991) Kinetics of fertilization in the sea urchin *Strongylocentrotus franciscanus*: interaction of gamete dilution, age and contact time. Biol Bull 181:371–378

Levitan DR, Sewell MA, Chia F-S (1992) How distribution and abundance influence fertilization success in the sea urchin *Strongylocentrotus franciscanus*. Ecology 73:248–254

Levitan DR, Fukami H, Jara J, Kline D, McGovern TM, McGhee KE, Swanson CA, Knowlton N (2004) Mechanisms of reproductive isolation among sympatric broadcast-spawning corals of the *Montastraea annularis* species complex. Evolution 58:308–323

Luning K, Müller DG (1978) Chemical interaction in sexual reproduction of several Laminariales (Phaeophyta): release and attraction of spermatozoids. Zeitschr Pflanzenphysiol 89:333–341

Maier I, Müller DG (1986) Sexual pheromones in algae. Biol Bull 170:145–175

Manríquez PH, Hughes RN, Bishop JDD (2001) Age-dependent loss of fertility in water-borne sperm of the bryozoan *Celleporella hyalina*. Mar Ecol Prog Ser 224:87–92

Marshall DJ (2002) In situ measures of spawning synchrony and fertilization success in an intertidal, free-spawning invertebrate. Mar Ecol Prog Ser 236:113–119

Marshall DJ, Semmens D, Cook C (2004) Consequences of spawning at low tide: limited gamete dispersal for a rockpool anemone. Mar Ecol Prog Ser 266:135–142

Mead KS, Denny MW (1995) The effects of hydrodynamic shear stress on fertilization and early development of the purple sea urchin *Strongylocentrotus purpuratus*. Biol Bull 188:46–56

Meidel SK, Yund PO (2001) Egg longevity and time-integrated fertilization in a temperate sea urchin (*Strongylocentrotus droebachiensis*). Biol Bull 201:84–94

Miller RL (1982) Sperm chemotaxis in ascidians. Am Zool 22:827–840

Miller RL (1985) Demonstration of sperm chemotaxis in Echinodermata: Asteroidea, Holothuroidea, Ophiuroidea. J Exp Zool 234:383–414

Mitchell JG, Okubo A, Fuhrman JA (1985) Microzones surrounding phytoplankton form the basis for a stratified marine microbial ecosystem. Nature 316:58–59

Mortensen T (1938) Contributions to the study of the development and larval forms of echinoderms. Kongl Danske Vidensk Selsk Naturvid Math Ser 9 7:1–59

Olive PJW (1995) Annual breeding cycles in marine-invertebrates and environmental-temperature—probing the proximate and ultimate causes of reproductive synchrony. J Thermal Biol 20:79–90

Palumbi SR (1999) All males are not created equal: fertility differences depend on gamete recognition polymorphisms in sea urchins. Proc Natl Acad Sci USA 96:12632–12637

Pearson GA, Brawley SH (1996) Reproductive ecology of *Fucus distichus* (Phaeophyceae): an intertidal alga with successful external fertilization. Mar Ecol Prog Ser 143:211–223

Pearson GA, Serrão EA (2006) Revisiting synchronous gamete release by fucoid algae in the intertidal zone: fertilization success and beyond? Integr Comp Biol 46:587–597

Pearson GA, Serrão EA, Brawley SH (1998) Control of gamete release in fucoid algae: sensing hydrodynamic conditions via carbon acquisition. Ecology 79:1725–1739

Pearson GA, Serrão EA, Dring M, Schmid R (2004) Blue- and green-light signals for gamete release in the brown alga, *Silvetia compressa*. Oecologia 138:193–201

Pemberton AJ, Hughes RN, Manríquez PH, Bishop JDD (2003) Efficient utilization of very dilute aquatic sperm: sperm competition is more likely than sperm limitation when eggs are retained. Proc R Soc Lond B 270:S223–S226

Pennington JT (1985) The ecology of fertilization of echinoid eggs: the consequences of sperm dilution, adult aggregation, and synchronous spawning. Biol Bull 169:417–430

Phillippi A, Hamann E, Yund PO (2004) Fertilization in an egg-brooding colonial ascidian does not vary with population density. Biol Bull 206:152–160

Podolsky RD (2001) Evolution of egg target size: an analysis of selection on correlated characters. Evolution 55:2470–2478

Podolsky RD (2004) Life-history consequences of investment in free-spawned eggs and their accessory coats. Am Nat 163:735–753

Riffell JA, Krug PJ, Zimmer RK (2004) The ecological and evolutionary consequences of sperm chemoattraction. Proc Natl Acad Sci USA 101:4501–4506

Rothschild L, Swann MM (1951) The fertilization reaction in the sea urchin: the probability of a successful sperm-egg collision. J Exp Biol 28:403–416

Santelices B (2002) Recent advances in fertilization ecology of macroalgae. J Phycol 38:4–10

Serrão EA, Kautsky L, Brawley SH (1996a) Distributional success of the marine seaweed *Fucus vesiculosus* L. in the brackish Baltic Sea correlates with osmotic capabilities of gametes. Oecologia 107:1–12

Serrão EA, Pearson G, Kautsky L, Brawley SH (1996b) Successful external fertilization in turbulent environments. Proc Natl Acad Sci USA 93:5286–5290

Serrão EA, Brawley SH, Hedman J, Kautsky L, Samuelson G (1999) Reproductive success of *Fucus vesiculosus* (Phaeophyceae) in the Baltic Sea. J Phycol 35:254–269

Slattery M, Hines GA, Starmer J, Paul VJ (1999) Chemical signals in gametogenesis, spawning, and larval settlement and defense of the soft coral *Sinularia polydactyla*. Coral Reefs 18:75–84

Sparck R (1927) Studies on the biology of the oyster (*Ostrea edulis*) IV. On fluctuations in the oyster stock in the Limfjord. Rep Danish Biol Stat 33:60–65

Starr M, Himmelman JH, Therriault JC (1990) Direct coupling of marine invertebrate spawning with phytoplankton blooms. Science 247:1071–1074

Styan CA (1998) Polyspermy, egg size, and the fertilization kinetics of free-spawning marine invertebrates. Am Nat 152:290–287

Styan CA, Byrne M, Franke E (2005) Evolution of egg size and fertilisation efficiency in sea stars: large eggs are not fertilised more readily than small eggs in the genus *Patiriella* (Echinodermata: Asteroidea). Mar Biol 147:235–242

Svane I, Havenhand JN (1993) Spawning and dispersal in *Ciona intestinalis* (L). Mar Ecol 14:53–66

Thomas FIM (1994) Physical properties of gametes in three sea-urchin species. J Exp Biol 194:263–284

Thorson G (1946) Reproduction and larval development of Danish marine bottom invertebrates. Meddr Kommn Danm Fisk- Havunders Ser Plankton 4:1–523

Togashi T, Cox PA (2001) Tidal-linked synchrony of gamete release in the marine green alga, *Monostroma angicava* Kjellman. J Exp Mar Biol Ecol 264:117–131

Togashi T, Motomura T, Ichimura T (1998) Gamete dimorphism in *Bryopsis plumosa* phototaxis, gamete motility and pheromonal attraction. Botanica Marina 41:257–264

Vacquier VD (1998) Evolution of gamete recognition proteins. Science 281:1995–1998

Vogel H, Czihak G, Chang P, Wolf W (1982) Fertilization kinetics of sea urchin eggs. Math Biosci 58:189–216

Vollmer SV, Palumbi SR (2002) Hybridization and the evolution of reef coral diversity. Science 296:2023–2025

Wahle RA, Gilbert AE (2002) Detecting and quantifying male sea urchin spawning with time-integrated fertilization assays. Mar Biol 140:375–382

Watson GJ, Bentley MG, Gaudron SM, Hardege JD (2003) The role of chemical signals in the spawning induction of polychaete worms and other marine invertebrates. J Exp Mar Biol Ecol 294:169–187

Williams ME, Bentley MG (2002) Fertilization success in marine invertebrates: the influence of gamete age. Biol Bull 202:34–42

Yund PO (1990) An in situ measurement of sperm dispersal in a colonial marine hydroid. J Exp Zool 253:102–106

Yund PO (1995) Gene flow via the dispersal of fertilizing sperm in a colonial ascidian (*Botryllus schlosseri*)—the effect of male density. Mar Biol 122:649–654

Yund PO (2000) How severe is sperm limitation in natural populations of marine free-spawners? Trends Ecol Evol 15:10–13

Yund PO, Meidel SK (2003) Sea urchin spawning in benthic boundary layers: are eggs fertilized before advecting away from females? Limnol Oceanogr 48:795–801

Chapter 11
Larval Supply and Dispersal

Dustin J. Marshall, Craig Styan, and Christopher D. McQuaid

11.1 Introduction

Most marine organisms have planktonic larvae that spend between minutes and months in the water column before settlement. For over 50 years, marine ecologists have recognised that the number of larvae that are produced, disperse and recruit successfully is extremely variable (Thorson 1950). More recently, it has been realised that variation in larval supply can drive the dynamics of marine populations and communities (Underwood and Fairweather 1989). To understand how marine populations and communities vary in time and space, we must first understand how propagule supply and dispersal are influenced. This chapter examines the causes of variability in larval production, survival in the plankton, and scales of dispersal and their consequences for marine organisms living on hard substrata.

11.2 Variability in the Production of Larvae

The enormous variability in production of larvae by any species can come from a variety of sources with two basic elements: (1) variation in fecundity (the production of gametes) and (2) variation in fertilisation success (the production of zygotes).

11.2.1 Variation in Fecundity

Despite the realisation that variation in egg production helps determine larval supply, surprisingly few patterns of variation in fecundity have been identified for marine organisms. This may be because fecundity is often difficult to quantify (Ramirez-Llodra 2002) but the paucity of identifiable patterns may also arise because data are scattered across studies on individual species. Studies of variation in fecundity show that there can be dramatic differences among populations and individuals (Ramirez-Llodra 2002). For example, the number of eggs produced by individual

M. Wahl (ed.), *Marine Hard Bottom Communities*, Ecological Studies 206,
DOI: 10.1007/978-3-540-92704-4_11,

females within a population of crabs ranged between 10 and 1,443 in the study of Sampedro et al. (1997).

The production of eggs is energetically expensive and, overall, reproductive output is limited by access to resources and the costs of somatic maintenance and/or growth. Accordingly, most factors affecting resource availability influence fecundity in marine organisms (Ramirez-Llodra 2002). Many factors affect fecundity but maternal size, nutritional history and age are major intrinsic factors. Importantly, maternal size and fecundity are not isometrically related—i.e. the ratio of body mass to fecundity increases with size such that larger mothers produce many more offspring per unit of body weight than do smaller mothers. Allometric reproduction has important implications for the relative contributions of small and large individuals to the next generation and, of course, exploitation by fisheries (Birkeland and Dayton 2005). The major extrinsic influences on fecundity seem to be environmental stress, pollution and both inter- and intra-specific competition (Ramirez-Llodra 2002). Overall, fecundity is highly plastic and variation across orders of magnitude should be expected both within and among marine populations.

Like egg quantity, the quality of eggs can also vary. The average within-population coefficient of variation in egg size across 102 species of marine invertebrates is ~9%, yielding a twofold size range across the largest and smallest 5% of eggs (Marshall and Keough 2008). Variation in egg size is important because offspring size can be a good proxy for offspring quality; bigger eggs typically have a better chance of being fertilised, developing in the plankton and recruiting successfully (Marshall and Keough 2008). Although variation in egg quality may be as important to recruitment as variation in their numbers, this is rarely considered.

11.2.2 Variation in Fertilisation

The production and release of eggs is not sufficient to ensure the supply of larvae into a population—these eggs must also be fertilised. In internally fertilising species, fertilisation success is largely assured; so, in this case fecundity is a good estimate of larval production. However, most marine animals reproduce by broadcast spawning, with eggs and sperm shed into the environment and external fertilisation. In such species, fertilisation is not assured, so fecundity and larval production may be decoupled.

Thorson was the first to consider the fertilisation rates of broadcast spawned eggs and concluded that, because many animals aggregate to spawn, the vast majority of eggs are fertilised. This assumption was largely unchallenged until Timothy Pennington examined the fertilisation success of sea urchin eggs in the field in 1985. Pennington found that when eggs were spawned further than 20 cm away from spawning males, less than 15% were successfully fertilised (Pennington 1985). Since this pioneering work, studies on a range of species have shown that fertilisation success is highly variable and rarely 100%. This has led to speculation that the dynamics of broadcast spawning populations are limited by fertilisation rates, though this remains controversial (Levitan and Petersen 1995; Yund 2000).

While many factors affect fertilisation success in broadcast spawners, the local concentration of sperm is the most important, as it determines the probability of egg/sperm encounters. The dynamic nature of the marine environment can result in rapid dilution of sperm to ineffective concentrations. Given that the average distance between spawning individuals is often inversely proportional to population density (assuming that aggregation during spawning is not always perfect), fertilisation success is positively related to population density. This relationship between fertilisation success and population density is termed an Allee effect and has interesting implications for the production of larvae.

Benthic marine organisms typically show a negative relationship between population density and fecundity, as increasing intra-specific competition can decrease individual reproductive output. Thus, we expect high-density populations to have lower per capita fecundity than do low-density populations. However, for broadcast spawners, fertilisation success increases with population density, suggesting higher fertilisation success in these high-density populations. Potential trade-offs between fecundity and fertilisation success obscure the relationship between population density and larval production for broadcast spawners. The precise relationship between sperm concentration (and its proxy population density) and fertilisation success varies with local hydrodynamics and the properties of the gametes per se, yielding a highly idiosyncratic relationship. Generally, more work is needed on the effects of population density on larval production.

11.3 Mortality in the Plankton

Most marine larvae are less than a few millimetres long and swim slowly. Their small size and poor swimming ability make these larvae vulnerable to a range of predators and to ocean currents, and the larval planktonic period has long been viewed as the greatest source of mortality for organisms with biphasic lifecycles (Thorson 1950). However, small larval size and the sheer scale of larval dispersal make estimating real rates of mortality in the field enormously challenging. This is important because planktonic mortality not only affects the number of larvae that can recruit into a population but it can also affect how far larvae can disperse (i.e. very high larval mortality rates will result in very few larvae surviving long-distance dispersal).

11.3.1 Estimates of Mortality in the Field

Estimating larval mortality in the field is extremely difficult (see reviews by Rumrill 1990 and Morgan 1995) and direct observations are restricted to large, easily observed ascidian larvae, which show ~70% survival (Stoner 1990). This suggests low mortality but such larvae spend less than 10 min in the plankton, so that daily mortality could be very high. Furthermore, the colonial ascidians studied produce large, conspicuous larvae that are good swimmers and may be chemically defended. Seeing that these are our only direct estimates of field mortality, it is unrealistic to extrapolate to more typical marine larvae.

Most studies examining larval mortality in the sea have used indirect methods or experiments (Morgan 1995; Johnson and Shanks 2003). Intuitively, given the high fecundity of most marine organisms with planktonic larvae, high levels of mortality must occur at some stage. Studies that track individual cohorts of larvae estimate that fewer than 0.01% of larvae survive in the plankton (Rumrill 1990; Morgan 1995). Again, it is difficult to determine how reliable such measures are and, overall, planktonic mortality rates remain one of the greatest 'black boxes' in marine ecology.

11.3.2 Sources of Planktonic Mortality

The issues that hamper estimates of mortality also hamper the identification of sources of mortality—but there are exceptions. Thorson (1950) considered planktonic predation as the major source of larval mortality in the plankton and many reviews support this view (Morgan 1995). However, in situ larval 'corrals' suggest that mortality through predation is typically less than 1% (Johnson and Shanks 2003). Larvae are also small, poorly protected and vulnerable to environmental stresses such as pollution, salinity changes and UV exposure. For larvae that feed in the plankton, starvation may be important but the evidence for such effects is variable (Olson and Olson 1989). Probably one of the greatest causes of mortality is advection: larvae may be transported away from suitable habitats and never reach a place where metamorphosis is successful. Again, whilst this is probably common, we have limited evidence for it in the field.

11.3.3 Phenotypic Degradation of Larvae in the Field

Larvae are not inert particles but living organisms that experience a range of conditions. The dramatic changes associated with metamorphosis encourage viewing the larval and adult stage as separate; however, their phenotypes are strongly linked. Consequently, events in the larval stage can affect performance in the adult stage (i.e. adult phenotype), and are termed 'carry-over' or 'latent' effects (Pechenik 2006). This link between life-history stages has some important consequences for understanding planktonic mortality. If larval experience degrades the larval phenotype so that they cannot survive as adults, then these larvae are effectively the 'swimming dead' (Pechenik 1990): although alive, they cannot contribute to recruitment. Many experiences can degrade larval phenotype, including pollution, metamorphic delay, increased swimming activity and poor nutrition (reviewed in Pechenik 2006). This could be an important driver of recruitment success because it becomes apparent only in larvae that have survived all other causes of mortality and would otherwise have recruited into the population. Unfortunately, the incidence of phenotypic degradation in the field is unknown, as most studies that manipulate larval phenotype are done in the laboratory.

11.4 Scales of Dispersal and Larval Supply

It is difficult to overestimate the importance of understanding larval supply, due to its influence on both gene flow and community structure. In evolutionary terms, scales of larval dispersal may be implicated in speciation rates, species longevity and geographic ranges (Jeffery and Emlet 2003). Ecologically, larval supply affects the degree to which communities or populations are shaped by competition for space (Gaines and Roughgarden 1985), spatial distribution patterns and how we define metapopulations (Ellien et al. 2000). Consequently, understanding larval supply has profound importance for the design of marine reserves (e.g. Stobutzki 2001; Jessopp and McAllen 2007), particularly where distinct source and sink populations exist (Bode et al. 2006).

Larval supply is affected by multiple factors operating at different spatial and temporal scales, and Pineda (1994) offers an extremely useful model of the hierarchical nature of the control of larval supply. Essentially, this scales down from the control of rates of larval arrival at a site (determined by the size of the larval pool and their physical transport) to small-scale factors including substratum availability, micro-hydrodynamics and behaviour. The advantage of this model is that it incorporates issues of large-scale dispersal occurring over scales of 10–100s km and differential nearshore delivery to places separated by 100s m (e.g. Porri et al. 2006). Importantly, large-scale offshore processes affect greater numbers of larvae and, so, have a disproportionally important influence on population fluctuations (Pineda 2000), while small-scale factors can operate only on the pool of larvae provided by offshore processes. This finds clear expression in the suggestion by Abrams et al. (1996) that large-scale oceanography affects material subsidies to coastal ecosystems, including propagule supply, and thereby dramatically affects the pace of community dynamics and the intensity of species interactions.

It is well known that different zooplankton, including larvae, occur in different water masses (Wing et al. 1998) and that larvae tend to accumulate at fronts (Roughgarden et al. 1991). Many physical processes affect offshore transport, including factors that influence cross-shelf transport, such as upwelling (e.g. Mace and Morgan 2006), surface slicks (Shanks 1986) and internal tidal bores (Pineda 1991), as well as longshore transport mechanisms such as wind-driven surface currents (Barkai and McQuaid 1988).

Coastal topography may have indirect and very powerful effects on larval dispersal and supply through its interaction with nearshore oceanography (Gaylord and Gaines 2000; Roughan et al. 2005; Webster et al. 2007). Bay and open coast sites differ in settlement rates, because of the interaction of topography with wind and tides (Gaines and Bertness 1992), they support different larval assemblages

(Jessopp and McAllen 2007) and, in South Africa, they support mussel populations that differ not only in their densities (unpublished data) but also in their genetic structure (Nicastro et al. 2008). The importance of zones of retention, where larval exchange with adjacent waters is reduced by patterns of water flow, is increasingly recognised (Largier 2004).

Given the small physical size of most larvae, it is tempting to assume that they essentially function as passive particles transported entirely through hydrological processes and, indeed, there is support for this (e.g. Barkai and McQuaid 1988; Griffin et al. 2008). However, growing evidence suggests that larval behaviour can influence dispersal, so that models based on passive transport define potential, not necessarily actual dispersal (Roberts 1997).

Some larvae can postpone settlement if conditions are not ideal (Seed and Suchanek 1992), and this carries direct (predation) and indirect costs (depletion of energetic resources). The desperate larva hypothesis (Knight-Jones 1951) predicts that larval discrimination with regard to settlement cues will decrease as their energy stores are diminished, so that they will accept a wider range of cues. For species with non-feeding larvae, dispersal potential can effectively be manipulated through larval size, which is assumed to correlate with energy stores, and influences readiness to settle and responsiveness to settlement cues; essentially, smaller or hungrier larvae become desperate more quickly (Botello and Krug 2006). However, models suggest that, because they can feed, the settlement behaviour of planktotrophic larvae should be strongly affected by the quality of potential settlement sites and local food availability; essentially, given enough food in the water column, they are less likely to become desperate (Elkin and Marshall 2007).

To what degree (if at all) can we regard planktonic larvae as the passive particles that many models (e.g. Man et al. 1995) assume these organisms to be? Clearly, this will differ among taxa, and will depend on larval behaviour (e.g. Shanks 1995), swimming abilities and duration (Stobutzki and Bellwood 1997). For many taxa, larval duration in turn depends on sea temperature, with implications for potential survival and dispersal, so that temperature effects can help to explain differences in recruitment among years (O'Connor et al. 2007).

Unfortunately, the situation is made even more complex by the fact that behaviour can differ within a single taxon. For example:

- Different larval stages of the same species may differ in their behaviour (Gallager et al. 1996) and position in the water column (Tapia and Pineda 2007).
- Conspecific larvae from different populations can show different vertical migration behaviour (Manuel and O'Dor 1997).

There is no doubt that larval dispersal occurs across a wide range of spatial scales (Shanks et al. 2003) that differ among taxa within the same community. Algae generally disperse over relatively small distances and fish over much larger distances; invertebrates, including many taxa, unsurprisingly show a wide range of dispersal scales (Shanks et al. 2003). Scales of connectivity are important in understanding whether reproductive output is correlated with recruitment, i.e. how closed are populations? Early work highlighted the ability of larvae to disperse over large distances (e.g. Scheltema and Williams 1983), so that populations were perceived as being extremely open (Scheltema 1971). Later work has tended to emphasise that dispersal is often less than anticipated (e.g. Barkai and McQuaid 1988) and there is considerable evidence for self-recruitment that exists across life-history types and geographic regions. Thus, marine populations may be much less open than was previously believed (Sponaugle et al. 2002), though there is extreme variability in the dispersal ranges observed in marine propagules (Kinlan and Gaines 2003) and there may be a bias in the reporting of results (Levin 2006).

Sponaugle et al. (2002) suggest that any departure from unidirectional flow that is uniform across depths will offer the possibility of larval retention. They make a useful separation here between physical retention, which operates on all passive particles and requires no behaviour of the larvae, and biophysical retention, which operates through the interaction of physical effects and larval behaviour. Much of the recent evidence emphasising the importance of self-recruitment comes from fish. Jones et al. (2005) marked individual larvae in a population of clown fish using tetracycline and found remarkably little dispersal, despite a 9–12 day larval period. Older fish larvae have considerable swimming potential and behavioural plasticity but the relative importance of behaviour and oceanography in their dispersal remains controversial (e.g. Colin 2003; Taylor and Helleberg 2003). Even for simpler invertebrate larvae, there is evidence that behaviour can have significant effects on dispersal distances (e.g. Shanks and Brink 2005), as well as on patterns of settlement (Bierne et al. 2003).

Even where self-recruitment is important, populations still depend to some degree on allochthonous inputs of propagules and this too can vary (Cowan et al. 2000). Models of the English Channel suggest that wind forcing increases the role of advection and strongly influences larval dispersal (Ellien et al. 2000), so that interannual variation in wind forcing leads to variability in settlement rates, and in the degree to which populations receive "external" or "local" larvae (Barnay et al. 2004).

There is evidence that large-scale oceanographic features and events can have cross-taxon effects (Gaylord and Gaines 2000) but, because dispersal scales can reflect the interaction of behaviour and hydrodynamics, these are likely to be governed by different taxon-specific combinations of factors. Mace and Morgan (2006) found that the crab *Cancer magister* in Bodega Bay settled mainly during relaxation conditions, while *C. antennarius/productus* settled during upwelling conditions, indicating that there may be different taxon-specific delivery mechanisms even for related taxa in the same small area.

Scales of larval dispersal depend on local conditions and hydrodynamics, phylogenetic constraint in terms of behaviour and potential propagule longevity, as well as on possible taxon-specific delivery shaped by the interaction of behaviour and oceanography. Dispersal scale may also differ among conspecific individuals, and a useful approach is to identify dispersal kernels, which describe the possibility of individual propagules dispersing over a given distance before settlement (Kinlan and Gaines 2003). Ultimately, the best method for studying larval dispersal depends on the question asked. In terms of connectivity and the structure of metapopulations, genetic approaches are suitable. In terms of population dynamics, an understanding of the interactions between behaviour and oceanography is required.

11.5 Genetic Consequences of Variation in Larval Production and Dispersal

Molecular genetic approaches are often used to understand gene flow and, indirectly, estimate larval dispersal/connectivity within marine metapopulations. The basic premise is that, assuming selective neutrality of molecular markers, genetic differences among populations suggest that larval exchange among these is limited (Hedgecock 1986). Early studies indicated that organisms with a short or no larval stage showed genetic differentiation on relatively small scales, while those with long-lived larvae showed differentiation across broader scales (Hedgecock 1986). The advent of polymerase chain reaction techniques led to an explosion of studies over the last 10 years using more variable and, thus, statistically more powerful markers, which have uncovered genetic discontinuities over spatial scales smaller than expected (Sotka et al. 2004; Temby et al. 2007). New simulation frameworks for predicting/analysing larval dispersal and gene flow in coastal marine populations (Sotka and Palumbi 2006) are also an exciting prospect, particularly because such models can potentially be adapted to account for realistic oceanographic patterns and large variances in larval output, quality and dispersal among populations.

Given the many processes influencing reproductive and larval success, reproductive skews in marine populations could be larger than previously thought. Large skews in reproductive success imply that the effective genetic sizes of populations may be orders of magnitude smaller than suggested by adult numbers, so that populations may function genetically as though much smaller (Hedrick 2005). Concern about the conservation consequences of small genetic population sizes in marine species (i.e. accelerated genetic drift, inbreeding, etc.) has generally been disregarded, presumably because marine populations are typically numerically enormous, but this lack of attention may be unwarranted (Waples 2002). Understanding reproductive skews is necessary to determine whether they are large or consistent enough to influence the effective size of populations, enabling us to interpret differences among populations that are related to their effective genetic size (N_e) and migration rates for each generation (m). The latter is usually the parameter of interest, because it relates to larval exchange, but it cannot be determined without knowing N_e.

Although data are scarce, initial molecular studies suggest that effective population sizes in sessile marine invertebrate populations are surprising low. Work on two species of oysters indicates that their effective population size may be many orders of magnitude less than their numerical size (Hedgecock et al. 2007). Other studies also show temporal differences in the genetic structure of recruits (Moberg and Burton 2000; Lee and Boulding 2007), a genetic pattern called "chaotic genetic patchiness" by Johnson and Black (1984). Similarly, Veliz et al. (2006) recently found evidence of kin relationships among barnacles with a relatively long planktonic period. These patterns are consistent with larval production by a limited subset of potential parents in a population at different times—i.e. very large skews in reproductive success among individual spawners. What affects effective population size, however, is large skews in the lifetime reproductive success of individuals.

Thus, longer temporal studies across time/multiple reproductive events are needed to understand whether these putative skews are consistent among individuals or change from one spawning event to the next.

11.6 Conclusions

Variation in the number of larvae produced in a population arises because some individuals produce/release fewer propagules in the first place, because gametes do not fertilise, or because the larvae die or are not transported to a suitable place. Transport is influenced by hydrodynamics interacting with larval behaviour. Even when delivered to appropriate habitats, larvae may count as 'living dead' because they are physiologically compromised or are reproductively isolated as foreign immigrants. Consequently, larval supply can and does vary enormously, with critical consequences for the ecology and evolution of marine species. Determining which processes are the primary drivers in the variation in larval production and dispersal is a key quest in marine ecology.

References

Abrams PA, Menge BA, Mittelbach GG, Spiller D, Yodzis P (1996) The role of indirect effects in food webs. In: Polis G, Winemiller K (eds) Food webs: dynamics and structure. Chapman and Hall, New York, pp 371–395

Barkai A, McQuaid C (1988) Predator-prey role reversal in a marine benthic ecosystem. Science 242:62–64

Barnay A, Ellien C, Gentil F, Thiebaut (2004) A model study on variations in larval supply: populations of the polychaete *Owenia fusiformis* in the English Channel open or closed? Helgol Mar Res 56:229–237

Bierne N, Bonhomme F, David P (2003) Habitat preference and the marine-speciation paradox. Proc R Soc Biol Sci Ser B 270:1399–1406

Birkeland C, Dayton PK (2005) The importance in fishery management of leaving the big ones. Trends Ecol Evol 20:356–358

Bode M, Bode L, Armsworth PR (2006) Larval dispersal reveals regional sources and sinks in the Great Barrier Reef. Mar Ecol Prog Ser 308:17–25

Botello G, Krug PJ (2006) Desperate larvae revisited: age, energy and experience affect sensitivity to settlement cues in larvae of the gastropod *Alderia sp.* Mar Ecol Prog Ser 312:149–159

Colin PL (2003) Larvae retention: genes or oceanography? Science 300:1657–1659

Cowan RK, Lwiza MM, Sponaugle S, Paris CB, Olson DB (2000) Connectivity of marine populations: open or closed? Science 287:857

Elkin CM, Marshall DJ (2007) Desperate larvae: the influence of deferred costs and habitat requirements on habitat selection. Mar Ecol Prog Ser 335:143–153

Ellien C, Thiebaut E, Barnay AS, Dauvin J-C, Gentil F, Salomon J-C (2000) The influence of variability in larval dispersal on the dynamics of a marine metapopulation in the eastern Channel. Oceanol Acta 23:423–442

Gaines SD, Bertness MD (1992) Dispersal of juveniles and variable recruitment in sessile marine species. Nature 360:579–580

Gaines S, Roughgarden J (1985) Larval settlement rate: a leading determinant of structure in an ecological community of the marine intertidal zone. Proc Natl Acad Sci USA 82:3707–3711
Gallager SM, Manuel JL, Manning DA, O'Dor R (1996) The vertical distribution of scallop larvae *Placopecten magellanicus* in 9 m deep mesocosms as a function of light, food and temperature. Part I. Ontogeny of vertical migration behaviour. Mar Biol 124:679–692
Gaylord B, Gaines SD (2000) Temperature or transport? Range limits in marine species mediated solely by flow. Am Nat 155:769–789
Griffin JN, de la Haye K, Hawkins SJ, Thompson RC, Jenkins SR (2008) Predator diversity and ecosystem functioning: density modifies the effect of resource partitioning. Ecology 89:298–305
Hedgecock D (1986) Is gene flow from pelagic larval dispersal important in the adaptation and evolution of marine-invertebrates. Bull Mar Sci 39:550–564
Hedgecock D, Launey S, Pudovkin AI, Naciri Y, Lapegue S, Bonhomme F (2007) Small effective number of parents (N-b) inferred for a naturally spawned cohort of juvenile European flat oysters Ostrea edulis. Mar Biol 150:1173–1182
Hedrick P (2005) Large variance in reproductive success and the N-e/N ratio. Evolution 59:1596–1599
Jeffery CH, Emlet RB (2003) Macroevolutionary consequences of developmental mode in temnopleurid echinoids from the tertiary of southern Australia. Evolution 57:1031–1048
Jessopp MJ, McAllen RJ (2007) Water retention and limited larval dispersal: implications for short and long distance dispersers in marine reserves. Mar Ecol Prog Ser 333:27–36
Johnson MS, Black R (1984) Pattern beneath the chaos–the effect of recruitment on genetic patchiness in an intertidal limpet. Evolution 38:1371–1383
Johnson KB, Shanks AL (2003) Low rates of predation on planktonic marine invertebrate larvae. Mar Ecol Prog Ser 248:125–139
Jones GP, Planes S, Thorrold SR (2005) Coral reef fish larvae settle close to home. Curr Biol 15:1314–1318
Kinlan BP, Gaines SD (2003) Propagule dispersal in marine and terrestrial environments: a community perspective. Ecology 84:2007–2020
Knight-Jones EW (1951) Gregariousness and some other aspects of the settling behaviour of *Spirobis*. J Mar Biol Assoc UK 30:201–222
Largier J (2004) The importance of retention zones in the dispersal of larvae. In: Proc American Fish Society Symp, pp 105–122
Lee HJ, Boulding EG (2007) Mitochondrial DNA variation in space and time in the northeastern Pacific gastropod, *Littorina keenae*. Mol Ecol 16:3084–3103
Levin LA (2006) Recent progress in understanding larval dispersal: new directions and digressions. Integ Comp Biol 46:282–297
Levitan DR, Petersen C (1995) Sperm limitation in the sea. Trends Ecol Evol 10:228–231
Mace AJ, Morgan SG (2006) Biological and physical coupling in the lee of a small headland: contrasting transport mechanisms for crab larvae in an upwelling region. Mar Ecol Prog Ser 324:185–196
Man A, Law R, Polunin NVC (1995) Role of marine reserves in recruitment to reef fisheries: a metapopulation model. Biol Conserv Restor Sustain 71:197–204
Manuel JL, O'Dor RK (1997) Vertical migration for horizontal transport while avoiding predators: I. A tidal/diel model. J Plankton Res 19:1929–1947
Marshall DJ, Keough MJ (2008) The evolutionary ecology of offspring size in marine invertebrates. Adv Mar Biol 53:1–60
Moberg PE, Burton RS (2000) Genetic heterogeneity among adult and recruit red sea urchins, *Strongylocentrotus franciscanus*. Mar Biol 136:773–784
Morgan SG (1995) Life and death in the plankton: larval mortality and adaptation. In: McEdward L (ed) Ecology of Marine Invertebrate Larvae, vol 1. CRC Press, Boca Raton, FL, pp 279–322
Nicastro KR, Zardi GI, McQuaid CD, Teske PR, Barker NP (2008) Topographically driven genetic structure in marine mussels. Mar Ecol Prog Ser (in press)
O'Connor MI, Bruno JF, Gaines SD, Halpern BS, Lester SE, Kinlan BP, Weiss JM (2007) Temperature control of larval dispersal and the implications for marine ecology, evolution, and conservation. Proc Natl Acad Sci USA 104:1266–1271

Olson RR, Olson MH (1989) Food limitation of planktotrophic marine invertebrate larvae: does it control recruitment success? Annu Rev Ecol Syst 20:225–247
Pechenik JA (1990) Delayed metamorphosis by larvae of benthic marine-invertebrates—does it occur—is there a price to pay. Ophelia 32:63–94
Pechenik JA (2006) Larval experience and latent effects—metamorphosis is not a new beginning. Integ Comp Biol 46:323–333
Pennington JT (1985) The ecology of fertilization of echinoid eggs: the consequences of sperm dilution, adult aggregation and synchronous spawning. Biol Bull 169:417–430
Pineda J (1991) Predictable upwelling and shoreward transport of planktonic larvae by internal tidal bores. Science 253:548–551
Pineda J (1994) Spatial and temporal patterns in barnacle settlement rate along a southern California rocky shore. Mar Ecol Prog Ser 107:125–138
Pineda J (2000) Linking larval settlement to larval transport: assumptions, potentials, and pitfalls. Oceanogr East Pacific 1:84–105
Porri F, McQuaid CD, Radloff S (2006) Spatio-temporal variability of larval abundance and settlement of *Perna perna*: differential delivery of mussels. Mar Ecol Prog Ser 315:141–150
Ramirez-Llodra ER (2002) Fecundity and life-history strategies in marine invertebrates. Adv Mar Biol 43:88–172
Roberts CM (1997) Connectivity and management of Caribbean coral reefs. Science 278:1454–1457
Roughan M, Mace AJ, Largier JL, Morgan SG, Fisher JL, Carter ML (2005) Subsurface recirculation and larval retention in the lee of a small headland: a variation on the upwelling shadow theme. J Geophys Res Oceans 110:C10
Roughgarden J, Pennington JT, Stoner D, Alexander S, Miller K (1991) Collisions of upwelling fronts with the intertidal zone: the cause of recruitment pulses in barnacle populations of central California. Acta Oecol 12:35–51
Rumrill SS (1990) Natural mortality of marine invertebrate larvae. Ophelia 32:163–198
Sampedro MP, Fernandex L, Freire J, Gonzalez-Gurriaran E (1997) Fecundity and reproductive output of *Pisifia longicornis* (Decapoda, Anomura) in the Ria de Arousa (Galicia, NW Spain). Crustaceana 70:95–109
Scheltema RS (1971) Larval dispersal as a means of genetic exchange between geographically separated populations of shallow-water benthic marine gastropods. Biol Bull 140:284–322
Scheltema RS, Williams IP (1983) Long-distance dispersal of planktonic larvae and the biogeography and evolution of some Polynesian and western Pacific mollusks. Bull Mar Sci 33:545–565
Seed R, Suchanek TH (1992) Population and community ecology of Mytilus. In: Gosling E (ed) The mussel Mytilus: ecology, physiology, genetics and culture. Development in Aquaculture and Fisheries Sciences, vol 25. Elsevier, Amsterdam, pp 87–169
Shanks AL (1986) Tidal periodicity in the daily settlement of intertidal barnacle larvae and an hypothesized mechanism for the cross-shelf transport of cyprids. Biol Bull 170:429–440
Shanks AL (1995) Mechanisms of cross-shelf dispersal of larval invertebrates and fish. In: McEdward LR (ed) Ecology of Marine Invertebrate Larvae, vol 1. CRC Press, Boca Raton, FL, pp 324–367
Shanks AL, Brink L (2005) Upwelling, downwelling, and cross-shelf transport of bivalve larvae: test of a hypothesis. Mar Ecol Prog Ser 302:1–12
Shanks AL, Grantham BA, Carr MH (2003) Propagule dispersal distance and the size and spacing of marine reserves. Ecol Appl 13:S159–S169
Sotka EE, Palumbi SR (2006) The use of genetic clines to estimate dispersal distances of marine larvae. Ecology 87:1094–1103
Sotka EE, Wares JP, Barth JA, Grosberg RK, Palumbi SR (2004) Strong genetic clines and geographical variation in gene flow in the rocky intertidal barnacle *Balanus glandula*. Mol Ecol 13:2143–2156
Sponaugle S, Cowen RK, Shanks A, Morgan SG, Leis JM, Pineda J, Boehlert GW, Kingsford MJ, Lindeman K, Grimes C, Munro JL (2002) Predicting self-recruitment in marine populations: biophysical correlates and mechanisms. Bull Mar Sci 70:341–375
Stobutzki IC (2001) Marine reserves and the complexity of larval dispersal. Rev Fish Biol Fish 10:515–518

Stobutzki IC, Bellwood DR (1997) Sustained swimming abilities of the late pelagic stages of coral reef fishes. Mar Ecol Prog Ser 149:35–41

Stoner DS (1990) Recruitment of a tropical colonial ascidian: relative importance of presettlement vs post-settlement processes. Ecology 71:1682–1690

Tapia FJ, Pineda J (2007) Stage-specific distribution of barnacle larvae in nearshore waters: potential for limited dispersal and high mortality rates. Mar Ecol Prog Ser 342:177–190

Taylor MS, Helleberg ME (2003) Response. Science 300:1657–1658

Temby N, Miller K, Mundy C (2007) Evidence of genetic subdivision among populations of blacklip abalone (*Haliotis rubra* Leach) in Tasmania. Mar Freshw Res 58:733–742

Thorson G (1950) Reproductive and larval ecology of marine bottom invertebrates. Biol Rev 25:1–45

Underwood AJ, Fairweather PG (1989) Supply-side ecology and benthic marine assemblages. Trends Ecol Evol 4:16–21

Veliz D, Duchesne P, Bourget E, Bernatchez L (2006) Genetic evidence for kin aggregation in the intertidal acorn barnacle (*Semibalanus balanoides*). Mol Ecol 15:4193–4202

Waples RS (2002) Evaluating the effect of stage-specific survivorship on the N-e/N ratio. Mol Ecol 11:1029–1037

Webster MS, Osborne-Gowey JD, Young TH, Freidenburg TL, Menge BA (2007) Persistent regional variation in populations of a tidepool fish. J Exp Mar Biol Ecol 346:8–20

Wing SR, Botsford LW, Ralston SV, Largier JL (1998) Meroplanktonic distribution and circulation in coastal retention zone of the N California upwelling system. Limnol Oceanogr 43:1710–1721

Yund PO (2000) How severe is sperm limitation in natural populations of marine free-spawners? Trends Ecol Evol 15:10–13

Chapter 12
Settlement and Recruitment

Stuart R. Jenkins, Dustin Marshall, and Simonetta Fraschetti

12.1 Introduction

The algae and invertebrates living on marine hard substrata cover an enormous variety of life forms and life histories. However, one common theme displayed in benthic algae and in the majority of benthic invertebrates is a planktonic phase within the lifecycle. Larvae of benthic invertebrates and the propagules of algae must survive a planktonic period before beginning their benthic way of life on hard substrata. The transition from pelagic to benthic habitat, which occurs through larval and propagule settlement, and survival through to recruitment, is one of the most important points in the life history of marine organisms. In this chapter, we review patterns of settlement and recruitment over a variety of scales, and examine biological and physical interactions which occur at settlement before looking at the challenges facing propagules at this critical period and assessing the consequences of this process for the dynamics of adult benthic populations.

12.2 Definitions of Settlement and Recruitment

Planktonic larvae of benthic invertebrates spend a variable period in the pelagic realm, from minutes to months, during which time they mature and at some point attain the competence to metamorphose. Metamorphosis is then usually delayed until a suitable substratum is found. On making contact with a hard bottom, larvae undergo a range of often complex behaviours involving exploration and inspection, followed by either acceptance of the site or rejection and release to enter the water column once again. Once accepted, larvae may permanently attach (in the case of sessile organisms), and metamorphosis generally ensues. This complex process leads to a range of definitions of settlement, generally dependent on whether metamorphosis is included. However, given the wide range of behaviour displayed by marine benthic organisms (for example, some taxa metamorphose before settlement), very specific definitions are probably counterproductive, and settlement is generally taken to mean the termination of the pelagic phase and assumption of

M. Wahl (ed.), *Marine Hard Bottom Communities*, Ecological Studies 206,
DOI: 10.1007/978-3-540-92704-4_12, © Springer-Verlag Berlin Heidelberg 2009

a sedentary life. Hence, settlement, rather than a single act, can be considered a process which generally includes reversible contact with the substratum, exploratory behaviour, orientation and metamorphosis (see Pawlik 1992 for discussion). This definition can also be used in considerations of algal settlement, although behaviour of algal propagules is less complex, and spores and zygotes generally do not have an obligate planktonic period (Santelices 1990). Hence, settlement of algal propagules may occur immediately following release.

The process of recruitment is very ill-defined and is not based on any specific life history stage or process. It reflects a combination of settlement and a post-settlement period of arbitrary length, determined by the length of time before the settler is counted by an observer (Keough and Downes 1982). This length of time varies enormously between studies. For example, in algal populations this period may be unusually long, where settled algal propagules delay or prolong development under unfavourable conditions (Hoffmann and Santelices 1991). Keough and Downes (1982) pointed out the importance of distinguishing between settlement and recruitment over a quarter of a century ago, yet the terms still appear to be used interchangeably. The main reasons for this are the logistical difficulty of measuring true settlement rates, especially where larvae are small and/or settle in cryptic habitats, and the generally arbitrary nature of recruitment definitions. In some taxa, clear developmental stages may be used to differentiate settlement from recruitment. For example, settled cyprids and metamorphosed spat may be used as indicators of settlement and recruitment respectively in intertidal barnacles (e.g. Jenkins et al. 2000). Whilst the terms settlement and recruitment will continue to be somewhat ambiguous because of the nature of the concepts, difficulties in interpretation will be avoided where the time periods following true settlement, or life stages where measurement is made, are defined.

12.3 Patterns of Settlement and Recruitment on Hard Substrata

Observations of settlement and recruitment on hard substrata invariably document huge variability at a wide range of spatial and temporal scales in both invertebrates (e.g. Hunt and Scheibling 1996; Hughes et al. 1999; Jenkins et al. 2000) and algae (e.g. Reed and Foster 1984; Aberg and Pavia 1997). One motivation for making quantitative observations of settlement and recruitment is that, through understanding of the scales at which variability occurs, insight will be gained into the processes controlling larval and propagule settlement and early post-settlement events. At a local scale, recruitment variability can be caused by interactions occurring at settlement or soon thereafter. At larger scales, spatial variability in recruitment may be related to patterns in the regional larval pool (e.g. Barnes 1956) and regional physical transport processes (Hughes et al. 1999). A number of large-scale rocky intertidal studies using hierarchical sampling designs have documented recruitment variability at all scales from regional (100s km) to highly localised (metres; Caffey 1985; Jenkins et al. 2000;

O'Riordan et al. 2004), indicating the difficulty in making generalisations about the influence of particular scales. However, although formal meta-analyses have not been conducted, it is likely that as in adult assemblages (see Fraschetti et al. 2005 for review), variability in recruitment at the scale of metres and less will be important across most, if not all taxa because of the ubiquitous importance of small-scale processes.

Propagules of algae generally do not have an obligate planktonic period and, so, may settle immediately following release. Although there are numerous examples of algal propagules being transported over large distances (e.g. Reed et al. 1988), the general pattern, especially in perennial long-lived species, is one of dispersal over short distances of the order of a few metres (Santelices 1990). Aberg and Pavia (1997) investigated the variation in abundance of new recruits of the macroalga *Ascophyllum nodosum* over a range of spatial scales from 100s of kilometres to centimetres. They found variation in recruit abundance at small but not large spatial scales, in agreement with the contention that algal assemblages form more closed populations than is the case for invertebrates with long-lived planktonic larvae. While propagule characteristics suggest limited dispersal, other factors can influence spatial patterns of settlement. For example, variation in environmental conditions such as ice cover over large scales (e.g. Aberg 1992; McCook and Chapman 1997) can affect overall reproductive output and, hence, the supply of propagules and recruitment over large spatial scales.

Temporal variability is another well-characterised feature of benthic invertebrate settlement, with huge differences documented at scales from days (Wethey 1983) to years (Barnes 1956). The barnacle *Semibalanus balanoides* is known to undergo occasional 'failure' years in recruitment, when settlement is at least an order of magnitude lower than normal (Barnes 1956; Hawkins and Hartnoll 1982; Jenkins et al. 2000). Such differences between years in pulse recruiters such as *S. balanoides* which recruit within a narrow window are ascribed to mismatches between development of the spring phytoplankton bloom and planktonic larval growth (Barnes 1956).

Integrating spatial and temporal variability, research in Chile has recorded patterns of settlement of intertidal organisms which are spatially synchronised (Lagos et al. 2007). Synchronous settlement implies that processes determining larval arrival to adult habitats and/or post-settlement survival operate simultaneously over a particular spatial scale. Lagos et al. (2007) showed that common invertebrate species with planktonic stages showed recruitment synchrony at spatial scales of less than 30 km along a 120-km stretch of the Chilean coast, whereas those with direct development (and those with planktonic stages but low recruitment levels) showed spatially heterogeneous recruitment dynamics. These results were consistent with mesoscale physical processes affecting delivery of larvae to the shore. In searching for generalities of settlement patterns, consistency among years in the ranking of sites by recruitment density has been recognised at a variety of spatial scales (Kendall et al. 1985; Connolly et al. 2001). Such consistency may reflect the influence of different orientations of coastlines relative to prevailing winds and currents, and the prevalence of large-scale upwelling regimes.

12.4 Behaviour at Settlement

At settlement, pelagic larvae must make contact with a suitable substratum on which to attach. In marine environments, hard substrata are far less common and more patchily distributed than soft sediments. Hence, the specific selection of settlement site is paramount, and invertebrate larvae have developed complex patterns of behaviour and finely tuned discriminatory abilities to ensure that settlement occurs in a habitat which is conducive to survival, growth and, ultimately, reproduction. The behaviour of invertebrate larvae at settlement has perhaps best been characterised in intertidal acorn barnacles. Work by Crisp and co-workers has shown that the exploratory behaviour of the barnacle *S. balanoides*, when seeking a favourable site for settlement, occurs in three consecutive phases—broad exploration, close exploration and inspection. With each advancing stage, the scale of exploration declines until, during 'inspection', the larva simply rotates at the scale of its own body size. This phase precedes permanent attachment and metamorphosis (Knight-Jones 1953; Crisp and Meadows 1963).

Larval behaviour is generally characterised as responding to positive cues to stimulate settlement behaviour (see section below). However, rejection of unsuitable substrata by settling invertebrates is an important mechanism in ensuring that settlement occurs in an appropriate environment. For hard bottom invertebrates, the lack of settlement on sedimentary material, for example, is undoubtedly a function of rejection. As well as ensuring settlement, rejection may also be an important driver ensuring aggregated settlement (Berntsson et al. 2004). Although settlers do reject poor-quality habitats, settlement behaviour is anything but uniform and settlers will sometimes colonise what appear to be very poor-quality habitats. Why does this variation occur? Why do some settlers accept one cue over another? Raimondi and Keough (1990) suggested that producing offspring which have uniform settlement behaviour is not necessarily the best strategy for parents to maximise their own fitness. If settlement cues are an imperfect estimator of subsequent habitat quality (either because the cues as such are unreliable or because the habitat changes through time), then producing colonisers which behave identically will mean that, occasionally, all of these will colonise the 'wrong' habitat. Raimondi and Keough (1990) speculated that, by avoiding putting all of their offspring into one 'habitat basket', parents may minimise the chances of a catastrophic loss of all of their offspring. Theoretical studies support this speculation and, more recently, some studies have shown that mothers produce offspring with intrinsically variable dispersal/settlement behaviour (Krug 2001; Toonen and Pawlik 2001a).

Intrinsic variability is not the only source of variation in settlement behaviour. In recent years, a 50-year-old observation of larval behaviour has been revived and re-examined—the 'desperate larvae hypothesis' (Toonen and Pawlik 2001b). In the 1950s, a number of studies found that as larvae age, they begin to accept lower-quality (or, at least, lower-ranked) settlement cues (Knight-Jones 1951, 1953). Recent work suggests that this change in settlement behaviour is typically found only in species with non-feeding larvae, interpreted as larvae becoming 'more desperate' to settle as they deplete their nutritional reserves (Botello and Krug 2006; Elkin and Marshall 2007).

For feeding larvae, the situation is more complex: the benefits of delaying metamorphosis, in the absence of appropriate settlement sites, depends strongly on food availability in the plankton (Elkin and Marshall 2007). Overall then, larval age and nutritional state are two of the most important elements of the larval phenotype to affect settlement behaviour.

In situ studies of settlement of the propagules of algae are few in comparison to invertebrates. This is due to the often complex life histories of algae and the very small size of many algal propagules (Schiel and Foster 2006). For example, the flagellated spores of kelps, probably along with fucoids the most important group of macroalgae on hard bottoms in temperate areas, are only approximately 7 μm long. However, fucoid propagules are considerably larger; newly settled embryos and germlings can be identified in the field under low magnification and may be visible to the naked eye after a few weeks or months. Unlike invertebrate larvae, the propagules of algae behave as passive particles at scales greater than a few mm, owing to their immotility or extremely low powers of locomotion. The swimming speeds of propagules (maximum of 80–300 $\mu m\ s^{-1}$) may be beneficial at the point of settlement in the slow-moving boundary layer close to the substratum, and potentially aid in selection of microhabitat (Amsler et al. 1992). For example, kelp spores can be positively chemotactic to nutrients and stimulated to settle by these (Amsler and Neushul 1990). Many algal propagules have adhesive properties formed by a high-viscosity mucilage to aid in attachment to settlement surfaces. The 'stickability' of propagules varies considerably, both among taxa and over time within a particular species. The extent to which propagules adhere to surfaces will have a considerable effect on settlement success in turbulent habitats, such as the intertidal and shallow subtidal (Vadas et al. 1992).

12.5 Biological and Physical Interactions at Settlement

Whilst some models of hard substratum community structure assume that the level of settlement and recruitment is a direct function of larval supply (e.g. Gaines et al. 1985) and the amount of free space available (Minchinton and Scheibling 1993), a vast range of physical and biological interactions dictate that reality is far more complex. This was elegantly demonstrated by Minchinton (1997), who showed that the level of recruitment of the tubeworm *Galeolaria caespitose* into patches of free space, within beds of the adult, was related not to patch area but to the length of the patch perimeter. These data indicated that recruitment is related to the proximity of conspecifics, rather than availability of space per se.

For sessile species the choice of settlement site is critical, since the exact location at which the benthic stage will spend its entire life is dictated by this one act. Location in relation to adequate food supplies and environmental conditions, and to potential mates, competitors and predators is in all cases dictated by the adequacy of the choice made at settlement. Hence, there is a huge selective pressure to ensure that settlement is dictated by adequate cues.

At the time of settlement, larvae are exposed to a vast range of novel physical and biological cues to which they have not been exposed in the pelagic environment. One very important indicator of the suitability of habitat for the settlement of larvae is the presence of conspecific adults. Numerous laboratory-based studies have demonstrated the importance of cues from conspecifics in promoting settlement of larvae (e.g. Knight-Jones 1953; Crisp and Meadows 1962), potentially leading to gregarious settlement. Field-based work has shown how such cues may have important effects on natural patterns of settlement and recruitment (e.g. Raimondi 1988b; Hills and Thomason 1996). Gregarious settlement among conspecifics carries many advantages but also the potential for enhanced mortality through cannibalism of larvae settling within beds of suspension-feeding adults. Demonstrations of these effects in the field are relatively rare and somewhat equivocal, although Navarrete and Wieters (2000) used field manipulations to show clearly the strong effects of the large intertidal barnacle, *Semibalanus cariosus*, on recruitment of other sessile species.

Another important biological cue is the microbial film, formed predominantly of diatoms and cyanobacteria, which coats all hard substratum surfaces immersed in water. This biofilm provides an interface between the substratum and the water column, and so is the point of first attachment for settling larvae and propagules. Microbial films have been shown to promote settlement in a range of invertebrate larvae (Todd and Keough 1994; Qian et al. 2007), with settlers responding to cues such as age of the film, specific taxa such as diatoms, and film characteristics related to tidal height (Strathmann et al. 1981).

Experimental work on the influence of independent biological cues has taught us much about the discriminatory abilities of invertebrate larvae, and provided insight into understanding natural settlement patterns. However, it is through observations of larval responses to multiple cues that the best progress may now be made. For example, Thompson et al. (1998) showed that while settling barnacle cyprids could discriminate among differently aged microbial films and those from different tidal heights in the laboratory environment, such discrimination was irrelevant under field conditions. Here, the presence of conspecific cues (either live adults or those recently detached) appeared to overrule cues from micro-biota within the film. This work raises interesting questions regarding the applicability of behavioural mechanisms deduced from controlled laboratory experiments on larval settlement behaviour.

Physical factors associated with the water column, such as light, temperature, salinity, hydrostatic pressure and flow, influence larval behaviour prior to settlement and will undoubtedly influence settlement patterns. One excellent example is the work of Larsson and Jonsson (2006), who showed that cyprid larvae of *Balanus improvisus* actively rejected high-flow environments which were suboptimal for suspension feeding in the early post-settlement phase. This finding suggests that larval choice can be adaptively connected to a specific part of the lifecycle—in this case, the very critical period in the early post-settlement phase. Physical characteristics of the substratum are important determinants of settlement. In the case of invertebrate larvae, settlers may make an active choice dependent on micro-topography (Berntsson et al. 2000), surface contour (Wethey 1986) and rock type (Raimondi 1988a). These factors may interact; the mineral composition of the substrate determines its hardness and resistance to weathering and can, therefore, influence surface topography and heterogeneity over considerable spatial scales (Herbert and Hawkins 2006).

12.6 Early Post-Settlement Survival

The period immediately following settlement is an extremely important time in the life history of benthic organisms. This transition from a pelagic to benthic form is associated with a high degree of mortality in both invertebrates and algae. Settlers are generally extremely small, have to cope with a dramatic change in morphology as they metamorphose and, for intertidal organisms, must withstand exposure to air within a few hours of settlement. In a review of juvenile mortality of benthic marine invertebrates, Gosselin and Qian (1997) showed that high levels of mortality soon after settlement were common across a range of taxa, with 20 of 30 studies reporting levels of over 90% mortality during the juvenile period. Examples of high mortality include loss of over 90% of settlers of the ascidian *Diplosoma similis* (Stoner 1990) over a 16-day period and a 50–80% reduction in the urchin *Strongylocentrotus purpuratus* over 24 days (Rowley 1990). In a fouling community, Osman et al. (1992) showed that two species of tiny gastropods, *Mitrella lunata* and *Anachis avara*, are able to prey selectively on newly settled colonial ascidians and eliminate almost all individuals. In algae, post-settlement survival is equally low, if not lower. For example, Schiel (1988) showed that only 5% of *Sargassum sinclairii* settlers survived to a visible size, Dudgeon and Petraitis (2005) that survival of fucoid embryos to juvenile stages was only between 0.3–3.8% over 400 days, and Wright et al. (2004) that only 2 of 5,395 embryos of *Fucus gardneri* survived to become visible recruits. While survival rates for these fucoid algae are low, they are likely to be orders of magnitude lower for the microscopic stages of kelp (Schiel and Foster 2006).

Early post-settlement stages of algae and invertebrates are generally at risk from the same sources of physical and biological mortality, presumably since both are vulnerable to abiotic and biotic environmental stresses not experienced in the pelagic realm. Much of the work on early post-settlement mortality has taken place in the rocky intertidal where physical stresses are particularly limiting factors. Stress associated with emersion during low-water periods increases along a unidirectional stress gradient as height on the shore increases. Evidence suggests that desiccation is a major source of mortality in the early stages of a variety of intertidal species, including patellid limpets (Branch 1975), barnacles (Denley and Underwood 1979) and littorinids (Behrens 1972). Strong wave action can lead to enormous losses of newly attached invertebrate and algal settlers, which have not developed sufficient strength of attachment. Some algae, such as those of the large canopy-forming species *Ascophyllum nodosum*, suffer from dislodgement under even very moderate wave conditions (Vadas et al. 1990). Artificially settled zygotes, when exposed to less than ten low-energy waves, suffered between 85 and 99% mortality. Biological sources of mortality include grazing (Hawkins and Hartnoll 1983), whiplash from canopy-forming algae (Jenkins et al. 1999) and 'bulldozing' by grazers (Miller 1989).

Grazing accounts for a high level of early post-settlement mortality in algae. A vast range of grazer types (e.g. molluscs, echinoids, fish, amphipods) have been shown to limit algal recruitment. Microphagous grazers such as patellid limpets, many littorinids and sea urchins, which feed upon the microalgal film growing on hard substrata, have an important role in limiting algal development through consumption of algal propagules (reviewed by Hawkins and Hartnoll 1983). This was first dramatically demonstrated by Jones (1948) in the Isle of Man; a large-scale removal experiment showed that exposed rocky shores were relatively devoid of macroalgae because of the grazing activities of patellid limpets, rather than through physical effects of wave action. Molluscan grazers can also cause mortality in vulnerable settled invertebrate larvae through physical dislodgement or 'bulldozing' (e.g. Miller 1989). Whiplash effects of large-canopy macroalgae are well documented. In the intertidal and shallow subtidal, water movement causes the fronds of macroalgae to scour across the substratum, leading to high mortality in both invertebrate larvae (e.g. Jenkins et al. 1999) and macroalgal propagules (e.g. Santelices and Ojeda 1984).

Rather than focusing on sources of mortality, another approach to understanding patterns of post-settlement survival is the determination of the conditions under which survival of new settlers is favoured. For example, complex micro-topography may not only enhance settlement but also provide developing propagules and larvae with refugia from grazers (Lubchenco 1983), physical stress and algal whiplash (Jenkins et al. 1999). Canopy algae, whilst negatively affecting settlement through scouring, can have positive effects on larval recruitment through mitigation of environmental extremes. The balance between positive and negative effects may be determined by the level of environmental stress experienced (Leonard 1999, 2000).

12.7 Consequences of Variation in Settlement and Recruitment

In open marine systems, the decoupling of local reproduction and input of new recruits has led to a long history of debate regarding the importance of variation in settlement relative to post-settlement density-dependent processes in determining adult density (see Caley et al. 1996 for review). One argument, termed the recruitment limitation hypothesis (sensu Doherty 1981), states that larval supply is insufficient for the total population size to reach a carrying capacity, and increases in recruitment will lead to increases in adult population size (e.g. Connell 1985; Menge 1991). Hence, there is a direct relationship between the level of supply/settlement and adult abundance, meaning that spatial and temporal variation in settlement has a strong influence on the density and distribution of benthic juveniles and adults. The implication here is that planktonic processes preceding settlement into the benthic habitat play a role in determining adult population structure. Alternatively, settlement may simply provide the input of new individuals, whose numbers are modified through strong post-settlement density-dependent interactions. Here, the input of larvae and, hence, patterns of settlement have little or no bearing on adult density.

A strong body of evidence suggests populations can be 'recruit limited' at low recruit densities, whereas at high recruit densities, density-dependent post-settlement processes predominate and recruitment variability has little impact on adult abundance. Under this recruit-adult hypothesis (sensu Menge 2000), the relative importance of recruitment declines with increasing density of recruits (Connell 1985; Roughgarden et al. 1985; Sutherland 1990; Menge 2000). However, recent work on the intertidal barnacle *S. balanoides* has shown that recruitment may be a useful predictor of adult density across all recruit densities, from very low to very high (Jenkins et al. 2008). In that study, the relationship between recruitment and adult density switched from positive, at low levels, to negative at high levels, owing to strong over-compensatory density-dependent mortality. Clearly, differences in the level of recruitment from place to place and time to time can be a strong determinant of adult density patterns but the strength of this relationship depends on context and varies with species and their lifecycles, and can be strongly modified by post-recruitment processes (Menge 2000; Penin et al. 2007). Evidently, recruitment limitation and density-dependent interactions are not mutually exclusive but, instead, act jointly to determine the densities of marine benthic populations and assemblages (Chesson 1998; Caselle 1999; McQuaid and Lindsay 2007).

Much of the research on the relationship between recruitment intensity and adult population structure has been carried out in the rocky intertidal on unitary sessile organisms, whereas less information is available for mobile and colonial organisms. In mobile taxa, post-settlement movement can relieve overcrowding due to high settlement intensity (Wahle and Incze 1997). Furthermore, the relationship between recruitment and adult abundance has been rarely explored at the community level. However, work on the Pacific coast of North America has shown how consistent large-scale differences in recruitment intensity can lead to differences in the strength of inter-specific interactions (Connolly et al. 2001). In Oregon and Washington, where recruitment is high, experimental determination of community organisation has emphasised the role of benthic processes such as competition and predation, in contrast to studies in central California where variations in larval supply and settlement have been stressed.

In our discussions of recruitment thus far, the focus has been on the *number* of propagules which reach and settle into a habitat. However, as implied by the desperate larva hypothesis described above, not all settling larvae are of the same quality. Compared to younger larvae, older larvae may be more likely to settle in poorer-quality habitats and, therefore, may have a lower chance of survival overall. In other words, older larvae with fewer nutritional reserves are of lower quality than younger larvae with more nutritional reserves. The dramatic changes in tissue, habitat and trophic mode associated with metamorphosis suggest that any differences among larvae before metamorphosis will be nullified after metamorphosis. However, in the 1980s Jan Pechenik and others found that events during the larval phase (for example, extended periods of swimming) affected performance in the adult stage (Pechenik and Eyster 1988). This link between larval experience and adult performance has been demonstrated in a range of organisms, and a range of larval experiences appear to affect post-metamorphic performance including pollution, salinity stress and the length of the larval period (Pechenik 2006). For example, Marshall et al. (2003) showed that ascidian larvae which were forced to swim for 3 h had lower post-metamorphic growth rates than larvae which were allowed to settle immediately. In addition to larval experience, larval size has also been shown to affect post-metamorphic performance, and a range of studies now suggest that bigger settlers have higher chances of surviving and growing than do smaller settlers (e.g. Marshall et al. 2006). These effects are not limited to species with non-feeding larvae, since larval nutrition in planktotrophs can also affect post-metamorphic performance (Phillips 2002).

12.8 Summary

The processes occurring immediately before, during and after the settlement of benthic invertebrate larvae and macroalgal propagules are fundamental to determining the distribution, abundance and dynamics of adult populations on marine hard substrata. Complex behavioural mechanisms, especially in invertebrate larvae, and the interaction with a range of physical and biological factors determine where settlement occurs and at what intensity. High levels of mortality immediately after settlement may modify these patterns. In this way, settlement and recruitment set the scene for the complexity of adult interactions which follow.

References

Aberg P (1992) Size based demography of the seaweed *Ascophyllum nodosum* in stochastic environments. Ecology 73:1488–1501

Aberg P, Pavia H (1997) Temporal and multiple scale variation in juvenile and adult abundance of the brown alga *Ascophyllum nodosum*. Mar Ecol Prog Ser 158:111–119

Amsler CD, Neushul M (1990) Nutrient stimulation of spore settlement in the kelps *Pterygophora california* and *Macrocystis pyrifera*. Mar Biol 107:297–304

Amsler CD, Reed DC, Neushul M (1992) The microclimate inhabited by macroalgal propagules. Br Phycol J 27:253–270

Barnes H (1956) *Balanus balanoides* (L.) in the Firth of Clyde: the development and annual variation in the larval population and the causative factors. J Anim Ecol 25:72–84

Behrens S (1972) The role of wave impact and desiccation on the distribution of *Littorina sitkana* Philippi. Veliger 15:129–132

Berntsson KM, Jonsson PR, Lejhall M, Gatenholm P (2000) Analysis of behavioural rejection of micro-textured surfaces and implications for recruitment by the barnacle *Balanus improvisus*. J Exp Mar Biol Ecol 251:59–83

Berntsson KM, Jonsson PR, Larsson AI, Holdt S (2004) Rejection of unsuitable substrata as a potential driver of aggregated settlement in the barnacle *Balanus improvisus*. Mar Ecol Prog Ser 275:199–210

Botello G, Krug PJ (2006) Desperate larvae revisited: age, energy and experience affect sensitivity to settlement cues in larvae of the gastropod *Alderia sp*. Mar Ecol Prog Ser 312:149–159

Branch G (1975) Ecology of *Patella* species from the Cape Peninsula. South Africa. IV. Desiccation. Mar Biol 32:179–188

Caffey HM (1985) Spatial and temporal variation in settlement and recruitment of intertidal barnacles. Ecol Monogr 55:313–332

Caley M, Carr M, Hixon M, Hughes T, Jones G, Menge B (1996) Recruitment and the local dynamics of open marine populations. Annu Rev Ecol Syst 27:477–500

Caselle J (1999) Early post-settlement mortality in a coral reef fish and its effect on local population size. Ecol Monogr 69:177–194

Chesson P (1998) Recruitment limitation: a theoretical perspective. Aust J Ecol 23:234–240

Connell JH (1985) The consequences of variation in initial settlement vs. post-settlement mortality in rocky intertidal communities. J Exp Mar Biol Ecol 93:11–43

Connolly SR, Menge BA, Roughgarden J (2001) A latitudinal gradient in recruitment of intertidal invertebrates in the northeast Pacific Ocean. Ecology 82:1799–1813

Crisp DJ, Meadows PS (1962) The chemical basis of gregariousness in cirripedes. Proc R Soc B 156:500–520

Crisp DJ, Meadows PS (1963) Adsorbed layers: the stimulus to settlement in barnacles. Proc R Soc B 158:364–387

Denley EJ, Underwood AJ (1979) Experiments on factors influencing settlement, survival, and growth of two species of barnacles in New South Wales. J Exp Mar Biol Ecol 36:269–293

Doherty P (1981) Coral reef fishes: recruitment limited assemblages? In: Proc Int Coral Reef Symp 4:465–470

Dudgeon S, Petraitis PS (2005) First year demography of the foundation species, *Ascophyllum nodosum*, and its community implications. Oikos 109:405–415

Elkin C, Marshall DJ (2007) Desperate larvae: influence of deferred costs and habitat requirements on habitat selection. Mar Ecol Prog Ser 335:143–153

Fraschetti S, Terlizzi A, Benedetti-Cecchi L (2005) Patterns of distribution of marine assemblages from rocky shores: evidence of relevant scales of variation. Mar Ecol Prog Ser 296:13–29

Gaines S, Brown S, Roughgarden J (1985) Spatial variation in larval concentration as a cause of spatial variation in settlement for the barnacle, *Balanus glandula*. Oecologia 67:267–272

Gosselin LA, Qian P (1997) Juvenile mortality in benthic marine invertebrates. Mar Ecol Prog Ser 146:265–282

Hawkins SJ, Hartnoll RG (1982) Settlement patterns of *Semibalanus balanoides* (L.) in the Isle of Man (1977-1981). J Exp Mar Biol Ecol 62:271–283

Hawkins SJ, Hartnoll RG (1983) Grazing of intertidal algae by marine invertebrates. Oceanogr Mar Biol Annu Rev 21:195–282

Herbert RJH, Hawkins SJ (2006) Effect of rock type on the recruitment and early mortality of the barnacle *Chthamalus montagui*. J Exp Mar Biol Ecol 334:96–108

Hills JM, Thomason JC (1996) A multi-scale analysis of settlement density and pattern dynamics of the barnacle *Semibalanus balanoides*. Mar Ecol Prog Ser 138:103–115

Hoffmann AJ, Santelices B (1991) Banks of algal microscopic forms: hypotheses on their functioning and comparisons with seed banks. Mar Ecol Prog Ser 79:1–2

Hughes TP, Baird AH, Dinsdale EA, Moltschaniwskyj NA, Pratchett MS, Tanner JE, Willis BL (1999) Patterns of recruitment and abundance of corals along the Great Barrier Reef. Nature 397:59–63

Hunt HL, Scheibling RE (1996) Physical and biological factors influencing mussel (*Mytilus trossulus, M. edulis*) settlement on a wave-exposed rocky shore. Mar Ecol Prog Ser 142:1–3

Jenkins SR, Norton TA, Hawkins SJ (1999) Settlement and post-settlement interactions between *Semibalanus balanoides* (L.) (Crustacea: Cirripedia) and three species of fucoid canopy algae. J Exp Mar Biol Ecol 236:49–67

Jenkins SR, Aberg P, Cervin G, Coleman RA, Delany J, Della Santina P, Hawkins SJ, LaCroix E, Myers AA, Lindegarth M, Power AM, Roberts MF, Hartnoll RG (2000) Spatial and temporal variation in settlement and recruitment of the intertidal barnacle *Semibalanus balanoides* (L.) (Crustacea: Cirripedia) over a European scale. J Exp Mar Biol Ecol 243:209–225

Jenkins SR, Murua J, Burrows MT (2008) Temporal changes in the strength of density-dependent mortality and growth in intertidal barnacles. J Anim Ecol 77:573–584

Jones NS (1948) Observations and experiments on the biology of *Patella vulgata* at Port St. Mary, Isle of Man. Proc Trans Lpool Biol Soc 56:60–77

Kendall MA, Bowman RS, Williamson P, Lewis JR (1985) Annual variation in the recruitment of *Semibalanus balanoides* on the North Yorkshire coast. J Mar Biol Assoc UK 65:1009–1030

Keough MJ, Downes BJ (1982) Recruitment of marine invertebrates: the role of active larval choices and early mortality. Oecologia 54:348–352

Knight-Jones EW (1951) Gregariousness and some other aspects of settling behaviour of *Spirobis*. J Mar Biol Assoc UK 30:201–222

Knight-Jones EW (1953) Laboratory experiments on gregariousness during settling in *Balanus balanoides* and other barnacles. J Exp Biol 30:584–598

Krug P (2001) Bet-hedging dispersal strategy of a specialist marine herbivore: a settlement dimorphism among sibling larvae of *Alderia modesta*. Mar Ecol Prog Ser 213:177–192

Lagos NA, Tapia FJ, Navarrete SA, Castilla JC (2007) Spatial synchrony in the recruitment of intertidal invertebrates along the coast of central Chile. Mar Ecol Prog Ser 350:29–39

Larsson AI, Jonsson PR (2006) Barnacle larvae actively select flow environments supporting post-settlement growth and survival. Ecology 87:1960–1966

Leonard GH (1999) Positive and negative effects of intertidal algal canopies on recruitment and survival of barnacles. Mar Ecol Prog Ser 178:241–249

Leonard GH (2000) Latitudinal variation in species interactions: a test in the New England rocky intertidal zone. Ecology 81:1015–1030

Lubchenco J (1983) *Littorina* and *Fucus*: effects of herbivores, substratum heterogeneity, and plant escapes during succession. Ecology 64:1116–1123

Marshall D, Pechenik J, Keough M (2003) Larval activity levels and delayed metamorphosis affect post-larval performance in the colonial, ascidian *Diplosoma listerianum*. Mar Ecol Prog Ser 246:153–162

Marshall DJ, Cook CN, Emlet RB (2006) Offspring size effects mediate competitive interactions in a colonial marine invertebrate. Ecology 87:214–225

McCook LJ, Chapman ARO (1997) Patterns and variations in natural succession following massive ice-scour of a rocky intertidal seashore. J Exp Mar Biol Ecol 214:121–147

McQuaid CD, Lindsay TL (2007) Wave exposure effects on population structure and recruitment in the mussel *Perna perna* suggest regulation primarily through availability of recruits and food, not space. Mar Biol 151:2123–2131

Menge BA (1991) Relative importance of recruitment and other causes of variation in rocky intertidal community structure. J Exp Mar Biol Ecol 146:69–100

Menge BA (2000) Recruitment vs. postrecruitment processes as determinants of barnacle population abundance. Ecol Monogr 70:265–288

Miller KM (1989) The role of spatial and size refuges in the interaction between juvenile barnacles and grazing limpets. J Exp Mar Biol Ecol 134:157–174

Minchinton TE (1997) Life on the edge: conspecific attraction and recruitment of populations to disturbed habitats. Oecologia 111:45–52

Minchinton TE, Scheibling RE (1993) Free space availability and larval substratum selection as determinants of barnacle population structure in a developing rocky intertidal community. Mar Ecol Prog Ser 95:233–244

Navarrete SA, Wieters EA (2000) Variation in barnacle recruitment over small scales: larval predation by adults and maintenance of community pattern. J Exp Mar Biol Ecol 253:131–148

O'Riordan RM, Arenas F, Arrontes J, Castro JJ, Cruz T, Delany J, Martinez B, Fernandez C, Hawkins SJ, McGrath D, Myers AA, Oliveros J, Pannacciulli FG, Power AM, Relini G, Rico JM, Silva T (2004) Spatial variation in the recruitment of the intertidal barnacles *Chthamalus montagui* southward and *Chthamalus stellatus* (Poli) (Crustacea: Cirripedia) over an European scale. J Exp Mar Biol Ecol 304:243–264

Osman RW, Whitlatch RB, Malatesta RJ (1992) Potential role of micropredators in determining recruitment into a marine community. Mar Ecol Prog Ser 83:35–43

Pawlik JR (1992) Chemical ecology of the settlement of benthic marine invertebrates. Oceanogr Mar Biol Annu Rev 30:273–335

Pechenik J (2006) Larval experience and latent effects—metamorphosis is not a new beginning. Integr Comp Biol 46:323–333
Pechenik J, Eyster L (1988) Influence of delayed metamorphosis on growth and metabolism of young *Crepidula fornicata* (Gastropoda) juveniles. Biol Bull 176:14–24
Penin L, Adjeroud M, Pratchett MS, Hughes TP (2007) Spatial distribution of juvenile and adult corals around Moorea (French Polynesia): implications for population regulation. Bull Mar Sci 80:379–389
Phillips NE (2002) Effects of nutrition-mediated larval condition on juvenile performance in a marine mussel. Ecology 83:2562–2574
Qian PY, Lau SCK, Dahms HU, Dobretsov S, Harder T (2007) Marine biofilms as mediators of colonization by marine macroorganisms: implications for antifouling and aquaculture. Mar Biotechnol 9:399–410
Raimondi PT (1988a) Rock type affects settlement, recruitment, and zonation of the barnacle *Chthamalus anisopoma* Pilsbury. J Exp Mar Biol Ecol 123:253–267
Raimondi PT (1988b) Settlement cues and determination of the vertical limit of an intertidal barnacle. Ecology 69:400–407
Raimondi P, Keough M (1990) Behavioural variability in marine larvae. Aust J Ecol 15:427–437
Reed DC, Foster MS (1984) The effect of canopy shading on algal recruitment and growth in a giant kelp forest. Ecology 65:937–948
Reed DC, Laur DR, Ebeling AW (1988) Variation in algal dispersal and recruitment, the importance of episodic events. Ecol Monogr 58:321–335
Roughgarden J, Isawa Y, Baxter C (1985) Demographic theory for an open marine population with space limited recruitment. Ecology 66:54–67
Rowley RJ (1990) Newly settled sea-urchins in a kelp bed and urchin barren ground—a comparison of growth and mortality. Mar Ecol Prog Ser 62:229–240
Santelices B (1990) Patterns of reproduction, dispersal and recruitment in seaweeds. Oceanogr Mar Biol Annu Rev 28:177–276
Santelices B, Ojeda F (1984) Recruitment, growth and survival of *Lessonia nigrescens* (Phaeophyta) at various tidal levels in exposed habitats of central Chile. Mar Ecol Prog Ser 19:73–82
Schiel DR (1988) Algal interactions on shallow subtidal reefs in northern New Zealand: a review. N Z J Mar Freshw Res 22:481–489
Schiel DR, Foster MS (2006) The population biology of large brown seaweeds: ecological consequences of multiphase life histories in dynamic coastal environments. Annu Rev Ecol Evol S 37:343–372
Stoner DS (1990) Recruitment of a tropical colonial ascidian—relative importance of pre-settlement vs post-settlement processes. Ecology 71:1682–1690
Strathmann RR, Branscomb ES, Vedder K (1981) Fatal errors in set as a cost of dispersal and the influence of intertidal flora on set of barnacles. Oecologia 48:13–18
Sutherland JP (1990) Recruitment regulates demographic variation in a tropical intertidal barnacle. Ecology 71:955–972
Thompson R, Norton T, Hawkins S (1998) The influence of epilithic microbial films on the settlement of *Semibalanus balanoides* cyprids—a comparison between laboratory and field experiments. Hydrobiologia 376:203–216
Todd CD, Keough MJ (1994) Larval settlement in hard substratum epifaunal assemblages: a manipulative field study of the effects of substratum filming and the presence of incumbents. J Exp Mar Biol Ecol 181:159–187
Toonen R, Pawlik J (2001a) Foundations of gregariousness: a dispersal polymorphism among the planktonic larvae of a marine invertebrate. Evolution 55:2439–2454
Toonen R, Pawlik J (2001b) Settlement of the gregarious tube worm *Hydroides dianthus* (Polychaeta: Serpulidae). II. Testing the desperate larva hypothesis. Mar Ecol Prog Ser 224:115–131
Vadas RL, Wright WA, Miller SL (1990) Recruitment of *Ascophyllum nodosum*: wave action as a source of mortality. Mar Ecol Prog Ser 61:263–272
Vadas RL, Johnson S, Norton TA (1992) Recruitment and mortality of early post-settlement stages of benthic algae. Br Phycol J 27:331–351

Wahle RA, Incze LS (1997) Pre- and post-settlement processes in recruitment of the American lobster. J Exp Mar Biol Ecol 217:179–207

Wethey DS (1983) Spatial pattern in barnacle settlement—daily changes during the settlement season. Am Zool 23:1021–1021

Wethey DS (1986) Ranking of settlement cues by barnacle larvae: influence of surface contour. Bull Mar Sci 39:393–400

Wright JT, Williams SL, Dethier MN (2004) No zone is always greener: variation in the performance of *Fucus gardneri* embryos, juveniles and adults across tidal zone and season. Mar Biol 145:1061–1073

Chapter 13
Seasonal Dynamics

Josep-Maria Gili and Peter S. Petraitis

13.1 Introduction

Seasonality is a characteristic of nearly all ecosystems and, thus, organisms have evolved a bewildering array of adaptations to sense and cope with the environmental variation imposed by seasonal changes. Seasonality is ultimately driven by the declination of the earth, which in turn determines the annual variation in day length and amount of solar radiation arriving at the surface (which is not only visible light and is measured in kW/m^2). Annual cycles of oceanographic and atmospheric conditions—what we normally associate with the change in seasons—are powered by the input of energy from solar radiation.

These seasonal patterns can affect organisms and, thus, populations and communities in two fundamentally different ways. First, organisms can sense and respond to these annual cycles. This includes sensing changes not only in photoperiod, which are directly linked to declination, but also in conditions such as water and air temperatures and wave surge, which are driven by variation in total solar radiation. Based on sensory input, organisms alter biochemical and physiological processes, reallocate resources among storage, maintenance, growth and reproduction, set the timing of reproduction, and adjust behaviours.

Second, mass rates of processes such as carbon fixation and per capita rates of mortality and population growth can change seasonally without any adjustment of physiology or behaviour occurring. For example, the rate of production by photoautotrophs responds to changes in the amount of photosynthetically active light and the variation in nutrient levels, which in many high-latitude areas of the ocean are in excess in winter due to wind-driven circulations, and low in summer due to the establishment of thermoclines and pycnoclines. The kinetics of carbon fixation are almost completely driven by nutrient concentration and availability of light, and the seasonal pattern of primary production is thus not caused by responses to sensory cues. Moreover, seasonal changes in rates may not reflect a simple link directly from cause to effect. For example, the barnacle *Chthamalus stellatus* shows lower rates of mortality during the winter than in the spring and summer because it

M. Wahl (ed.), *Marine Hard Bottom Communities*, Ecological Studies 206,
DOI: 10.1007/978-3-540-92704-4_13,

is crowded out by its competitor *Semibalanus balanoides*, which grows more rapidly during the spring phytoplankton bloom (Connell 1961).

Generally, organisms react to a set of environmental conditions, rather than to a single factor, and so studies that address a combination of environmental factors, instead of single factors one at a time, are more likely to provide successful approaches to understanding effects of seasonality on individuals, populations and communities. However, it is often difficult to integrate the effects of multiple factors, as well as their interactive effects into a single study or experiment.

Clearly, it is impossible for us to provide a review of seasonality that is both brief and comprehensive, given the daunting problems of interactive effects, co-variation of environmental cues, and the idiosyncratic characteristics of different communities. Below, we address some of these seasonal patterns and bring to the forefront several overlooked aspects of seasonality in marine communities.

13.2 Causes, Cues and Clocks

The ecological effects of seasonality depend in part on understanding both the proximate and ultimate causes for seasonal shifts in an organism's biochemistry, physiology, timing of reproduction and behaviour. Investigations of proximate causes are usually seen as "how" questions (Alcock 1998). For example, how does rising temperature in the spring cause the development and release of barnacle larvae? In contrast, studies of ultimate causes address "why" questions (Alcock 1998)—why does rising temperature in spring cause the release of barnacle larvae? While the distinction between how and why questions is well understood, ignoring the issues raised by this distinction can lead to mixing up cause and effect, and this is especially relevant for ecological studies of seasonality. For example, there are dozens, if not hundreds of examples of annual cycles in the physiology and growth of marine organisms that appear to be cued by temperature. The common interpretation is that temperature is the cause, and the physiology or growth is the response. Yet, it is just as likely that the annual cycles are endogenous and temperature is a cue that resets the clock (i.e. Zeitgeber). Annual endogenous cycles, or circannual rhythms, are well studied in insects (e.g. Danks 2005) but have been largely overlooked in marine invertebrates (but see Brock 1975; Fong and Pearse 1992; but see Last and Olive 2004; Matrai et al. 2005). In particular, Brock's (1975) work on the hydroid *Campanularia flexuosa* is quite stunning; colonies that were held in constant darkness for 4 years and at three temperatures exhibited circannual rhythms in growth, development and colony longevity. While it is true that circannual cycles in marine systems have been overlooked, certainly part of the problem is the logistical difficulty of doing studies of circannual rhythms, which require following organisms for at least two cycles in a constant environment (Alcock 1998).

13.3 Identifying Drivers and Responses

Seasonal fluctuations in light, nutrients, temperature and food availability are intertwined as crucial environmental factors affecting the dynamics of hard-bottom communities. For photoautotrophs, variation in light and nutrients are key factors, while for heterotrophs it is temperature and food availability that are critical. Seasonal availability of light, mixing depth and nutrients are often mismatched. At high latitudes, for example, these mismatches cause the typical spring bloom in phytoplankton, followed in some systems by a second smaller increase in production in the late summer and autumn, fuelled by regenerated nutrients. Since solar radiation drives both the availability of photosynthetically active light and mixing depth, the boom and crash cycle of primary production varies with latitude. Mixing depths vary little over the year near the equator where there is little seasonal variation in solar radiation, while at high latitudes mixing depths vary considerably. In addition, the dynamics of algae from different groups and at different depths often depend on the seasonal cycle of different limiting factors. For example, nutrients tend to be the primary limiting factor for shallow-water algae, whereas light is the most limiting factor for deep-water algae. In addition, algal production peaks in spring in shallow-water communities and in summer in deep-water communities (Ballesteros 1991). Other sources of nutrients can also come into play—for example, the input of nutrients from terrestrial runoff (Valiela 1995).

Heterotrophs ultimately depend on the input from primary producers and, so, the seasonal patterns of the two groups are linked. In addition, rates of physiological processes vary seasonally with temperature cycles, especially for organisms in cold temperate waters. Below, we will discuss various effects of seasonal changes in temperature and food levels on physiology, behaviour and species interactions. Notably, an intrinsic feature of many temperate ecosystems is that temperature and food availability tend to be positively correlated. Thus, it is difficult to distinguish the relative importance of these two effects without undertaking experiments that control for both (e.g. Menge et al. 2008). To the extent that food quality varies seasonally, it often is not possible to manipulate both food availability and temperature at the same time, since the same level of food in different seasons may vary dramatically in quality.

Physiological processes, such as respiration rates of benthic sessile species, usually follow closely the seasonal pattern of the water temperature, rising during summer as temperature increases (e.g. Coma et al. 2002). In bivalves, the relationship between clearance rate and respiration has been long recognized (Jörgensen et al. 1986), and it has been shown that the increase in clearance rate with temperature may depend on a decline in viscosity (Petersen and Riisgård 1986). In contrast, Coma et al. (2002) observed that the respiration rate of gorgonian species, which are passive suspension feeders, did not exhibit any significant response to temperature. They found that the highest respiratory demand occurred in spring when temperature was not at its highest, and low respiration rates in summer when temperatures were at maximum. The high rates in the spring appear to be related to the seasonal pattern of investment in gonad development that overlaps with the period of greatest growth (Coma et al. 1995). This pattern of secondary production investment is related to the seasonal pattern of food uptake. The low respiration rate in summer is related to the reduction in respiration due to polyp contraction.

Regardless of the seasonal effects of temperature itself, rates of ingestion, assimilation and, ultimately, population growth depend on food availability. For suspension feeders, rates of ingestion and assimilation vary not only with seston concentration but also with quality and feeding strategy (Jordana et al. 2001; Rossi and Gili 2005). Generally, in cool temperate waters the spring bloom in primary production is followed by a summer increase in secondary production. In contrast, the abundances of ephemeral and perennial organisms in warm temperate seas, such as the Mediterranean, have been shown to respond to summer food shortages in the water column (Rossi et al. 2006). Usually, the reproductive period coincides with increases in secondary production (Coma et al. 1998).

Linkages between seston characteristics and availability and seasonal cycles in benthic organisms are common phenomena in littoral communities (Grémare et al. 1997), and seasonal variation in the biochemical composition of the tissue of a benthic organism can reveal a "record" of seasonal cycles in water column productivity. For example, Rossi et al. (2006) have shown that the protein, carbohydrate and lipid tissue concentrations in the common gorgonian *Paramuricea clavata* reflect seasonal fluctuations in seston quantity and quality. In *Heteroxenia fuscescens*, seasonal changes in biochemical composition were seen not only in zooxanthellae concentration, somatic tissue variations, and planulae release but also in the coral tissue itself (Ben-David-Zaslow and Benayahu 1999).

Patterns of seasonal dormancy are also common in temperate seas (Coma et al. 2000). For example, 3 years of data on the biochemical composition of Mediterranean gorgonians show periodic summer minima in energy storage of lipids and proteins (Rossi et al. 2006). These observations agree with the expected physiological changes associated with a summer energy shortage, and with an annual pattern of investment in growth and reproduction of the gorgonian species. Seasonal dormancy phenomena are often associated not only with energetic constraints but also with physiological adaptations to seasonal water temperature extremes (Sulak et al. 2007).

Many fish and mobile invertebrates move in response to food levels, temperature and shelter. For example, near-shore crabs in the North Atlantic move inshore in the spring and summer to the intertidal zone where mating occurs, often in dense aggregations, and then move into deeper offshore water during winter to avoid wave action and severe decreases in near-shore water temperatures (e.g. Rebach 1987; Hunter and Naylor 1993). In contrast, other species migrate offshore in the summer to avoid high temperatures, and into shallow habitats in the winter for food (SanVicente and Sorbe 1995).

Larval stages of many invertebrates and fish also use transitory seasonal habitats for shelter and feeding. During periods of high growth of foliage and thalli, meadows of macroalgae provide shelter and protection for the life history stages of many marine species that are most vulnerable to predation. Because such habitats are essential for the success of larval stages, seasonal variation in these habitats could drive the seasonal and inter-annual variations in adult populations that do not live in the larval habitat.

Seasonal patterns of recruitment can also affect community structure. For example, Underwood and Anderson (1994) found that fouling plates set out in summer were always dominated by Sydney Rock oysters (*Saccostrea commercialis*) within 5 months, while panels placed out in spring showed several alternative outcomes.

Seasonal variation in competitive interactions have been investigated extensively in intertidal systems and, in one of the best studies, Underwood (1984) showed that levels of intra- and interspecific competition among gastropods were more intense in spring–summer months than in autumn–winter months. The effects were correlated with chlorophyll a concentrations, which reflected microalgal abundance.

There are also seasonal aspects to top-down control by predators (e.g. Beal et al. 2001). For example, herbivorous fishes have an important effect on the organization, both in space and time, of shallow marine benthic communities (Horn 1989), although the pattern is highly variable. In tropical waters, fish control seaweed development and are essential for maintaining the temporal patterns (Hay 1991). In contrast, in temperate waters fish often are not as important and have less effect on algal abundance and seasonal variability, which is often controlled by the activity of other groups such as sea urchins (Verlaque 1990). In the Mediterranean, overgrazing by fishes has led to the formation of an assemblage seasonally dominated by an alga with chemical defences that disguise natural seasonal community patterns (Sala and Boudouresque 1997).

13.4 From Intertidal Habitats to Deep-Sea Communities

The relative importance of seasonal oceanographic and atmospheric conditions shifts with changes in depth. In intertidal communities, seasonal changes in atmospheric conditions have the largest impact because the seasonality of total solar radiation drives winds, waves, surface currents, temperature, upwelling of nutrients and other such environmental conditions that are tightly linked with atmospheric conditions. The direct effects of atmospheric conditions diminish with depth. This decline can be gradual, such as the exponential extinction of light with depth, or abrupt as with thermoclines and pycnoclines. With increasing depth, oceanographic processes become dominant and, although seasonal variation is quite low (e.g. water temperature throughout most of the deep sea remains between 1 and 2°C), there is one striking exception, which is the seasonal pattern of downward fluxes of organic material.

There is also a depth gradient in how the water column and benthos are seasonally linked. In shallow waters, the seasonal cycles of water column and benthos are closely linked, with processes such as sedimentation, resuspension, vertical mixing and nutrient fluxes coupling the planktonic and benthic systems. Benthic organisms also influence the dynamics of the water column by providing nutrients through regeneration in shallow and estuarine areas (Valiela 1995). In deeper water, benthic–pelagic coupling is asymmetrical, with the ecological processes of the deep-sea benthos and water column being fed by carbon fluxes from production in the surface layers.

Below, we summarize some of these seasonal patterns moving from the intertidal zone to deep-sea shelf communities. The overarching patterns are driven by light and the type of resource. At shallow locations where light can reach, the bottom-up processes of primary production are entrained by the seasonal input of light and nutrients. As the availability of light declines with depth, the seasonal input of detritus becomes the limiting factor in communities that are primarily composed of heterotrophic organisms.

Communities inhabiting intertidal shores face diurnal shifts between oceanographic and atmospheric conditions because the shore alternates between marine and terrestrial habitats as the tide rises and falls. Marine organisms living higher on the shore are faced not only with a more terrestrial environment than their lower shore counterparts, but also a set of seasonal conditions strongly driven by solar radiation and atmospheric conditions. The interaction between diurnal and seasonal cycles across the intertidal zone can lead to complex patterns of distribution and abundance along shorelines in a particular region. Nevertheless, some general seasonal patterns are evident at a regional scale. First, seasonal conditions in conjunction with the timing of low tides can have a profound effect on organisms by exposing these to extreme conditions. For example, the lowest tides in the Gulf of Maine, USA tend to occur near dusk or dawn, and so organisms are rarely exposed to midday sun in the summer but are often exposed to below freezing temperatures on winter mornings. In contrast, summer low tides in south-eastern Australia occur around midday and expose organisms to extraordinarily high temperatures. Second, wave surge varies seasonally and affects the types of organisms found on the shore and their distributions. Wave surge tends to expand the extent of the intertidal zone by lowering the rate of desiccation throughout the intertidal zone, enabling species to extend farther up onto the shore. Yet, wave surge can also force mobile animals to seek refuge and can limit the distribution of slow-moving species. The force of breaking waves or debris carried by waves (e.g. logs)—which varies seasonally—can damage and sweep away organisms on shores exposed to large waves and open new patches for colonization (Dayton 1971; Paine and Levin 1981).

As in the water column, most intertidal habitats in temperate regions exhibit seasonal peaks in standing stocks of algae in late spring through summer, with minima in winter (Murray and Littler 1984). In most communities, seasonal patterns also involve changes in community structure (Boudouresque 1971; Mann and Lazier 2006), and there is marked seasonality in growth rates in intertidal pools or shallow sublittoral habitats (Leukart 1994). In contrast, seasonal changes in community structure are usually small in tropical areas, compared to temperate and cold regions (Lubchenco et al. 1984).

Sublittoral habitats tend to be characterized by either seaweed-dominated communities or animal-dominated communities that are largely composed of encrusting colonial invertebrates. Seaweed-dominated communities often show a clear seasonal change from spring to summer, corresponding to the increase of light availability. However, production and biomass peaks take place in the summer, rather than in the spring, as is often found in shallower habitats. In most animal-dominated communities, there are fewer seasonal changes in biomass, production and composition and, often, the only detectable seasonal variation occurs in physiological processes and the timing of reproductive periods (Turón and Becerro 1992).

Shelf and deep-sea communities are not free of seasonal change even though the ocean's floor is generally a highly constant environment. Carbon and sediment inputs vary seasonally, and deep-sea organisms also show seasonal rhythms. Deep-sea benthic communities are fuelled by primary productivity in the surface waters and subsequent food transport to the deep-sea floor (Billett et al. 1983; Tyler 1988; Gooday 2002). Large carbon and particle fluxes to the deep have been reported in several areas such as the North Atlantic sponge banks (Witte et al. 1997), deep-coral banks (Roberts et al. 2006) and the Antarctic deep shelf (Smith et al. 2006). In addition, seasonal episodes of surface-water sinking, such as the winter cascades that deliver food-rich surface waters, have been linked to the development of benthic communities in submarine canyons (Canals et al. 2006). Deep hard-bottom communities are frequently seen at sites with locally accelerated currents and in areas of the continental slope where internal tidal waves enhance seabed food supply. Thus, seasonal trends in deep-water hydrodynamic patterns influence benthic temporal processes. Because of these seasonal inputs of carbon and particulates, there have been several suggestions that these fluxes could act as cues for breeding in benthic organisms (Tyler 1988; Gooday 2002). There is some evidence that this may be the case. For example, in the NE Atlantic it has been observed that gamete production and subsequent spawning in the deep coral *Lophelia pertusa* follows phytodetrital food fall (Roberts et al. 2006).

Finally, one of the long-lasting paradoxes in marine ecology is the occurrence of the pronounced seasonality on the Antarctic shelf, given that the seawater temperature there remains practically constant year round (Clarke 1988). Even though recruitment of many sessile invertebrates shows strong annual cycles, it is commonly thought that the high Antarctic fauna undergoes a period of quiescence in winter as a consequence of reduced food availability. The first inkling that this might not be entirely accurate arose after the discovery of the rich marine fauna that dwells on the continental shelves in the high Antarctic, with species richness comparable to that of temperate and tropical rocky shores (Gili et al. 2001). In addition, many species have rates of reproduction similar to those in more temperate regions, while other species can quickly occupy areas scraped clean by icebergs, presumably due to higher than expected growth rates (Teixidó et al. 2004). Experimental observations have furnished solid evidence supporting the view that species may not be as inactive as hitherto thought during the long Antarctic winter. Resuspension by tidal currents and the high nutritional quality of the seabed sediment after the summer (Isla et al. 2006) form "food banks" that extend over hundreds of kilometres (Mincks et al. 2005), enabling benthic trophic transfers to remain constant throughout the year.

13.5 Future Directions

Many aspects of seasonality have been fully investigated by marine ecologists, and the examples we have provided are but a fraction of the published literature on the subject. However, the interaction between circannual rhythms and long-term environmental change is one area that needs more study and that could provide new insights for marine ecology in general. Circannual rhythms are difficult to detect because studies of these endogenous cycles must be at least 2 years long and done under constant conditions. Thus, while it has been suggested that climate change could affect timing of reproduction, keystone predation and community structure in marine communities (e.g. Sanford 2002), it will be extremely difficult to separate cause and effect if a keystone predator or one or more competing species rely on circannual rhythms. To make matters even more complex, annual cycles in reproduction and behaviour are in many cases not only endogenous but also rely on Zeitgebers such as photoperiod, which will be unaffected by climate change. In these cases, organisms may commence reproduction, begin to migrate or switch feeding modes at the correct moment, based on photoperiod, but at the wrong moment based on the new conditions imposed by climate change. It is clear that disentangling circannual rhythms from environmentally driven cycles is critical but will be difficult and will require careful consideration of alternative hypotheses.

References

Alcock J (1998) Animal behavior: an evolutionary approach. Sinauer, Sunderland, MA

Ballesteros E (1991) Structure and dynamics of North-Western Mediterranean phytobenthic communities: a conceptual model. Oecol Aquat 10:223–242

Beal BF, Parker MR, Vencile KW (2001) Seasonal effects of intraspecific density and predator exclusion along a shore-level gradient on survival and growth of juveniles of the soft-shell clam, *Mya arenaria* L., in Maine, USA. J Exp Mar Biol Ecol 264:133–169

Ben-David-Zaslow R, Benayahu Y (1999) Temporal variation in lipid, protein and carbohydrate content in the Red Sea soft coral *Heteroxenia fuscescens*. J Mar Biol Assoc UK 79:1001–1006

Billett DSM, Lampitt RS, Rice AL, Mantoura RFC (1983) Seasonal sedimentation of phytoplankton to the deep-sea benthos. Nature 302:520–522

Boudouresque CF (1971) Contribution à l'étude phytosociologique des peuplements algaux des côtes varoises. Végétatio 22:83–184

Brock MA (1975) Circannual rhythms. 2. Temperature-compensated free-running rhythms in growth and development of marine cnidarian, *Campanularia flexuosa*. Comp Biochem Physiol A 51:385–390

Canals M, Puig P, Durrieu de Madron X, Heussner S, Palanques A, Fabres J (2006) Flushing submarine canyons. Nature 444:354–357

Clarke A (1988) Seasonality in the Antarctic marine ecosystem. Comp Biochem Physiol 90B:461–473

Coma R, Ribes M, Zabala M, Gili JM (1995) Reproduction and cycle of gonadal development in the Mediterranean gorgonian *Paramuricea clavata*. Mar Ecol Prog Ser 117:173–183

Coma R, Ribes M, Gili JM, Zabala M (1998)An energetic approach to the study of life-history traits of two modular colonial benthic invertebrates. Mar Ecol Prog Ser 162:89–103

Coma R, Ribes M, Gili JM, Zabala M (2000) Seasonality in coastal benthic ecosystems. Trends Ecol Evol 15:448–453

Coma R, Ribes M, Gili JM, Zabala M (2002) Seasonality of in situ respiration rate in three temperate benthic suspension feeders. Limnol Oceanogr 47:324–331

Connell JH (1961) Influence of interspecific competition and other factors on distribution of barnacle *Chthamalus stellatus*. Ecology 42:710–723

Danks HV (2005) How similar are daily and seasonal biological clocks? J Insect Physiol 51:609–619

Dayton PK (1971) Competition, disturbance, and community organization—provision and subsequent utilization of space in a rocky intertidal community. Ecol Monogr 41:351–389

Fong PP, Pearse JS (1992) Evidence for a programmed circannual life-cycle modulated by increasing daylengths in *Neanthes limnicola* (Polychaeta, Nereidae) from Central California. Biol Bull 182:289–297

Gili JM, Coma R, Orejas C, López-González PJ, Zavala M (2001) Are Antarctic suspension feeding communities different from those elsewhere in the world? Polar Biol 24:273–485

Gooday AJ (2002) Biological responses to seasonally varying fluxes of organic matter to the ocean floor: a review. J Oceanogr 58:305–332

Grémare A, Amouroux JM, Charles F, Dinet A, Riaux-Gobin C, Baudart J, Medernach L, Bodiou JY, Vètion G, Colomines JC, Albert P (1997) Temporal changes in the biochemical composition and nutritional value of the particulate organic matter available to surface deposit-feeders: a two year study. Mar Ecol Prog Ser 150:195–206

Hay ME (1991) Fish-seaweed interactions on coral reefs: effects of herbivorous fishes and adaptations of their prey. In: Sale PF (ed) The ecology of fishes on coral reefs. Academic Press, San Diego, CA, pp 96–119

Horn MH (1989) Biology of marine herbivorous fishes. Mar Biol Annu Rev 27:167–272

Hunter E, Naylor E (1993) Intertidal migration by the shore crab *Carcinus maenas*. Mar Ecol Prog Ser 101:131–138

Isla E, Rossi S, Palanques A, Gili JM, Gerdes D, Arntz WE (2006) Biochemical composition of marine sediments from the eastern Weddell Sea (Antarctica): high nutritive value in a high benthic-biomass environment. J Mar Syst 60:255–267

Jordana E, Grémare A, Lantoine F, Courties C, Charles F, Amoroux JM, Vètion G (2001) Seasonal changes in the grazing of coastal picoplankton by the suspension-feeding polychaete *Ditrupa arietina* (O.F. Müller). J Sea Res 46:245–259

Jörgensen C, Mühlenberg F, Sten-Knudsen O (1986) Nature of relation between ventilation and oxygen consumption in filter feeders. Mar Ecol Prog Ser 29:73–88

Last KS, Olive PJW (2004) Interaction between photoperiod and an endogenous seasonal factor in influencing the diel locomotor activity of the benthic polychaete *Nereis virens Sars*. Biol Bull 206:103–112

Leukart P (1994) Field and laboratory studies on depth dependence, seasonality and light requirement of growth in three species of crustose coralline algae (Corallinales, Rhodophyta). Phycologia 33:281–290

Lubchenco J, Menge BA, Garrity SD, Lubchenco PJ, Ashkenas LA, Gaines SD, Emlet R, Lucas J, Strauss S (1984) Structure, persistence and the role of consumers in a tropical rocky intertidal community (Taboguilla Island, Bay of Panama). J Exp Mar Biol Ecol 78:23–73

Mann KH, Lazier JRN (2006) Dynamics of marine ecosystems. Biological-physical interactions in the oceans. Blackwell, Malden, MA

Matrai P, Thompson B, Keller M (2005) Circannual excystment of resting cysts of *Alexandrium* spp. from eastern Gulf of Maine populations. Deep-Sea Res II 52:2560–2568

Menge BA, Chan F, Lubchenco J (2008) Response of a rocky intertidal ecosystem engineer and community dominant to climate change. Ecol Lett 11:151–162

Mincks SL, Smith CR, DeMaster DJ (2005) Persistence of labile organic matter and microbioal biomass in Antractic shelf sediments: evidence of a sediment ';food bank'. Mar Ecol Prog Ser 300:3–19

Murray SN, Littler MM (1984) Analysis of seaweed communities in a disturbed rocky intertidal environement near Whites Point, Los Angeles, California, USA. Hydrobiologia 116/117:374–382

Paine RT, Levin SA (1981) Intertidal landscapes—disturbance and the dynamics of pattern. Ecol Monogr 51:145–178
Petersen JK, Riisgård HV (1986) Filtration capacity of the ascidian *Ciona intestinalis* (L) and its grazing impact in a shallow fjord. Mar Ecol Prog Ser 88:9–17
Rebach S (1987) Entrainment of seasonal and non-seasonal rhythms by the rock crab *Cancer irroratus*. J Crustacean Biol 7:581–594
Roberts JM, Wheeler AJ, Freiwald A (2006) Reefs of the deep: the biology and geology of cold-water coral ecosystems. Science 312:543–547
Rossi S, Gili JM (2005) Composition and temporal variation and of near-bottom seston in a Mediterranean coastal area. Estuar Coast Shelf Sci 65:385–395
Rossi S, Gili JM, Coma R, Linares C, Gori A, Vert N (2006) Temporal variation in protein, carbohydrate and lipid concentrations in *Paramuricea clavata* (Anthozoa, Octocorallia): evidence for summer-autumn feeding constraints. Mar Biol 149:643–665
Sala E, Boudouresque CF (1997) The role of fishes in the organization of a Mediterranean sublittoral community. 1. Algal communities. J Exp Mar Biol Ecol 212:25–44
Sanford E (2002) Community responses to climate change: links between temperature and keystone predation in a rocky intertidal system. In: Schneider SH, Root TL (eds) Wildlife responses to climate change: North American case studies. Island Press, Covelo, CA, pp 165–200
SanVicente C, Sorbe JC (1995) Biology of the suprabenthic mysid *Schistomysis spiritus* (Norman, 1860) in the southeastern part of the Bay of Biscay. Sci Mar 59:71–86
Smith CR, Mincks S, DeMaster DJ (2006) A synthesis of bentho-pelagic coupling on the Antarctic shelf: food banks, ecosystem inertia and global climate change. Deep-Sea Res II 53:875–894
Sulak KJ, Brooks RA, Randall MT (2007) Seasonal refugia and trophic dormancy in gulf sturgeon: test and refutation of the thermal barrier hypothesis. Am Fish Soc Symp 56:19–49
Teixidó N, Garrabou J, Gutt J, Arntz WE (2004) Recovery in Antarctic benthos after iceberg disturbance: trends in benthic composition, abundance and growth forms. Mar Ecol Prog Ser 278:1–16
Turón X, Becerro MA (1992) Growth and survival of several ascidian species from Northwestern Mediterranean. Mar Ecol Prog Ser 82:235–247
Tyler PA (1988) Seasonality in the deep sea. Oceanogr Mar Biol Annu Rev 26:227–258
Underwood AJ (1984) Vertical and seasonal patterns in competition for microalgae between intertidal gastropods. Oecologia 64:211–222
Underwood AJ, Anderson MJ (1994) Seasonal and temporal aspects of recruitment and succession in an intertidal estuarine fouling assemblage. J Mar Biol Assoc UK 74:563–584
Valiela I (1995) Marine ecological processes. Springer, Berlin Heidelberg New York
Verlaque M (1990) Relations entre *Sarpa salpa* (Linnaeus, 1758) (Téléostéens, Sparidae), les autres poissons brouteurs et le phytobenthos algal méditerranéen. Oceanol Acta 13:373–388
Witte U, Brattegard T, Graf G, Springer B (1997) Particle capture and deposition by deep-sea sponges from the Norwegian-Greenland Sea. Mar Ecol Prog Ser 154:241–252

Chapter 14
Disruption, Succession and Stochasticity

J. Timothy Wootton, Mathieu Cusson, Sergio Navarrete, and Peter S. Petraitis

Ecological disruptions play an important role in generating spatial and temporal patterns of species occurrence and coexistence. Disruptions are particularly obvious in marine benthic communities, where the scale of disruption and recovery can be particularly amenable to experimental investigation and quantification.

14.1 Definitions

Disruptions can be defined as interference with the orderly course of a process. Several different general mechanisms can be considered as ecological disruptions. Perhaps the most discussed is the concept of disturbance. Two general perspectives of disturbance have emerged in the literature. A view generally taken by terrestrial ecologists is that disturbance is a pulsed perturbation from an equilibrium state (White and Pickett 1985). This definition follows more closely the dictionary definition of disturbance, and is similar to the idea of ecological disruption. Implicit in this definition is that an identifiable equilibrium state of a system exists, and that the system tends to converge to that equilibrium state in the absence of disturbance.

A second definition of disturbance is a physical process that removes living biomass from an ecosystem (Sousa 1984b). This definition requires neither that an equilibrium reference point be identified, nor that the system must reach the equilibrium point in the absence of disturbance, and it avoids the difficulty of identifying whether a perturbation is pulsed or not, which depends largely on the temporal perspective adopted by an investigator. Additionally, this definition enables empirical quantification through demographic analysis of known individuals in conjunction with detailed measurements of the physical environment, or observations of signs associated with a physical process (e.g., impressions left by ice scour, changed orientation of boulders, spatial patterning of mortality, characteristic patterns of body part remnants). Experimental studies can also be implemented in which investigators impose (e.g., Petraitis and Dudgeon 1999), or in some cases remove (e.g., Sousa 1979b), mortality patterns on the system to infer effects of disturbance. For these reasons, we will consider disturbance a physical process that removes living biomass from an ecosystem, and use this definition for the remainder of the chapter.

M. Wahl (ed.), *Marine Hard Bottom Communities*, Ecological Studies 206,
DOI: 10.1007/978-3-540-92704-4_14, © Springer-Verlag Berlin Heidelberg 2009

14.1.1 Anthropogenic Versus Natural Disturbance

Another perspective on disturbance makes a distinction between perceived origins. Disturbance can be considered anthropogenic (derived from human activity) or "natural" (derived without human intervention). In some cases, making this distinction leads to a fundamental conflict between the perturbation-based vs. mortality-based definitions described above; "natural" disturbance, when viewed over a sufficiently long timescale, may be considered as a chronic process, whereas anthropogenic changes may be considered a recent novel event. Hence, a human activity (e.g., building a breakwater) may be considered a disturbance that moves a system from a prior equilibrium, by reducing a process (e.g., wave impact) considered a disturbance in the mortality-based framework.

14.1.2 Physical Disturbance Versus Consumers

By focusing on physical processes as drivers of disturbance, we are explicitly excluding consumer–resource interactions as a type of disturbance. A key difference between the two is that negative feedback is expected between consumers and their resources, but not for disturbance; eating prey contributes to increased fitness of the consumer, thereby increasing consumer pressure, whereas the physical processes creating disturbance are generally unaffected by the populations they impact. However, under some circumstances, it may be practical to treat consumer–resource interactions as another source of disturbance (e.g., Menge and Sutherland 1976). For instance, the two processes would be similar when the effects of prey on consumers are weak, such as when consumers are extreme generalists, when pronounced interference interactions strongly regulate predator populations, or when dispersal scales of predators are much greater than the scale of the study. Several models, however, propose that the relative importance of predation and disturbance will be observable at different ends of a stress gradient (Menge and Sutherland 1976).

Organisms are more clearly agents of disturbance when they are ecosystem engineers (Jones et al. 1994) with activities changing the physical environment in ways that kill other organisms. For example, basking pinnipeds or human visitors crush intertidal organisms as they move across the shoreline, or kick past a reef while diving (Hawkins and Roberts 1993; Keough and Quinn 1998). In other cases, physical disturbance is closely linked to the feeding mechanism employed by the predator while foraging, such as rays, sea otters, and gray whales that burrow, dig, or turn over rocks while foraging, smothering or crushing non-prey species in the process (Oliver and Slattery 1985), and species such as gulls that actively rip algae from the rocks while searching for hiding prey, or building nests. Canopy algae can also be a source of disturbance through abrasion or "whiplash" as they are moved around by water motion, and dislodge or kill other organisms (Dayton 1971). In all these cases, there is no clear benefit for the individual species creating the mortality.

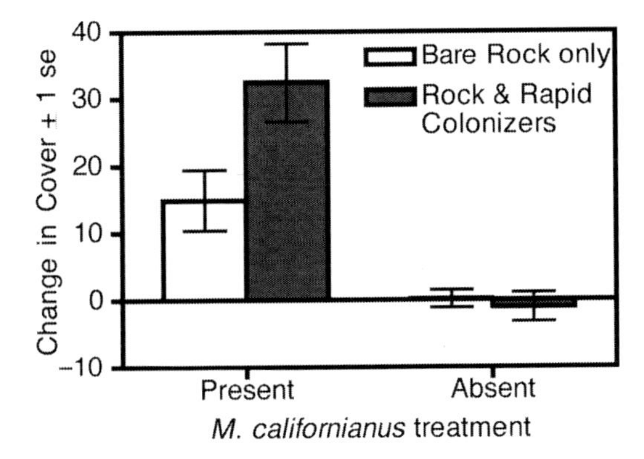

Fig. 14.1 Effect of wave disturbance on rocky intertidal plots with the dominant mussel *Mytilus californianus* present, and on immediately adjacent experimental plots where all *M. californianus* were chronically removed (see Wootton 2004). Experimental plots were dominated by coralline algae (*Corallina vancouveriensis*), acorn barnacles (*Semibalanus cariosus*), and goose barnacles (*Pollicipes polymerus*). Data are change in (%) cover of bare rock, or of bare rock and species that rapidly colonize bare rock (ephemeral algae and *Balanus glandula*) in plots between censuses spanning a disturbance event. Strong disturbance mortality of mussels is not reflected in mortality of other species in the same area

Aside from biotic sources of disturbance, numerous physical mechanisms commonly create disturbances in marine environments (reviewed in Sousa 1984b, 2001).

A critical aspect of disturbance that is sometimes ignored is that, within the same area and at the same time, the disturbance agent may affect species or individuals differentially. This is distinct from the abilities of different species to recover in specific disturbance regimes. For example, wave disturbance can strongly impact populations of large mussels, yet leave other species essentially intact (Fig. 14.1), and typhoons more strongly affect mature, highly developed coral reefs (Connell et al. 1997). This feature can have profound impacts on ecosystems by introducing species differences that may facilitate coexistence (Wootton 1998), affect recovery patterns (Connell and Slatyer 1977), and contribute to spatial patterning (Wootton 2001). On a practical level, broad gradients of disturbance assumed in many theoretical treatments (e.g., Connell 1978) may be difficult to identify empirically, because they do not act uniformly on all species. Thus, the observed responses to disturbance might in part reflect the fact that some species or types of individuals are not really disturbed, instead of reflecting their ability to better recover from disturbance.

14.1.3 Other Ecological Disruptions

Although disturbance is most often considered within the broader topic of environmental disruptions, other ecological processes might also be considered disruptions. Changes in environmental stress can generally cause disruptions. If disturbance is defined as a physical process that induces mortality, then stress can be defined as a physical process that reduces physiological performance without immediately causing death, although it may do so when combined with other factors such as food limitation over longer time frames (e.g., Wootton 1998). In some cases, disturbance and stress may have the same general effects on ecosystems, particularly when stress operates in a density-independent manner on organisms (e.g., by changing metabolic rates). Stress impacts may differ fundamentally, however, if they change physiological rates in ways that alter interactions among species. For example, high temperatures could rapidly kill predators by preventing these from maintaining water balance, or could change physiological rates at which predators capture and digest food, fundamentally altering interactions with their prey. Because the latter alters feedbacks within the food web, its consequences could be quite different than those of simple reductions in abundance of the predator.

Changes in water flow regimes are another ecological disruption of particular relevance to marine systems. There has been much recent interest in this topic, as models suggest that global warming may substantially alter large-scale ocean currents and storm patterns (IPCC 2001; ACIA 2004). Some aspects of water flow are easily related to disturbance and stress concepts: changes in storm intensity may alter wave disturbance, and changes in water temperatures associated with different currents can affect physiological stress levels and some types of disturbance (e.g., the prevalence of ice scour or freeze events). Changes in water movement can also disrupt key marine processes in ways that are less easily linked to direct mortality or physiological performance, because of its central role in material transport. Disruption of current and upwelling patterns may change the rate of recruitment of organisms to different areas (Farrell et al. 1991), gene flow, or the aggregation and delivery of nutrients and food resources to local shorelines (Bustamante et al. 1995; Menge et al. 1997).

14.1.4 Stochasticity

If disruption can be defined as interference in the orderly course of a process, then it is also relevant to consider the concept of stochasticity. Stochasticity can be defined as a process lacking any particular order or plan—hence, it underlies the concept of disruption. There is an emerging body of theoretical literature on stochastic processes and their implications that may help to understand the role of disruptions in marine systems. Stochasticity arises from two fundamental sources (Lande 1993). First, demographic stochasticity and related concepts (e.g., genetic drift) arise as a sampling phenomenon involving a limited number of discrete events, analogous to flipping a coin. For example, the annual probability of survival by sea otters is some fractional value between 0 and 1, but an individual sea otter either lives or dies. A similar situation holds for the probability of a physical event occurring (e.g., a volcanic eruption, or a tsunami). Although such stochasticity may not usually be important to ecosystems, because it is a factor only when the event is rare, in some instances, such as in the examples provided, the impacts of rare components are sufficiently strong that their variability can serve as an important disruption. Second, environmental stochasticity arises from variation over time in factors external to a system, which cause deviations from mean values of parameters such as demographic rates, interaction strengths among species, and supply rates of recruits and nutrients. Because it is externally derived, environmental stochasticity is typically expected to be independent of local system conditions. The perception of environmental stochasticity depends on the scale with which a system is viewed. The source of stochasticity can be spatially external to the system, such as physical forcing by large-scale climatic processes or movement of organisms from other spatial locales, or from changes in local abiotic or biotic conditions that have not been included in the conceptualization of the system. Hence, changing the spatial scale, and the suite of variables considered as a system can affect the perceptions of the role of stochastic processes.

More generally, it is important to appreciate that the scale of observation affects the perception of disruptions, and how they are studied. A seemingly persistent set of conditions in ecological time (e.g., present sea-level conditions, chronic nutrient runoff) may be perceived as pulsed disruptions over geological time. Similarly, wave disturbances, mudslides or lava flows represent massive unpredictable disturbances at local spatial scales (<km), but are regular features of the seascape at larger scales (e.g., >100s km). Hence, perceptions of the same system viewed at different scales may vary radically, with small-scale studies characterizing a system as being in a highly stochastic non-equilibrium state, whereas large-scale observation suggests that the same system is in a highly deterministic equilibrium state (De Angelis and Waterhouse 1987; Petraitis et al. 1989). The study of disruptions, particularly disturbances, has taken two polar perspectives on the problem, which have both contributed to our understanding of ecological communities: disruptions as unique events, and disruptions as chronically occurring processes. In the next sections we review these two perspectives.

14.2 Disruptions as Unique Events

Disruptions are most notable when they occur as intense events that wreak large-scale destruction on a system, such as hurricanes, earthquakes, landslides, volcanic eruptions, or ice ages. Frequently, these events have recurrence times on the scale of a human lifespan or longer, and hence draw particular attention because they are unusual situations. Such circumstances naturally lead to treating disturbances as unique events that introduce a historical aspect to the compositions of communities and ecosystem processes, resulting in a focus on the patterns and mechanisms of recovery from the disruption back to "normal" conditions. These issues underpin one of the earliest conceptual pillars of ecology, the study of ecological succession and ecosystem development (Cowles 1899; Gleason 1926; Clements 1936), and provide an important basic touchstone for considering the restoration of ecosystems following human disruption.

Succession has a variety of definitions. The narrow view of succession suggests that it is a highly deterministic pattern of species replacements following a disturbance event. A broader view defines succession as a non-random pattern of species replacement following disturbance. Although early investigators clearly appreciated that stochastic elements were present during recovery from disturbance (Cowles 1899), adoption of the narrow view of succession led to intense debate between advocates (e.g., Clements 1936), who tended to view patterns at large spatial scales, and detractors, who were impressed by the variable patterns they observed at local scales (e.g., Gleason 1926). Intensive study of recovery patterns now indicate that the broader view is compatible with both perspectives in the debate, as patterns of recovery are not strongly deterministic, but are clearly not random either (Paine and Levin 1981). In benthic marine systems, unpredictable recruitment from the plankton, and priority effects arising from indeterminate growth of many marine species generate an element of stochasticity in the recovery, while differences in life history traits, tolerances to physical conditions, and responses to interactions with other species lend some predictability to the recovery.

Given that recovery patterns are non-random, what are the mechanisms important in succession? One organizing framework for shedding light on this question was proposed by Connell and Slatyer (1977), who advanced a trichotomy of scenarios based on the relationship between species characteristic of early and later stages of succession. They noted that three qualitative relationships were possible: early species could have positive (facilitation), negligible (tolerance), or negative (inhibition) effects on later-stage species, and these relationships serve to identify mechanisms that are important in recovery from disturbance. This clear qualitative contrast also serves as a framework that enables succession to be readily probed with field experiments, a feature that benthic ecologists have used advantageously. Multiple patterns of recovery have been documented in marine systems, and any successional sequence may contain elements of multiple mechanisms (e.g., facilitation at one stage, inhibition at another).

Facilitation embodies the classical view of succession, and implies that beneficial interactions play a key role in disturbance recovery. For this reason, facilitation is often expected when physical conditions are harsh in the aftermath of a disturbance, where early species have both a high physiological tolerance to stress, and are capable of modifying the environment when present. For example, barnacles can be important for facilitating establishment of mussels and algae on rocky shores (e.g., Johnson 1992; Berlow 1999).

Tolerance patterns of succession imply an important role of life history differences in dispersal, growth, maturation, and reproductive rates among species, as neither negative nor positive effects of early species on later species are detectable. Many marine species exhibit distinct differences in life history characteristics (Thorson 1950), ranging from benthic diatoms that have lifecycles on the order of days, to brown algae, bivalves, and corals that take several years to mature, and have distinct annual reproductive cycles. Hence, it would not be surprising if, for instance, the early dominance of diatoms in many benthic successional sequences arises primarily from their fast lifecycle. Nevertheless, direct evidence for tolerance succession is scant in marine benthic systems. It is uncertain why this should be, and further study of this mechanism is required. One possible explanation is that recruitment in many species is dependent on transport via ocean currents, which may be highly unpredictable, thereby disrupting any clear successional pattern based on life history differences.

Inhibitory effects of early species on later-stage species in succession present a quandary: how can succession proceed to later species under these conditions? One solution to this problem is that other mortality factors may differentially affect early compared to later-stage species. For example, consumers may differentially prefer feeding upon early-stage species, releasing later species from inhibition. Consumers can be particularly effective at promoting succession following disturbance, because of a natural tendency for predator populations to lag behind prey populations over time. Their subsequent arrival and population increase may then release later species from inhibition, creating a non-random pattern of species replacements over time. Connell and Slatyer (1977) noted that inhibitory interactions are common among sessile organisms in general, and suggested that this mechanism of succession might be more common than previously suspected, because studies of succession often do not consider recovery patterns of mobile consumers following disturbance. Experimental studies, in which populations of early species were reduced and in which populations of consumers that feed on early species were reduced, have now demonstrated that inhibition patterns of succession occur frequently in marine benthic systems. Examples include grazers removing rapidly growing ephemeral algae, and favoring more resistant perennial algae (Sousa 1979a; Lubchenco 1983), and predators removing earlier-colonizing animals that inhibit later species (Wootton 2002).

Although understanding the relationship between early and later successional species has proven fruitful, other conceptual frameworks may also aid in identifying key mechanisms operating during recovery from disturbance. For example, if different phases in the recovery are characterized by dominance of different species, it may be profitable to ask what processes are important in the rise to dominance of particular species. This involves first colonization (life history traits, physical transport mechanisms) and then establishment (competition, facilitation, physical stress) processes, and the demise from dominance of species, which generally involves changes in mortality regimes, or recruitment failure (Farrell 1991; Wootton 2002). This rise-demise perspective identifies a richer suite of mechanisms that may be important in recovery from disturbance than does the Connell-Slatyer framework, and further emphasizes that key mechanisms may vary throughout the recovery process.

14.3 Disruption as a Chronically Recurring Process

When ecosystems are observed over sufficiently long time periods, even large intense disruptions occur multiple times. Given such observations, ecologists have more recently started asking how different disturbance regimes can affect community and ecosystem characteristics. In taking this perspective, it is useful to identify general characteristics of disruptions to understand their consequences better. Disruptions introduce at least three features to ecosystem structure and dynamics. First, adding disruptions generally leads to increases in average mortality experienced by organisms within the system, which can alter population dynamics and long-term persistence of interacting species by altering the potential equilibrium points of the system (Wootton 1998). Second, disruptions are usually pulsed events; hence, they add temporal variability to the system, which makes it less likely that ecosystems will strictly attain an equilibrium condition (Connell 1978; Huston 1979). Third, disruptions are generally limited in aerial extent; hence, they introduce spatial variability into the landscape, which can affect population viability and species coexistence (Paine and Levin 1981; Petraitis et al. 1989; Chesson 2000). A related approach to understanding disruptions involves characterizing their intensity (related to average mortality and spatial extent) and frequency (related to temporal variability; Sousa 1984a). The generality of these categorizations is scale-dependent, as a frequent disturbance event from the perspective of a long-lived organism, such as a seagull, might be quite different from that of a species with rapid population turnover, such as an amphipod.

The effects of recurring disturbance are most extensively explored in the intermediate disturbance hypothesis (IDH; Dayton 1971; Horn 1975; Connell 1978). This theory posits that species diversity is maximized at intermediate disturbance levels, with high disturbance eliminating species, and low-disturbance situations being characterized by high competitive exclusion. Benthic marine systems have played an important role in generating and experimentally testing aspects of the hypothesis. Observations of rocky intertidal areas disturbed by waves and floating logs revealed that species that would otherwise be displaced via competition persisted in these areas (Dayton 1971; Paine and Levin 1981). Long-term studies of Australian coral reefs emphasized the recurrent nature of tropical storm disturbances on shorelines, and identified unimodal relationships between coral diversity and time since disturbance (Connell 1978). Experiments and observations in coastal boulder fields supported the intermediate disturbance hypothesis by documenting increasing disturbance rates via rolling by waves with decreasing rock size, maximal richness of sessile species on intermediate-sized rocks, and increases in richness when small rocks were experimentally stabilized (Sousa 1979b).

Although empirical patterns in benthic marine communities support the IDH, more detailed studies are needed to fully understand the phenomenon, and when it occurs. Specifically, what mechanisms and components of disturbance are responsible for the patterns? Initially, an emphasis was placed on the introduction by disturbance of non-equilibrium conditions (Connell 1978; Huston 1979). Recent theoretical analyses indicate, however, that simply introducing stochastic non-equilibrium conditions while holding other components of disturbance constant either does not affect species coexistence, or reduces it, because the risk of stochastic extinction increases (Wootton 1998; Chesson 2000). In contrast, increases in average mortality rates by disturbance can promote species coexistence under some conditions. In general, disturbance needs to act more strongly on competitively dominant species to promote coexistence, which tends to equalize the fitness of different species, and its effectiveness also requires that a stabilizing mechanism be present, such as differentiation in resource use, external immigration, or disturbance effects that are density-dependent. The introduction of spatial heterogeneity by disturbance provides perhaps the most powerful mechanism for promoting coexistence, both by creating temporary refuges from dominant species in space if tradeoffs between colonization and competitive dominance exist, and by promoting spatial heterogeneity among sessile species, which can enable niche differentiation by species that use the sessile species as habitat. Detailed empirical studies are needed to evaluate which of these mechanisms are most important in natural communities.

Although good empirical support for the IDH exists from communities of sessile marine species, there are likely to be many situations where the theory does not hold. For example, the IDH is based on the assumption that systems are organized by interspecific competition, but there is ample evidence that consumer-resource interactions and positive interactions play important roles in marine ecosystems (Paine 1994; Bertness and Leonard 1997). Theoretical analysis indicates that in multi-trophic situations, the IDH often does not hold, particularly for mobile animals, and that it is highly dependent on where in the food web disturbance has its strongest effects (Wootton 1998). These results may explain why the best evidence for the IDH comes from studies of sessile species, but more extensive study of disturbance in a food web context is required.

Further understanding of the implications of disturbance on species diversity may come from distinguishing between its causes and effects, and considering potential feedbacks related to disturbance. For example, species may interfere with each other in a manner that is relevant only in the presence of disturbance. In benthic marine systems, cracks in the rock often provide refuges from disturbance caused by strong wave shear, scouring debris, and desiccation. As the space available in these cracks can be limited, increasing abundance may prevent all population members from taking refuge in these cracks, thereby increasing mortality at high densities in the presence of disturbance, and creating negative density dependence. Density of organisms may also affect susceptibility to wave disturbance by affecting adhesion ability or drag coefficients. In cases where density limits attachment area or strength, negative density or biomass dependence may arise. For example, as barnacles or mussels become abundant, the area of rock on which they can attach remains the same, but attachment area per unit biomass declines, making it easier for wave dislodgement to occur. Not all interactions confer increased susceptibility to disturbance, however. For example, increases in sessile species density can increase water trapping, reduce evaporation, and increase the boundary layer, thereby reducing disturbance from desiccation, salinity stress, and waves, and lowering the chances of species loss to disturbance (Palumbi 1985; Bertness and Shumway 1993).

Aside from affecting overall mortality levels, disturbances are a key component in generating large-scale landscape pattern. In many cases, spatial heterogeneity is created from spatially restricted intense physical conditions, such as when multiple waves converge on a local area, or a piece of floating log impacts the shore. In other cases, landscape patterning arises from the interplay of physical factors promoting disturbance, and interactions among organisms. For example, when organisms attach to each other and one individual is disturbed, the disturbance can be transmitted to neighbors, creating a locally enlarged area of disturbance that is dependent on organism density (Wootton 2001). Locally propagated disturbances may also arise in marine systems when organisms change the local hydrodynamic or physical environment, thereby increasing susceptibility of their neighbors to disturbance when they are removed (Paine and Levin 1981). Such localized processes can introduce density dependence into disturbance impacts, and create spatial patterning that self-organizes in predictable ways (Wootton 2001).

Ecological disruptions in general can also play an important role in shaping marine ecosystems in the presence of strong non-linear interactions among organisms. In such situations, multiple basins of attraction may exist to which the ecosystem tends to converge depending on initial conditions. Strong differences in body size or other traits among individuals of a species, coupled with strong trait-based shifts in species interactions can generate such non-linearities. Disruptions play an essential role in these situations by shifting the system from one basin of attraction to another, which can create strong spatial heterogeneity and facilitate coexistence (Wilson 1992). Within marine ecosystems, strong individual variation within species is the norm as a result of indeterminate growth and variation in larval-adult life forms, which may favor the existence of multiple basins of attraction (Berlow and Navarrete 1997). Although demonstrating regime shifts between alternative stable states is difficult, because of the long-term studies and experiments required (Connell and Sousa 1983; Bertness et al. 2002), several situations from marine systems suggest that they may occur. In South Africa, spiny lobsters once dominated nearshore benthic communities, but were reduced to low populations following anoxia events linked to harbor construction. Associated with the decline, predatory snails were released from lobster predation, and increased to become dominant species in the system. Following this ecological disruption, managers attempted to reestablish lobster dominance by releasing adults, but this program failed when the snails at high abundance were able to quickly overwhelm and eat their former predators (Barkai and McQuaid 1988). Another possible example occurs in New England rocky intertidal shores, where disturbance by ice and other sources creates patches of bare space on the rocks. Here, the community dominance of the area appears to depend on whether mussels or rockweed colonize the area first; established mussel beds abrade rockweed, and harbor grazers that prevent establishment of the seaweed, whereas rockweed "whiplashes" the rocks and harbors mussel predators, inhibiting the establishment of small mussels (Petraitis and Dudgeon 1999, 2005). Further research is needed into these types of disruption-induced regime shifts, as they may be very hard to predict a priori.

In summary, disruptions play an important role in shaping the patterns of ecosystems, and understanding their impacts is critical to developing appropriate responses to human impacts on the environment. Marine benthic communities have provided key insights into the mechanisms associated with disruptions and recovery, and are likely to play an important role in future investigations, because of their experimental tractability and relatively rapid dynamics through time.

References

ACIA (2004) Impacts of a warming Arctic: Arctic Climate Impact Assessment. Cambridge University Press, Cambridge
Barkai A, McQuaid C (1988) Predator-prey role reversal in a marine benthic ecosystem. Science 242:62–64
Berlow EL (1999) Strong effects of weak interactions in ecological communities. Nature 398:330–334
Berlow EL, Navarrete S (1997) Spatial and temporal variation in rocky intertidal community organization: lessons from repeating field experiments. J Exp Mar Biol Ecol 214:195–229
Bertness MD, Leonard GH (1997) The role of positive interactions in communities: lessons from intertidal habitats. Ecology 78:1976–1989
Bertness MD, Shumway SW (1993) Competition and facilitation in marsh plants. Am Nat 142: 718–724
Bertness MD, Trussell GC, Ewanchuk PJ, Silliman BR (2002) Do alternate stable community states exist in the Gulf of Maine rocky intertidal zone? Ecology 83:3434–3448
Bustamante RH, Branch GM, Eekhout S (1995) Maintenance of an exceptional intertidal grazer biomass in South Africa: subsidy by subtidal kelps. Ecology 76:2314–2329
Chesson P (2000) Mechanisms of maintenance of species diversity. Annu Rev Ecol Syst 31:343–366
Clements FE (1936) Nature and structure of the climax. J Ecol 24:252–284
Connell JH (1978) Diversity in tropical rain forest and coral reefs. Science 199:1302–1310
Connell JH, Slatyer RO (1977) Mechanisms of succession in natural communities and their role in community stability and organization. Am Nat 111:1119–1144
Connell JH, Sousa WP (1983) On the evidence needed to judge ecological stability and persistence. Am Nat 121:789–824
Connell JH, Hughes TP, Wallace CC (1997) A 30-year study of coral abundance, recruitment, and disturbance at several scales in space and time. Ecol Monogr 67:461–488
Cowles HC (1899) The ecological relations of the vegetation on the sand dunes of Lake Michigan. Bot Gaz 27:95–117, 167–202, 281–308, 361–391
Dayton PK (1971) Competition, disturbance, and community organization: the provision and subsequent utilization of space in a rocky intertidal community. Ecol Monogr 41:351–389
De Angelis DL, Waterhouse JC (1987) Equilibrium and nonequilibrium concepts in ecological models. Ecol Monogr 57:1–22
Farrell TM (1991) Models and mechanisms of succession—an example from a rocky intertidal community. Ecol Monogr 61:95–113
Farrell TM, Bracher D, Roughgarden J (1991) Cross-shelf transport causes recruitment to intertidal populations in central California USA. Limnol Oceanogr 36:279–288
Gleason HA (1926) The individualistic concept of the plant association. Bull Torrey Bot Club 53:7–26
Hawkins JP, Roberts CM (1993) Effects of recreational scuba diving on coral reefs: trampling on reef-flat communities. J Appl Ecol 30:25–30
Horn HS (1975) Markovian properties of forest succession. In: Cody ML, Diamond JM (eds) Ecology and evolution of communities. Harvard University Press, Cambridge, MA, pp 196–211
Huston M (1979) General hypothesis of species-diversity. Am Nat 113:81–101
IPCC (2001) Climate Change 2001: impacts, adaptation, and vulnerability. In: MacCarthy JJ (ed) Intergovernmental Panel on Climate Change (IPCC) working group II. Cambridge University Press, Cambridge
Johnson LE (1992) Potential and peril of field experimentation: the use of copper to manipulate molluscan herbivores. J Exp Mar Biol Ecol 160:251–262
Jones CG, Lawton JH, Shachak M (1994) Organisms as ecosystem engineers. Oikos 69:373–386
Keough MJ, Quinn GP (1998) Effects of periodic disturbances from trampling on rocky intertidal algal beds. Ecol Appl 8:141–161

Lande R (1993) Risks of population extinction from demographic and environmental stochasticity and random catastrophes. Am Nat 142:911–927
Lubchenco J (1983) *Littorina and Fucus*—effects of herbivores, substratum heterogeneity, and plant escapes during succession. Ecology 64:1116–1123
Menge BA, Sutherland JP (1976) Species diversity gradients: synthesis of the roles of predation, competition, and temporal heterogeneity. Am Nat 110:351–369
Menge BA, Daley BA, Wheeler PA, Dahlhoff E, Sanford E, Strub PT (1997) Benthic-pelagic links and rocky intertidal communities: bottom-up effects on top-down control? Proc Natl Acad Sci USA 94:14530–14535
Oliver JT, Slattery PN (1985) Destruction and opportunity on the sea floor: effects of gray whale feeding. Ecology 66:1965–1975
Paine RT (1994) Marine rocky shores and community ecology: an experimentalist's perspective. Ecology Institute, Oldendorf/Luhe, Germany
Paine RT, Levin SA (1981) Intertidal landscapes: disturbance and the dynamics of pattern. Ecol Monogr 51:145–178
Palumbi SR (1985) Spatial variation in an alga-sponge commensalism and the evolution of ecological interactions. Am Nat 126:267–274
Petraitis PS, Dudgeon SR (1999) Experimental evidence for the origin of alternative communities on rocky intertidal shores. Oikos 84:239–245
Petraitis PS, Dudgeon SR (2005) Divergent succession and implications for alternative states on rocky intertidal shores. J Exp Mar Biol Ecol 326:14–26
Petraitis PS, Latham RE, Niesenbaum RA (1989) The maintenance of species diversity by disturbance. Q Rev Biol 64:393–418
Sousa WP (1979a) Experimental investigation of disturbance and ecological succession in a rocky intertidal algal community. Ecol Monogr 49:227–254
Sousa WP (1979b) Disturbance in marine intertidal boulder fields: the nonequilibrium maintenance of species diversity. Ecology 60:1225–1239
Sousa WP (1984a) Intertidal mosaics: patch size, propagule availability, and spatially variable patterns of succession. Ecology 65:1918–1935
Sousa WP (1984b) The role of disturbance in natural communities. Annu Rev Ecol Syst 15:353–391
Sousa WP (2001) Natural disturbance and the dynamics of marine benthic communities. In: Bertness MD, Gaines SD, Hay ME (eds) Marine community ecology. Sinauer, Sunderland, MA, pp 85–130
Thorson G (1950) Reproductive and larval ecology of marine bottom invertebrates. Biol Rev 25:1–45
White PS, Pickett STA (1985) Natural disturbance and patch dynamics: an introduction. In: Pickett STA, White PS (eds) The ecology of natural disturbance and patch dynamics. Academic Press, New York, pp 3–13
Wilson DS (1992) Complex interactions in metacommunities, with implications for biodiversity and higher levels of selection. Ecology 73:1984–2000
Wootton JT (1998) Effects of disturbance on species diversity: a multitrophic perspective. Am Nat 152:803–825
Wootton JT (2001) Local interactions predict large-scale pattern in an empirically-derived cellular automata. Nature 413:841–843
Wootton JT (2002) Mechanisms of successional dynamics: consumers and the rise and fall of species dominance. Ecol Res 17:249–260
Wootton JT (2004) Markov chain models predict the consequences of experimental extinctions. Ecol Lett 7:653–660

Chapter 15
Changes in Diversity and Ecosystem Functioning During Succession

Laure M.-L.J. Noël, John N. Griffin, Paula S. Moschella, Stuart R. Jenkins, Richard C. Thompson, and Stephen J. Hawkins

15.1 Introduction

Biological succession occurs following disturbance when new space becomes available for colonisation. A sequence ensues during which assemblage composition and diversity change through time. This chapter summarises spatial and temporal changes of processes described by Wahl (1989), in addition to the modes of succession proposed by Connell and Slatyer (1977) and subsequent studies (e.g. Farrell 1991; Benedetti-Cecchi 2000a). Processes creating space for colonisation are described. We examine the early microbial phase of colonisation on natural rock and subsequent colonisation by macrobiota, including the changes in species diversity and the associated consequences for ecosystem functioning.

15.2 Concepts and Terminology

The term succession, first coined by Clement (1916), is currently defined as species replacement during recolonisation following a disturbance or the creation of new substrata (Sousa and Connell 1992). In marine habitats, it is predominantly driven by the availability of larvae and propagules, the migration of mobile organisms, and vegetative growth of neighbouring communities. Disturbance can be defined as any event which affects community structure by 'freeing' resources. On hard substrata, the limiting resource affected is usually space. Through disturbance, the impact of which depends on its frequency and magnitude, competitive species are prevented from monopolising space. In this way, a mosaic of patches of different successional age may be generated (Dethier 1984). Disturbance may lead to either primary or secondary succession. Primary succession is the colonisation of completely virgin substrata (e.g. associated with volcanic or erosive activity, displacement of hard man-made substrata or appearance of new biogenic substrata) or completely denuded substrata (e.g. after ice scour, ice melt or shifts in sediments covering rock). Secondary succession occurs on substrata which may still be partially occupied following physical disturbance (waves, ice or sediment scouring), or after the relaxation of grazing or predation pressure.

M. Wahl (ed.), *Marine Hard Bottom Communities*, Ecological Studies 206,
DOI: 10.1007/978-3-540-92704-4_15, © Springer-Verlag Berlin Heidelberg 2009

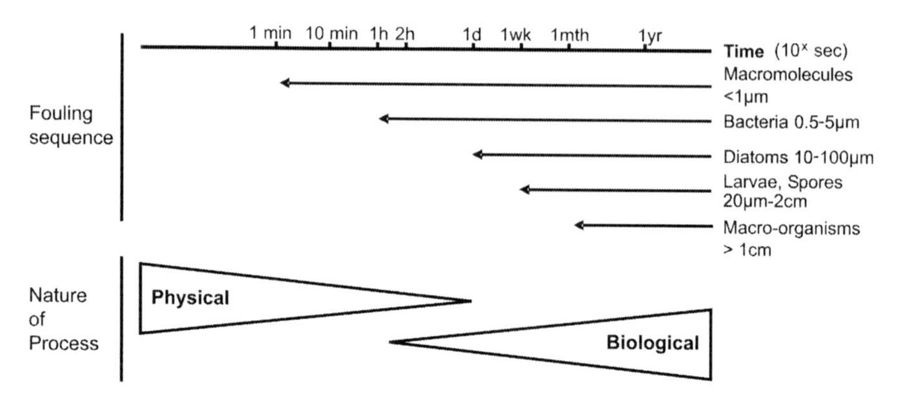

Fig. 15.1 A generalised schematic model of the colonisation sequence in a fouling community on submerged surfaces indicating the nature of the processes involved (modified from Wahl 1989 with permission). The typical size range of organisms is shown

Wahl (1989) summarised the initial events involved in succession and their timescales. Many of the early physicochemical events occur over very short time periods (seconds) and very small spatial scales (microns), and are followed by biological colonisation by bacteria and diatoms within 24 h (Fig. 15.1). Connell and Slatyer (1977) described three models of succession: *facilitation*, where early stages are necessary for subsequent development; *inhibition*, where early stages block the colonisation of later stages; and *tolerance*, where succession progresses due to the life history characteristics of the organisms, rather than positive or negative interactions. Following this seminal paper, many experimental studies demonstrated inhibition (see Sousa and Connell 1992 for review) and showed the importance of biological interactions in enabling succession to proceed by interrupting inhibitory processes (Sousa and Connell 1992). More recently, the role of facilitation has been reemphasised as both a process operating during succession and as a term to define positive interactions involved in structuring assemblages (Bertness and Callaway 1994). The models of Connell and Slatyer have been criticised for their oversimplified nature, and for offering little predictive power (Benedetti-Cecchi 2000a) but, despite limitations, the models are a very useful classification of the modes of succession and have had an enormous influence on subsequent experimental studies of succession.

In any successional sequence, different processes may act at different times (Farrell 1991). Facilitation can be important in the very early stages (Harris et al. 1984) but has also been demonstrated in late succession where long-lived habitat-forming or environment-ameliorating species are involved (e.g. Jenkins et al. 1999a). Inhibition can also occur in the earlier microbial phase (e.g. Wieczorek and Tood 1998) but may be more common in mid-succession. Both direct and indirect interactions (Benedetti-Cecchi 2000a; see Chap. 16 by Molis and da Gama) can modify the course of succession such that apparently simple systems can display a variety of successional trajectories. Berlow (1997) classified succession as “canalized”, when trajectories are predictable and convergent, or “contingent” when colonisation processes are more variable and divergent. Historically, succession was considered as a linear sequence moving towards a climax community (Clement 1916). However, experimental work on marine hard substrata has contributed significantly towards understanding succession as a continuous process with changing trajectories deflected by both physical and biological processes.

15.3 Creation of New Space

The creation of new space on marine hard substrata occurs through a variety of physical and biological means. Volcanic activity provides the most spectacular example of space creation, associated with the formation of completely new shorelines (e.g. Surtsey Island, Iceland; Frederiksen et al. 2001) and, hence, initiating primary succession. A similar process occurs when man-made structures are first immersed in the sea. On a smaller but more widespread scale, cleaving of sedimentary or metamorphic rocks due to erosional processes can also provide virgin surface for primary succession. Biogenic substrata can also provide new space and may be locally important (e.g. whale bones, inner side of bivalve shell after their death, newly moulted carapace, or seagrass blade protruding from its sheath during growth).

Secondary succession typically follows any disturbance event which partially removes existing biota. Such events can be physically generated due to wave action (Dethier 1984), heat stress (Underwood and Jernakoff 1984), transient anoxia (Lehvo and Back 2001), scouring by ice (Petraitis and Dudgeon 2005), rock dismantling by destructive fisheries (e.g. Bevilacqua et al. 2006), and by sediment (Airoldi 1998) or waterborne stones, logs or cobbles (Dayton 1971; Benedetti-Cecchi and Cinelli 1996). Biological disturbance can be caused by grazing, predation and algal sweeping (Hawkins and Hartnoll 1985; Jenkins et al. 1999c). Loss of dominant consumers can also lead to secondary succession, by providing the opportunity for colonisation by organisms which would otherwise be consumed. For example, numerous studies undertaken since the 1940s have shown how localised loss of patellid limpets can lead to macroalgal succession on intertidal rocky shores (see below).

15.4 Early Colonisation by Microorganisms

The microbial phase of colonisation in natural environments has been studied much less than subsequent colonisation by macrobiota (e.g. Sousa and Connell 1992), and more on artificial (e.g. Wahl 1989) than on natural substrata (e.g. Williams et al. 2000). Wahl (1989) proposed a model of temporal sequences of biofilm formation in seawater (Fig. 15.1). The initial phase consists of surface adsorption of organic compounds within seconds after submersion. This film modifies physicochemical properties at the surface/liquid interface, affecting the composition of primary bacteria which colonise the substratum within an hour. Subsequent microbial succession can follow different pathways, depending on environmental and biological factors such as propagule availability, seasonal variation, nutrient supply, competition and grazing (e.g. Callow and Callow 2006). In general, bacteria and cyanobacteria are the first colonisers, followed by diatoms and then other microorganisms (protozoans, fungi, yeasts). In freshwater habitats, a mature microbial community can develop in 3–4 weeks (Hamilton and Duthie 1984), whereas on intertidal rocky shores 10 months may be necessary (Underwood 1984; R.C. Thompson, personal observation).

The colonisation process is characterised by a diverse and apparently non-structured assemblage during the first week. After this period, the assemblage becomes dominated by one or only a few species (Niell and Varela 1984). After several days to weeks, the microbial community is enriched by the settlement of larvae and algal propagules, eventually leading to colonisation by macrobiota (Characklis 1981). Grazing (e.g. Thompson et al. 2004) and algal sweeping (e.g. Jenkins et al. 1999c) can limit the succession by macrobiota and lead to persistence of a microbial "lawn".

Biofilms can have both positive and negative interactions with settling macrobiota (see reviews by Callow and Callow 2000; Dobretsov et al. 2006). Diatoms can directly interact with macroalgae, enhancing or inhibiting growth. Diatoms can also be affected by the shading from macroalgae (Thompson et al. 2004). The relative importance of these effects can change depending on the composition of the diatom component in the biofilm (Huang and Boney 1985). Biofilms can provide cues for settlement of many invertebrate species (e.g. Satuito et al. 1997) but can also have an inhibitory effect (e.g. Wieczorek and Tood 1998). Barnacle cyprids may be attracted differently according to the tidal height at which the microbial film develops, to the age of the biofilm (e.g. Thompson et al. 1998), and to the interplay between biofilm formation and substrate nature (Faimali et al. 2004). However, other factors (e.g. the proximity of adult conspecifics) may overpower these subtle effects in field settings (e.g. Thompson et al. 1998).

5.5 Macrobiotic Succession on Rocky Shores

Following the microbial phase, diatoms, and filamentous and foliose algae rapidly monopolise cleared substrata exposed to natural light. Then, turf-forming algae or sessile invertebrates become established (e.g. *Corallina* spp., *Gelidium* spp., barnacles). Finally, late colonists including slow-growing perennial species with complex morphologies, such as canopy-forming fucoids and kelps, or large macroinvertebrates such as mussels appear (Benedetti-Cecchi 2000a). Although this is the usual pattern of colonisation, fucoids and other perennials can also arrive earlier in successional sequences (Hawkins 1981; Noël 2007).

Patterns of colonisation are highly variable and are influenced by a variety of physical and biological factors (Benedetti-Cecchi 2000b). The distance between propagule source and the site of recruitment can directly dictate the pattern of succession (Reed et al. 2000). Colonists typically arrive by settlement of propagules from remote sources (Dethier 1984) but also by vegetative reproduction from borders or by migration of animals (Underwood 1977). Recruitment of species with short-range dispersal relies on the local pool of larvae and propagules (Dayton 1973; Paine 1979). When space for colonisation is created during the main reproductive period of a given species, their propagules may recruit and dominate the subsequent succession (Benedetti-Cecchi and Cinelli 1996). In more diverse assemblages, the outcomes of competition between organisms will vary depending on their relative size and density: large species have an inhibiting effect by overgrowing, shading or bulldozing other species. At high densities, even small organisms may pre-empt space, leading to non-hierarchical competition (Benedetti-Cecchi 2000b).

Herbivores can have a major effect on succession (see Sousa and Connell 1992 for review). Depending on their intensity, grazers may remove most of the settling organisms (Dethier 1984) or can mediate the interaction between early and late colonisers, thereby accelerating or slowing succession. In this way, grazing effects can match all three models proposed by Connell and Slatyer (Farrell 1991). Seasonality in grazer abundance and feeding rate (e.g. Jenkins et al. 2001) can also influence the successional trajectory (Hawkins 1981). Gastropods and other grazers feed preferentially on ephemeral algae and, by removing these algae, they can reduce inhibitory effects (van Tamelen 1987). Some algae (e.g. calcareous forms) are better adapted to resist consumer pressure (Steneck and Watling 1982), whilst others escape grazing by rapid growth in order to attain a size-related refuge (Lubchenco 1983), or through chemical defence (Padilla and Allen 2000). Protection from consumers also occurs when non-palatable species provide associational defences for palatable species (Bertness and Callaway 1994). Herbivores can promote the establishment of non-palatable species (Benedetti-Cecchi 2000a) which otherwise would be out-competed by more palatable algae (Steneck 1982). Thus, consumer pressure can lead to different successional sequences, often interacting directly and indirectly with other species in the assemblage, and being modulated by the various life history traits of the species involved (Sousa and Connell 1992; Benedetti-Cecchi 2000a).

The inhibition model of Connell and Slatyer has been reported to occur frequently, usually as a consequence of ephemeral algae inhibiting the establishment of later colonisers (Sousa and Connell 1992). Algae can directly inhibit the establishment of barnacles by pre-empting space and overgrowing barnacles, reducing their growth and increasing their mortality (e.g. Hawkins 1983; Farrell 1991). By grazing these ephemerals, herbivores can accelerate succession, with indirect positive effects on barnacle settlement (e.g. Hawkins 1983; van Tamelen 1987) and on later algal colonists (Benedetti-Cecchi 2000b). Strong consumer pressure can stop algal succession (e.g. Lubchenco 1983), leaving only barnacles and lichens to persist in time (e.g. Noël 2007). Barnacles can also have an inhibitory effect on ephemeral algae (Hawkins 1981), perhaps due to the filtering activity of the barnacles on propagules (Benedetti-Cecchi 2000a). Later colonists such as fucoids generally have a negative effect on early algal colonists by out-competing these for light and space (Jenkins et al. 1999b), and by sweeping, which can also affect barnacles (Hawkins 1983; Petraitis and Dudgeon 2005).

Facilitation often occurs in strongly grazed rockpools (Noël et al. 2009). Grazer-resistant species such as encrusting and articulated calcareous algae usually develop and dominate succession, since palatable ephemeral algae are removed by high consumer pressure (Lubchenco 1982; Benedetti-Cecchi 2000a). In later stages of succession, canopy species can escape grazing via associational defences (e.g. *Sargassum muticum* is protected when developing amongst *Corallina* spp.; Noël 2007). Ephemeral species can avoid herbivory when growing as epiphytes on perennial species in late succession (Noël 2007). These associational defences typically increase the species richness in rockpools (Noël 2007).

15.6 Succession, Species Diversity and Ecosystem Processes

15.6.1 Diversity

Disturbance events have been directly related to species diversity—for example, in the intermediate disturbance hypothesis (Connell 1978). At intermediate levels of disturbance, competitive exclusion is prevented, enabling coexistence of both early colonisers and later-colonising, competitively superior species. The model predicts that species richness will be maximised at intermediate disturbance frequency, or intensity, both within and across patches which are disturbed at different times. Evidence in support of this hypothesis has been derived from work on rocky shore assemblages in the 1970s and 1980s. Lubchenco (1978) demonstrated that intermediate densities of grazers maximised macroalgal diversity; Sousa (1979) found a unimodal relationship between boulder size, disturbance frequency and the diversity of principal space-occupiers. Subsequently, in a long-term field investigation, Dethier (1984) showed the importance of disturbance in preventing competitive exclusion in rockpools. Despite the concerted use of rocky shores as a model for successional species interactions during the 1980s and 1990s (e.g. Farrell 1991), changes in diversity were seldom described in sufficient detail to enable testing of the intermediate disturbance hypothesis. More recently, unimodal disturbance–diversity relationships have been reported in subtidal fouling assemblages (e.g. Svensson et al. 2007), lending further support to the intermediate disturbance hypothesis.

Research in a range of other systems has provided at best mixed support for the intermediate disturbance hypothesis (see Mackey and Currie 2001 for review). The intermediate disturbance hypothesis can only be expected to apply where competition for a limiting resource is intense, dominant species are affected by disturbance, and a competition–colonisation trade-off is in evidence (Mackey and Currie 2001). Within hard-bottomed coastal assemblages, where primary space is often limiting, these conditions are often met, perhaps explaining the strong empirical support for the intermediate disturbance hypothesis in these communities (but see Mook 1981 for counterexample).

Even within hard-bottomed communities, the intermediate disturbance hypothesis may not apply in some cases. Rare and small-bodied organisms are favoured by increasing habitat complexity in later stages of succession (e.g. Suchanek 1994), as created by 'foundation species' (Dayton 1973), or 'ecosystem engineers' (Jones et al. 1997) such as slower-growing invertebrates and structurally complex macroalgae. Taking these into account may result in a plateau (Dean and Connell 1987) or in a continuous increase (McKindsey and Bourget 2001) in diversity. Changes may also occur in disturbance–diversity relationships when displaced, so-called early-successional species find refuge on—or within—assemblages of later successional species, especially as epibionts (Benedetti-Cecchi 2000a; Noël 2007).

15.6.2 Functional Consequences

Research over the last decade has shown that species richness within a local assemblage can have a positive effect on the magnitude and stability of ecosystem processes, such as primary productivity and nutrient cycling (Hooper et al. 2005). This relationship has recently been demonstrated in marine macroalgal assemblages (e.g. Bruno et al. 2006). In such "biodiversity–ecosystem functioning" studies, there is often no differentiation between early- and late-successional species. Thus, the extent to which findings from biodiversity–ecosystem functioning research applies to changes in diversity during natural successional sequences is not clear. To integrate successional concepts with current theories, we need to understand both (1) the extent to which diversity contributes to differences in ecosystem functioning during succession and (2) whether the effect of diversity depends on the stage of succession.

To our knowledge, there are no studies which have explicitly tested the functional effect of changes in species diversity through succession. However, the available evidence suggests that diversity effects may be swamped by changes in the identity of dominant species. In marine macroalgal assemblages, the effect of species richness on productivity has been shown to be relatively weak, compared to the effect of species identity (Bruno et al. 2006). The role of species identity is likely to be particularly marked during succession in macroalgae, as species-specific productivity decreases in a predictable way with increasing successional status (Littler and Littler 1980). Differences in the productivity of component species, measured per unit biomass, do not necessarily scale to the productivity of the assemblage as a whole, as standing biomass can also vary markedly during succession. For instance, across a 14-year chronosequence of man-made tide pools, Griffin (unpublished data) found an increase in total standing-stock throughout succession, which acted to equilibrate productivity across pools of differing age, despite a decrease in mass-specific productivity of component species with pool age. Martins et al. (2007) also found no difference in community productivity between early-successional and undisturbed control pools; this may also be a result of higher accumulated biomass in mature pools.

An increase in biomass during succession may modify the effect of species diversity as succession proceeds. Density determines the intensity of interactions within and between species, in part controlling the strength of diversity effects driven by resource partitioning (Griffin et al. 2008). Although likely to be weak compared to species identity effects, the effect of diversity as mediated through resource partitioning would be expected to become more pronounced as succession proceeds and the total density of organisms in a community increases (Weis et al. 2007).

Species composition may also change the effect of species richness on ecosystem functioning, and this will also depend upon the successional stage. Functional diversity, rather than species richness per se, is most likely to drive emergent effects of diversity through niche complementarity (Petchey and Gaston 2006). Succession on rocky shores typically involves a transition from a suite of ephemeral species which display minimal variation in functional traits, to a more functionally diverse mature community, often characterised by canopy-forming algae and associated epiphytes, together with an under-storey of crustose or turf-forming coralline algae with greater functional complementarity (Méndez 2007).

The stability of ecosystem processes may increase during succession. Again, species identity may well have a far greater influence on stability than does species diversity. Intuitively, the temporal stability of productivity will increase during succession as ephemeral species are replaced by well-defended, resistant perennials. Species richness may also act as a biological 'insurance' in the case of dominant species being lost (Yachi and Loreau 1999). The relative roles of species identity and diversity have yet to be tested within a successional context in marine systems (see Steiner et al. 2005 for an aquatic microcosm example).

15.7 Overview and Concluding Remarks

Rocky shore communities typically consist of mosaics of assemblage patches at differing successional stages initiated by their histories of physical and biological disturbance. These processes determine the species composition and diversity within and between habitats which, in turn, influences ecosystem functioning. Developing our understanding of the functional consequences of changes of species diversity during succession clearly requires consideration of changes in density and species composition, as well as changes in species richness itself. Existing evidence suggests that the combined effects of species identity and biomass explain a large proportion of the variation in ecosystem functioning during succession in rockpool communities (Méndez 2007; Griffin, unpublished data). Explicit consideration of how changes in diversity and identity through succession affect ecosystem functioning is required if a full understanding of the role of biodiversity in highly disturbed environments such as marine hard substrata is to be gained.

Acknowledgements This work was supported by a NERC funded Fellowship at The Marine Biological Association of the UK, NERC standard grant NE/B504649/1 to SJH, RCT and SRJ, MARBEF, NERC's Oceans 2025 Theme 4 (Biodiversity and Ecosystem Functioning) and The European Marie Curie Research Training Fellowship MAS3-CT98-5055.

References

Airoldi L (1998) Roles of disturbance, sediment stress, and substratum retention on spatial dominance in algal turf. Ecology 79:2759–2770

Benedetti-Cecchi L, Cinelli F (1996) Patterns of disturbance and recovery in littoral rock pools: nonhierarchical competition and spatial variability in secondary succession. Mar Ecol Prog Ser 135:145–161

Benedetti-Cecchi L (2000a) Predicting direct and indirect interactions during succession in a mid-littoral rocky shore assemblage. Ecol Monogr 70:45–72

Benedetti-Cecchi L (2000b) Priority effects, taxonomic resolution, and the prediction of variable patterns of colonisation of algae in littoral rock pools. Oecologia 123:265–274

Berlow EL (1997) From canalization to contingency: historical effects in a successional rocky intertidal community. Ecol Monogr 67:435–460

Bertness MD, Callaway R (1994) Positive interactions in communities. Trends Ecol Evol 9:191–193

Bevilacqua S, Terlizzi A, Fraschetti S, Russo GF, Boero F (2006) Mitigating human disturbance: can protection influence trajectories of recovery in benthic assemblages? J Anim Ecol 75:908–920

Bruno JF, Lee SC, Kertesz JS, Carpenter RC, Long ZT, Duffy JE (2006) Partitioning the effects of algal species identity and richness on benthic marine primary production. Oikos 115:170–178

Callow ME, Callow JA (2000) Substratum location and zoospore behaviour in the fouling alga *Enteromorpha*. Biofouling 15:49–56

Callow JA, Callow ME (2006) Biofilms. In: Fusetani N, Clare AS (eds) Antifouling compounds. Progress in Molecular and Subcellular Biology, Sub-series Marine Molecular Biotechnology. Springer, Berlin Heidelberg New York, pp 141–169

Characklis WG (1981) Fouling biofilm development: a process analysis. Biotechnol Bioeng 23:1923–1960

Clement FE (1916) Plant succession: an analysis of the development of vegetation. Carnegie Institution Publ vol 242, Washington, DC

Connell JH (1978) Diversity in tropical rain forest and coral reefs—high diversity of trees and coral is maintained only in a non-equilibrium state. Science 199:1302–1310

Connell JH, Slatyer RO (1977) Mechanisms of succession in natural communities and their role in community stability and organization. Am Nat 111:1119–1144

Dayton PK (1971) Competition, disturbance and community organisation: the provision and subsequent utilization of space in a rocky intertidal community. Ecol Monogr 41:351–389

Dayton PK (1973) Dispersion, dispersal and persistence of the annual intertidal alga, Postelsia *palmaeformis* Ruprecht. Ecology 54:433–438

Dean RL, Connell JH (1987) Marine invertebrates in an algal succession. 1. Variations in abundance and diversity with succession. J Exp Mar Biol Ecol 109:195–215

Dethier MN (1984) Disturbance and recovery in intertidal pools—maintenance of mosaic patterns. Ecol Monogr 54:99–118

Dobretsov S, Dahms HU, Qian PY (2006) Inhibition of biofouling by marine microorganisms and their metabolites. Biofouling 22:43–54

Faimali M, Garaventa F, Terlizzi A, Chiantore M, Cattaneo-Vietti R (2004) The interplay of substrate nature and biofilm formation in regulating *Balanus amphitrite* Darwin, 1854 larval settlement. J Exp Mar Biol Ecol 306:37–50

Farrell TM (1991) Models and mechanisms of succession: an example from a rocky intertidal community. Ecol Monogr 61:95–113

Frederiksen HB, Kraglund HO, Ekelund F (2001) Microfaunal primary succession on the volcanic island of Surtsey, Iceland. Polar Res 20:61–73

Griffin JN, De la Haye KL, Hawkins SJ, Thompson RC, Jenkins SR (2008) Predator diversity and ecosystem functioning: density modifies the effect of resource partitioning. Ecology 89:298–305

Hamilton PB, Duthie H (1984) Periphyton colonization of rock surfaces in a boreal forest stream studied by scanning electron microscopy and track autoradiography. J Phycol 20:525–532

Harris LG, Ebling AW, Laur DR, Rowley RJ (1984) Community recovery after storm damage: a case of facilitation in primary succession. Science 224:1336–1338

Hawkins SJ (1981) The influence of season and barnacles on the algal colonization of *Patella vulgata* exclusion areas. J Mar Biol Assoc UK 61:1–15

Hawkins SJ (1983) Interactions of patella and macroalgae with settling *Semibalanus balanoides* (L). J Exp Mar Biol Ecol 71:55–72

Hawkins SJ, Hartnoll RG (1985) Factors determining the upper limits of intertidal canopy forming algae. Mar Ecol Prog Ser 20:265–271

Hooper DU, Chapin FS, Ewel JJ, Hector A, Inchausti P, Lavorel S, Lawton JH, Lodge DM, Loreau M, Naeem S (2005) Effects of biodiversity on ecosystem functioning: a consensus of current knowledge. Ecol Monogr 75:3–35

Huang R, Boney AD (1985) Individual and combined interactions between littoral diatoms and sporelings of red algae. J Exp Mar Biol Ecol 85:101–111

Jenkins SR, Hawkins SJ, Norton TA (1999a) Direct and indirect effects of a macroalgal canopy and limpet grazing in structuring a sheltered inter-tidal community. Mar Ecol Prog Ser 188:81–92

Jenkins SR, Hawkins SJ, Norton TA (1999b) Interaction between a fucoid canopy and limpet grazing in structuring a low shore intertidal community. J Exp Mar Biol Ecol 233:41–63

Jenkins SR, Norton TA, Hawkins SJ (1999c) Settlement and post-settlement interactions between *Semibalanus balanoides* (L.) (Crustacea: Cirripedia) and three species of fucoid canopy algae. J Exp Mar Biol Ecol 236:49–67

Jenkins SR, Arenas F, Arrontes J, Bussell J, Castro J, Coleman RA, Hawkins SJ, Kay S, Martinez B, Oliveros J (2001) European-scale analysis of seasonal variability in limpet grazing activity and microalgal abundance. Mar Ecol Prog Ser 211:193–203

Jones CG, Lawton JH, Shachak M (1997) Positive and negative effects of organisms as physical ecosystem engineers. Ecology 78:1946–1957

Lehvo A, Back S (2001) Survey of macroalgal mats in the Gulf of Finland, Baltic Sea. Aquat Conserv Mar Freshw Ecosyst 11:11–18

Littler MM, Littler DS (1980) The evolution of thallus form and survival strategies in benthic marine macroalgae: field and laboratory test of a functional form model. Am Nat 116:25–44

Lubchenco J (1978) Plant species-diversity in a marine intertidal community: importance of herbivore food preference and algal competitive abilities. Am Nat 112:23–39

Lubchenco J (1982) Effects of grazers and algal competitors on fucoid colonization in tide pools. J Phycol 18:544–550

Lubchenco J (1983) *Littorina and Fucus*—effects of herbivores, substratum heterogeneity, and plant escapes during succession. Ecology 64:1116–1123

Mackey RL, Currie DJ (2001) The diversity-disturbance relationship: is it generally strong and peaked? Ecology 82:3479–3492

Martins GM, Hawkins SJ, Thompson RC, Jenkins SR (2007) Community structure and functioning in intertidal rockpools: effects of pool size and shore height at different successional stages. Mar Ecol Prog Ser 329:43–55

McKindsey CW, Bourget E (2001) Diversity of a northern rocky intertidal community: the influence of body size and succession. Ecology 82:3462–3478

Méndez V (2007) The effect of functional diversity on ecosystem functioning: an experimental test in a macroalgal assemblage. MSc Thesis, University of Plymouth

Mook DH (1981) Effects of disturbance and initial settlement on fouling community structure. Ecology 62:522–526

Niell FX, Varela M (1984) Initial colonization stages on rocky coastal substrates. Mar Ecol Publ Stn Zool Napoli 1:45–56

Noël LM-LJ (2007) Species interactions during succession in rockpools: role of herbivores and physical factors. PhD Thesis, University of Plymouth

Noël LM-LJ, Hawkins SJ, Jenkins SR, Thompson RC (2009) Grazing dynamics in intertidal rockpools: connectivity of microhabitats. J Exp Mar Biol Ecol 370:9–17

Padilla DK, Allen BJ (2000) Paradigm lost: reconsidering functional form and group hypotheses in marine ecology. J Exp Mar Biol Ecol 250:207–221
Paine RT (1979) Disaster, catastrophe and local persistence of the sea palm *Postelsia palmaeformis*. Science 205:685–687
Petchey OL, Gaston KJ (2006) Functional diversity: back to basics and looking forward. Ecol Lett 9:741–758
Petraitis PS, Dudgeon SR (2005) Divergent succession and implications for alternative states on rocky intertidal shores. J Exp Mar Biol Ecol 326:14–26
Reed DC, Raimondi PT, Carr MH, Goldwasser L (2000) The role of dispersal and disturbance in determining spatial heterogeneity in sedentary organisms. Ecology 81:2011–2026
Satuito CG, Shimizu K, Fusetani N (1997) Studies on the factors influencing larval settlement in *Balanus amphitrite* and *Mytilus galloprovincialis*. Hydrobiologia 358:275–280
Sousa WP (1979) Experimental investigations of disturbance and ecological succession in a rocky intertidal algal community. Ecol Monogr 49:227–254
Sousa WP, Connell JH (1992) Grazing and succession in marine algae. In: John DM, Hawkins SJ, Price JH (eds) Plant-animal interactions in the marine benthos. Oxford University Press, New York, pp 425–441
Steiner CF, Long ZT, Krumins JA, Morin PJ (2005) Temporal stability of aquatic food webs: partitioning the effects of species diversity, species composition and enrichment. Ecol Lett 8:819–828
Steneck RS (1982) A limpet coralline algal association: adaptations and defences between a selective herbivore and its prey. Ecology 63:507–522
Steneck RS, Watling L (1982) Feeding capabilities and limitation of herbivorous molluscs: a functional group approach. Mar Biol 68:299–319
Suchanek TH (1994) Temperate coastal marine communities: biodiversity and threats. Am Zool 34:100–114
Svensson JR, Lindegarth M, Siccha M, Lenz M, Molis M, Wahl M, Pavia H (2007) Maximum species richness at intermediate frequencies of disturbance: consistency among levels of productivity. Ecology 88:830–838
Thompson RC, Hawkins SJ, Norton TA (1998) The influence of epilithic microbial films on the settlement of *Semibalanus balanoides* cyprids—a comparison between laboratory and field experiments. Hydrobiologia 376:203–216
Thompson RC, Norton TA, Hawkins SJ (2004) Physical stress and biological control regulate the producer-consumer balance in intertidal biofilms. Ecology 85:1372–1382
Underwood AJ (1977) Movements of intertidal gastropods. J Exp Mar Biol Ecol 26:191–201
Underwood AJ (1984) Vertical and seasonal patterns in competition for microalgae between intertidal gastropods. Oecologia 64:211–222
Underwood AJ, Jernakoff P (1984) The effects of tidal height, wave-exposure, seasonality and rock-pools on grazing and the distribution of intertidal macroalgae in New-South-Wales. J Exp Mar Biol Ecol 75:71–96
van Tamelen PG (1987) Early successional mechanisms in the rocky intertidal: the role of direct and indirect interactions. J Exp Mar Biol Ecol 112:39–48
Wahl M (1989) Marine epibiosis. I. Fouling and antifouling: some basic aspects. Mar Ecol Prog Ser 58:1–2
Weis JJ, Cardinale BJ, Forshay KJ, Ives AR (2007) Effects of species diversity on community biomass production change over the course of succession. Ecology 88:929–939
Wieczorek SB, Tood CD (1998) Inhibition and facilitation of settlement of epifaunal marine invertebrate larvae by micobial biofilm cues. Biofouling 12:81–118
Williams GA, Davies M, Nagarkar S (2000) Primary succession on a seasonal tropical rocky shore: the relative roles of spatial heterogeneity and herbivory. Mar Ecol Prog Ser 203:81–94
Yachi S, Loreau M (1999) Biodiversity and ecosystem productivity in a fluctuating environment: the insurance hypothesis. Proc Natl Acad Sci USA 96:1463–1468

Chapter 16
Simple and Complex Interactions

Markus Molis and Bernardo A.P. da Gama

16.1 Introduction

While the role of abiotic factors in governing species interactions is dealt with in various chapters of this book (e.g. Chap. 7 by Terlizzi and Schiel, Chap. 9 by Benedetti-Cecchi and Chap. 13 by Gili and Petraitis), we will focus here on the biotic factors that affect species interactions. Due to the large number of examples on biotic interactions, we can not and do not attempt to give a complete overview on this topic. Rather, we will present selected examples, mainly from competitive and trophic interactions among macroscopic individuals, describing the principal mechanisms that turn simple into complex interactions. One gradient of complexity concerns the number of interacting species. In this regard, we define the simplest level of species interactions as (1) among conspecific individuals and populations (intraspecific level), followed by interactions (2) between species (interspecific level), and how this reflects on (3) larger sets of species (community level), as the highest level of complexity. Orthogonal to this cline of complexity based on the number of participating species, a number of non-mutually exclusive factors further affect and complicate species interactions, including (1) context specificity, (2) variability, (3) modulation and (4) simultaneous action of several interactions (Fig. 16.1).

16.2 Intraspecific Interactions

Interference competition among conspecifics may represent one of the simplest forms of biotic interaction. Here, individuals of the same species directly affect each other. Antagonistic behaviours between conspecifics represent a commonly observed mechanism of intraspecific interference competition, which may lead to mortality rates as high as 10% of production (Cerda and Wolff 1993). From an ecological and evolutionary perspective, it is favourable to reduce injury or mortality rates among conspecific competitors, as this increases survival and fitness of each antagonist and, thus, benefits the species as a whole. An elaborated example of avoiding conspecific rivals to reduce aggressive encounters has been observed in

M. Wahl (ed.), *Marine Hard Bottom Communities*, Ecological Studies 206,
DOI: 10.1007/978-3-540-92704-4_16,

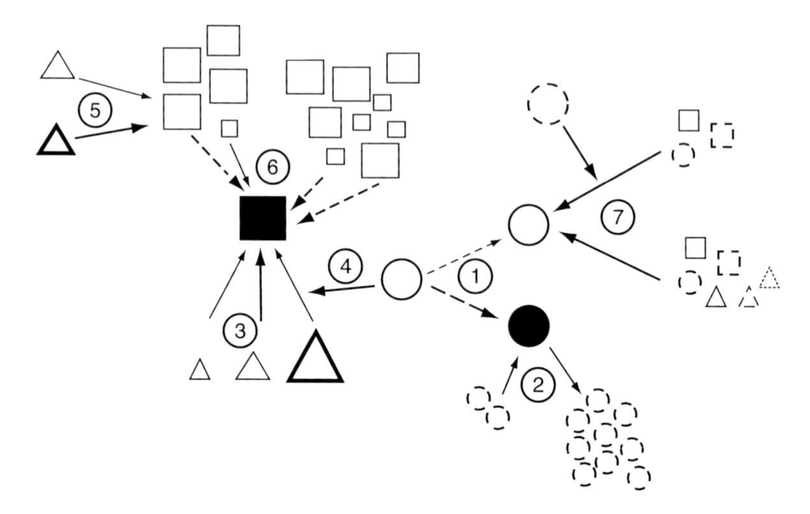

Fig. 16.1 Schematic interaction web showing different types of context-specific interaction (sex-dependent cannibalism (1), density-dependent predator–prey reversal (2), size-dependent optimal foraging (3)), modulation of interaction (trait-mediated indirect effects on, e.g. trophic interaction (4) and community structure (7)), variability of interaction (inducible responses (5)), and simultaneous actions of multiple interactions (size- and density-dependent effects (6)). *Circled numbers* refer to studies presented in the text. *Triangle* photoautotrophs, *square* herbivores, *circle* predators, *black* females, *white* males. Different line patterns of symbols indicate different species, thickness of symbol contours indicates different physiological or phenotypic states, thickness and direction of arrows indicate strength of interaction and flow of energy/competitive superiority respectively, and stippled and solid arrows indicate intra- and interspecific interactions respectively

populations of the sea anemone *Anthopleura elegantissima* (Ayre and Grosberg 2005). Clonal aggregations of this species expand and encounter other clones, among which distinctive anemone-free bands several centimetres wide are formed. These bands are a demarcation line, reflecting the current status of intraspecific interference competition between two clones. Competing clones exist of morphologically distinct casts of polyps, including scouts, warriors and reproductive individuals. Scout polyps explore the border area and, depending on their fate (death or repulsion), the scout-sending clone can organize heavily armed warrior polyps to coordinate the next attack, with the aim to repress the antagonistic clone.

Intraspecific interactions gain complexity if they are context-specific—e.g. by the density-dependency of effects. Positive density-dependent effects can result from beneficial effects under gregarious conditions, like protection against abiotic stress. For instance, crowding of barnacles (*Chthamalus anisopoma*) increased survivorship, due to increased resistance to desiccation (Lively and Raimondi 1987). In contrast, in several fucoid algae an increase in density showed detrimental effects on plant growth (Choi and Norton 2005), with the result that intraspecific competition led to the death of conspecifics (self-thinning). Species traits may modify intraspecific interactions. For instance, sex can determine competitive superiority. The intensity of cannibalism is higher in male than in female crabs (Cerda and Wolff 1993).

Moreover, success of cannibals depends on age, with older individuals of *Carcinus maenas* foraging more effectively on conspecifics than do younger individuals (Moksnes et al. 1998).

Multiple context-specific interactions may simultaneously drive intraspecific interactions. For instance, Shuster and Wade (1991) demonstrated that species traits may interact with density-dependent effects in intraspecific competition (see 6 in Fig. 16.1). In their study, equal male mating success among three differently sized (=trait) morphs of the isopod *Paracerceis sculpta* was dependent on their relative densities. Large males defend harems within intertidal sponges and ensure paternity at low densities of conspecific males. However, by mimicking female behaviour and morphology, medium-sized males successfully enter harems and mate. Invasion of the dwarf males further increases male densities in sponges and alters the relative share to sire. Thus, within one species alternating density-dependent effects may occur, suggesting high context specificity in species responses that make predictions about their population structure and dynamics ultimately difficult.

16.3 Interspecific Interactions

Competition and consumption are direct and, therefore, relatively simple ways in which individuals of different species can interact. Exploitation competition is one form of competition in which detrimental effects on one organism are caused by the depletion of a resource through another organism that also requires this resource. The differential success of species in the depletion of a shared resource will cause asymmetrical competition. For this to happen, the superior competitor must use a limiting resource more effectively than does the other species. The competitive outcome may, however, vary with environmental conditions (Dudgeon et al. 1990) and, ultimately, with the physiological status of rivals, making predictions on the outcome in biotic interactions more difficult. Besides resource depletion, exploitation competition may also occur as a result of antagonistic behaviours. For instance, species may display antagonistic behaviours when competing for access to resources, like suitable substratum for settlement, which is often the most limiting resource for sessile organisms on rocky shores. Antagonistic behaviours among competitors may result in hierarchical competition webs if competitive abilities are fixed. However, the variability in interactions makes interspecific interactions more complex. For instance, competitive superiority may alternate between rivals, if the physiological status of opponents is variable. The stony coral *Oculina patagonica* is the superior competitor for space in the interaction with the bryozoan *Watersipora* sp. between bleaching events but becomes inferior during bleaching events (Fine and Loya 2003).

The study by Raimondi and Lively (1986) provides an example of how simultaneous positive and negative direct effects in species interactions complicate interspecific interactions. In their study, the herbivorous gastropod *Nerita funiculata* adversely affects the distribution of the hermit crab *Clibanarius digueti* by removal of hermit crab food. While this exploitation competition has a negative effect on the distribution of hermit crabs, gastropods were disproportionately more beneficial to hermit crabs, as they are the only source of small-sized shells that small hermit crabs need for successful recruitment. Consequently, negative effects at the level of individuals are not reflected at the population level, strongly indicating that the outcome of interspecific competition is dependent on the relative importance of different resources that competitors require, like food and habitat (=shells) in the case of hermit crabs.

Relatively simple interspecific interactions are also known from trophic interactions. The direct effect of consumers on their prey leads to a sometimes dramatic removal of prey biomass. For instance, labrid fish consumed up to 95% of mussel recruits from experimental plots within 1 day (Rilov and Schiel 2006), and sea urchins are known to deforest kelp beds (Scheibling et al. 1999). The pronounced removal of habitat-forming species can have an important influence on the structure and functioning of benthic food webs at the level of seascapes (see Sect. 16.4). Besides the effects of larger consumers like fish and urchins, the smaller (<2.5 cm) meso-herbivores may also severely affect prey population structure when, for instance, early life stages are consumed (Chapman 1990).

Context specificity also occurs in trophic interspecific interactions, which depend on a large number of traits of the interacting species, for which we present here five examples, starting with size-dependent effects. The Optimal Foraging Theory predicts that consumers will prefer prey that maximises the ratio of energy uptake relative to energy used for foraging, handling and feeding (Stephens and Krebs 1986). Besides the identity, morphology or behaviour of the predator, prey characteristics like shell thickness (Brousseau et al. 2001), epibiosis (Laudien and Wahl 1999), induction of byssus thread production (Cote 1995) or the presence of shell-boring species (Buschbaum et al. 2006) can further complicate prey size selections by predators in opposing directions and with different magnitudes.

Additional traits known to complicate interspecific interactions include age, sex, dietary status and defensive status of interacting species. The age of prey has been shown to alter trophic interactions. For instance, juvenile dogwhelks (*Nucella lapillus*) take fewer risks when foraging in the presence of predatory cues (*Carcinus maenas*) than do adults (Vadas et al. 1994). Presumably, this behavioural trait change reflects the higher susceptibility of thin-shelled juveniles to crab predation. This effect was further dependent on the dietary status of the juveniles, with hungry whelks being less risk-sensitive than fed conspecifics (Vadas et al. 1994). A predator-induced reduction in grazing may have strong indirect effects on other parts of the food web (this chapter, 'trait-mediated indirect effects'). Finally, the defensive status of an organism will affect species interactions. An extensive theory on defensive species responses exists (summarized in Cronin 2001), like the Optimal Defence Theory, which predicts a positive correlation between the level of defence and the fitness value of an algal part (Rhoades 1979).

Besides species traits, numerical responses can complicate interspecific interactions in quite different ways. In benthic invertebrate and macroalgal populations, density-dependent effects are likely to occur at early life stages as a result of mass spawning. The importance of density-dependent effects includes (1) modification of competitive hierarchies among species (e.g. macroalgal seed banks; Worm et al. 2001),

(2) establishment of negative feedback loops in species interactions (Zabin and Altieri 2007), (3) simultaneous alteration of intra- and interspecific interactions (Clark et al. 1999) and (4) reversal in predator-prey relationships (Barkai and McQuaid 1988).

The mechanisms and processes we have considered thus far assume constancy in the responses of individuals, at least for a given life stage. Yet, formal models have been developed (Karban et al. 1999) and experimental evidence gathered that demonstrate that species traits change not only among life stages but also ‘on demand’ within one life stage. This temporal variability in species traits will ultimately modulate species interactions because patterns that might exist today may not be valid tomorrow, e.g. when defences have been switched on in response to increased consumption pressure and/or risk of consumption (Yun et al. 2007, see 5 in Fig. 16.1). Phenotypic plasticity in the responses of individuals from rocky shores, mainly alga/meso-herbivore interactions, have received increasing attention over the last 15 years (Toth and Pavia 2007). In this regard, benthic ecologists lag behind terrestrial colleagues, where more information exists on the biological responses, mechanisms and responsible chemical substances that trigger induced responses (Karban and Baldwin 1997).

Induced responses may concern morphological, behavioural and/or chemical traits, with many studies investigating the induction of chemical traits in several species of algae (Toth and Pavia 2007). The production of secondary metabolites may be induced to serve directly as anti-herbivory defences or these may be used as waterborne infochemicals to warn adjacent conspecific algae about actual or even potential grazer attacks (Yun et al. 2007). As an intrinsic part of inducible defence theory, defences can be switched off when threat-related cues diminish, thereby further complicating species interactions (Rohde et al. 2004). The ability of an organism to induce, rather than permanently exhibit defences carries several important ecological consequences for its bearer. First, Molis et al. (2006) observed consumer specificity in the induction of defences, indicating an ability of the defender to discriminate between the magnitude of consumption pressure exerted by different consumer species and, thus, optimise resource allocation (Cronin 2001). Second, inducible responses increase trait variability. For instance, the palatability of macroalgae in which anti-herbivory defences have been induced is variable in space and time, resulting in increased feeding dispersal (Borell et al. 2004) that, in turn, may lead to an as yet to be experimentally tested increased visibility of grazers to their predators. Third, intraspecific variation in palatability of seaweed species can lower herbivore fitness and, thus, indirectly reduce grazing impact (Toth et al. 2005). Finally, induced responses may affect indirect interactions with consequences on community structure (see Sect. 16.4).

The addition of a third species makes indirect effects possible, adding a new quality to interspecific interactions, which strongly complicates even the simplest interaction web. Indirect effects occur when the interaction between two species depends on the presence of a third species. Associative defences (bodyguard hypothesis) are an example of mutualistic indirect effects that may be generated by epibionts (Enderlein et al. 2003) or occur at larger spatial scales (Pfister and Hay 1988). The latter study demonstrated that mutualistic indirect effects may even occur between competitors but also revealed the conditional nature of mutualisms. Indirect effects are highly context-dependent. The study by Coleman et al. (2007) of predator-mediated consumption of *Ascophyllum nodosum* by the herbivorous snail *Littorina obtusata* provides a clear example. In this tri-trophic interaction, crabs function as bodyguards. The release of waterborne cues by grazed algae is perceived by green crabs and used as infochemicals, guiding these organisms to their prey, i.e. *L. obtusata*. This effect is predator-specific, in that crabs responded to all cues, while fish responded only to cues released from snail-damaged but not from artificially damaged algae. Thus, the bodyguard function was tailored by and dependent on the quality of the 'emergency call' of grazer-attacked seaweeds. Informing the enemy of one's own enemy may be an effective way to foster indirect species interactions, as this may maximise predation success of higher-ranked consumers, especially if signals serve as guides to preferred, inconspicuous prey (Hay et al. 2004).

Between-species associations may also result in negative indirect effects in species interactions (shared doom, sensu Wahl and Hay 1995). For example, barnacle-fouled mussels share a higher predation risk than do clean conspecifics (Enderlein et al. 2003). This pattern was also displayed when barnacle mimics were used, suggesting that improved handling to open mussels, rather than additional food gain may drive crab predation. Furthermore, the Enderlein et al. (2003) study enabled a relative weighing of the ecological importance of indirect effects against other theoretical aspects in trophic interactions. Prey size was the primary control of mussel selection by crabs, corroborating the predictions of the Optimal Foraging Theory, while the indirect effects of epibionts were responsible for the fine tuning of predation success by crabs for a given prey size class.

Jormalainen et al. (2001) demonstrate that indirect effects in species interactions may be sex-dependent. Their study revealed within-alga (*Fucus vesiculosus*) differences in consumption according to gender in isopods (*Idotea baltica*), with males and females grazing apical and basal tissues respectively. As both tissue types exhibit different predation risks for isopods, males face a higher risk of being consumed by predators than do females. This ecological configuration may also pose evolutionary constraints, as males performed better on apical tissues—which pose a higher risk of predation—than on basal algal tissues, while performance (weight gain, intermoult duration) of females was comparable when reared on apical and basal tissues. Because size is more important for mating success of males than of females, males fed in a high predation-risk microhabitat to increase fitness, whereas females adapted to utilise the low-quality food of the safer basal microhabitat as efficiently as the high-quality food of the apical plant parts.

Indirect effects can be classified as density- or trait-mediated (Abrams et al. 1996). The former is a function of a numerical response of the intermediary species, while the latter involves the modification of the interaction between two species by a third species. This interaction modification can arise in two ways. First, indirect effects can result from an environment-mediated modification of interactions. Here, one species changes the environmental context that affects the interaction between two other species. For example, the brown seaweed *Dictyota menstrualis* chemically deters omnivorous fish, providing shelter from fish predation to the amphipod

Ampithoe longimana, which resists the fish-deterring secondary metabolites of the alga (Duffy and Hay 1994). Second, indirect effects may occur from 'trait-mediated indirect interactions' (TMII; Abrams et al. 1996). The principal mechanism in TMIIs results from the non-lethal effect of species C on a trait of species A from an interaction pair A–B. Due to the altered trait of species A, the outcome of its interaction with species B will change (see 4 in Fig. 16.1). TMIIs seem to be more important in aquatic than terrestrial ecosystems (Preisser et al. 2005). Studies addressing TMIIs in rocky shore communities have emerged recently, with a strong emphasis on trophic interactions. Field studies by Trussell et al. (2002) indicated that the mere presence of crabs (*Carcinus maenas*) can reduce the activity level of their prey, the periwinkle *Littorina littorea*, which in turn relaxes grazing pressure on *Fucus* recruits. Interestingly, this *Fucus–Littorina–Carcinus* food chain has been a classical textbook example of a density-mediated indirect interaction (DMII). Furthermore, the sign and magnitude of TMII effects may change between safe and risky habitats, and the magnitude of TMII effects being comparable with or higher than that of DMIIs (Trussell et al. 2006). There are several reasons why the ecological relevance of TMIIs should exceed that of DMIIs (Peacor and Werner 2001; Preisser et al. 2005). First, TMIIs have immediate effects as soon as a modifying species enters a community, while DMIIs will not be immediately effective. Second, TMIIs affect entire populations. For instance, waterborne cues from green crabs influenced the behaviour of nearby snails; further assuming an even distribution of crabs in a given habitat, TMIIs should operate across a broad spatial scale (Trussell et al. 2002). In contrast, DMIIs will affect only that part of a snail population that falls prey to crabs. Third, TMIIs act over the entire period when cues are present. The disappearance of differences in snail density after green crabs were removed from experimental plots in the study by Trussell et al. (2002) clearly shows this immediate function of TMIIs. In contrast, DMIIs operate only at times when crabs kill prey. Finally, DMIIs attenuate through food chains, while TMIIs continue to be strong, as indicated by a more pronounced contrast in effect size between DMIIs and TMIIs when trophic cascades were considered in the meta-analysis by Preisser et al. (2005).

16.4 Community Interactions

The above examples from competitive and trophic interactions among up to three species form the basis for even more complex interactions, when these relatively simple interaction webs merge with each other at the community level. The review by Worm and Duffy (2003) highlights the importance of joining food web theory with biodiversity research, as consumers can modify the directionality of biodiversity–productivity–stability relationships that are derived from isolated studies of simple species interactions. Present empirical and theoretical knowledge of the effects of more species, adding trophic links to a community, has been reviewed by Duffy et al. (2007). These authors emphasise the role of multi-trophic interactions across trophic levels, rather than within one trophic level, as an important driver increasing the variety of diversity-functioning relationships in ecosystems. Such multi-trophic interactions will partly depend on (1) the numbers of consumer and prey, (2) food chain length, (3) relative importance of top-down vs. bottom-up effects within a food web and (4) the level of plasticity of individual species-species interactions within a food web.

16.4.1 Multiple Predator and Prey Effects

Predation is one of the key factors governing patterns in natural communities but is usually understood from the perspective of a single predator species, rather than from a multi-species perspective. However, in nature each prey species is usually exposed to multiple predators, rather than to a single specialized one. Recent studies demonstrated that predator richness can have a strong effect on the efficiency of resource capture and, thus, ecosystem functioning (Griffin et al. 2008). This is particularly important in marine hard-bottom communities, where feeding specialization rarely—if ever—occurs (Hay 1992), in contrast to terrestrial habitats.

Although many trait-mediated indirect interactions (TMIIs) are caused by changes in prey behaviour, less is known about the effects of changes in predator behaviour, such as prey switching, or multiple predator effects (MPEs) on indirect interactions, especially in marine systems. Thus, understanding emergent MPEs is a critical issue for marine community ecology. Few works have studied MPEs in marine communities (reviewed by Sih et al. 1998). Siddon and Witman (2004) tested for the presence of behaviourally mediated indirect effects in a multi-predator system. Here, the effects of crab (*Cancer borealis*) as well as crab and lobster (*Homarus americanus*) predation (=MPE treatment) on sea urchins (*Strongylocentrotus droebachiensis*) were quantified in three habitats (algal beds dominated by the green seaweed *Codium fragile*, barrens, and mussel beds), representing differing combinations of food and shelter, to examine the effects of prey switching by crabs. The study revealed that the presence of lobsters modifies crab behaviour, thereby dampening changes in community structure. These results illustrate the importance of predator behaviour and habitat context in modifying consumer pressure and community structure, and argue for the consideration of these factors in other multi-predator systems where habitats represent food and/or shelter.

The diversity of prey can also influence consumer impact. Hillebrand and Cardinale (2004) conducted a meta-analysis of 172 laboratory and field experiments that manipulated consumer presence to assess their effects on freshwater and marine periphyton, and concluded that grazer effects on algal biomass tend to decrease as algal diversity increases, indicating that periphyton communities characterized by higher species diversity are less prone to consumption by grazers. Diversity of prey may enhance the probability of inedibility and/or of positive interactions. The presence of non-edible prey is generally expected to reduce the efficiency of consumer–prey interactions. Less known, however, are the effects of prey defences on predator–predator interactions. When prey have non-specific defences (i.e. those that can be moderately effective against more than one predator), the addition of a second predator is expected to result in an increase in anti-predator behaviour (or perhaps in chemical defences) and a reduction in predation by both predators, i.e. risk reduction (Sih et al. 1998).

Consumers may also have pervasive indirect impacts on community organization when prey strongly interact with other species in the community. Depending on the timing of the interaction, the carnivorous whelk *Acanthina angelica* kills the barnacle *Chthamalus anisopoma* or induces a predation-resistant morph, which resulted in strong differences in species composition of intertidal communities between sites where the predation-induced morph was present or absent (Raimondi et al. 2000). Indirect effects can thus extend well beyond the particular prey taxa consumed. One classic demonstration of this phenomenon is the fundamental change in community structure after the removal of the starfish *Pisaster ochraceous* from a northeast Pacific rocky intertidal habitat (Paine 1974). Despite its relatively low abundance, *Pisaster* removal resulted in a dramatic reduction in species diversity because its main prey, the mussel *Mytilus edulis*, was competitively dominant and excluded other species when released from predation.

By linking two or more direct interactions together via intermediate species involved in two interactions, 'trophic cascades' emerge in which, e.g. the impact of a top predator indirectly affects the biomass of photoautotrophs. Perhaps the best documented example of a trophic cascade, ranging over four trophic levels, comes from the northeast Pacific and includes killer whales (*Orcinus orca*), sea otters, urchins and kelp (*Macrocystis*) beds (Estes and Duggins 1995). Sudden increases in killer whale predation on otters were correlated with pronounced decreases in otter densities, increases in sea urchin biomass and grazing intensity, and dramatic declines in kelp abundance. In fact, humans may have long been causing periodic shifts between urchin barrens and kelp communities by acting as top predators in the role played by killer whales, by overharvesting sea otters and then allowing these to recover, resulting in what Simenstad et al. (1978) called alternate stable state communities.

In contrast, if prey have conflicting predator-specific defences, then predators would have mutualistic effects on each other (reviewed by Sih et al. 1998). Complex interactions such as these are poorly known in the marine environment but are expected to occur. Many benthic marine invertebrates and algae have physical (i.e. sclerites, spicules, spines, hard exoskeletons, tunicae, etc.) and chemical adaptations against predators, generally as non-specific defences (e.g. Amsler 2008). We do know whether these defences can, in some cases, have conflicting effects, on one hand protecting against generalist predators but, on the other hand, exposing the organisms to more specialized consumers that may even use defences as cues signalling the presence of prey (Avila 2006). These defences can even be sequestered and employed by consumers as a defence against their own predators by a variety of consumers, such as molluscs and some crustaceans.

However, the effects of defences against predators at the community level are hardly known in marine communities, although they have long been postulated to be important in maintaining high species diversity in marine benthic communities.

An example given to illustrate this is associational defence in communities dominated by one or more chemically defended macroalga species. Hay (1986) shows that increasing abundance of one or a few species does not necessarily lead to decreased species richness within the community. In seasons with higher fish abundance, the in situ frequency of species associated with *Sargassum filipendula* and *Padina vickersiae* increased. Similar patterns occurred in microcosm experiments. In fish-inclusion treatments, a significant positive correlation between the cover of *Sargassum* and *Padina* and the number of other species present was found, so that species richness increased as the community became dominated by *Sargassum* and *Padina*. This seemed to result from the unpalatable species creating microhabitats of lowered herbivory that then facilitated the invasion of palatable species, which were excluded by herbivores if these refuges were not available. TMIIs may also be considered in this context, as changes in the density of one species that are caused by induced changes in one or more traits of an intervening species can affect community structure by altering the nature of indirect effects that are mediated through intervening species (e.g. Raimondi et al. 2000). The role of environmentally induced polymorphisms on species interactions and, ultimately, on the structure of hard-bottom communities will be one logical next step in community ecology studies.

In conclusion, we emphasised in this chapter that, in addition to the number of interacting species, context specificity, variability, modulation and simultaneous actions of multiple interactions complicate interactions among individuals. Furthermore, phenotypic plasticity within individuals generates temporal variation in environmentally triggered species traits, and the only recently recognized role of trait-mediated indirect interactions in ecological systems suggests ubiquity of and far-ranging effects on species interactions due to this plasticity. Single experiments may be of limited help in assessing species interactions, as they can sample only a subset of possible species configurations. Thus, the resulting evidence on species interactions from one site or season does not enable extraction of generalities about a studied phenomenon. Due to the strong context dependency of species interactions, future studies should strive for replication in space and time to better understand and predict the causes and consequences of simple and complex species interactions for the functioning of ecological systems.

References

Abrams PA, Menge BA, Mittelbach GG, Spiller D, Yodzis P (1996) The role of indirect effects in food webs. In: Polis G, Winemiller K (eds) Food webs: dynamics and structure. Chapman and Hall, New York, pp 371–395

Amsler CD (2008) Algal chemical ecology. Springer, Berlin Heidelberg New York

Avila (2006) Molluscan natural products as biological models: chemical ecology, histology, and laboratory culture. In: Cimino G, Gavagnin M (eds) Molluscs. Springer, Berlin Heidelberg New York, pp 1–23

Ayre DJ, Grosberg RK (2005) Behind anemone lines: factors affecting division of labour in the social cnidarian *Anthopleura elegantissima*. Anim Behav 70:97–110

Barkai A, McQuaid C (1988) Predator-prey role reversal in a marine benthic ecosystem. Science 242:62–64
Borell EM, Foggo A, Coleman RA (2004) Induced resistance in intertidal macroalgae modifies feeding behaviour of herbivorous snails. Oecologia 140:328–334
Brousseau DJ, Filipowicz A, Baglivo JA (2001) Laboratory investigations of the effects of predator sex and size on prey selection by the Asian crab, *Hemigrapsus sanguineus*. J Exp Mar Biol Ecol 262:199–210
Buschbaum C, Buschbaum G, Schrey I, Thieltges DW (2006) Shell-boring polychaetes affect gastropod shell strength and crab predation. Mar Ecol Prog Ser 329:123–130
Cerda C, Wolff M (1993) Feeding ecology of the crab *Cancer polyodon* in La Herradura Bay, northern Chile. II. Food spectrum and prey consumption. Mar Ecol Prog Ser 100:119–125
Chapman AS (1990) Effects of grazing, canopy cover and substratum type on the abundances of common species of seaweeds inhabiting littoral fringe tide pools. Botanica Marina 33:319–326
Choi HG, Norton TA (2005) Competitive interactions between two fucoid algae with different growth forms, *Fucus serratus* and *Himanthalia elongata*. Mar Biol 146:283–291
Clark ME, Wolcott TG, Wolcott DL, Hines AH (1999) Intraspecific interference among foraging blue crabs *Callinectes sapidus:* interactive effects of predator density and prey patch distribution. Mar Ecol Prog Ser 178:69–78
Coleman RA, Ramchunder SJ, Davies KM, Moody AJ, Foggo A (2007) Herbivore-induced infochemicals influence foraging behaviour in two intertidal predators. Oecologia 151:454–463
Cote IM (1995) Effects of predatory crab effluent on byssus production in mussels. J Exp Mar Biol Ecol 188:233–241
Cronin G (2001) Resource allocation in seaweeds and marine invertebrates: chemical defense patterns in relation to defense theories. In: McClintock JB, Baker BJ (eds) Marine chemical ecology. CRC Press, New York, pp 325–353
Dudgeon SR, Davison IR, Vadas RL (1990) Freezing tolerance in the intertidal red algae *Chrondrus crispus* and *Mastocarpus stellatus:* relative importance of acclimation and adaptation. Mar Biol 106:427–436
Duffy JE, Hay ME (1994) Herbivore resistance to seaweed chemical defense: the roles of mobility and predation risk. Ecology 75:1304–1319
Duffy JE, Carinale BJ, France KE, McIntyre PB, Thebault E, Loreau M (2007) The functional role of biodiversity in ecosystems: incorporating trophic complexity. Ecol Lett 10:522–538
Enderlein P, Moorthi S, Rohrscheidt H, Wahl M (2003) Optimal foraging versus shared doom effects: interactive influence of mussel size and epibiosis on predator preference. J Exp Mar Biol Ecol 292:231–242
Estes JA, Duggins DO (1995) Sea otters and kelp forests in Alaska—generality and variation in a community ecological paradigm. Ecol Monogr 65:75–100
Fine M, Loya Y (2003) Alternate coral-bryozoan competitive superiority during coral bleaching. Mar Biol 142:989–996
Griffin JN, de la Haye K, Hawkins SJ, Thompson RC, Jenkins SR (2008) Predator diversity and ecosystem functioning: density modifies the effect of resource partitioning. Ecology 89:298–305
Hay ME (1986) Associational plant defenses and the maintenance of species-diversity—turning competitors into accomplices. Am Nat 128:617–641
Hay ME (1992) The role of seaweed chemical defenses in the evolution of feeding specialization and in the mediation of complex interactions. In: Paul VJ (ed) Ecological roles of marine natural products. Comstock, Ithaca, NY, pp 93–118
Hay ME, Parker JD, Burkepile DE, Caudill CC, Wilson AE, Hallinan ZP, Chequer AD (2004) Mutualisms and aquatic community structure: the enemy of my enemy is my friend. Annu Rev Ecol Syst 35:175–197
Hillebrand H, Cardinale BJ (2004) Consumer effects decline with prey diversity. Ecol Lett 7:192–201

Jormalainen V, Honkanen T, Makinen A, Hemmi A, Vesakoski O (2001) Why does herbivore sex matter? Sexual differences in utilization of *Fucus vesiculosus* by the isopod *Idotea baltica*. Oikos 93:77–86
Karban R, Baldwin IT (1997) Induced responses to herbivory. University of Chicago Press, Chicago, IL
Karban R, Agrawal AA, Thaler JS, Adler LS (1999) Induced plant responses and information content about risk of herbivory. Trends Ecol Evol 14:443–447
Laudien J, Wahl M (1999) Indirect effects of epibiosis on host mortality: seastar predation on differently fouled mussels. Mar Ecol PSZNI 20:35–47
Lively CM, Raimondi PT (1987) Desiccation, predation, and mussel-barnacle interactions in the northern Gulf of California. Oecologia 74:304–309
Moksnes PO, Pihl L, van Montfrans J (1998) Predation on postlarvae and juveniles of the shore crab *Carcinus maenas:* importance of shelter, size and cannibalism. Mar Ecol Prog Ser 166:211–225
Molis M, Körner J, Ko YW, Kim JH, Wahl M (2006) Inducible responses in the brown seaweed *Ecklonia cava:* the role of grazer identity and season. J Ecol 94:243–249
Paine RT (1974) Intertidal community structure: experimental studies on the relationship between a dominant competitor and its principal predator. Oecologia 15:93–120
Peacor SD, Werner EE (2001) The contribution of trait-mediated indirect effects to the net effects of a predator. Proc Natl Acad Sci USA 98:3904–3908
Pfister CA, Hay ME (1988) Associational plant refuges: convergent patterns in marine and terrestrial communities result from differing mechanisms. Oecologia 77:118–129
Preisser EL, Bolnick DI, Benard MF (2005) Scared to death? The effects of intimidation and consumption in predator-prey interactions. Ecology 86:501–509
Raimondi PT, Lively CM (1986) Positive abundance and negative distribution effects of a gastropod on an intertidal hermit crab. Oecologia 69:213–216
Raimondi PT, Forde SE, Delph LF, Lively CM (2000) Processes structuring communities: evidence for trait-mediated indirect effects through induced polymorphisms. Oikos 91:353–361
Rhoades D (1979) Evolution of plant chemical defenses against herbivores. In: Rosenthal GA, Janzen DH (eds) Herbivores. Academic Press, New York, pp 4–54
Rilov G, Schiel DR (2006) Trophic linkages across seascapes: subtidal predators limit effective mussel recruitment in rocky intertidal communities. Mar Ecol Prog Ser 327:83–93
Rohde S, Molis M, Wahl M (2004) Regulation of anti-herbivore defence by *Fucus vesiculosus* in response to various cues. J Ecol 92:1011–1018
Scheibling RE, Hennigar AW, Balch T (1999) Destructive grazing, epiphytism, and disease: the dynamics of sea urchin–kelp interactions in Nova Scotia. Can J Fish Aquat Sci 56:2300–2314
Shuster SM, Wade MJ (1991) Equal mating success among male reproductive strategies in a marine isopod. Nature 350:608–610
Siddon CE, Witman JD (2004) Behavioral indirect interactions: multiple predator effects and prey switching in the rocky subtidal. Ecology 85:2938–2945
Sih A, Englund G, Wooster D (1998) Emergent impacts of multiple predators on prey. Trends Ecol Evol 13:350–355
Simenstad CA, Estes JA, Kenyon KW (1978) Aleuts, sea otters, and alternate stable-state communities. Science 200:403–411
Stephens DW, Krebs JR (1986) Foraging theory. Princeton Academic Press, Princeton, NJ
Toth GB, Pavia H (2007) Induced herbivore resistance in seaweeds: a meta-analysis. J Ecol 95:425–434
Toth GB, Langhamer O, Pavia H (2005) Inducible and constitutive defenses of valuable seaweed tissues: Consequences for herbivore fitness. Ecology 86:612–618
Trussell GC, Ewanchuk PJ, Bertness MD (2002) Field evidence of trait-mediated indirect interactions in a rocky intertidal food web. Ecol Lett 5:241–245
Trussell GC, Ewanchuk PJ, Matassa CM (2006) Habitat effects on the relative importance of trait- and density-mediated indirect interactions. Ecol Lett 9:1245–1252

Vadas RL, Burrows MT, Hughes RN (1994) Foraging strategies of dogwhelks, *Nucella lapillus* (L.): interacting effects of age, diet and chemical cues to the threat of predation. Oecologia 100:439–450

Wahl M, Hay ME (1995) Associational resistance and shared doom: effects of epibiosis on herbivory. Oecologia 102:329–340

Worm B, Duffy JE (2003) Biodiversity, productivity and stability in real food webs. Trends Ecol Evol 18:628–632

Worm B, Lotze HK, Sommer U (2001) Algal propagule banks modify competition, consumer and resource control on Baltic rocky shores. Oecologia 128:281–293

Yun HY, Cruz J, Treitschke M, Wahl M, Molis M (2007) Testing for the induction of anti-herbivory defences in four Portuguese macroalgae by direct and water-borne cues of grazing amphipods. Helgoland Mar Res 61:203–209

Zabin CJ, Altieri A (2007) A Hawaiian limpet facilitates recruitment of a competitively dominant invasive barnacle. Mar Ecol Prog Ser 337:175–185

Part IV
Changing Biodiversity

Coordinated by Angus C. Jackson and M. Gee Chapman

Introduction

Angus C. Jackson and M. Gee Chapman

During the past few decades, changes to biodiversity (the variety of life at all levels of biological organization; Gaston and Spicer 2004) have been dominated by anthropogenic influences and activities. The term *anthropogenic* describes effects, processes, objects and materials that are caused or produced from human activities, in contrast with those occurring in natural environments that are not influenced by humans. Recent major impacts to ecosystems have had serious social and economic consequences. For example, the 1990 collapse of the fishery for cod in the western Atlantic left tens of thousands of fisherman out of work. Consequently, such changes to biota and their habitats are increasingly of concern among scientists, politicians, managers and the public alike. In general, however, awareness and understanding of changes in marine habitats lag far behind those of changes on land (Jackson et al. 2001).

Changes to diversity caused by humans are not new phenomena by any means. Many impacts pre-date modern expansion and relocation of populations with associated urban and industrial development, pollution and exploitation of resources. For instance, intensive irrigation in Mesopotamia up to 6,000 years ago profoundly altered terrestrial and freshwater habitats by changing salinity and rates of sedimentation (Garbrecht 1983). Major impacts on hard substrata in marine environments may be less ancient but, nonetheless, have been around for some time. Over-harvesting is also a long-standing problem. Fishing of *Semicossyphus pulcher* (Sheephead), which can regulate populations of sea urchins, may be one example (Tegner and Dayton 1981; Cowen 1983). Decreases over the last 4,000 years in the size of bones in middens suggest that historical overfishing of this species was a cause of increases in populations of sea urchins in California (Erlandson et al. 1996), with consequences for associated species such as sea otters, lobsters and kelp. Waterfront development dates back more than 2 millennia, with natural habitats being replaced by built infrastructure for shipping and fortification (Mann 1988). Jackson et al. (2001) emphasise the need for historical perspectives such as these, alongside modern impacts, to enable the full understanding of causes of contemporary changes.

In Part IV of this book, the authors examine several prevailing aspects of change in diversity on rocky marine substrata around the world and describe up-to-the-minute studies and classic examples in their field of expertise. Chapter 17 is a brief overview

M. Wahl (ed.), *Marine Hard Bottom Communities*, Ecological Studies 206,
DOI: 10.1007/978-3-540-92704-4_IV, © Springer-Verlag Berlin Heidelberg 2009

of the main types of anthropogenic activity, providing some definitions and raising the importance of the scale at which impacts are considered. Human activities not only affect organisms directly but also often alter the abiotic environment in which organisms live. This may have indirect consequences for many species. Chapter 18 describes patterns and processes of change in abiotic variables at global, regional and local scales and proposes a framework for understanding anthropogenic impacts from a catchment perspective. Changes to coastal habitats and some of their causes (e.g. poor management of fisheries or addition of artificial hard substrata) are described in Chapter 19. Mortality is a natural process for all populations but changes to physical or biotic conditions caused by human activities are strongly implicated in recent increases in mass mortalities. Chapter 20 highlights the importance of multiple concurrent or sequential stressors, and the need to account for synergistic or multiplicative effects in research, management and conservation. The occurrence (frequency, extent and taxonomy) and ecological and economic consequences of mass mortalities in hard-bottom marine habitats are reviewed in Chapter 21. Chapters 22 and 23 deal with the introduction of non-indigenous species, which is a major contemporary issue in ecology. Although species naturally spread and change their range, the rate at which this now occurs (for example, because of mass transportation, changes in ice cover associated with climatic change and creation of new waterways such as the Suez canal) is unprecedented. The final two chapters approach the often emotive but increasingly important subjects of conservation of biota and habitat via the use of protected areas (Chap. 24) and the rehabilitation of degraded habitat (Chap. 25).

Conclusion

Part IV illustrates clearly that the depth and detail of research on and knowledge about anthropogenic impacts to hard-bottomed marine habitats and associated changes to biodiversity are proceeding apace. Here, we provide some very brief, general conclusions, alongside some interpretations about important future directions for research into changing biodiversity.

Perhaps the most important message is the need to appreciate the temporal and spatial scales, extents and complexities of impacts and subsequent changes to diversity. Many impacts on ecosystems have now become global in scale (e.g. declines in major fisheries; Brander 2007); others are limited to a local sphere of influence. There is a tendency to associate current changes in diversity with contemporary causes such as pollution, global warming and invading species. In reality, impacts may be temporally disconnected from their causes, and long-term or historical activities may have precipitated chains of events, causing what we see today. We need to know about these historical influences, in addition to modern impacts, to fully understand any current associations between species and habitats.

Anthropogenic influences and their consequences may also be disconnected spatially (Chap. 19). Changes to hard substrata in marine environments do not only stem from activities and processes that occur in the sea but can also be affected strongly by human activities in other environments. The consequences of terrestrial activities can find their way, via the atmosphere, water tables and rivers, to distant estuaries, coastal waters and even entire seas on the other side of the globe. A well-known example is the human release of halogenated compounds causing pronounced reductions to the ozone layer in polar regions (Solomon et al. 1986). Trying to manage and conserve marine habitats in isolation from multiple terrestrial processes that affect these may be a fruitless pursuit. For instance, coastal habitats are often influenced by input of freshwater, nutrients and terrigenous sediment from rivers. In Australia, the Murray-Darling river system contains large loads of suspended sediment and has a variable flow regime (Walker 1985; Prosser et al. 2001). The system has several human influences, the most important of which are extraction of surface waters (approximately 80% of the mean annual flow is diverted for consumptive use; MDBMC 1995), land clearing, loss of riparian vegetation, and agricultural runoff. These processes will have impacts in the coastal zone, which will receive less and more variable supplies of freshwater, nutrients, organic matter and sediments (Meybeck 2002).

Until fairly recently, the Gordian nature of human stressors has been underestimated and studies have tended to focus on the consequences of single aspects of human activities. With greater appreciation of the enormous scales of some activities or processes (e.g. rising sea levels) and the complexity of responses by organisms and habitats, it is becoming increasingly apparent that anthropogenic stressors seldom act in simple ways and that simple definitions of disturbances (e.g. press, pulse) are not sufficient (Chap. 17). Impacts seldom occur in isolation and often arise through the confluence of multiple stressors (Chap. 20). Coral reefs provide several prime examples (Hughes and Connell 1999). Corals can be damaged directly or indirectly by human activities at sea (e.g. boating, collecting, fishing) or on land (e.g. forestry, agriculture and urbanisation). Climatic change will inevitably cause additional impacts. Increases in CO_2 levels are predicted to reduce the pH of ocean water and decrease the availability of aragonite, hindering calcification (Chap. 18). Coral bleaching can be caused by elevated temperatures, reduced salinity or increases in suspended particulate matter. Impacts are also often cumulative, occurring in response to repeated events (Chap. 17). For example, repeated small amounts of damage by boat anchors or divers to slow-growing corals can result in large and long-lasting damage (Rouphael and Inglis 1997). Repeated impacts may elicit non-linear responses, potentially causing ramp responses when disturbances exceed some ecological threshold (Lake 2000; Chap. 17). For instance, predicted increases in frequency and intensity of storms may have increasingly large effects on reefs as damage from successive storms eventually reaches a level where reefs have lost structural integrity.

Consequently, studies of single activities or processes may give unrealistic or unrepresentative impressions of the consequences of stressors. Growing recognition of cumulative impacts of stresses and the widespread occurrence of different stresses requires concurrent evaluation of multiple stresses at different locations. When multiple stressors co-occur, there may be synergistic, multiplicative or antagonistic effects, impeding our ability to predict overall changes in diversity (Folt et al. 1999).

We are currently faced with the great challenge of solving or at least ameliorating some of the changes that threaten our planet. Species or habitats per se cannot be managed; rather, we need to regulate human activities that impact on these. For some activities (e.g. fishing, mineral extraction), this is straightforward but other processes are not so easy to regulate. For example, stresses wrought by changes in climate will not 'respect' boundaries imposed by humans (e.g. countries, cities, nature reserves). Thus, setting aside protected natural habitat and assuming this is adequate is not sufficient to protect diversity.

Assessment about how much change in diversity is important and the development of policy for managing marine habitats face two tricky issues. The first is that change is natural and must be allowed to occur. Many current attitudes to conservation focus on a static picture, to find an undisturbed area and to try and keep it unchanged into the future. The second issue is that of 'shifting baselines' (i.e. our perception of what is 'natural' is altered by our activities; Pauly 1995; Jackson 2001). There is no such thing as an original, 'natural' condition and, until recently, historical changes due to human activity have been forgotten or have gone unrecorded. The issue of change, restoration and rehabilitation is conceptually and practically difficult but it is reality. Humans are an integral part of the planet and we need to accept our role in changes to diversity. Consequently, conservation and management practise (Chaps. 24 and 25) must be tailored to account for our changing human perspective and be sufficiently realistic to work in a world dominated by people (Bawa et al. 2004; Palmer et al. 2005).

Loss of habitat has long been claimed to be the primary threat to loss of diversity (Wilson 1988) but this may no longer be true. Climatic change, with major predicted shifts in air and sea temperature, rises in sea level, changes in ocean chemistry, and increasing intensity and variability of storm events (Chap. 18), is an enormous threat that may elicit mass extinctions (Thomas et al. 2004; Chap. 19). The increasing potential for and facilitation of the spread of invasive species already have major impacts on diversity (Chap. 22). These impacts are undoubtedly set to increase and will have enormous consequences for species and habitats. Invasion pressure is, however, by no means spatially uniform, with hotspots of invasion (e.g. sheltered hard substrata) reflecting the provenance of the invasive species and their means of transport (Chap. 23). Teasing out the relative contributions of different stressors, understanding the actual responses of populations and habitats, and managing their use in a sustainable way will be very difficult and will require much more thought and careful experimentation. In a world increasingly impacted by humans, changes in biodiversity will assume ever greater importance and we must improve our ability to understand, predict and deal with future conditions.

References

Bawa KS, Seidler R, Raven PH (2004) Reconciling conservation paradigms. Conserv Biol 18:859–860

Brander KM (2007) Global fish production and climate change. Proc Natl Acad Sci USA 104:19709–19714

Cowen RK (1983) The effect of sheephead (*Semicossyphus pulcher*) predation on red sea urchin (*Strongylocentrotus franciscanus*) populations—an experimental analysis. Oecologia 58:249–255

Erlandson JM, Kennett DJ, Ingram BL, Guthrie DA, Morris DP, Tveskov MA, West GJ, Walker PL (1996) An archaeological and palaeontological chronology for Daisy Cove (CA-SMI-261), San Miguel Island, California. Radiocarbon 38:355–373
Folt CL, Chen CY, Moore MV, Burnaford J (1999) Synergism and antagonism among multiple stressors. Limnol Oceanogr 44:864–877
Garbrecht G (1983) Ancient water works—lessons from history. UNESCO, Paris, p 8
Gaston KJ, Spicer JI (2004) Biodiversity: an introduction. Blackwell, Oxford
Hughes TP, Connell JH (1999) Multiple stressors on coral reefs: a long-term perspective. Limnol Oceanogr 44:932–940
Jackson JBC (2001) What was natural in the coastal oceans? Proc Natl Acad Sci USA 98:5411–5418
Jackson JBC, Kirby MX, Berger WH, Bjorndal KA, Botsford LW, Bourque BJ, Bradbury RH, Cooke R, Erlandson J, Estes JA, Hughes TP, Kidwell S, Lange CB, Lenihan HS, Pandolfi JM, Peterson CH, Steneck RS, Tegner MJ, Warner RR (2001) Historical overfishing and the recent collapse of coastal ecosystems. Science 293:629–638
Lake PS (2000) Disturbance, patchiness, and diversity in streams. J N Am Benth Soc 19:573–592
Mann RB (1988) Ten trends in the continuing renaissance of urban waterfronts. Landsc Urban Plan 16:177–199
MDBMC (1995) An audit of water use in the Murray-Darling basin. Murray-Darling Basin Ministerial Council, Canberra
Meybeck M (2002) Riverine quality at the Anthropocene: propositions for global space and time analysis, illustrated by the Seine River. Aquat Sci 64:376–393
Palmer MA, Bernhardt ES, Chornesky EA, Collins SL, Dobson AP, Duke CS, Gold BD, Jacobson RB, Kingsland SE, Kranz RH, Mappin MJ, Martinez ML, Micheli F, Morse JL, Pace ML, Pascual M, Palumbi SS, Reichman O, Townsend AR, Turner MG (2005) Ecological science and sustainability for the 21st century. Frontiers Ecol Environ 3:4–11
Pauly D (1995) Anecdotes and the shifting base-line syndrome of fisheries. Trends Ecol Evol 10:430–430
Prosser IP, Rutherfurd ID, Olley JM, Young WJ, Wallbrink PJ, Moran CJ (2001) Large-scale patterns of erosion and sediment transport in river networks, with examples from Australia. Mar Freshw Res 52:81–99
Rouphael AB, Inglis GJ (1997) Impacts of recreational scuba diving at sites with different reef topographies. Biol Conserv 82:329–336
Solomon S, Garcia RR, Rowland FS, Wuebbles DJ (1986) On the depletion of Antarctic ozone. Nature 321:755–758
Tegner MJ, Dayton PK (1981) Population-structure, recruitment and mortality of 2 sea urchins (*Strongylocentrotus franciscanus* and *S. purpuratus*) in a kelp forest. Mar Ecol Prog Ser 5:255–268
Thomas CD, Cameron A, Green RE, Bakkenes M, Beaumont LJ, Collingham YC, Erasmus BFN, de Siqueira MF, Grainger A, Hannah L, Hughes L, Huntley B, van Jaarsveld AS, Midgley GF, Miles L, Ortega-Huerta MA, Peterson AT, Phillips OL, Williams SE (2004) Extinction risk from climate change. Nature 427:145–148
Walker KF (1985) A review of the ecological effects of river regulation in Australia. Hydrobiologia 125:111–129
Wilson EO (1988) Biodiversity. National Academy Press, Washington, DC

Chapter 17
Anthropogenic Changes in Patterns of Diversity on Hard Substrata: an Overview

Brianna G. Clynick, David Blockley, and M. Gee Chapman

17.1 Introduction

Whatever processes determine natural patterns in distribution and abundance of biota on rocky reefs, there is little doubt that these have been greatly altered by human activities in many parts of the world. These include, but are not restricted to, extraction of resources, especially seafood but also sediments, addition of contaminants and solid waste, intentional and accidental spread of species and widespread changes to habitat. Yet, relative to changes that humans have wrought on land, those that people have imposed on the ocean have been largely ignored. This is, in part, due to an "out of sight, out of mind" mentality but also possibly to the fact that we have been altering nearshore coastal habitats over very long periods, so that it now appears to be the "norm" (Jackson 2001).

On land, changes to habitat and, associated with this, changes to diversity of species are often extreme, in terms of both the spatial scales over which conditions have been altered and the length of time during which humans have had major impacts. In the sea, changes are often assumed to be smaller, more localized or more easily reversed, although this may not be the case—e.g. large fishing grounds that have been exploited over long periods have been extremely damaged and may take decades to recover (Thrush and Dayton 2002). Complete loss of entire areas of marine habitat is generally restricted to intertidal areas. Although large areas of soft sedimentary habitats may be destroyed by dumping or disposal or waste, or eutrophication, these may be transitory effects and when the disturbance is removed, the habitats tend to recover. Many changes to intertidal areas are more permanent. For example, tidal wetlands have been extensively destroyed by land-reclamation and direct exploitation of resources (such as culling mangroves for wood) has destroyed large areas of habitat. Once reclaimed land is built upon, there is generally little chance of reversal. In addition, built structures destroy many shorelines—e.g. aquaculture facilities, docks and wharves. Hard rocky or coral substrata may be damaged by excessive use, e.g. by trampling or removal of species, and also over large spatial scales, by activities such as dynamite fishing. One obvious loss of subtidal habitat is the decimation of oyster reefs in many parts of the USA (Rothschild et al. 1994; Jackson 2001), which is discussed in more detail in Chapter 19

M. Wahl (ed.), *Marine Hard Bottom Communities*, Ecological Studies 206,
DOI: 10.1007/978-3-540-92704-4_17, © Springer-Verlag Berlin Heidelberg 2009

by Airoldi et al. This not only reduced habitat created by reefs but, because oysters are very effective filter-feeders, also caused major changes to water-quality.

Direct impacts on species, rather than habitat, are often less visible, although marine flora and fauna have long been exploited for food, industry, agriculture or for cultural reasons. Biological diversity, especially at higher taxonomic levels, is large in the sea but, as is generally the case, most faunal diversity is among small, often cryptic species. These tend to "fall off the radar" when it comes to conservation (Roberts and Hawkins 1999; Underwood and Chapman 1999). Marine extinctions due to humans are still considered rare, leading to complacency when addressing the capacity of the oceans to withstand increasing human interference. Yet, people have major influences on distribution and abundance of species, through their capacity to exploit selected species heavily, the speed and ease with which organisms are moved around the world and the changes to habitat that we impose on nearshore environments.

17.2 Scales of Disturbances Affecting Distributions and Abundances

When assessing anthropogenic changes to patterns of distribution and abundance of organisms, it is important to consider the scale (in time and space) of the perturbation itself and of the biological response to it (Glasby and Underwood 1996). Perturbations can be short pulse events (e.g. a chemical spill; Fig. 17.1a) or longer press events (e.g. extensive shading on the substratum by marinas; Fig. 17.1b). Generally, pulse disturbances are followed by a pulse response, i.e. the population recovers soon after the disturbance ceases. Similarly, a press disturbance is usually followed by a long-term ecological change. Pulse or press perturbations do not, however, necessarily result in pulse or press responses respectively. If a chemical spill removes a breeding population, then a species may not recover, even after the spill is removed. Similarly, a species may exploit novel conditions created by press disturbances, such as using artificial structures as habitat (Glasby and Connell 1999), so that loss of natural habitat may have a deleterious effect only until species have had time to recover.

Simple definitions, such as these, may not apply to many situations because disturbances seldom arrive one at a time. Cumulative effects may occur when a second disturbance occurs before a population has recovered from the previous one—e.g. organisms may be affected by repeated exposures to low levels of contaminants (Barreiro et al. 1993). Recreational foraging for food may also be a cumulative disturbance. At any given time, foraging may be limited to a few people taking a few animals but, if this continues year after year, populations of exploited species can not recover because new arrivals are removed from the shore. Another form of cumulative disturbance is a "ramp" disturbance (Lake 2000), where the effects of repeated disturbance gradually increase over time, even if the size of the disturbance itself is the same each time. Proliferation of built habitat in urbanized estuaries or

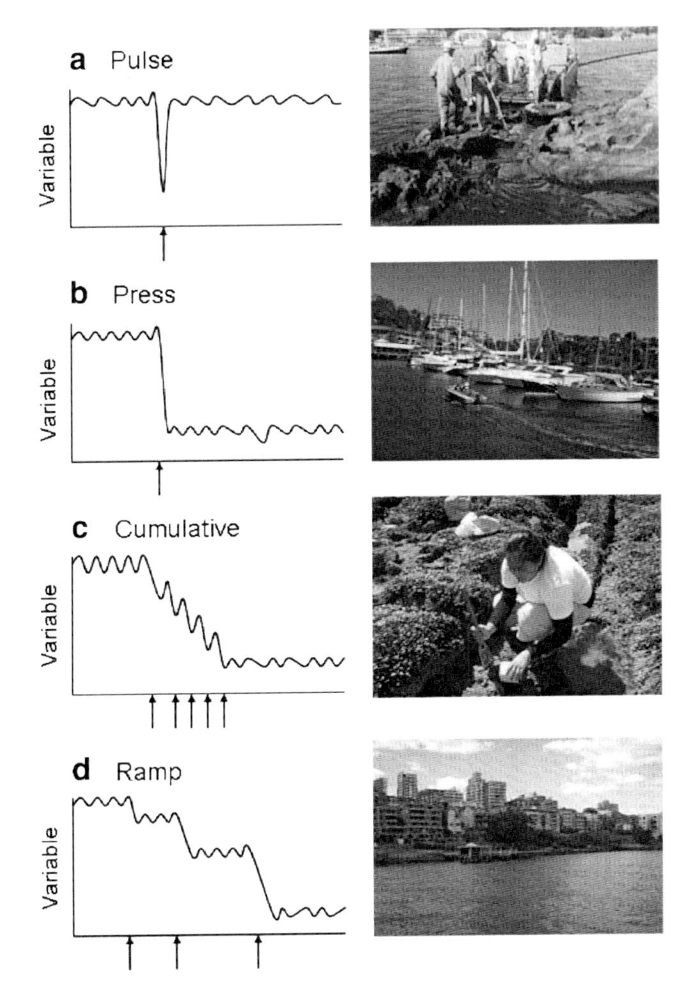

Fig. 17.1 **a** *Pulse disturbances* are short-term and assemblages recover naturally—e.g. many accidental spills. **b** *Press disturbances*, such as major changes to shorelines, have long-term press effects because the disturbance makes a long-term change to the environment. **c** *Cumulative effects* may occur when populations are subjected to repeated short-term impacts, such as small amounts of harvesting, if they cannot recover before the next impact occurs. **d** *Ramp disturbances*—e.g. continued additional replacement of patches of shores with seawalls, with the strength of the impact increasing through time because the environment is becoming increasingly more stressed

the spread of exotic species may be ramp disturbances because each additional disturbance (e.g. each newly-built wall, wharf, marina, etc.), even when the same as previous ones, may have a more severe effect because it impacts on an increasingly stressed environment.

Anthropogenic activities can thus have major influences on marine biota on hard substrata, both subtidal reefs and intertidal shores, through short-term, long-term, one-off and cumulative impacts. Many of these stresses occur simultaneously and may act antagonistically or synergistically, making impacts difficult to predict. They include changes to habitat such as loss, fragmentation, degradation and the proliferation of novel habitat, contamination and pollution and major direct impacts on populations of flora and fauna. Some major changes are briefly reviewed here.

17.2.1 Effects of Disposal of Waste Material and Spills

Contamination and pollution have long been of concern and, with continued development, there are increasing potential sources of contamination. Although the terms are often confused, contamination and pollution have different meanings, which should be kept in mind for logical evaluation of impacts (GESAMP 1980). Contamination is the introduction of anthropogenic substances into the environment and is directly measured by amounts of substances. Pollution, in contrast, refers to their effects on biota and implies a biological response to the contaminants. Contamination can be spatially or temporally limited (i.e. a pulse perturbation), or chronic (i.e. single or multiple stressors acting over larger temporal and/or spatial scales—a press or ramp perturbation) and, like other disturbances, responses to contamination can be short- or long-term (see Chap. 25 by Goodsell and Underwood).

Highly visible pulse disturbances receive most attention (e.g. oil spills; Myers et al. 1980), although there are many chronic sources of pollution associated with anthropogenic activities. Solid and chemical contaminants, including nutrients, heavy metals, litter or terrestrial sediments, can enter the marine environment accidentally—e.g. from runoff—or be deliberately discharged—e.g. stormwater or sewage discharge (e.g. Stark 1998), or disposal of waste (Stark et al. 2004). Although many contaminants may not have serious short-term effects, they can be chronic pollutants due to their persistence in marine systems (Stark 1998), particularly in bays and estuaries, where there is limited flushing. Although there is a tendency to try to convert waste into wetlands to replace lost habitat (Zedler et al. 1998), their value is not always assessed.

Most research on pollution has been on vegetated or unvegetated soft sediments, other than effects of disposal of sewage onto rocky reefs and effects of major spills onto intertidal shores. Sewage can directly affect the structure of marine assemblages living on or associated with hard substrata—e.g. fish (Smith et al. 1999), invertebrates and algae (Lopez Gappa et al. 1990)—but does not always have measurable effects (Chapman et al. 1995). Abundances of consumers may be greater near sewage outfalls, possibly because of greater quality or quantity of food (Lopez Gappa et al. 1990), although sessile invertebrates and macro-algae may be absent or show reduced coverage near sewage discharges (Gorgula and Connell 2004). Although contaminants, such as heavy metals, can accumulate in biota, this will alter patterns of distribution and abundance only if it changes mortality, recruitment, or outcomes of biological interactions. Poor competitors, but rapid colonizers can become more abundant in the presence of heavy metals—e.g. introduced bryozoans (Piola and Johnston (2006).

Solid contaminants in the form of silt, litter or other waste from runoff, stormwater and active disposal can smother benthic assemblages—e.g. smothering of corals by sediments derived from the land (Wesseling et al. 1999). Sedimentation has been linked to assemblages becoming dominated by turfing algae (Gorgula and Connell 2004). Marine litter can cover extensive areas of the seafloor, affecting benthic biota, although accumulations of hard waste can also form subtidal reefs, which provide habitat for hard-bottom assemblages (Chapman and Clynick 2006). Such impacts are often (illogically) viewed as positive when waste is intentionally discarded to form reefs for the enhancement of selected biota (usually fish) at the expense of other biota (see also Chap. 24 by Goodsell and Chapman).

It is important to remember that contamination seldom occurs without other disturbances, e.g. boating, spread of non-native species, alterations of shorelines, loss and fragmentation of habitat. In addition, although contamination itself may be polluting, getting rid of it may be equally problematic and cleaning up after spills can have severe and long-lasting effects on biota on hard substrata (Gelin et al. 2003).

17.2.2 Changes to Habitat Provided by Hard Substrata

Many changes to marine habitat are intentional—e.g. mining for sediment or minerals. These are largely confined to nearshore waters, although fishing or extraction of resources can occur many 100s of kilometres offshore. Coastal areas tend to be heavily populated and the prediction is that coastal urbanization will increase disproportionately compared to inland areas. A large human population requires considerable infrastructure and, because many goods are moved by sea, this entails building ports and dredging channels (which each produce wastes that need to be disposed of). Although impacts of land-reclamation and dredging are often associated with soft sediments (e.g. in seagrass meadows, mangroves and wetlands), they can also affect hard substrata, either directly by removal of habitat or indirectly by, for example, dredging, which increases suspended material in the water column which is then deposited on the substratum.

Loss of natural habitats is one of the major causes of local extinction of species, although there are few documented extinctions of the taxa that constitute the vast majority of marine fauna and flora (i.e. invertebrates and algae). These taxa may be more resilient to major disturbances or, perhaps, our poor knowledge of marine biodiversity means that local extinctions go unnoticed (Roberts and Hawkins 1999). No-one knows how many previously undescribed species may have disappeared, nor how much change most marine species can tolerate before populations can no longer recover.

As well as direct loss of habitat, anthropogenic impacts often result in habitat-fragmentation. Although fragmentation generally affects large stretches of habitat, it can also affect naturally patchy habitats, e.g. rocky reefs separated by mangrove forests (Goodsell et al. 2007). Fragmentation of the natural shoreline in Sydney Harbour by seawalls reduced diversity on adjacent intertidal shores, possibly due to changes in size of the shores, their isolation or the surrounding matrix. This occurred even though most species disperse through the water column, which is thought to make marine populations less vulnerable to habitat-fragmentation. Changes to water currents may also modify distributions of marine species by disrupting connectivity among populations. This may occur through changes to shorelines, adding structures into nearshore waters, changing depths of water, etc. Climate change is likely to have major permanent influences on large offshore water-currents, as already occurs with alternating El Niño/La Niña phenomena and these changes may affect patterns of recruitment over large areas of coast (see Chap. 18 by Harley and Connell).

Anthropogenic modification of natural marine habitats (see Chap. 19 by Airoldi et al.) also adds new hard substrata to areas (Glasby and Connell 1999), which can change distributions of species. Although not specifically built to provide habitat, many artificial substrata are used as habitat by many taxa. Because modified or constructed habitats differ from natural habitat in a number of ways—e.g. the material from which they are constructed, orientation, surface-topography—they may not support the same range of taxa that live on natural substrata (Chapman 2003). In addition, many structures form completely novel habitat—e.g. pontoons are large, floating, shaded structures that do not have a natural analogue in coastal estuaries (Glasby and Connell 2001). Artificial structures can also add hard substrata to areas where these are normally absent (e.g. groynes on sandy beaches). This will inevitably introduce taxa into areas where they do not normally occur, increasing the range of many species or acting as stepping stones to connect naturally separated populations.

Artificial structures often provide habitat for introduced species (Floerl and Inglis 2003). The strong association of invasive species with artificial habitat may be related to the absence or paucity of cover by native species, or the resilience of many invasive species to disturbance (e.g. Piola and Johnston 2006; King et al. 2007). A major impact associated with changes to habitat is homogenization of biota, which occurs via two pathways. First, vulnerable species that cannot survive in highly disturbed habitats tend to disappear from assemblages. These are often habitat-specialists, which are not particularly widespread and are those that tend to make local assemblages more unique, thus increasing diversity over large areas. Second, invasive species tend to be similar worldwide and so their introduction into widely separated assemblages also tends to make local assemblages more similar. Loss, fragmentation and alteration of habitat are likely to increase homogenization via each of these pathways. Losing habitat or replacing it with habitat that does not provide suitable conditions for all biota will increase local loss of species. Providing habitat that is suitable for invasive species will aggravate the situation.

17.2.3 Direct Effects on Species

It is impossible to separate the multiple effects of disturbances that change distributions of species; for example, changes to habitat often include direct removal or additions of species and contamination is often associated with other disturbances to the environment, e.g. coastal infrastructure (see Chap. 20 by Schiel for multiple stressors). Many changes to distributions of species thus occur indirectly as a consequence of other disturbances. People do, however, have direct effects on many marine biota, by exploitation, aquaculture or by moving species around.

Fish and invertebrates are directly exploited for food or bait through commercial, recreational or subsistence fisheries (Suchanek 1994). Numerous species are also collected for their ornamental value and for use in aquaria. Algae are collected for production of agar, alginates and, to a lesser extent, for food or fertiliser. Although commercial exploitation is considered much more intensive and, thus, more of a threat to biodiversity, the cumulative effects of recreational fishing can be considerable. Even subsistence fishing can be very destructive, resulting in almost complete removal of prized species over large areas of shores, or critical reductions and localized extinctions of species (e.g. Duran and Castilla 1989).

Associated with reductions in targeted species are shifts in associated assemblages. As populations have been depleted, the relative abundances of other species have increased (Moreno 2001). Russ and Alcala (1989) found major changes in the assemblage of coral reef fish after an increase in fishing in a previously protected marine reserve, with dramatic decreases in the abundance of favoured targets and increases in abundance of non-targeted species. Fishing of top predators may affect trophic chains; e.g. excessive exploitation of predators may have led to increased populations of urchins and, thence, to major changes to kelp forests (Tegner and Dayton 2000). Harvesting of herbivorous and carnivorous gastropods on rocky shores can strongly increase their source of food (Moreno 2001), which can, in turn, decrease diversity of other species (Duran and Castilla 1989).

Disturbances associated with fishing and extraction of resources can severely alter the physical structure of habitat and, thus, indirectly affect associated populations. Although most studies on disturbances associated with fishing gear have focused on soft sediments (Kaiser et al. 2006), many reefs are destroyed to remove fish. Exploitation of the eastern oyster (a multimillion dollar fishery) in the US led to the destruction of nearly 50% of habitat formed by oyster-reefs (Lenihan and Peterson 1998). Extraction of *Lithophaga lithophaga* off the coast of Italy destroyed large areas of reefs (Fanelli et al. 1994).

A pervasive and irreversible impact of human activity on natural marine ecosystems is introduction of non-indigenous species (see Chap. 22 by McQuaid and Arenas). This is not a new phenomenon but the opportunities for species introductions have steadily increased over recent centuries (Hewitt and Campbell 2007). Many have been deliberately introduced, particularly for aquaculture and fish stocking. The vast majority, however, arrive accidentally through human-mediated transport, including fouling on ships' hulls or in ballast water, transport of species for aquaculture, or accidental releases into the wild. Non-indigenous species are now common inhabitants of most geographic regions of the world. For example, up to 230 introduced species have been documented for a single estuary and about 400 are established in marine and estuarine habitats in the US alone (Ruiz et al. 1997).

Despite coastal habitats being most frequently invaded, the ecological effects of introduced species to native assemblages are not clearly understood and, unless (and until) the introduced species become a dominant component of the system, have rarely been investigated. They may, however, pose a significant stress to marine communities, particularly in areas that are already stressed by loss of habitat or high levels of contamination. The kelp *Undaria pinnatifida* is native to northeast Asia and its introduction to temperate coasts worldwide is associated with a dramatic decrease in species richness and diversity of native seaweeds (Casas et al. 2004).

Some introduced species, in contrast, increase abundances of native species. For example, the American slipper limpet, *Crepidula fornicata*, was accidentally introduced into Europe and soon labelled an "oyster pest". Nevertheless, it decreases predation of mussels by sea stars and may increase abundance, biomass and numbers of species in the benthos (de Montaudouin and Sauriau 1999). It is important to recognise, however, that all introduced species are environmental impacts and it is inherently illogical to decide that some impacts are "good" and others "bad", depending on one's anthropocentric view of their effects.

Whether the arrival of non-native species is due to humans or not, their success in establishing viable populations is often accelerated by anthropogenic disturbances. For example, the addition of artificial structures to urbanized estuaries provides favourable conditions for non-indigenous species, enhancing their spread in areas with little natural hard substrata (Ruiz et al. 1997). Glasby et al. (2007) showed that numbers of introduced species were 1.5–2.5 times greater on artificial structures than on nearby rocky reefs, which contrasted with native species that were much more common on natural reefs. Invasive organisms appear to be able to take advantage of chemically disturbed environments and a greater resistance to heavy metals may be an important characteristic of invading species (Piola and Johnston 2006).

17.3 Conclusions

This introduction is not meant to be a definitive summary of all of the anthropogenic changes to distributions and abundances of biota living on rocky reefs, which will be dealt with in more detail in the following chapters. It does, however, demonstrate the large scale of changes to rocky reefs that are being brought about by human activities and the cumulative effects of multiple disturbances that disrupt habitat in various ways, remove and/or introduce species, or add large amounts of chemical contaminants to waterways. Even when impacts are considered singly, it is important to remember the multiplicity of disturbances and their cumulative effects. No disturbance of marine habitats occurs in isolation from others, making predictions of impacts extremely difficult.

It is also extremely important to consider what is likely to change in the next century. Few, if any of the many current disturbances will decrease. In addition, climatic change will introduce new ones: increases in sea level, changes in major oceanic currents, acidification of the oceans and increased storm activities along many coasts. Some of these will change distributions of species over large areas—e.g. due to changes in currents or water temperature. Other changes will increase loss of habitat. So far, most public concern regarding sea-level rise has focused on loss of habitat for people but, in addition, large areas of intertidal habitat will be lost along coasts where intertidal habitats cannot retreat inland. There will be added incentive to "armour" the shoreline to keep the sea at bay, leading not only to loss of intertidal habitat but also to proliferation of more artificial habitat, with potentially dire consequences for diversity. Key research needs in the near future are to identify synergistic and antagonistic effects of different disturbances, so that they are not considered in isolation—this will be essential for better management and conservation of marine biodiversity.

Acknowledgements We thank numerous colleagues for extensive discussions on many topics raised here. This contribution was supported by funds from the Australian Research Council Special Research Centre's Scheme and the University of Sydney.

References

Barreiro R, Real C, Carballeira A (1993) Heavy-metal accumulation by *Fucus ceranoides* in a small estuary in north-west Spain. Mar Environ Res 36:39–61

Casas G, Scrosati R, Piriz ML (2004) The invasive kelp *Undaria pinnatifida* (Phaeophyceae, Laminariales) reduces native seaweed diversity in Nuevo Gulf (Patagonia, Argentina). Biol Invas 6:411–416

Chapman MG (2003) Paucity of mobile species on constructed seawalls: effects of urbanization on biodiversity. Mar Ecol Prog Ser 264:21–29

Chapman MG, Clynick BG (2006) Experiments testing the use of waste material in estuaries as habitat for subtidal organisms. J Exp Mar Biol Ecol 338:164–178

Chapman MG, Underwood AJ, Skilleter GA (1995) Variability at different spatial scales between a subtidal assemblage exposed to the discharge of sewage and two control assemblages. J Exp Mar Biol Ecol 189:103–122

de Montaudouin X, Sauriau PG (1999) The proliferating Gastropoda *Crepidula fornicata* may stimulate macrozoobenthic diversity. J Mar Biol Assoc UK 79:1069–1077

Duran LR, Castilla JC (1989) Variation and persistence of the middle rocky intertidal community of Central Chile with and without human harvesting. Mar Biol 103:555–562

Fanelli G, Piraino S, Belmonte G, Geraci S, Boero F (1994) Human predation along Apulian rocky coasts (SE Italy): desertification caused by *Lithophaga lithophaga* (Mollusca) fisheries. Mar Ecol Prog Ser 110:1–8

Floerl O, Inglis GJ (2003) Boat harbour design can exacerbate hull fouling. Austral Ecol 28:116–127

Gelin A, Gravez V, Edgar GJ (2003) Assessment of Jessica oil spill impacts on intertidal invertebrate communities. Mar Pollut Bull 46:1377–1384

GESAMP (1980) Monitoring biological variables related to marine pollution. Group of Experts in Marine Pollution reports and studies no 12. UNESCO, Geneva

Glasby TM, Connell SD (1999) Urban structures as marine habitats. Ambio 28:595–598

Glasby TM, Connell SD (2001) Orientation and position of substrata have large effects on epibiotic assemblages. Mar Ecol Prog Ser 214:127–135

Glasby TM, Underwood AJ (1996) Sampling to differentiate between pulse and press perturbations. Environ Monitor Assess 42:241–252

Glasby TM, Connell SD, Holloway MG, Hewitt CL (2007) Nonindigenous biota on artificial structures: could habitat creation facilitate biological invasions ? Mar Biol 151:887–895

Goodsell PJ, Chapman MG, Underwood AJ (2007) Differences between biota in anthropogenically fragmented and in naturally patchy habitats. Mar Ecol Prog Ser 351:15–23

Gorgula SK, Connell SD (2004) Expansive covers of turf-forming algae on human-dominated coast: the relative effects of increasing nutrient and sediment loads. Mar Biol 145:613–619
Hewitt CL, Campbell ML (2007) Mechanisms for the prevention of marine bioinvasions for better biosecurity. Mar Pollut Bull 55:395–401
Jackson JBC (2001) What was natural in the coastal oceans ? Proc Natl Acad Sci 98:5411–5418
Kaiser MJ, Clarke KR, Hinz H, Austen MCV, Somerfield PJ, Karakassis I (2006) Global analysis of response and recovery of benthic biota to fishing. Mar Ecol Prog Ser 311:1–14
King RS, Deluca WV, Whigham DF, Marra PP (2007) Threshold effects of coastal urbanization on *Phragmites australis* (common reed) abundance and foliar nitrogen in Chesapeake Bay. Estuaries Coasts 30:469–481
Lake PS (2000) Disturbance, patchiness, and diversity in streams. J N Am Benth Soc 19:573–592
Lenihan HS, Peterson CH (1998) How habitat degradation through fishery disturbance enhances impacts of hypoxia on oyster reefs. Ecol Appl 8:128–140
Lopez Gappa JJL, Tablado A, Magaldi NH (1990) Influence of sewage pollution on a rocky intertidal community dominated by the mytilid *Brachidontes rodriguezi*. Mar Ecol Prog Ser 63:163–176
Moreno CA (2001) Community patterns generated by human harvesting on Chilean shores: a review. Aquat Conserv 11:19–30
Myers AA, Southgate T, Cross TF (1980) Distinguishing the effects of oil pollution from natural cyclical phenomena on the biota of Bantry Bay, Ireland. Mar Pollut Bull 11:204–207
Piola RF, Johnston EL (2006) Differential resistance to extended copper exposure in four introduced bryozoans. Mar Ecol Prog Ser 311:103–114
Roberts CM, Hawkins JP (1999) Extinction risk in the sea. Trends Ecol Evol 14:241–246
Rothschild BJ, Ault JS, Goulletquer P, Heral M (1994) Decline of the Chesapeake Bay oyster population—a century of habitat destruction and overfishing. Mar Ecol Prog Ser 111:23–29
Ruiz GM, Carlton JT, Grosholz ED, Hines AH (1997) Global invasions of marine and estuarine habitats by non-indigenous species: mechanisms, extent, and consequences. Am Zool 37:621–632
Russ GR, Alcala AC (1989) Effects of intense fishing pressure on an assemblage of coral reef fishes. Mar Ecol Prog Ser 56:13–28
Smith AK, Ajani PA, Roberts DE (1999) Spatial and temporal variation in fish assemblages exposed to sewage and implications for management. Mar Environ Res 47:241–260
Stark JS (1998) Heavy metal pollution and macrobenthic assemblages in soft sediments in two Sydney estuaries, Australia. Mar Freshw Res 49:533–540
Stark JS, Riddle MJ, Smith SDA (2004) Influence of an Antarctic waste dump on recruitment to nearshore marine soft-sediment assemblages. Mar Ecol Prog Ser 276:53–70
Suchanek TH (1994) Temperate coastal marine communities biodiversity and threats. Am Zool 34:100–114
Tegner MJ, Dayton PK (2000) Ecosystem effects of fishing in kelp forest communities. ICES J Mar Sci 57:579–589
Thrush SF, Dayton PK (2002) Disturbance to marine benthic habitats by trawling and dredging: implications for marine biodiversity. Annu Rev Ecol Syst 33:449–473
Underwood AJ, Chapman MG (1999) Problems and practical solutions for quantitative assessment of biodiversity of invertebrates in coastal habitats. In: Ponder W, Lunney D (eds) The other 99%: the conservation and biodiversity of invertebrates. Trans R Zool Soc New South Wales, Sydney, pp 19–25
Wesseling I, Uychiaoco A, Alino P, Aurin T, Vermaat J (1999) Damage and recovery of four Philippine corals from short-term sediment burial. Mar Ecol Prog Ser 176:11–15
Zedler JB, Fellows MQ, Trnka S (1998) Wastelands to wetlands: links between habitat protection and ecosystem science. In: Pace ML, Groffman PM (eds) Successes, limitations and frontiers in ecosystem science. Springer, Berlin Heidelberg New York, pp 69–112

Chapter 18
Shifts in Abiotic Variables and Consequences for Diversity

Christopher D.G. Harley and Sean D. Connell

18.1 Introduction

Throughout Earth's history, abiotic conditions in the oceans have changed naturally at timescales ranging from days to decades to millennia. Recently, this natural variation has been overlaid by directional climate change as a result of human activities, such as the burning of fossil fuels. Humans are also changing local environmental conditions by altering (usually increasing) the flow of nutrients, sediments and pollutants. In this chapter, we discuss how these abiotic factors vary in time and space, and briefly address some of the ecological consequences of that variation. We break abiotic changes down according to the scales at which they operate; for a start, we describe the global phenomenon of anthropogenic climate change, then we explore several of the important 'natural' interannual and interdecadal cycles which have strong impacts at the regional scale. Finally, we discuss local changes which are driven by human activities.

18.2 Global-Scale Change

Anthropogenic climate change affects the oceans at the global scale through a variety of pathways (Fig. 18.1). The root cause of anthropogenic climate change is the emission of greenhouse gases—particularly carbon dioxide—into the atmosphere through the burning of fossil fuels and other activities such as deforestation. Once in the atmosphere, greenhouse gases have important consequences for the heat budget of the planet. Specifically, some of the Sun's thermal energy, which would otherwise re-radiate into space, is trapped in the atmosphere, resulting in warming. This phenomenon is commonly known as the greenhouse effect, and it is expected to increase global surface temperatures by several degrees during the current century (IPCC 2007). Rising temperatures are causing sea levels to rise due to the thermal expansion of the oceans and the addition of melt water from terrestrial glaciers and ice caps. Furthermore, because warming will not be uniform over the Earth's surface, there will be changes in storm patterns and in ocean circulation. Greenhouse gases will also have important chemical effects in the atmosphere and in the oceans.

M. Wahl (ed.), *Marine Hard Bottom Communities*, Ecological Studies 206,
DOI: 10.1007/978-3-540-92704-4_18, © Springer-Verlag Berlin Heidelberg 2009

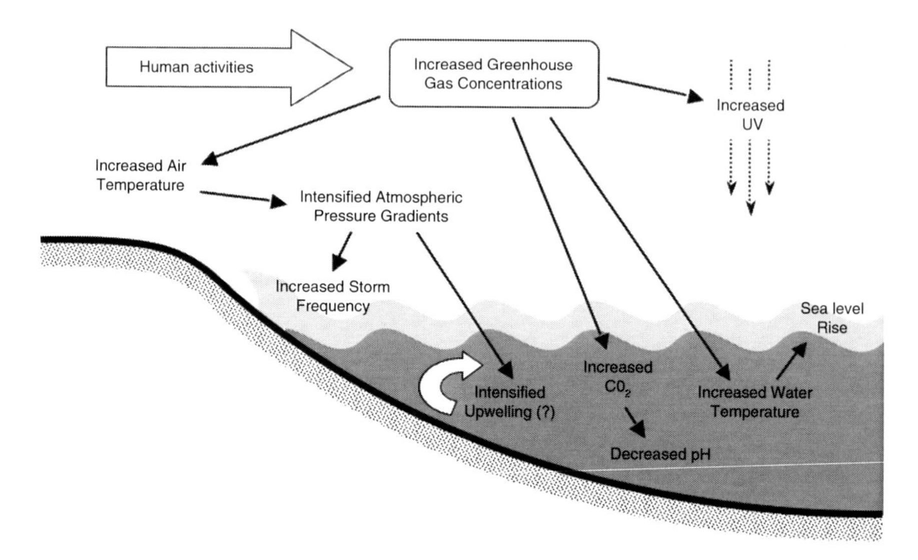

Fig. 18.1 Abiotic changes associated with climate change may lead to an array of potential stressors, some of which are well known and others uncertain (?). The interplay between these abiotic changes and the response of the biota is shaded by thick uncertainty. Reprinted with permission from Harley et al. (2006)

For example, they are thought to interfere with the ozone layer, resulting in increased levels of ultraviolet radiation at the Earth's surface. Perhaps more importantly, excess carbon dioxide in the atmosphere is being absorbed by the oceans, which are becoming more acidic as a result. All of these physical and chemical changes have important implications for the ecology of benthic ecosystems.

18.2.1 Changes in Water Temperature

Temperature is one of the most important components of an organism's environment. It affects metabolic rate, growth, reproduction and survival. Because of the overriding importance of temperature, patterns of oceanic temperature change and the ecological effects of warming seas have received a great deal of scientific attention.

Sea surface temperatures have increased, and are expected to continue to increase, in virtually every part of the planet (IPCC 2007). As a result of warming trends, the distribution of many species is shifting towards the poles (see, e.g. Southward et al. 2005). Rising temperatures are also changing the ways in which species interact with one another. For example, the predatory sea star *Pisaster ochraceus*, which controls diversity on rocky shores in the Northeast Pacific, consumes prey more quickly when the water is warmer (Sanford 1999). The effects of many diseases are also exacerbated by warming (Harvell et al. 2002).

More severe effects of warming are expected in the tropics, particularly on coral reefs, despite the fact that warming in the tropics is expected to be less pronounced than that at higher latitudes. Coral reef species have evolved in relatively stable thermal environments, and even small changes in temperature may exceed the range of conditions to which they have adapted. For example, many corals already live very close to their maximal thermal tolerance, and seemingly minor increases in temperature can, and have, resulted in widespread mortalities (Hughes et al. 2003).

Temperature is also predicted to have important consequences on the phenology (timing of life-history transitions) of marine species. For example, in northern Europe, the timing of larval release by the bivalve *Macoma balthica* has shifted to earlier in the spring due to warming. It now occurs earlier than would be optimal for larvae to feed on the spring plankton bloom (Philippart et al. 2003). Mortality due to predation has also increased for *Macoma*, as seasonally abundant predatory shrimp now arrive earlier in the year when juvenile clams are smaller and more vulnerable to predators (Philippart et al. 2003).

18.2.2 Changes in Sea Level

Mean sea level is currently rising at roughly 2–3 mm per year, although actual rates of sea-level rise vary regionally depending on uplift or subsidence of the coastline (IPCC 2007). Predictions for sea-level rise in the current century range from approximately 20 to 50 cm (IPCC 2007). This has important implications for the distribution of shallow-water benthic marine species, which typically live in a particular depth range relative to the ocean's surface or, in the intertidal zone, relative to the positions of high and low tide. As sea level rises, intertidal species will, of necessity, shift their zones upwards to keep pace. This will be fairly straightforward for most species. There are, however, two general cases in which sea-level rise will have important ecological consequences. The first is for slow-growing species such as corals, which can increase their vertical position in the water column only by the slow process of accretion. If sea-level rise outpaces accretion rates, then many shallow-water reefs may "drown" (Knowlton 2001). The second important case applies to shorelines which are not topographically uniform. For example, the high intertidal zone on many rocky shores is a broad, nearly horizontal, wave-cut bench bordered on the shoreward side by steep cliffs. Thus, a species which is a high intertidal specialist may currently enjoy a very extensive habitat within its vertical range. Sea-level rise could displace the vertical range of this species above the broad horizontal benches to a very narrow ribbon along the vertical cliffs. This will greatly reduce the available habitat for such species, assuming that they can survive on vertical cliffs at all.

18.2.3 Increasing Frequency and Intensity of Storms

Because of the large thermal inertia of water compared to land, the oceans are not warming quite as quickly as the continents (IPCC 2007). This is resulting in stronger temperature gradients between land and sea. Because these temperature gradients control wind patterns, stronger thermal gradients will result in stronger wind fields and, thus, more frequent and more intense storms. Indeed, wind speeds and significant wave heights have already increased in coastal regions, particularly in the northern hemisphere (IPCC 2007). Tropical cyclones, which are fuelled by warm surface waters, are also predicted to increase in intensity during the current century (IPCC 2007).

Storms are a major source of disturbance and play an important role in structuring hard-bottomed communities. Storm disturbance is not necessarily a 'bad thing', as it prevents competitively dominant species from excluding other species and can thus serve to enhance biodiversity (Sousa 1979). If storms become too frequent, or if damage is too extensive, there may be insufficient time for recovery between storm events and biodiversity may decline.

18.2.4 Changes in Upwelling and Circulation

Ocean currents and upwelling are driven by the winds. As wind patterns change, so too will patterns of ocean circulation. Future patterns of upwelling are particularly difficult to predict. On the one hand, increasing coastal winds may lead to enhanced upwelling, thereby bringing cool, nutrient-rich waters to the surface and negating some of the effects of warming (Bakun 1990). On the other hand, rising temperatures could lead to increasing stratification of the surface ocean and, thus, inhibit upwelling (Roemmich and McGowan 1995). Although the exact future of regions of upwelling remains a matter of debate, there is little doubt that ocean circulation is an important determinant of benthic community structure (Connolly and Roughgarden 1998). Where upwelling is strong, seaweeds and kelps tend to perform well because of the abundance of nutrients. Conversely, species with planktonic larvae, such as invertebrates and fish, may suffer from reduced recruitment if their larvae are washed offshore away from the shallow-water habitats of the adults. Patterns of recruitment can play a central role in determining the distribution and abundance of benthic species (Underwood and Keough 2001).

18.2.5 Ocean Acidification

As carbon dioxide concentrations increase, excess CO_2 is shifting the carbonate equilibrium in the surface ocean (Feely et al. 2004). Because this results in a decrease in pH, it is commonly referred to as ocean acidification. The important outcome of this shifting chemical balance is a reduction in available carbonate ions, which are the building blocks of the shells and structural defences of plants and animals, ranging from crustose algae to corals to oysters to sea urchins. The more CO_2 that is absorbed by the ocean, the more difficult it is for these taxa to build calcium carbonate skeletons (Feely et al. 2004). In many cases, reduced calcification will translate directly into reduced growth, as has been shown for mussels (Berge et al. 2006). Although researchers are beginning to understand which species will suffer reduced growth and by how much, the ways in which ocean acidification will change benthic community structure (e.g. competitive dominance hierarchies and predator–prey dynamics) and ecosystem function (e.g. primary production and nutrient cycling) remain essentially unknown (Fabry et al. 2008). Our lack of understanding should not, however, downplay the importance of ocean acidification. Unlike global warming, predicted changes in seawater pH have not been experienced in tens of millions of years (Feely et al. 2004). In other words, predicted acidification is well beyond the evolutionary experience of most species. Furthermore, there are no equivalents of cooler regions to which species could migrate to avoid acidification. In the oceans, the effects of acidification may ultimately become more important than those of warming.

18.2.6 Increasing UV Radiation

Human activities have resulted in a general thinning of the ozone layer which serves to block much of the ultraviolet (UV) radiation reaching the Earth from the Sun. UV radiation damages DNA and can be lethal to marine organisms in shallow waters. The organisms most vulnerable to UV damage include those which require bright sunlight for other reasons (e.g. intertidal and shallow-water photosynthesizers, such as seaweeds and many corals), and small life-history stages (e.g. larvae and juveniles) which have not developed UV-protective mechanisms such as thick exoskeletons or heavy pigmentation (Harley et al. 2006 and references therein). Although the effects of UV radiation could become more important through time, the impact of current levels of UV radiation on marine benthic communities appears to be relatively minor (Wahl et al. 2004).

18.3 Regional-Scale Change

There are several regional-scale climate cycles which influence the ecology of the marine benthos. The El Niño Southern Oscillation (ENSO) is perhaps the best known but interdecadal oscillations in the North Atlantic and North Pacific also play important roles in those ocean basins. Unlike anthropogenic climate change, which is directional, these climate cycles oscillate between one phase and another. They can thus produce rapid accelerations of abiotic change when the direction is the same as the global warming trend. Certain regions may also experience periodic reversals when, for example, the switch to a cool phase of an interdecadal cycle temporarily overrides the more gradual trend of global warming.

18.3.1 El Niño–Southern Oscillation (ENSO)

The El Niño-Southern Oscillation refers to changes in the equatorial Pacific Ocean which vary in a 3–7 year cycle. During an El Niño event, unusually warm equatorial surface waters extend from the coast of South America westwards for thousands of kilometres. Warm surface waters and a bulge of elevated sea level also propagate north and south along the west coast of the Americas, and have strong effects in Australia and Southeast Asia. The warm water during El Niño events leads to brief poleward range expansions of many marine species; fish and other mobile taxa were found north of their usual ranges in California during the 1997–1998 El Niño (Hayward 2000). ENSO events also change patterns of larval transport, with important implications for benthic species such as crabs and sea urchins (Botsford 2001). El Niños are often characterized by unusually powerful storms, which can cause extensive damage to nearshore ecosystems such as kelp beds. Changes in other abiotic patterns, particularly rainfall, are also noteworthy as they can change the salinity and hydrography of estuaries. The frequency of El Niño events has increased in recent years, and it has been suggested that this increase may be due to anthropogenic climate change.

The alternate phase of the ENSO cycle is known as a La Niña. During La Niñas, the surface waters of the eastern equatorial Pacific are unusually cold, and many of the weather anomalies observed during an El Niño are reversed.

18.3.2 Other Interannual Oscillations

The atmospheric circulation patterns in the North Atlantic region vary in an ~8 year cycle in what is known as the North Atlantic Oscillation (NAO). The cycle is defined by changes in the position of high and low atmospheric pressure systems over the region, and variation in these atmospheric parameters leads to changes in circulation and water temperature at sites in the North Atlantic Ocean. Through this cascade of abiotic changes, the NAO can have important implications for hard-bottom ecology. For example, benthic diversity was inversely correlated with the NAO index over a multi-decadal observation period in Svalbard, and the shift from the positive to the negative mode of the NAO in the mid-1990s coincided with a decline in sea anemones and an increase in brown algae (Beuchel et al. 2006).

The Pacific Decadal Oscillation (PDO) similarly refers to shifting positions of high and low pressure systems over the North Pacific. The PDO alternates between ‘warm’ and ‘cool’ phases, each of which lasts for 20–30 years. During the warm phase, sea surface temperatures are warmer along the west coast of North America, and cooler in the central and western North Pacific. Productivity in Alaska is typically enhanced, and productivity in the California Current is typically depressed (Francis et al. 1998). During the cool phase, the inverse patterns prevail. Although PDO impacts on pelagic fisheries are well established, less is known about their effects on the benthos. Large declines in the diversity of mussel-bed dwelling fauna may be attributable to changes in the PDO (Smith et al. 2006) but the mechanistic linkages remain unclear.

18.4 Local-Scale Change

A broad spectrum of land- and marine-based activities, coupled with continued growth of the human population and migration to coastal areas are driving unprecedented and complex changes in water chemistry at the scale of individual coastlines (i.e. 100s m to 10s km). These abiotic shifts are important to understand because it is at these smaller scales that most organisms interact with their environment. Local conditions not only represent the cumulative effect of broad- through small-scale conditions but variation in local influences may also rival climate change in their impacts on ecological diversity. Whilst climate change is of emerging concern, this necessary focus is insufficient per se for forecasting future abiotic environments. We need to understand the manifestation of climate change in local settings.

18.4.1 Permanent Abiotic Shifts: a Catchment Perspective

Permanent change in abiotic environments is created by the continuous and increasing amounts of terrestrially derived discharge through point and non-point sources (Carpenter et al. 1998). The diverse nature of human activities which influence ocean chemistry means that it is useful to understand spatial and temporal shifts in abiotic environments from a perspective of various 'catchment types'. This provides a focus on the collective abiotic environment, rather than on individual parameters which can vary in concentration and type over very short timescales (i.e. hours, days, weeks). Understanding land-sea linkages among different types of catchments (i.e. land–use practices) provides a useful framework for understanding human-driven changes at local through middle scales. Although non-permanent changes in abiotic environments created by short-term activities (e.g. dredging, drilling and oil spills) can cause intergenerational change to ecological systems, this section focuses on permanent changes to the abiotic environment caused by long-term activities (i.e. coastal urbanisation and agriculture).

Abiotic shifts are most noticeable along the land–sea fringe where human populations have increased in the coastal zone and have intensified the use of land for agriculture and urban living. These activities can alter adjacent marine habitats and their ecological functions (Vitousek et al. 1997). Throughout the Tertiary, human inhabitation and expansion in the coastal zone and destruction of native vegetation (e.g. Flannery 1994) has altered land-sea linkages. In the last few centuries, humans have accelerated coastal development and subsequent nutrient input into coastal seas through agriculture, widespread land clearing, extensive use of fertilisers and pesticides, and the introduction of herbivores (e.g. sheep and cattle). Recent urbanisation has produced large and densely populated cities and a bewildering array of coastal inputs, arising from industrial (e.g. heavy metal and industrial chemicals) and residential activity (e.g. runoff of lawn fertilisers, waste from pets, pesticides) to pharmaceutical and recreational activities (e.g. oestrogen and synthetic steroid present in sewage effluent).

During the last few decades, diversification of novel coastal inputs, as well as increasing demand for and intensification of use of coastal land mean that identifying ecological changes caused by any one human activity is problematic (see Chap. 20 by Schiel). Rather than understanding how the dynamics of each abiotic parameter links with ecological change, it is particularly instructive to assess the type and magnitude of ecological change in association with the broad nature of land-based activities, i.e. catchment-type (McClelland and Valiela 1998). Of particular concern has been the increase of land-derived material from soil erosion, runoff and sewage release (Bouwman et al. 2005). This discharge contains large amounts of suspended material, soluble nutrients and toxicants which alter the physical environment of coastal seas through increasing deposition of sediments (sedimentation), and increasing levels of nutrients (eutrophication) and toxicants (contamination).

The most noticeable changes to the land–sea fringe centre on accelerated soil erosion, heavy and inefficient use of fertilisers in rural catchments (Bouwman et al. 1997), and elevated nutrients and sediments associated with point and non-point sources in urban catchments. Natural forest catchments export few nutrients, relative to cleared catchments (see review by Harris 2001), and urban and agriculture catchments generally have a significantly higher δ^{15}N value than seawater. The strength of this signal is directly related to the proportion of human influence (McClelland and Valiela 1998). It is important to recognise that this signal is related to the proportion of human *influence*, rather than the proportion of human *population*. The historical idea that size of human populations is proportional to the size of ecological impact is misleading because it does not account for key contingencies. Variation in treatment of discharge, historical and ambient concentrations of nutrients, and presence of key functional groups of consumers (e.g. herbivores) modify the influence of human populations. For example, benthic herbivores can counter sudden nutrient events in oligotrophic systems (i.e. increased rates of consumption of opportunistic algae in low nutrient systems) but not in eutrophic systems (Russell and Connell 2007). Hence, in this example, size of ecological impact is less a product of size of human population and more a product of human influence (e.g. influence of heavy fishing vs. marine protected areas on herbivores) as contingent on the disparity between resource availability between donor (concentration of nutrient input from land) and recipient system (concentration of nutrients in sea; Connell 2007).

18.4.2 Regional and Middle-Scale Contingencies of the Catchment Perspective

Understanding ambient and historical abiotic conditions within their regional contexts is important. In Australia, marine ecologists often fail to detect simple correlations between size of human populations and magnitude of ecological impacts caused by nutrient inputs. The influence of nutrient pollution may be greater for smaller coastal populations where ambient levels are naturally low (i.e. oligotrophic coasts). For example, the effects of nutrients adjacent to Australia's largest city (Sydney) have relatively minor effects (Russell et al. 2005; S.D. Connell, unpublished data). In this regard, it may be more informative to assess the ratio between nutrient delivery (e.g. concentration of human inputs) and productivity (e.g. natural nutrient concentrations) as a framework to reconcile regional differences in ecological responses (i.e. predicting larger responses where the ratio is largest).

At middle scales, coastal geomorphology represents a useful feature to consider both global and local-scale shifts in abiotic parameters. The inshore of coral reefs and the sheltered gulfs and bays of rocky coasts appear more vulnerable to terrestrial runoff from developed catchments, and tend to have less flushing and smaller volumes of water. Declining water quality in semi-enclosed water (e.g. bays, gulfs, inshore reefs) is typically associated with declines in foundation species (e.g. macrophytes and corals), due to increases in structurally simple and opportunistic functional groups (Rabalais 2002).

18.4.3 Departures: Abiotic Shifts Can Be Subtle and Disconnected from Their Source

The study of shifting abiotic conditions in the ecology of rocky coasts has a relatively short history. Such shifts are often insidious, particularly when compared to the often visually impressive effects of invasive species, consumer outbreaks and fishing. Many of these biotic impacts are well documented and common to most subtidal rocky habitats worldwide. In contrast, shifts in abiotic environments have been neglected. The effects of declining water quality may be subtle, and sometimes may be secondary to other impacts. For example, in systems under strong influence of herbivores, nutrient effects are likely to become important when herbivores are removed (Worm et al. 2002) because fast-growing, opportunistic algae can then dominate space when released from intense rates of consumption (Eriksson et al. 2007). In many places, however, water quality may combine with other impacts to be a primary source of habitat degradation, as considered for historical shifts in dominance from large and slow-growing perennial algae (e.g. kelp forests) to low-lying, fast-growing ephemeral algae (Eriksson et al. 2002; Connell et al. 2008).

Whilst the 'catchment-type' approach appears to provide a progressive framework for research, it is worth recognising that some parameters may not be best understood at a catchment scale. For example, organochlorine chemicals, including chlorinated pesticides, are found at appreciable concentrations in the polar regions, presumably as a result of long-range atmospheric transport (Wania and Mackay 1993). Additionally, it is perplexing that terrestrial runoff to a bay (Monterey Bay) can create spectacular changes (albeit non-permanent), including diatom blooms, which have killed over 400 California sea lions (*Zalophus californianus*) but the diatoms could not be observed even in trace amounts in the benthos (i.e. blue mussels, *Mytilus edulus*; Scholin et al. 2000). In coastal systems, there remains much to discover about the conditions in which land-sea linkages are uncoupled from benthic ecology.

18.5 Conclusions

This chapter recognises that the abiotic conditions of the world's oceans have changed naturally at timescales ranging from days to decades to millennia, but we highlight the directional changes that overlay this natural variation as a result of human activities. Whilst we find that these shift as a function of their scale of operation, we conclude that they can combine in alternate ways which depend on local conditions. This cross-scale interaction means that the abiotic conditions at any one place will reflect the combined influence of broader-scale (e.g. temperature and storm frequency) and local-scale interactions (e.g. catchment type × coastal morphology).

The ecological response to these cross-scale interactions is contingent on historical and ambient abiotic conditions (e.g. productivity) and the biotic conditions which create ecological resistance and resilience (e.g. functional types of consumers; Connell and Irving 2008). For example, the novel environmental conditions generated by climate change (e.g. increases in frequency and intensity of storms) create bigger abiotic shifts (e.g. nutrient and sediment runoff) on coasts adjacent to developed catchments (i.e. urban and rural catchments) in semi-enclosed waters (e.g. gulfs and bays). The largest abiotic shifts are, therefore, a product of their local–regional settings. The largest ecological shifts are contingent on the capacity of the biological systems to withstand (resistance), or absorb and recover from (resilience) the multiple assaults of natural and human origin. Resilience is particularly relevant for subtidal systems in which natural disturbance and turnover is frequent, and the capacity for normal recovery in human-altered environments is uncertain (Hughes and Connell 1999).

While the link between human activity and abiotic changes appears obvious, ecologists continue to be surprised by the ecological outcomes—indeed, "ecological surprises" are set to continue (Paine et al. 1998). These surprises occur as a product of synergies between abiotic and biotic drivers which cannot be predicted by simply 'adding up' the effects of single drivers. As progress is made in identifying the conditions leading to major abiotic shifts, there will be a need to assess the parameters most responsible for this ecological change and the scales at which they operate. These assessments will need to focus on meaningful combinations ensuring that predictions of future shifts in ecological diversity are not underestimated by any false impression created by summing single drivers. By improving predictive power, such an approach will help guide conservation and management responses to abiotic change.

References

Bakun A (1990) Global climate change and intensification of coastal ocean upwelling. Science 247:198–201

Berge JA, Bjerkeng B, Pettersen O, Schaaning MT, Oxnevad S (2006) Effects of increased sea water concentrations of CO2 on growth of the bivalve *Mytilus edulis* L. Chemosphere 62:681–687

Beuchel F, Gulliksen B, Carrol ML (2006) Long-term patterns of rocky bottom macrobenthic community structure in an Arctic fjord (Kongsfjorden, Svalbard) in relation to climate variability (1980–2003). J Mar Syst 63:35–48

Botsford LW (2001) Physical influences on recruitment to California Current invertebrate populations on multiple scales. ICES J Mar Sci 58:1081–1091
Bouwman AF, Lee DS, Asman WAH, Dentener FJ, Van der Hoek KW, Olivier JGJ (1997) A global high-resolution emission inventory for ammonia. Global Biogeochem Cycles 11:561–587
Bouwman AF, Van Drecht G, Knoop JM, Beusen AHW, Meinardi CR (2005) Exploring changes in river nitrogen export to the world's oceans. Global Biogeochem Cycles 19 GB1002 doi:10.1029/2004GB002314
Carpenter SR, Caraco NF, Correll DL, Howarth RW, Sharpley AN, Smith VH (1998) Nonpoint pollution of surface waters with phosphorus and nitrogen. Ecol Appl 8:559–568
Connell SD (2007) Water quality and the loss of coral reefs and kelp forests: alternative states and the influence of fishing. In: Connell SD, Gillanders BM (eds) Marine ecology. Oxford University Press, Melbourne, pp 556–568
Connell SD, Irving AD (2008) Integrating ecology with biogeography using landscape characteristics: a case study of subtidal habitat across continental Australia. J Biogeogr 35:1608–1621
Connell SD, Russell BC, Turner DJ, Shepherd SA, Kildae T, Miller DJ, Airoldi L, Cheshire A (2008) Recovering a lost baseline: missing kelp forests on a metropolitan coast. Mar Ecol Prog Ser 360:63–72
Connolly SR, Roughgarden J (1998) A latitudinal gradient in Northeast Pacific intertidal community structure: evidence for an oceanographically based synthesis of marine community theory. Am Nat 151:311–326
Eriksson BK, Johansson G, Snoeijs P (2002) Long-term changes in the macroalgal vegetation of the inner Gullmar Fjord, Swedish Skagerrak coast. J Phycol 38:284–296
Eriksson BK, Rubach A, Hillebrand H (2007) Dominance by a canopy forming seaweed modifies resource and consumer control of bloom-forming macroalgae. Oikos 116:1211–1219
Fabry VJ, Seibel BA, Feely RA, Orr JC (2008) Impacts of ocean acidification on marine fauna and ecosystem processes. ICES J Mar Sci 65:414–432
Feely RA, Sabine CL, Lee K, Berelson W, Kleypas J, Fabry VJ, Millero FJ (2004) Impact of anthropogenic CO_2 on the $CaCO_3$ system in the oceans. Science 305:362–366
Flannery TF (1994) "The future eaters." New Holland Publishers, Sydney
Francis RC, Hare SR, Hollowed AB, Wooster WS (1998) Effects of interdecadal climate variability on the oceanic ecosystems of the NE Pacific. Fish Oceanogr 7:1–21
Harley CDG, Hughes AR, Hultgren KM, Miner BG, Sorte CJB, Thornber CS, Rodriguez LF, Tomanek L, Williams SL (2006) The impacts of climate change in coastal marine systems. Ecol Lett 9:228–241
Harris GP (2001) Biogeochemistry of nitrogen and phosphorous in Australian catchments, rivers and estuaries: effects of land use and flow regulation and comparisons with global patterns. Mar Freshw Res 52:139–149
Harvell CD, Mitchell CE, Ward JR, Altizer S, Dobson AP, Ostfeld RS, Samuel MD (2002) Climate warming and disease risks for terrestrial and marine biota. Science 296:2158–2162
Hayward TL (2000) El Niño 1997–98 in the coastal waters of southern California: a timeline of events. CalCOFI Rep 41:98–116
Hughes TP, Connell JH (1999) Multiple stressors on coral reefs: a long-term perspective. Limnol Oceanogr 44:932–940
Hughes TP, Baird AH, Bellwood DR, Card M, Connolly SR, Folke C, Grosberg R, Hoegh-Guldberg O, Jackson JBC, Kleypas J, Lough JM, Marshall P, Nystrom M, Palumbi SR, Pandolfi JM, Rosen B, Roughgarden J (2003) Climate change, human impacts, and the resilience of coral reefs. Science 301:929–933
IPCC (2007) Climate change 2007: the physical science basis. Contribution of Working Group I to the Fourth Assessment Report of the Intergovernmental Panel on Climate Change. Cambridge University Press, Cambridge
Knowlton N (2001) The future of coral reefs. Proc Natl Acad Sci 98:5419–5425
McClelland JW, Valiela I (1998) Linking nitrogen in estuarine producers to land-derived sources. Limnol Oceanogr 43:577–585
Paine RT, Tegner MJ, Johnson EA (1998) Compounded perturbations yield ecological surprises. Ecosystems 1:535–545

Philippart CJM, van Aken HM, Beukema JJ, Bos OG, Cadée GC, Dekker R (2003) Climate-related changes in recruitment of the bivalve *Macoma balthica*. Limnol Oceanogr 48:2171–2185
Rabalais NN (2002) Nitrogen in aquatic ecosystems. Ambio 31:102–112
Roemmich D, McGowan JA (1995) Climatic warming and the decline of zooplankton in the California current. Science 267:1324–1326
Russell BD, Connell SD (2007) Response of grazers to sudden nutrient pulses in oligotrophic v. eutrophic conditions. Mar Ecol Prog Ser 349:73–80
Russell BD, Elsdon TE, Gillanders BM, Connell SD (2005) Nutrients increase epiphyte loads: broad scale observations and an experimental assessment. Mar Biol 147:551–558
Sanford E (1999) Regulation of keystone predation by small changes in ocean temperature. Science 283:2095–2097
Scholin CA, Gulland F, Doucette G et al. (2000) Mortality of sea lions along the central California coast linked to a toxic diatom bloom. Nature 403:80–84
Smith JR, Fong P, Ambrose RF (2006) Dramatic declines in mussel bed community diversity: response to climate change? Ecology 87:1153–1161
Sousa WP (1979) Disturbance in marine intertidal boulder fields: the nonequilibrium maintenance of species diversity. Ecology 60:1225–1239
Southward AJ, Langmead O, Hardman-Mountford NJ, Aiken J, Boalch GT, Dando PR, Genner MJ, Joint I, Kendall MA, Halliday NC, Harris RP, Leaper R, Mieszkowska N, Pingree RD, Richardson AJ, Sims DW, Smith T, Walne AW, Hawkins SJ (2005) Long-term oceanographic and ecological research in the western English Channel. Adv Mar Biol 47:1–105
Underwood AJ, Keough MJ (2001) Supply-side ecology: the nature and consequences of variations in recruitment of intertidal organisms. In: Bertness MD, Gaines SD, Hay ME (eds) Marine community ecology. Sinauer, Sunderland, MA, pp 183–200
Vitousek P, Mooney H, Lubchenco J, Melillo J (1997) Human domination of earth's ecosystems. Science 277:494–499
Wahl M, Molis M, Davis A, Dobretsov S, Dürr ST, Johansson J, Kinley J, Kirugara D, Langer M, Lotze HK, Thiel M, Thomason JC, Worm B, Zeevi ben-Yosef D (2004) UV effects that come and go: a global comparison of marine benthic community level impacts. Global Change Biol 10:1962–1972
Wania F, Mackay D (1993) Global fractionation and cold condensation of low volatile organochlorine compounds in polar regions. Ambio 22:10–18
Worm B, Lotze HK, Hillebrand H, Sommer U (2002) Consumer versus resource control of species diversity and ecosystem functioning. Nature 417:848–851

Chapter 19
The Loss of Natural Habitats and the Addition of Artificial Substrata

Laura Airoldi, Sean D. Connell, and Michael W. Beck

19.1 Human Changes to Coastal Habitats

Humans depend on coastal environments for food, energy, construction, transport, recreation and many other resources and services. The use of these resources will inevitably have an influence on the coastal environment and associated marine life. Few coastal environments (if any) can today be regarded as "natural" in the sense that the non-human components are independent from anthropogenic changes (Jackson and Sala 2001). Even a coast without human development is affected by regional processes on the land or at sea, such as freshwater inflow, sedimentation, excess input of upstream nutrients, and threats such as oil spills and pollutants as well as sea-level rise (see Chap. 18 by Harley and Connell). The novel environmental conditions created by human activities modify the habitats and the diverse assemblages of plants and animals which rely on these for food and living space. These changes usually result in loss of diversity, structural complexity, functional attributes and ecosystem services (Airoldi et al. 2008). Current considerations about future population growth identify the coastal temperate zones as shouldering much environmental change. As coastal populations continue to increase, changes to habitats and their inhabitants will occur more and more. Thus, mismanagement of potentially disruptive human activities on natural coastal habitats carries a large cost for future generations (Millennium Ecosystem Assessment 2005).

This chapter focuses on changes to the distribution of hard-bottom habitats which are directly or indirectly attributable to human activities. Hard-bottom habitats are among the most diverse and productive systems in temperate marine environments but they are rapidly being degraded by human activities which directly damage and change the habitat itself (see Chaps. 17 by Clynick et al. and 18 by Harley and Connell). We report some key examples about the past and current losses of hard-bottom habitats, particularly those of biogenic origin. We also discuss threats related to the addition of artificial hard or rock-armoured substrata as a consequence of coastal urbanization, in particular the significance of these changes to coastal areas and why we should be concerned about these. Finally, we also point out how much change has already passed unrecognized, and recommend greater effort in the maintenance and restoration of what remains of valuable natural hard-bottom habitats.

M. Wahl (ed.), *Marine Hard Bottom Communities*, Ecological Studies 206,
DOI: 10.1007/978-3-540-92704-4_19, © Springer-Verlag Berlin Heidelberg 2009

This chapter thus expands upon many of the topics introduced in the overview of this book part.

19.2 Causes of Habitat Loss

Here, "habitat" indicates a focus on the predominant geological or biological features which create structural complexity in the environment (e.g. rocky outcrops, kelp forests or oyster reefs), while "loss" indicates a focus on a measurable reduction in the amount of habitat (Airoldi and Beck 2007). "Loss" clearly occurs when, for example, intertidal rocky shores are changed into urban coastal landscapes, or when shellfish reefs are dredged. We also refer to a loss when more structurally complex habitats are converted to less complex ones (e.g. canopy-forming algae are replaced by turfs or urchin barrens), which generally leads to smaller abundances (biomasses) and often to declines in species richness. Habitat loss commonly occurs through a process of fragmentation, when previously continuous habitats become patchier and less connected.

Habitat degradation (e.g. caused by invasions of non-native species or by pollution) also represent a serious alteration which often acts as a precursor to the loss of natural habitats (see Chaps. 18 by Harley and Connell and 22 by McQuaid and Arenas). Degradation, however, represents a decrease in condition, rather than a change in distribution of habitats as discussed in other chapters in this book part.

Habitat loss is directly or indirectly a consequence of the exponential growth of human populations in coastal areas, and the related intensification and multiplication of their impacts (Lotze et al. 2006). The loss of hard-bottom habitats has occurred through a variety of means, most of which are associated with land reclamation, coastal development (including indirect effects on circulation patterns), degraded water quality, and destructive fishing and overexploitation (e.g. Thompson et al. 2002). Coastal settlements, industries, ports and harbours, military installations, mines, power plants, tourist and other infrastructures have often directly damaged hard-bottom habitats and deeply modified the nature of coastal habitats and landscapes (Fig. 19.1), particularly in intertidal and shallow water areas.

Urbanization may have greater indirect effects on hard-bottom habitats by affecting water clarity and quality, in addition to enhancing sediment loads. Many biogenic habitats are considered highly sensitive to deterioration in water quality, including shellfish reefs, kelps and other macro-algal canopy habitats (Airoldi and Beck 2007). For example, the expansion of turf-forming algae at the expense of canopy-forming habitats on human-dominated rocky coast is clearly facilitated by the synergic effects of increasing nutrient and sediment loads (Airoldi 1998; Gorgula and Connell 2004).

Biogenic and rocky habitats are also severely damaged by disruptive fishing techniques. The most obvious threats are from direct commercial extraction of commercially valuable habitats, such as shellfish reefs and maerl beds (see Chap. 22 by McQuaid and Arenas for a dramatic example). Severe damage can also come from disruptive fishing techniques targeting other species. One of the most blatant

Fig. 19.1 Aerial view of the harbour of Genova (northwest Italy; photo courtesy F. Boero) showing the severe habitat changes due to coastal development prevailing along this rocky coast

examples is the use of pneumatic hammers and explosives to harvest the date mussel *Lithophaga lithophaga*. This fishery has devastated thousands of square metres of rocky coasts around the Mediterranean Sea, directly and irreversibly altering the rocky environment and causing the permanent loss of canopy-forming seaweeds and the formation of barrens (Guidetti et al. 2003). Offshore hard-bottom formations, both living biogenic and fossil in nature, are also significantly altered by the use of dredges and bottom trawling nets (Thrush and Dayton 2002). These destructive forms of fishery remove the fauna and damage the bottom surface, as a consequence of which many biogenic reefs have been flattened and reduced in size.

19.3 Trends of Habitat Loss

19.3.1 A Case History: the Decline of Native Oyster Reefs in Europe[1]

Of all hard-bottom habitats, biogenic formations are among those which historically have been subject to the most dramatic losses (Airoldi and Beck 2007). However, many of these losses have passed unnoticed and direct quantitative data are relatively scarce. The European native oyster case history provides a time

[1]Material in this section is largely from Airoldi and Beck (2007)

course of how humans can irreversibly force hard-bottom habitats to virtual ecological extinction. This dramatic process is typical of many exploited hard-bottom habitats around the world.

The European flat oyster (*Ostrea edulis*) is a sessile, filter-feeding bivalve mollusc which used to be very abundant throughout its native range, from Norway to Morocco and across the coasts of the Mediterranean and Black seas. It has been an extremely popular food for centuries and, in some British estuaries, there are archaeological signs of overexploitation of native oyster beds dating back to Roman times. For centuries, *O. edulis* reefs supported a productive commercial fishery but their abundance declined significantly during the 19th and 20th centuries (Mackenzie et al. 1997). Regulations and fishery closures were imposed in some regions. The decline could not, however, be halted and, in the 20th century, catches collapsed. By the late 19th century, beds of *O. edulis* were commercially extinct along most European coasts. This was explained largely in terms of overfishing and wasteful exploitation, combined with outbreaks of diseases, habitat loss and change or destruction, reduction in water quality and other large-scale environmental alterations, adverse weather conditions, and the introduction of non-native oysters (and associated parasites and diseases, such as the protozoan *Bonamia ostreae*) for aquaculture and other non-native species (e.g. the invasive gastropod *Crepidula fornicata*). Whatever the cause, the collapse continued inexorably and, by the mid 20th century, wild native beds were virtually nonexistent in many regions of Europe (reviewed in Airoldi and Beck 2007).

Nowadays, aquaculture provides the main supply of native oysters in most European countries (Mackenzie et al. 1997). This industry has also been seriously affected by epidemic diseases in recent decades, with documented losses of commercial stocks above 80% in France, and most Mediterranean native oyster beds in such poor conditions that they are unable to support intensive culture. Although marketplace demand for native oysters remains strong, the introduced Pacific oyster *Crassostrea gigas*, which is easier to cultivate than the native oyster, now makes the major share of oyster production in Europe. The sparse remains of wild native oyster beds are probably one of the most endangered marine habitats in Europe, and management actions are urgently needed for addressing their decline through conservation, restoration, fishery management, and aquaculture.

19.3.2 Habitat Conversion: Switches from Canopy Habitats to Barrens/Turfs

Because of their physical nature, rocky coasts are less vulnerable to fragmentation and loss than biogenic reefs. Nevertheless, many rocky coasts around the world have been degraded, which has made these extremely susceptible to shifts in habitat structure. Perennial canopies of kelps, fucoids and other complex, erect macro-algae—some of the largest, most diverse and valuable biogenic marine habitats—are clearly becoming rarer at local, regional and global scales.

This may be increasingly considered as one of the main global ecological threats for temperate rocky coasts (Steneck et al. 2002; Connell 2007; Airoldi and Beck 2007). Lost canopy-forming algae tend to be replaced by species with less structural complexity, such as turf-forming, filamentous or other ephemeral seaweeds (Fig. 19.2), which seem to be favoured in conditions of decreased water quality (Benedetti-Cecchi et al. 2001; Connell et al. 2008), or by mussels, particularly at exposed sites with high loads of particulate organic matter (Thibaut et al. 2005). Macro-algal canopies have also been frequently replaced by "barrens", where outbreaks of urchins, possibly due to overfishing, may have been the primary cause of macro-algal loss (Steneck et al. 2002; Guidetti et al. 2003).

Canopy-forming algae, turfs and barrens have been suggested to represent alternative states on shallow temperate rocky coasts under different disturbance and stress regimes (Fig. 19.3; Airoldi 1998; Worm et al. 1999; Connell 2005, 2007). The proximate cause of these habitat shifts are often easily observed (e.g. a massive pulse of sediments) but the factors maintaining the shifts are notoriously more difficult to identify (Connell 2007). This is particularly true of the insidious effects

Fig. 19.2 View of a degraded, turf-dominated reef along the urban coasts of Adelaide (Gulf St. Vincent, South Australia) after the complete loss of macro-algal canopies (photo S.D. Connell)

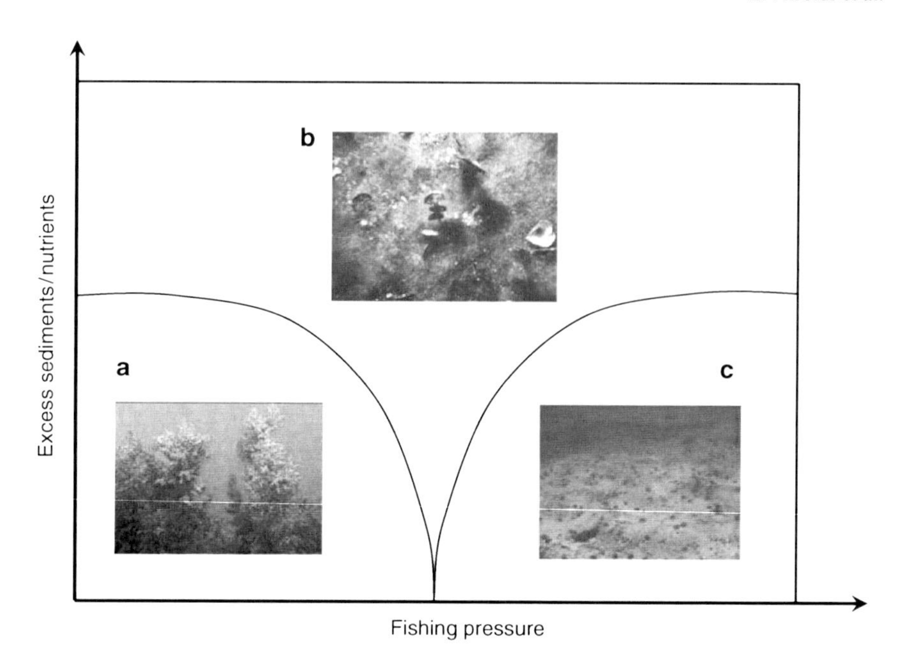

Fig. 19.3 Conceptual model of human-induced shifts between *A* algal canopies (photo F. Colosio), *B* algal turfs (photo L. Airoldi) and *C* urchin barrens (photo P. Guidetti). Recovery back from turfs is difficult because turfs and sediments, once established, seem to provide positive feedbacks for each other that deter grazers and inhibit recruitment of canopy-forming algae

of declining water quality, which often do not occur in a consistent and uniform manner and may be synergic with other local or global stressor (Gorgula and Connell 2004; see also Chap. 20 by Schiel). Thus, while there are examples of recovery of canopies from outbreak of urchins (e.g. in northern New Zealand, where the closure of fishing in marine protected areas has resulted in the rapid decline of sea urchin populations and the widespread recovery of kelp forests; Shears and Babcock 2002), turfs and sediments, once established, seem to provide positive feedbacks for each other which inhibit the recovery of canopy habitats (Airoldi 1998).

19.4 The Addition of Artificial Hard Substrata

One of the major consequences of coastal development and urbanization is the addition of a variety of artificial hard-bottom structures to coastal areas, including seawalls, dykes, breakwaters, groynes, jetties and offshore platforms. The extent and projected trend of this phenomenon are dramatic. A review of the status of European coastlines (Airoldi and Beck 2007) has shown that nowadays 22,000 km^2

of the coastal zone in Europe is covered in concrete or asphalt, and urban artificial surfaces have increased by nearly 1,900 km^2 between 1990 and 2000 alone. Similar examples occur in other parts of the world—e.g. California (Davis et al. 2002), Australia (Chapman 2003) and Japan (Koike 1996)—where hundreds of kilometres of coasts have been "hardened" to variable extents.

Artificial substrata pose severe threats to coastal areas (Airoldi et al. 2005). Their construction causes the loss of the natural habitats which they replace, and the disruption of the surrounding environments (Martin et al. 2005; Goodsell et al. 2007). Furthermore, these structures introduce new artificial hard substrata (Glasby and Connell 1999). These artificial substrata attract and support species typical of hard bottoms but are not analogues of natural rocky or biogenic habitats (Glasby 1999; Bulleri 2005). Even in the comparatively rare situations when artificial structures have been purpose designed to mimic natural habitats and enhance species of recreational, commercial or naturalistic value (e.g. artificial reefs), there has been no consistent evidence that these aims have been achieved (Lenihan 1999; Svane and Petersen 2001; see Chap. 24 by Goodsell and Chapman for more detail). In most instances, artificial hard structures are built for a variety of human uses in areas which otherwise have soft sediment habitats (e.g. breakwaters on beaches or oilrigs in the Gulf of Mexico), where they do not mimic natural habitats in any way, and where they create unnatural changes in species composition, abundance and diversity (Bacchiocchi and Airoldi 2003).

Extensive hardening of coasts can also have critical impacts on regional species diversity (Airoldi et al. 2005). High numbers of urban structures can act as stepping stones, disrupting natural barriers and facilitating the dispersal of nearby rocky coast species across habitats and regions which naturally would be poorly connected (Dethier et al. 2003), possibly affecting gene flow within individual species. The system of artificial structures can also provide new dispersal routes enabling the invasion of non-indigenous species, including pests (Glasby et al. 2007). For example, along the northeast coast of Italy, coastal-defence structures, which run almost uninterrupted for about 300 km, have promoted the expansion of numerous introduced species including *Codium fragile* ssp. *tomentosoides*, which finds particularly favourable growth conditions on the sheltered landward sides of breakwaters (Bulleri and Airoldi 2005).

The hardening of coastal areas and the resulting changes in species composition, abundance and diversity could have important consequences for the functioning of coastal ecosystems, modifying productivity, nutrient cycling and detrital pathways and, ultimately, affecting resources and services to humans. Because of their artificial nature, however, these substrata and their associated assemblages have not been as much a focus of marine science and management as has been the case for natural substrata, and our understanding of the ecology of urban marine habitats lags surprisingly far behind (Bulleri 2006). Urban structures and other artificial substrata are often uncritically claimed as reasonable mimics of natural hard-bottom habitats and valuable replacements for the habitats which they damage, while the consequences of the major unnatural changes in species and habitat distribution and diversity are ignored (Airoldi et al. 2005).

19.5 The Importance of Regional and Historical Contexts

Habitat change may be influenced by processes operating at larger scales than those at which they are commonly observed. Macro-ecological observations can complement local knowledge by placing it in a broader perspective which explicitly considers (1) spatial and temporal variation, (2) openness to exchange and (3) the influence of history (Brown 1995). Most studies of habitat loss or coastal urbanization, however, focus only on local scales and short periods over contemporary times. Whilst this focus is also necessary, the preoccupation with these local and present-day drivers means that we are often not aware of historical habitats and the regional context of local studies.

19.5.1 Regional Contexts of Habitat Change

Whether the drivers of ecological change are global or local, they are unlikely to act independently of regional differences in native species pool, environmental conditions (including climate and nutrient delivery), prevailing habitat type, and anthropogenic pressures on the coast (Connell and Irving 2008). For example, regions with few strongly interacting species (i.e. strong per capita effects of some species of urchin) may be more vulnerable to loss of canopy habitats through overfishing (Steneck et al. 2002), whereas oligotrophic regions may experience loss of canopies when excess nutrients are delivered via urban catchments (Connell 2007). Similarly, case studies in Europe have shown that the addition of urban structures can have interactive effects with regional impacts, such as eutrophication, harvesting and depletion of living resources, or high rates of introduction of exotic species, leading to different environmental consequences which require different solutions (Airoldi et al. 2005). Thus, considering the regional context is important in that regional differences in the biotic (e.g. diversity of functional groups and species) and abiotic environments (e.g. biochemical cycles) can lead to contrasting outcomes of human pressures on the environment (e.g. overfishing and pollution) and mitigation actions (e.g. marine protected areas; see Chap. 25 by Goodsell and Underwood, and water quality guidelines).

It is also important to consider the regional context because the consequences of habitat change can differ qualitatively and quantitatively at local vs. regional scales (Sax and Gaines 2003). For example, whilst the construction of individual urban structures may not pose particularly serious threats at local scales, the proliferation of urban structures has substantially changed the identity and nature of coastal landscapes in many regions of the world, unnaturally affecting the composition and distribution of biota at regional scales (Airoldi et al. 2005). Successful management of coastal areas can be achieved only by considering the whole coast as an integrated unit, where the ecological consequences of human-induced changes to natural habitats are evaluated in terms of a regional environmental and societal framework.

19.5.2 Historical Habitat Loss and the Shifting-Baseline Syndrome

The amount of habitat loss is contingent on the capacity of the biological systems to withstand (resistance) or absorb and recover from (resilience) the multiple assaults of natural and human origin. The magnitude of loss may depend not only on regional-scale phenomena but also on their history of past disturbances. Early changes can increase the sensitivity of coastal marine habitats to subsequent disturbances and, thus, precondition their eventual collapse (Jackson et al. 2001). The most well-understood examples are derived from coral reefs, where extensive coral covers can withstand chronic stressors (e.g. elevated nutrients and fishing) but fail to recover from subsequent acute disturbances (e.g. storm damage and freshwater runoff; Hughes 1994). Recognizing the historical context of habitat loss is, therefore, fundamental to clarifying underlying causes and rates of ecological change and for identifying achievable goals for restoration and management (Jackson et al. 2001). The unrecognized historical loss of habitats causes each generation to observe a progressively more degraded state as "natural", the so-called shifting-baseline syndrome (sensu Dayton et al. 1998), where we recognize declines in the natural environment relative only to the baseline of recent memory in each generation, and both scientists and managers continually reduce their expectations of the environment.

Our short-sightedness is particularly true for hard-bottom habitats. It should be apparent from the cited examples that the past and current losses of hard-bottom coastal habitats are alarming, and need immediate consideration. Nevertheless, alteration and threats to these habitats are still largely overlooked, with important implications for policy and conservation (Connell et al. 2008). Of all hard-bottom habitats, shellfish reefs are probably one of the most globally impacted, as shown by the case history of the native oyster reefs in Europe and by analogous examples in the US and Australia (Ocean Studies Board 2004). Many of these losses, however, have gone unrecognized, and nowadays these severely endangered marine habitats do not seem to be the focus of adequate protection measures, conservation legislation or conventions (Airoldi and Beck 2007).

19.6 The Case for Mitigating Habitat Loss

All hard-bottom habitats have been subject to substantial alterations throughout their history, and so are they today. These changes have been globally or locally driven, and have resulted from natural but especially from anthropogenic agents of environmental change and overexploitation. While many conservation biologists consider habitat loss the single greatest threat to species and ecosystems worldwide (Sih et al. 2000), there seems to be limited public, political and even scientific awareness of the extent, importance and consequences of the profound habitat changes and losses which have occurred during the history of marine coastal environments (Airoldi and Beck 2007). We have documented some striking alterations to the nature and distribution of hard-bottom habitats, where the systems have been driven through thresholds of degradation, and the losses have become long-lasting or even permanent. These alterations have caused not only large changes in the natural distribution of valuable coastal habitats and their associated species but also the loss of fundamental functional attributes, services and economic resources, including food production. Hard-bottom habitats support an impressive array of marine life as well as human livelihoods and wellbeing—alarmingly, they are declining globally. We need to provide better protection for existing natural hard-bottom habitats. This is in part also because the restoration science and practice for hard bottoms lags substantially behind that for soft-bottom habitats such as marshes, mangroves and seagrasses, and opportunities for restoration can be extremely expensive, challenging or even impossible (Lenihan 1999).

Acknowledgements We thank M. Wahl for the invitation to write this chapter. Work was supported by funding from Bologna University (Strategic Project ADRIABIO and Marco Polo Grant) to LA, funding from The Nature Conservancy to MWB, and funding from a QEII Fellowship to SDC. We are grateful to all the colleagues who provided their photographs.

References

Airoldi L (1998) Roles of disturbance, sediment stress, and substratum retention on spatial dominance in algal turf. Ecology 79:2759–2770

Airoldi L, Beck MW (2007) Loss, status and trends for coastal marine habitats of Europe. Oceanogr Mar Biol Annu Rev 45:345–405

Airoldi L, Abbiati M, Beck MW, Hawkins SJ, Jonsson PR, Martin D, Moschella PS, Sundelöf A, Thompson RC, Åberg P (2005) An ecological perspective on the deployment and design of low-crested and other hard coastal defence structures. Coast Eng 52:1073–1087

Airoldi L, Balata D, Beck MW (2008) The Gray zone: relationships between habitat loss and marine diversity and their applications in conservation. J Exp Mar Biol Ecol 366:8–15

Bacchiocchi F, Airoldi L (2003) Distribution and dynamics of epibiota on hard structures for coastal protection. Estuar Coast Shelf Sci 56:1157–1166

Benedetti-Cecchi L, Pannacciulli F, Bulleri F, Moschella PS, Airoldi L, Relini G, Cinelli F (2001) Predicting the consequences of anthropogenic disturbance: large-scale effects of loss of canopy algae on rocky shores. Mar Ecol Prog Ser 214:137–150

Brown JH (1995) Macroecology. The University of Chicago Press, Chicago, IL

Bulleri F (2005) Role of recruitment in causing differences between intertidal assemblages on seawalls and rocky shores. Mar Ecol Prog Ser 287:53–64

Bulleri F (2006) Is it time for urban ecology to include the marine realm? Trends Ecol Evol 21:658–659

Bulleri F, Airoldi L (2005) Artificial marine structures facilitate the spread of a non-indigenous green alga, *Codium fragile* ssp *tomentosoides*, in the North Adriatic Sea. J Appl Ecol 42:1063–1072

Chapman MG (2003) Paucity of mobile species on constructed seawalls: effects of urbanization on biodiversity. Mar Ecol Prog Ser 264:21–29

Connell SD (2005) Assembly and maintenance of subtidal habitat heterogeneity: synergistic effects of light penetration and sedimentation. Mar Ecol Prog Ser 289:53–61

Connell SD (2007) Water quality and the loss of coral reefs and kelp forests: alternative states and the influence of fishing. In: Connell SD, Gillanders BM (eds) Marine ecology. Oxford University Press, Melbourne, pp 556–568
Connell SD, Irving AD (2008) Integrating ecology with biogeography using landscape characteristics: a case study of subtidal habitat across continental Australia. J Biogeogr 35:1608–1621
Connell SD, Russell BC, Turner DJ, Shepherd SA, Kildae T, Miller DJ, Airoldi L, Cheshire A (2008) Recovering a lost baseline: missing kelp forests on a metropolitan coast. Mar Ecol Prog Ser 360:63–72
Davis JLD, Levin LA, Walther SM (2002) Artificial armored shorelines: sites for open-coast species in a southern California bay. Mar Biol 140:1249–1262
Dayton PK, Tegner MJ, Edwards PB, Riser KL (1998) Sliding baselines, ghosts, and reduced expectations in kelp forest communities. Ecol Appl 8:309–322
Dethier MN, McDonald K, Strathmann RR (2003) Colonization and connectivity of habitat patches for coastal marine species distant from source populations. Conserv Biol 17:1024–1035
Glasby TM (1999) Differences between subtidal epibiota on pier pilings and rocky reefs at marinas in Sydney. Estuar Coast Shelf Sci 48:281–290
Glasby TM, Connell SD (1999) Urban structures as marine habitats. Ambio 28:595–598
Glasby TM, Connell SD, Holloway MG, Hewitt CL (2007) Nonindigenous biota on artificial structures: could habitat creation facilitate biological invasions? Mar Biol 151:887–895
Goodsell PJ, Chapman MG, Underwood AJ (2007) Differences between biota in anthropogenically fragmented habitats and in naturally patchy habitats. Mar Ecol Prog Ser 351:15–23
Gorgula SK, Connell SD (2004) Expansive covers of turf-forming algae on human-dominated coast: the relative effects of increasing nutrient and sediment loads. Mar Biol 145:613–619
Guidetti P, Fraschetti S, Terlizzi A, Boero F (2003) Distribution patterns of sea urchins and barrens in shallow Mediterranean rocky reefs impacted by the illegal fishery of the rock-boring mollusc *Lithophaga lithophaga*. Mar Biol 143:1135–1142
Hughes TP (1994) Catastrophes, phase shifts and large-scale degradation of a Caribbean coral reef. Science 265:1547–1551
Jackson JBC, Sala E (2001) Unnatural oceans. Scientia Marina 65 suppl 2:273–281
Jackson JBC, Kirby MX, Berger WH, Bjorndal KA, Botsford LW, Bourque BJ, Bradbury RH, Cooke R, Erlandson J, Estes JA, Huges TP, Kidwell S, Lange CB, Lenhian HS, Pandolfi JM, Peterson CH, Steneck RS, Tegner MJ, Warner RR (2001) Historical overfishing and the recent collapse of coastal ecosystems. Science 293:629–638
Koike K (1996) The countermeasures against coastal hazards in Japan. GeoJournal 38:301–312
Lenihan HS (1999) Physical-biological coupling on oyster reefs: how habitat structure influences individual performance. Ecol Monogr 69:251–275
Lotze HK, Lenihan HS, Bourque BJ, Bradbury RH, Cooke RG, Kay MC, Kidwell SM, Kirby MX, Peterson CH, Jackson JBC (2006) Depletion, degradation, and recovery potential of estuaries and coastal seas. Science 312:1806–1809
Mackenzie CL Jr, Burrell VG Jr, Rosenfield A, Hobart WL (1997) The history, present condition and future of the molluscan fisheries of North and Central America and Europe. Vol 3, Europe. US Department of Commerce, Seattle, NOAA Tech Rep NMFS 129
Martin D, Bertasi F, Colangelo MA, de Vries M, Frost M, Hawkins SJ, Macpherson E, Moschella PS, Satta MP, Thompson RC, Ceccherelli VU (2005) Ecological impact of coastal defence structures on sediments and mobile infauna: evaluating and forecasting consequences of unavoidable modifications of native habitats. Coast Eng 52:1027–1051
Millennium Ecosystem Assessment (2005) Ecosystems and human well-being: synthesis. Island Press, Washington, DC
Ocean Studies Board (2004) Nonnative oysters in the Chesapeake Bay. National Academy Press, Washington, DC
Sax DF, Gaines SD (2003) Species diversity: from global decreases to local increases. Trends Ecol Evol 18:561–566
Shears NT, Babcock RC (2002) Marine reserves demonstrate top-down control of community structure on temperate reefs. Oecologia 132:131–142

Sih A, Jonsson BG, Luikart G (2000) Habitat loss: ecological, evolutionary and genetic consequences. Trends Ecol Evol 15:132–134

Steneck RS, Graham MH, Bourque BJ, Corbett D, Erlandson JM, Estes JA, Tegner MJ (2002) Kelp forest ecosystems: biodiversity, stability, resilience and future. Environ Conserv 29:436–459

Svane I, Petersen JK (2001) On the problems of epibioses, fouling and artificial reefs, a review. PSZNI Mar Ecol 22:169–188

Thibaut T, Pinedo S, Torras X, Ballesteros E (2005) Long-term decline of the populations of Fucales (*Cystoseira* spp. and *Sargassum* spp.) in the Alberes coast (France, North-western Mediterranean). Mar Pollut Bull 50:1472–1489

Thompson RC, Crowe TP, Hawkins SJ (2002) Rocky intertidal communities: past environmental changes, present status and predictions for the next 25 years. Environ Conserv 29:168–191

Thrush SF, Dayton PK (2002) Disturbance to marine benthic habitats by trawling and dredging: implications for marine biodiversity. Annu Rev Ecol Syst 33:449–473

Worm B, Lotze HK, Boström C, Engkvist R, Labanauskas V, Sommer U (1999) Marine diversity shift linked to interactions among grazers, nutrients and propagule banks. Mar Ecol Prog Ser 185:309–314

Chapter 20
Multiple Stressors and Disturbances: When Change Is Not in the Nature of Things

David R. Schiel

20.1 Introduction

Hard shore plants and animals show a high degree of adaptation to the dynamic environment of nearshore waters that includes physical forces acting over temporal scales from seconds (waves) to decades (e.g. Pacific Decadal Oscillation) and spatial scales from sub-millimetre (benthic boundary layer) to ocean-wide (currents, oceanic circulation). Overlaid on these are anthropogenic influences that impinge on one or more life history stages of organisms and, increasingly, affect benthic community structure along shorelines worldwide. Collectively, these forces are often called "stressors" (Vinebrooke et al. 2004; Adams 2005), which can be complex and act through many direct or indirect pathways, from effects on individual organisms to changed communities and trophic dynamics. Here, multiple stressors and their effects on hard shore communities are discussed, focusing primarily on algae and mussel communities that dominate many temperate rocky shores.

In the broadest sense, "stress" can be any factor that negatively affects the physiology, growth, reproduction and survival of an organism or that has consequences affecting populations and communities, such as shortages of light, water, nutrients or suboptimal temperatures that extend the responses and functions of organisms beyond their normal range (Grime 1977; Vinebrooke et al. 2004). Barton (1997) defines "stressor" as a stimulus and "stress response" as something that can be measured as indicative of an altered state. In this definition, stressors should exceed what is (or, historically, was) normally encountered by organisms; increasingly, this means the combination of anthropogenic components added to natural processes. Hughes and Connell (1999) define multiple stressors more broadly as "natural or man-made disturbances" including acute, cataclysmic or chronic disturbances causing declines in numbers of organisms, assemblages or ecosystems, or recruitment and regenerative processes. Sousa (2001), in his thorough review of natural disturbance to benthic marine communities, simply defines multiple stressors as multiple agents of disturbance resulting in displacement or mortality from physical agents or incidentally by biological agents. Here, I retain the broader terminologies recommended by Sousa (2001) and Hughes and Connell (1999), and focus on anthropogenic agents of disturbance that interact with natural processes.

M. Wahl (ed.), *Marine Hard Bottom Communities*, Ecological Studies 206,
DOI: 10.1007/978-3-540-92704-4_20,

20.2 A Framework of Disturbance by Multiple Stressors

Deciding what is "natural" in communities and what has been changed by anthropogenic influences may often be impossible. Some argue it is safe to assume that all marine communities have been altered in some way by human influences (Steneck and Carlton 2001; Jackson et al. 2001). The documented and hypothesized changes to coastal ecosystems over many centuries echo Connell's (1980) classic paper ("ghost of competition past") in ascribing the current state of coastal ecosystems as being a product of "disturbance past". Dayton et al. (1998) highlight the problem of "sliding baselines", in recognition that our reference points are increasingly altered states. Hughes and Connell (1999) argue that we cannot understand structure, relationships and causal mechanisms without understanding the history of change. Adams (2005) recommends that multiple lines of evidence are needed to understand relationships between stressors and their effects, including causal criteria based on epidemiological principles. This is often impossible, however, when effects are diffuse, over long times and large scales.

20.3 Types of Stressors and Responses

Sousa (2001) lists 19 agents of physical disturbance and five agents of biological disturbance, citing hundreds of references. These include storms, various effects of wave forces, ice scour, increased sedimentation, aerial exposure, temperature changes, hypoxia, and biological agents such as algal whiplash and suffocation by detached organic material. Human agents of disturbance include harvesting of species, chemical pollution, nutrient loading, oil spills, eutrophication, alteration of physical habitats, introductions of non-indigenous species and, potentially, climate change. The effects of these occur across a wide range of temporal and spatial scales, interact with the life histories, morphologies, susceptibilities and regenerative abilities of affected species, and within the context of biotic interactions such as competition and food web dynamics.

Central to all considerations of stressors and disturbance is the extent to which organisms, species and communities are vulnerable to damage. This can be quite complex because of the varying susceptibilities of the numerous species within assemblages, and the positive and negative effects they may have on each other as abundances change under various degrees of disturbance. For example, dense mussel beds may shield mussels from minor impacts but, when large impacts strike, entire clumps may be removed (Dayton 1971). Similarly, dense algal stands in the intertidal zone can ameliorate heat stress, and thinning causes far-reaching impacts on the rest of the community (Allison 2004).

Recent reviews provide definitions of terms and discussion of disturbances (see also Chaps. 17 by Clynick et al. and 30 by Underwood and Jackson). Resilience is the capacity of a system to recover from a disturbance (Stachowicz et al. 2007), the magnitude of disturbance that it can absorb before shifting into another state, or its capacity to absorb disturbance while undergoing change, yet retaining its structure and function (Folke et al. 2004). Resistance is the degree to which communities can resist changes (with reference to measurable effects) when disturbed. A key relationship, still contentious, is that between diversity and function (Naeem et al. 2002). Stachowicz et al. (2007) concluded that diversity often has a weak effect on production and biomass, compared to the strong effects exerted by some species.

Resistance and resilience depend to some degree on the nature of "redundancies" in communities, that is, whether the function of affected species can be replaced by that of others (Loreau 2004). Compensatory dynamics may occur whereby tolerant species succeed affected competitors and preserve overall function; nevertheless, this depends on the sign and magnitude of species' co-tolerance to different stressors (Vinebrooke et al. 2004). Hughes et al. (2005) point out that the degree to which damaged marine ecosystems can recover from repeated disturbances, the time trajectory of recovery and whether or not they shift into alternate states have considerable bearing on conservation and management practices. The symmetry of damage and recovery processes is referred to as hysteresis and, in marine systems, recovery follows a different trajectory than seen in decline. Localized or short-term reductions in stressors, therefore, do not ensure recovery, especially over short time periods.

20.4 Temporal Stressors

The temporal components of stressors, their duration and frequency, are critical factors that interact with the magnitude and intensity of stressors. For example, Connell's (1978) intermediate disturbance hypothesis related diversity directly to the frequency, intensity and time between disturbances. Diversity was lowest when there was either frequent intense disturbance, in which case most species were removed with little time for recovery, or when disturbance was infrequent or small, when one or a few species were competitive dominants. There is considerable support for this but it also relies on the types of species present and their vulnerabilities (Dial and Roughgarden 1998). The frequency of disturbance is also categorised as "pulse" (one-time or infrequent) or "press" (continual through time; Dayton 1971; see Chap. 17 by Clynick et al.). These recognize explicitly that the history and starting points of assessments of disturbance affect the trajectories of communities through time.

The temporal progression of life stages of affected species is crucial to assessments of stressor effects, and involves life histories, dispersal, population connectivity and nearshore oceanic regimes (Kinlan and Gaines 2003; Schiel 2004). Sousa (2001) lists four mechanisms by which populations become established after disturbance: vegetative re-growth, lateral encroachment from nearby undisturbed assemblages, and recruitment from either a remaining seed bank or via dispersive propagules. These depend on the types of species present, their modes of propagation and the severity of disturbance. For example, the adults of many large algae are readily removed by storms and most can re-establish from spores remaining on the substratum (Dayton et al. 1984) or from dispersive propagules. Small clearances in mussel beds, however, may be filled by lateral encroachment, while tough turfing algae may be resistant to complete removal and regenerate by vegetative growth (Sousa 1985). Effects of disturbance will vary, therefore, by the sizes of patches created and the nature and extent of remaining species.

Chronic effects of stressors may entail numerous regenerative processes including successional sequences of spore arrival (Sousa 1985), demographic lags due to lower reproductive output in unfavourable conditions, and altered competitive and trophic relationships and behaviours. An example is the change in sea urchin feeding behaviour from passive to active foraging after giant kelp is removed over large areas (Harrold and Reed 1985). Many authors point out that to manage and restore coastal communities, it is necessary to understand the history of disturbances and the mechanisms and causal relationships between stressors and effects (Hughes and Connell 1999; Adams 2005).

20.5 Spatial Patterns of Stressors

The severity of disturbance by multiple stressors can be gauged both by the magnitude of their effect on assemblages (e.g. composition, diversity) and by its areal extent. Generally, where the patch of disturbance is small, surrounding species tend to replenish it but, where disturbed patches are large, recolonization requires spores from distant sources. For example, multiple storms associated with El Niño events can remove giant kelp forests from most of a coastline and recolonization can take a few years (Dayton and Tegner 1984), possibly from refuge populations in quiet, protected bays (Schiel and Foster 1992).

There is considerable experimental evidence for mussels and large brown algae that patch sizes of disturbances affect recovery and community processes. For example, large clearings in dominant canopies of the fucoid *Ascophyllum* in New England had different recruitment by mussels and fucoids, and different levels of predation, compared to small clearings (Petraitis and Latham 1999). Large disturbances may trigger cascading effects, particularly in thermally stressed areas of mid- and upper intertidal zones where the loss of canopy algae can result in die-off of understory species, reduced diversity (Lilley and Schiel 2006) and altered patterns of grazing and predation.

20.6 Empirical Evidence of Stressor Effects

Here, I highlight examples of stressors acting on populations and communities. None acts in isolation of other stressors (i.e. multiple stressors are common) but their covariation and synergies in effects may not always be known.

20.6.1 Sedimentation

Increases in coastal sediments from urbanization, changes in land use such as intensive farming and silviculture, underwater mining, and construction of breakwaters and marinas have produced disturbances to marine communities worldwide (Airoldi 2003; Valiela 2006). Sediments tend to be more abundant and are commonly finer-grained in more protected areas where wave forces are diminished (Airoldi 2003; Schiel et al. 2006). Sediments can kill by smothering organisms, preventing effective feeding and photosynthesis, inhibiting recolonization by impeding settlement and recruitment, and altering light penetration. Effects can occur across a wide range of spatiotemporal scales, interacting with species tolerances and life histories, and with other stressors such as pollutants (Eriksson and Johansson 2005).

Early life stages of large algae seem to be particularly vulnerable. Schiel et al. (2006) showed that sediment cover differentially affected two fucoid species. A layer of fine sediment caused a 71% reduction in attachment of zygotes of *Durvillaea antarctica* and a 34% reduction in *Hormosira banksii*, and thicker sediment layers prevented any attachment in both species. On a larger scale, the interaction of sediments with other stressors, including eutrophication and increased turbidity since the 1940s, are believed to have caused a narrowing in the vertical distribution of fucoid algae, especially *Fucus vesiculosus*, along the Baltic coast (Kautsky et al. 1986). This upward movement of the lower distribution limit seems to have been accelerated by a reduction in light penetration and photosynthesis, eutrophication favouring epibionts, and by a degradation of herbivory defences under light limitation and warmer temperatures (Rohde et al. 2008). Eriksson and Johansson (2005) found that ephemeral green algae are highly tolerant of sediments, while perennial browns are not; *F. vesiculosus* recruited only where sediments were removed.

Sediments from land can be laden with toxic chemicals and trace metals (Valiela 2006). An example of effects on rocky coasts is the decline of a large kelp forest in California (Palos Verdes, Los Angeles) that started in the early 1950s when sewage discharge rates increased (Schiel and Foster 1992). After a period of warm water, the kelp forest was completely lost. Turbidity and benthic sludge may have prevented spore settlement and gametophyte development, while toxic chemicals in the sediment (DDT) and metals such as copper may have inhibited reproduction. Kelp returned in the 1970s soon after the outfall was moved further offshore and contaminated sediments went into deeper water.

Sediments on hard reefs seem to favour simplified communities dominated by low-lying turfing algae, such as corallines, and reduced diversity (Gorgula and Connell 2004). Quick turnover species, with reproduction throughout the year, seem better able to take advantage of the annual flux in sediments and also live on sediments (Kiirikki and Lehvo 1997; Eriksson and Johansson 2005), possibly due to enhanced nutrients, diminished competition and less grazing. Overall, increased sedimentation is considered a major threat to coastal ecosystems, triggering losses of habitat-engineering species and cascading effects on diversity and functional relationships.

20.6.2 Species Reductions

Multiple stressors can affect biomass, diversity, production and overall "function" (Schwartz et al. 2000). Not all species are equal in these respects. Stressors that lead to losses of foundation species (key species, structural species and ecosystem engineers) can exert a pervasive influence on the functioning of reef communities because they provide much of the biomass and many of the "services" on which other species rely. On rocky reefs, these species are often large brown algae.

We know a great deal about the consequences of losses of some large algae, such as *Macrocystis pyrifera*, but less so about others. The major stressor that removes large kelp is storms, which may be particularly intense in El Niño years (Dayton and Tegner 1984). Giant kelp loss increases light to areas below, and can lead to significant increases in recruitment of understory kelps and other species (Schiel and Foster 1986). Because many species of fish, invertebrates and other algae are intimately connected to the extent of kelp cover, kelp loss can trigger significant changes in primary production, diversity, food web dynamics, and most "functions" associated with kelp communities (Holbrook et al. 1997; Dayton et al. 1998).

The community consequences of removal of fucoid algae are less known but may be as pervasive as for kelps. Intertidal zones worldwide have been the focus of such studies because of their accessibility, amenability to experimental manipulations over physical gradients, and social and recreational importance. In mid- and upper intertidal areas, where aerial exposure and temperature can be harsh at low tide, canopies of fucoids are usually crucial to community structure and function. For example, in New Zealand the removal of *Hormosira banksii* from large patches triggered quick, cascading losses of most understory species through burn-off, with temperatures >35°C (Lilley and Schiel 2006). The community became dominated by small turfing coralline algae and bare space, and diversity was half that of controls even after 2 years. Despite the presence of six other fucoids, there was no functional replacement of *Hormosira* by other species (Schiel 2006). This species, and probably many others worldwide, are "key", "idiosyncratic" (Lawton 1994) or "singular" species (Naeem et al. 2002) with no functional redundancies. Furthermore, the community was largely structured by positive interactions that depended on the services (shade, moisture retention, reduction in exposure) provided by the fucoid canopy. In eastern Australia, the storm-induced loss of *Hormosira* triggered a series of direct and indirect effects (Underwood 1999). Grazing molluscs increased and delayed algal recruitment, barnacles recruited, and predatory whelk behaviour and abundance changed, none of which occurred in New Zealand. Even with the same canopy species, therefore, functional roles and interactions within communities can be context-dependent.

An influx of grazers and consequent effects of herbivory often follow removal of macroalgal canopies. On British shores, for example, the removal of *Ascophyllum* triggered changes in grazers and understory species that lasted 12 years (Jenkins et al. 2004). Turfing algae declined, limpets increased, and a mixed canopy of fucoids resulted. The slow recovery of *Ascophyllum* and the combination of direct and indirect effects of canopy removal highlight the differences between geographically separated systems and the long-term nature of changes. In the eastern USA, it has been argued both that the removal of *Ascophyllum* leads to alternate states between fucoid and mussel/barnacle dominance (Petraitis and Dudgeon 2004) and that removal leads deterministically to the replenishment of *Ascophyllum* beds (Bertness et al. 2004). Life histories clearly play a significant role. The variable recruitment of *Ascophyllum*, its long life (Aberg 1992), and resistance to grazing but slow growth mean that eventual dominance takes a long time after major disturbance. It is noteworthy that much of the debate about processes and outcomes involves indirect effects on the diversity and abundances of understory species, which seem less deterministic and potentially longer-lasting than for the canopy species.

Other stressors include direct removal by harvesting and trampling (Lindberg et al. 1998; Pinn and Rodgers 2005). In Australia, for example, high levels of trampling by pedestrians on reef platforms caused the fucoid *Hormosira* to be removed, followed by increased densities of herbivorous gastropods and reduced cover of coralline algae (Keough and Quinn 1998). In New Zealand, only ten steps at a given spot caused canopies to be damaged, and 25 steps or more led to burn-off of understory species (Schiel and Taylor 1999). Similar effects have been shown in many assemblages worldwide, with increasing examples of pulse effects coalescing into press impacts that alter communities and diversity over entire reefs.

Far more information is needed on the extent to which there are gradients or thresholds in population and community responses to stressors, which will have a direct bearing on understanding how variable degrees of impacts affect assemblages (Schiel and Lilley 2007), and on predictive models. One concern is that many species in assemblages will be reduced to densities below critical levels for effective reproduction, fertilization, subsequent recruitment and growth (Dayton 2003; Schiel and Foster 2006), with cascading effects throughout communities.

20.6.3 Extractions, Harvesting, Removals

Recent papers emphasize the history and effects of human impacts on coastal ecosystems through over-harvesting of many species, including predators, such as sea otters, fish and lobsters, herbivores such as gastropods, sea urchins and abalone, and filter-feeding invertebrates (Steneck and Carlton 2001; Jackson et al. 2001). It is believed that this has caused great changes in functional relationships, structure and resilience of coastal ecosystems worldwide (Jackson et al. 2001; Steneck et al. 2002; Hughes et al. 2005). Although the degree to which extractions have resulted in alternate states is debated, large-scale removals, reductions in abundance, and decreases in size have reduced many species to functionally minor roles (Dayton 2003).

In systems heavily stressed by extractions, there can be a shift towards fast-growing, "weedier" species (Hughes and Connell 1999). Habitat-dominant kelps, however, have weedy life histories (Schiel and Foster 2006), which can make these organisms resilient to numerous stressors.

20.6.4 Non-indigenous Species (NIS)

Non-native species herald potentially large changes to coastal ecosystems, altering functional relationships, and the character of native assemblages over large areas (see also Chap. 22 by McQuaid and Arenas). They do not "respect" borders or management and protected zones. Steneck and Carlton (2001) list five major vectors: shipping and boating; aquaculture, fishing and the aquarium trade; scientific research; canals; and other activities such as moving structures around. There are many examples of NIS affecting invaded communities. For example, the green alga *Caulerpa taxifolia* has disrupted seagrass communities along much of the Mediterranean coast; *Codium fragile tomentosoides* is the dominant alga around many regions of Cape Cod; the fucoid *Sargassum muticum* has spread throughout much of the northern temperate zone; the European crab *Carcinus maenas* has spread to North America and Australia; the Japanese kelp *Undaria pinnatifida* has spread through both hemispheres; and hundreds of species of most phyla have been spread worldwide (Steneck and Carlton 2001; Williams and Smith 2007). Not all NIS have had negative impacts on native species and communities. Ruiz et al. (1999) found that of the 196 known NIS in Chesapeake Bay, 20% were thought to have had an impact and 6% had a measurable impact. Williams and Smith (2007) found that only 6% of 400 global seaweed invasions had been studied for their effects on native biota but most effects were negative. The problem of NIS is exacerbated by many species not being readily detected, often cryptic, having uncertain taxonomy, or there simply being a shortage of monitoring and taxonomists to identify incursions (Drake and Lodge 2007).

Displacement of native species, invasion of new habitats, and changed relationships within communities may have historical components. For example, the European snail *Littorina littorea* has inhabited the New England shoreline for >100 years and is the numerically dominant herbivore (Bertness 1984), affecting algal diversity, recruitment of *Ascophyllum*, and modifying some habitats by decreasing sediment accumulation. There can be rapid evolution in response to invaders, such as shell thickening by native mussels within 15 years of invasion by the Asian shore crab *Hemigrapsus sanguineus* (Freeman and Byers 2006). Invaders can even affect other invaders indirectly. The clam *Gemma gemma*, which had been transported from the eastern USA to California, greatly increased in abundance over 6 years following the arrival of the European crab *C. maenas* because of selective predation and differing life histories (Grosholz 2005). The crab ate larger native clams (*Natricola*); by doing so, it consumed most of the reproductive females of this protandrous hermaphrodite and diminished subsequent recruitment, whereas the smaller, dioecious *Gemma* was able to maintain populations.

Propagule pressure is a major mechanism of spread of NIS into native habitats (von Holle and Simberloff 2005) but, given the vast numbers and types of invaders that have become established, there are clearly numerous life histories and modes of spread possible. Some aggressively make their way into native habitats, such as crabs, while others key into little-used areas, such as the cryptic tunicate *Styela clava* that has affected aquaculture areas in eastern North America (Locke et al. 2007). Others, such as the Asian kelp *Undaria*, are highly prolific annuals that take advantage of disturbed areas. For example, in New Zealand it recruits not only into bare patches but also into coralline turfs, which few other large algae can do (Schiel, unpublished data). Once localized patches are established, it then can "leapfrog" along the shoreline. Generally, disturbance tends to decrease resistance and increase invasibility of marine communities (Byers 2002; Williams and Smith 2007).

Given the ever-increasing volume of marine traffic between countries, it is difficult to see how the advent of NIS to new shores will be abated, despite considerable efforts to do so, including requirements for ballast water to be exchanged in open, rather than coastal waters, and other measures aimed at prevention and early detection (Wotton and Hewitt 2004). In terms of human-induced stressors, NIS are in many ways the most pernicious because, once established, they are probably impossible to eradicate. There is no hysteresis, only management and adaptation.

20.6.5 Climate Change

Climate change includes potentially numerous stressors such as increased temperature, sea-level rise, ocean acidification, increased ultraviolet-b radiation, altered nutrients, and increased frequency and intensity of storms (Harley et al. 2006; see Chap. 18 by Harley and Connell). Potential ecological responses include changes in zonational patterns, biogeographical shifts, and changes in community structure, dynamics and productivity (Parmesan 2006; Helmuth et al. 2006a, b). The mechanisms include the interaction of species' life histories, physiological tolerances and adaptations, ecological interactions within communities, altered functional relationships, and synergistic effects of multiple stressors associated with climate change.

The major indicator of climate change has been increased temperatures, although recent reviews highlight the need to consider temperature acting in concert with other stressors. For example, Smith et al. (2006) re-sampled 22 mussel-dominated sites that had previously been sampled in the 1960s and 1970s along the California coast, finding a 59% loss of invertebrate species. They suggested this was due to large-scale processes, such as the Pacific Decadal Oscillation, rather than local habitat changes. Helmuth et al. (2006a) present other evidence that local descriptors of weather were better predictors of change than were large-scale oceanographic events, and argued that both must be considered.

Most studies on climate change in marine ecosystems have assessed changes over varying periods of time and correlated these with changes in temperature.

Along the European coast, for example, where good long-term data exist, barnacles and other invertebrates have increased or declined in abundance as seawater temperatures have increased or decreased since the 1930s (Southward et al. 1995). Sagarin et al. (1999) reported an influx of warm-water species and a decline of cold-water species between the 1930s and the 1990s in central California, which was ascribed to rising seawater temperature. The robustness of the conclusion was compromised, however, by the limited, one-off samples of the 1930s. Helmuth et al. (2006b) criticize studies that attempt to correlate changes in abundance and distribution with a single environmental parameter; however, the loose correlations of most studies do just that. A large-scale test of temperature increase of around 3.5°C was provided by the thermal outfall of a power plant into a bay in central California over a period of 20 years (Schiel et al. 2004). There were significant changes in >150 species of algae and invertebrates throughout the intertidal and subtidal zones, and little replacement of cold-water species by warm-water southern species. Instead, the communities were greatly altered in apparently cascading responses to changes in abundance of several key taxa, particularly habitat-dominating kelps and foliose algae. This study showed the difficulties of predicting climate change effects because there was a complex interplay of physical, ecological and life history features of numerous species.

Experimental work sheds light on physiological tolerances of species in a wide range of interacting circumstances (Helmuth et al. 2006a). The effects of climate change, however, are only beginning to be understood, and our predictive abilities will rely on targeted long-term monitoring, innovative experiments, far greater understanding of cascading processes and indirect effects, and better knowledge of the life histories, vulnerabilities and ecological roles of species, particularly the key or keystone species on which community structure rely.

20.6.6 Other Stressors

Numerous other stressors have affected coastal ecosystems, with both acute and chronic effects. Hypoxia events ("dead zones") have been recorded in many places worldwide, affecting areas of continental shelf, estuaries and the open ocean (Grantham et al. 2004). Changes in the timing and strength of upwelling, high nearshore temperatures, depressed nutrient levels and lower recruitment rates of benthic invertebrates could be related to climate change (Barth et al. 2007).

Point-source pollution such as oil spills has had considerable long-lasting effects. In Prince William Sound, for example, the dominant intertidal alga *Fucus gardneri* was reduced for several years after the Exxon Valdez oil spill and subsequent clean-up (Stekoll and Deysher 2000), and oil residue persisted for at least 10 years (Irvine et al. 2006). Hawkins and Southward (1992) found in Britain that recovery of macrophytes took up to 15 years and was facilitated by the demise of grazers that recruited after an oil spill and clean-up operations.

Other pollutants, such as tributyl tin (TBT) used in anti-fouling paints before it was banned on small vessels, had severe toxic effects on a wide variety of invertebrates, altering shell morphologies and sexes (Bryan et al. 1986). There are many examples of pollutants affecting benthic communities, often with long-lasting effects (Hawkins et al. 2002).

20.7 Conclusions

It is always timely to remember that even global problems act locally and are manifested in many contexts. Impacts on diversity are usually greatest or more obvious when they affect dominant species on which much of community structure and function rely. The problems in understanding the direct and indirect pathways of change are exacerbated by the often diffuse and more generalized effects of many stressors acting simultaneously on key species, keystone predators and grazers, and the fact that loss or severe reduction of some species seems to leave few detectable vestiges that resound through the ecosystem.

There are increasing numbers of multidisciplinary publications, in recognition that multiple stressors require multiple levels of understanding of biology, ecology, species' demographies, physiology, genetics and oceanography. As forecasting models develop, and as societies act more to predict, solve and mitigate coastal problems, sociology will play a greater role in this mix of disciplines (Hughes et al. 2005).

Acknowledgements Thanks to The New Zealand Foundation for Research, Science and Technology, the Andrew W. Mellon Foundation of New York and the Marsden fund of the Royal Society of New Zealand, which supported much of the work on which this chapter is based.

References

Aberg P (1992) A demographic-study of 2 populations of the seaweed *Ascophyllum nodosum*. Ecology 73:1473–1487

Adams SM (2005) Assessing cause and effect of multiple stressors on marine systems. Mar Pollut Bull 51:649–657

Airoldi L (2003) The effects of sedimentation on rocky coast assemblages. Oceanogr Mar Biol 41:161–236

Allison G (2004) The influence of species diversity and stress intensity on community resistance and resilience. Ecol Monogr 74:117–134

Barth JA, Menge BA, Lubchenco J, Chan F, Bane JM, Kirincich AR, McManus MA, Nielsen KJ, Pierce SD, Washburn L (2007) Delayed upwelling alters nearshore coastal ocean ecosystems in the northern California current. Proc Natl Acad Sci USA 104:3719–3724

Barton BA (1997) Stress in finfish: past, present and future—a historical perspective. In: Iwama GK, Pickering AD, Sumpter JP, Schreck CB (eds) Fish stress and health in aquaculture. Soc Exp Biol Sem Ser vol 62. Cambridge University Press, Cambridge, pp 1–33

Bertness MD (1984) Habitat and community modification by an introduced herbivorous snail. Ecology 65:370–381

Bertness M, Silliman BR, Jefferies R (2004) Salt marshes under siege. Am Sci 92:54–61

Bryan GW, Gibbs PE, Hummerstone LG, Burt GR (1986) The decline of the gastropod *Nucella lapillus* around southwest England—evidence for the effect of tributyl tin from antifouling paints. J Mar Biol Assoc UK 66:611–640

Byers JE (2002) Impact of non-indigenous species on natives enhanced by anthropogenic alteration of selection regimes. Oikos 97:449–458

Connell JH (1978) Diversity in tropical rain forests and coral reefs. Science 199:1302–1310

Connell JH (1980) Diversity and the coevolution of competitors, or the Ghost of Competition Past. Oikos 35:131–138

Dayton PK (1971) Competition, disturbance, and community organization—provision and subsequent utilization of space in a rocky intertidal community. Ecol Monogr 41:351–389

Dayton PK (2003) The importance of the natural sciences to conservation. Am Nat 162:1–13

Dayton PK, Tegner MJ (1984) Catastrophic storms, El-Nino, and patch stability in a southern-California kelp community. Science 224:283–285

Dayton PK, Currie V, Gerrodette T, Keller BD, Rosenthal R, Ventresca D (1984) Patch dynamics and stability of some California kelp communities. Ecol Monogr 54:253–289

Dayton PK, Tegner MJ, Edwards PB, Riser KL (1998) Sliding baselines, ghosts, and reduced expectations in kelp forest communities. Ecol Appl 8:309–322

Dial R, Roughgarden J (1998) Theory of marine communities: the intermediate disturbance hypothesis. Ecology 79:1412–1424

Drake JM, Lodge DM (2007) Hull fouling is a risk factor for intercontinental species exchange in aquatic ecosystems. Aquat Invasions 2:121–131

Eriksson BK, Johansson G (2005) Effects of sedimentation on macroalgae: species-specific responses are related to reproductive traits. Oecologia 143:438–448

Folke C, Carpenter S, Walker B, Scheffer M, Elmqvist T, Gunderson L, Holling CS (2004) Regime shifts, resilience, and biodiversity in ecosystem management. Annu Rev Ecol Evol Syst 35:557–581

Freeman AS, Byers JE (2006) Divergent induced responses to an invasive predator in marine mussel populations. Science 313:831–833

Gorgula SK, Connell SD (2004) Expansive covers of turf-forming algae on human-dominated coast: the relative effects of increasing nutrient and sediment loads. Mar Biol 145:613–619

Grantham BA, Chan F, Nielsen KJ, Fox DS, Barth JA, Huyer A, Lubchenco J, Menge BA (2004) Upwelling-driven nearshore hypoxia signals ecosystem and oceanographic changes in the northeast Pacific. Nature 429:749–754

Grime JP (1977) Evidence for existence of three primary strategies in plants and its relevance to ecological and evolutionary theory. Am Nat 111:1169–1194

Grosholz ED (2005) Recent biological invasion may hasten invasional meltdown by accelerating historical introductions. Proc Natl Acad Sci USA 102:1088–1091

Harley CDG, Hughes AR, Hultgren KM, Miner BG, Sorte CJB, Thornber CS, Rodriguez LF, Tomanek L, Williams SL (2006) The impacts of climate change in coastal marine systems. Ecol Lett 9:228–241

Harrold C, Reed DC (1985) Food availability, sea-urchin grazing, and kelp forest community structure. Ecology 66:1160–1169

Hawkins SJ, Southward AJ (1992) The Torrey Canyon oil spill: recovery of rocky shore communities. In: Thayer GW (ed) Restoring the Nation's marine environment. Maryland Sea Grant, College Park, MD, pp 583–631

Hawkins SJ, Gibbs PE, Pope ND, Burt GR, Chesman BS, Bray S, Proud SV, Spence SK, Southward AJ, Southward GA, Langston WJ (2002) Recovery of polluted ecosystems: the case for long-term studies. Mar Environ Res 54:215–222

Helmuth B, Broitman BR, Blanchette CA, Gilman S, Halpin P, Harley CDG, O'Donnell MJ, Hofmann GE, Menge B, Strickland D (2006a) Mosaic patterns of thermal stress in the rocky intertidal zone: implications for climate change. Ecol Monogr 76:461–479

Helmuth B, Mieszkowska N, Moore P, Hawkins SJ (2006b) Living on the edge of two changing worlds: forecasting the responses of rocky intertidal ecosystems to climate change. Annu Rev Ecol Evol Syst 37:373–404
Holbrook SJ, Schmitt RJ, Stephens JS (1997) Changes in an assemblage of temperate reef fishes associated with a climate shift. Ecol Appl 7:1299–1310
Hughes TP, Connell JH (1999) Multiple stressors on coral reefs: a long-term perspective. Limnol Oceanogr 44:932–940
Hughes TP, Bellwood DR, Folke C, Steneck RS, Wilson J (2005) New paradigms for supporting the resilience of marine ecosystems. Trends Ecol Evol 20:380–386
Irvine GV, Mann DH, Short JW (2006) Persistence of 10-year old Exxon Valdez oil on Gulf of Alaska beaches: the importance of boulder-armoring. Mar Pollut Bull 52:1011–1022
Jackson JBC, Kirby MX, Berger WH, Bjorndal KA, Botsford LW, Bourque BJ, Bradbury RH, Cooke R, Erlandson J, Estes JA, Hughes TP, Kidwell S, Lange CB, Lenihan HS, Pandolfi JM, Peterson CH, Steneck RS, Tegner MJ, Warner RR (2001) Historical overfishing and the recent collapse of coastal ecosystems. Science 293:629–638
Jenkins SR, Norton TA, Hawkins SJ (2004) Long term effects of *Ascophyllum nodosum* canopy removal on mid shore community structure. J Mar Biol Assoc UK 84:327–329
Kautsky N, Kautsky H, Kautsky U, Waern M (1986) Decreased depth penetration of *Fucus vesiculosus* (L) since the 1940s indicates eutrophication of the Baltic Sea. Mar Ecol Prog Ser 28:1–8
Keough MJ, Quinn GP (1998) Effects of periodic disturbances from trampling on rocky intertidal algal beds. Ecol Appl 8:141–161
Kiirikki M, Lehvo A (1997) Life strategies of filamentous algae in the northern Baltic proper. Sarsia 82:259–267
Kinlan BP, Gaines SD (2003) Propagule dispersal in marine and terrestrial environments: a community perspective. Ecology 84:2007–2020
Lawton JH (1994) What do species do in ecosystems? Oikos 71:367–374
Lilley SA, Schiel DR (2006) Community effects following the deletion of a habitat-forming alga from rocky marine shores. Oecologia 148:672–681
Lindberg DR, Estes JA, Warheit KI (1998) Human influences on trophic cascades along rocky shores. Ecol Appl 8:880–890
Locke A, Hanson JM, Ellis KM, Thompson J, Rochette R (2007) Invasion of the southern Gulf of St. Lawrence by the clubbed tunicate (*Styela clava* Herdman): potential mechanisms for invasions of Prince Edward Island estuaries. J Exp Mar Biol Ecol 342:69–77
Loreau M (2004) Does functional redundancy exist? Oikos 104:606–611
Naeem S, Loreau M, Inchausti P (2002) Biodiversity and ecosystem functioning: the emergence of a synthetic ecological framework. In: Loreau M, Naeem S, Inchausti P (eds) Biodiversity and ecosystem functioning: synthesis and perspectives. Oxford University Press, Oxford, pp 3–17
Parmesan C (2006) Ecological and evolutionary responses to recent climate change. Annu Rev Ecol Evol Syst 37:637–669
Petraitis PS, Dudgeon SR (2004) Do alternate stable community states exist in the Gulf of Maine rocky intertidal zone? Comment Ecology 85:1160–1165
Petraitis PS, Latham RE (1999) The importance of scale in testing the origins of alternative community states. Ecology 80:429–442
Pinn EH, Rodgers M (2005) The influence of visitors on intertidal biodiversity. J Mar Biol Assoc UK 85:263–268
Rohde S, Hiebenthal C, Wahl M, Karez R, Bischof K (2008) Decreased depth distribution of *Fucus vesiculosus* (Phaeophyceae) in the Western Baltic: effects of light deficiency and epibionts on growth and photosynthesis. Eur J Phycol 43:143–150
Ruiz GM, Fofonoff P, Hines AH, Grosholz ED (1999) Non-indigenous species as stressors in estuarine and marine communities: assessing invasion impacts and interactions. Limnol Oceanogr 44:950–972
Sagarin RD, Barry JP, Gilman SE, Baxter CH (1999) Climate-related change in an intertidal community over short and long time scales. Ecol Monogr 69:465–490
Schiel DR (2004) The structure and replenishment of rocky shore intertidal communities and biogeographic comparisons. J Exp Mar Biol Ecol 300:309–342

Schiel DR (2006) Rivets or bolts? When single species count in the function of temperate rocky reef communities. J Exp Mar Biol Ecol 338:233–252

Schiel DR, Foster MS (1986) The structure of subtidal algal stands in temperate waters. Oceanogr Mar Biol Annu Rev 24:265–307

Schiel DR, Foster MS (1992) Restoring kelp forests. In: Thayer GW (ed) Restoring the Nation's marine environment. Maryland Sea Grant, College Park, MD, pp 279–342

Schiel DR, Foster MS (2006) The population biology of large brown seaweeds: ecological consequences of multiphase life histories in dynamic coastal environments. Annu Rev Ecol Evol Syst 37:343–372

Schiel DR, Lilley SA (2007) Gradients of disturbance to an algal canopy and the modification of an intertidal community. Mar Ecol Prog Ser 339:1–11

Schiel DR, Taylor DI (1999) Effects of trampling on a rocky intertidal algal assemblage in southern New Zealand. J Exp Mar Biol Ecol 235:213–235

Schiel DR, Steinbeck JR, Foster MS (2004) Ten years of induced ocean warming causes comprehensive changes in marine benthic communities. Ecology 85:1833–1839

Schiel DR, Wood SA, Dunmore RA, Taylor DI (2006) Sediment on rocky intertidal reefs: effects on early post-settlement stages of habitat-forming seaweeds. J Exp Mar Biol Ecol 331:158–172

Schwartz MW, Brigham CA, Hoeksema JD, Lyons KG, Mills MH, van Mantgem PJ (2000) Linking biodiversity to ecosystem function: implications for conservation ecology. Oecologia 122:297–305

Smith JR, Fong P, Ambrose RF (2006) Long-term change in mussel (*Mytilus californianus* Conrad) populations along the wave-exposed coast of southern California. Mar Biol 149:537–545

Sousa WP (1985) Disturbance and patch dynamics on rocky intertidal shores. In: Pickett STA, White PS (eds) The ecology of natural disturbance and patch dynamics. Academic Press, Orlando, FL, pp 101–124

Sousa WP (2001) Natural disturbance and the dynamics of marine benthic communities. In: Bertness MD, Gaines SD, Hay ME (eds) Marine ecological communities. Sinauer, Sunderland, MA, pp 85–130

Southward AJ, Hawkins SJ, Burrows MT (1995) 70 Years observations of changes in distribution and abundance of zooplankton and intertidal organisms in the western English Channel in relation to rising sea temperature. J Therm Biol 20:127–155

Stachowicz JJ, Bruno JF, Emmett Duffy J (2007) Understanding the effects of marine biodiversity on communities and ecosystems. Annu Rev Ecol Evol Syst 38:739–766

Stekoll MS, Deysher L (2000) Response of the dominant alga *Fucus gardneri* (silva) (Phaeophyceae) to the Exxon Valdez oil spill and clean-up. Mar Pollut Bull 40:1028–1041

Steneck RS, Carlton (2001) Human alterations of marine communities–students beware! In: Bertness MD, Gaines SD, Hay ME (eds) Marine community ecology. Sinauer, Sunderland, MA, pp 445–468

Steneck RS, Graham MH, Bourque BJ, Corbett D, Erlandson JM, Estes JA, Tegner MJ (2002) Kelp forest ecosystems: biodiversity, stability, resilience and future. Environ Conserv 29:436–459

Underwood AJ (1999) Physical disturbances and their direct effect on an indirect effect: responses of an intertidal assemblage to a severe storm. J Exp Mar Biol Ecol 232:125–140

Valiela I (2006) Global coastal change. . Blackwell, Malden, MA

Vinebrooke RD, Cottingham KL, Norberg J, Scheffer M, Dodson SI, Maberly SC, Sommer U (2004) Impacts of multiple stressors on biodiversity and ecosystem functioning: the role of species co-tolerance. Oikos 103:451–457

von Holle B, Simberloff D (2005) Ecological resistance to biological invasion overwhelmed by propagule pressure. Ecology 86:3212–3218

Williams SL, Smith JE (2007) A global review of the distribution, taxonomy, and impacts of introduced seaweeds. Annu Rev Ecol Syst 38:327–359

Wotton DM, Hewitt CL (2004) Marine biosecurity post-border management: developing incursion response systems for New Zealand. N Z J Mar Freshw Res 38:553–559

Chapter 21
Mass Mortalities and Extinctions

Carlo Cerrano and Giorgio Bavestrello

21.1 Introduction

During recent years, many researchers have predicted that diseases in terrestrial and marine ecosystems could increase due to climate warming. Thermal stress may directly affect the physiology of organisms (Zocchi et al. 2001) or reduce their resistance (Harvell et al. 2002), resulting in numerous diseases affecting natural populations. Mortality is a natural process but recent diseases seem to be ever more important in regulating population dynamics of several organisms (Boero 1996; Hayes and Goreau 1998; Epstein et al. 1998; Harvell et al. 1999). Sometimes, at a regional scale, sudden and diffused diseases give rise to a mass mortality of one or more species. Generally, these mass mortalities do not lead to extinction, although local populations, with their associated fauna, may disappear (Carlton et al. 1999). True documented extinction of benthic species are very rare and generally related to the destruction of their habitat. There is recent evidence of an increase in the frequency and intensity of mass mortalities of several species over the past 30–40 years in tropical (Hayes et al. 2001) and temperate areas (Cerrano et al. 2000; Garrabou et al. 2001), even if an increased awareness could also be partially responsible for this trend.

Here, we review major mass mortalities of the most frequently affected benthic organisms, summarizing (where possible) the causes, dynamics and documented effects of these phenomena.

21.2 Porifera

Sponges are considered simple and primitive metazoans, which play a key structural and functional ecological role (Wulff 2006a). They provide refuge for numerous small organisms (Cerrano et al. 2006a) and, through efficient filter-feeding, couple productivity in the water column with secondary productivity of benthic communities (Gili and Coma 1998). Sponges host a huge number of micro-organisms in their tissues, including contaminants and true symbionts, so that it is very difficult to separate the species used as food from symbionts or opportunistic and saprophytic species.

M. Wahl (ed.), *Marine Hard Bottom Communities*, Ecological Studies 206,
DOI: 10.1007/978-3-540-92704-4_21, © Springer-Verlag Berlin Heidelberg 2009

Carter (1878) first reported *Ircinia* species in the Indian Ocean being affected by filamentous fungi. Since then, several episodes of mortalities have been documented, often involving species with a skeleton of spongin fibres. In the Mediterranean, sponge diseases were reported in 1986 when catastrophic mortality of commercial sponges in the East Mediterranean reduced annual output to only a few tonnes (Vacelet 1994). A spongin fibre-perforating bacterium was hypothesised as the causative agent of the disease (Gaino and Pronzato 1989). During an extensive die-off of benthic suspension-feeders in 1999 in the North Mediterranean (Cerrano et al. 2000), the most seriously affected sponges were the Dictyoceratida, Dendroceratida and Verongida, all characterized by a skeleton of spongin fibres (Fig. 21.1a). An α-proteobacterium was considered to be responsible for the death of *Rhopaloeides odorabile* and the degradation of its spongin fibres on the Great Barrier Reef (Webster et al. 2002), suggesting that a similar organism may have contributed to the Mediterranean episodes. Sponge mass mortalities have increased dramatically in recent years, including in Papua New Guinea, the Great Barrier Reef and the reefs of Cozumel, Mexico (see Webster 2007 for a review).

Generally, diseases start as small necrotic areas on the body of the sponge, which increase in size until they affect the whole sponge. The progression of the disease is influenced by growth form, with massive sponges damaged faster than are branching forms (Wulff 2006b). The affected specimens are sometimes able to isolate the necrotic tissue by means of a layer of new pinacoderm between healthy and necrotic tissue, which aids recovery, as was seen during an episode in the West Mediterranean when a severe necrosis of the symbiotic sponge *Petrosia ficiformis* followed a bleaching event. This was concomitant with environmental stress (heavy rainfall, land runoff, high seawater temperature; Cerrano et al. 2001). From 1996–2000 in Papua New Guinea, populations of *Ianthella basta* suffered from a disease due to bacteria (*Bacillus* and *Pseudomonas*) closely related to species that are widely used as pesticides. This suggests a possible horizontal gene flow from terrestrial species used in wide-spectrum pesticides to related marine species that can then infect marine invertebrates (Cervino et al. 2006).

The loss of sponges has an unpredictable impact on the ecology of rocky benthos because of the contribution of their three-dimensional structure to habitat architecture, and the wide range of trophic and symbiotic relationships that link sponges to virtually all the other marine taxa (Wulff 2006a; Cerrano et al. 2006a). For example, the loss of sponges from reefs in Florida Bay resulted in the redistribution of juvenile lobsters, particularly an influx of lobsters into sites with artificial shelters and their decline at sites without shelters (Butler et al. 1995).

21.3 Cnidaria

21.3.1 Hexacorals—Hard Corals

Coral reefs are among the most biodiverse ecosystems of the world. They show optimal development where the annual mean minimum temperatures are about 23–25°C. Generally, coral reefs are found in oligotrophic waters and their diversity

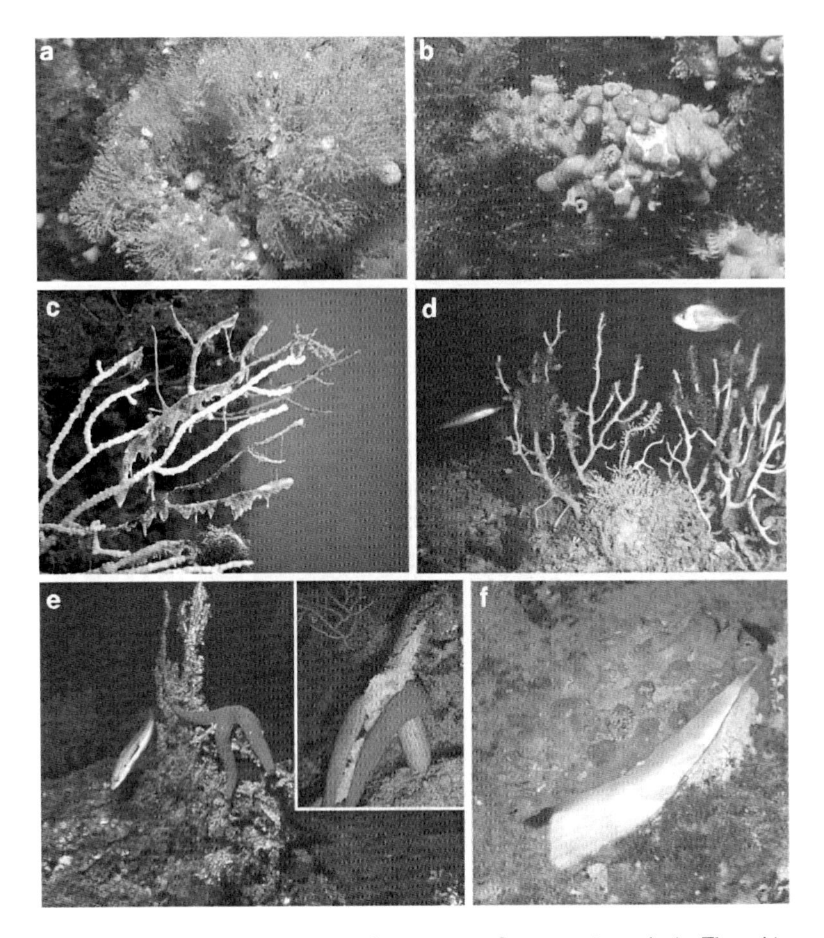

Fig. 21.1 **a** Sponge mortality: skeleton of the sponge *Cacospongia scalaris*. The *white spots* indicate dead barnacles usually living in the sponge tissue. **b** Mortality of the zoanthid *Parazonathus axinellae*: first, a purple cyanobacterial mat chokes the polyps (not shown) and then a white fungus leads to the detachment of colonies. **c** *Eunicella cavolinii* tissue sloughing off during a mortality event. **d** *Paramuricea clavata*: the coenechyme is still in situ but signs of necrosis are evident. **e** The starfish *Echinaster sepositus* feeding on organisms settled on a *Paramuricea clavata* skeleton. *Inset*: *Hacelia attenuata* feeding on the sponge *Plerapysilla spinifera* growing on a dead colony. **f** *Spondylus gaederopus*: a dead specimen with the upper valve still in situ and covered by the sponge *Crambe crambe*

and productivity are sustained by symbiotic relationships between zooxanthellae (*Symbiodinium* spp.) and madreporian corals.

During the past 30–40 years, madreporian corals in intertropical areas and the Mediterranean Sea have been prone to a wide spectrum of diseases that have decimated their populations. Nearly all of the world's major coral-reef regions (Caribbean/western Atlantic, eastern Pacific, central and western Pacific, Indian Ocean, Arabian Gulf, Red Sea) have experienced some degree of mortality since the 1970s. A coral disease was first documented in the Caribbean with local episodes of black band disease (Antonius 1973) but the first widespread episode was due to the white plague in 1975, affecting *Mycetophyllia ferox* (Dustan 1977).

Sixteen pathologies are known to affect hard corals in the Caribbean (Weil 2004) and six have been implicated in the Indopacific (Willis et al. 2004). To these, we have to add bleaching, first described in a wide area of the Pacific Ocean by Glynn (1984). Bleaching consists of a major decline in densities of zooxanthellae (e.g. Hoegh-Guldberg and Smith 1989) and/or loss of photosynthetic pigments (e.g. Kleppel et al. 1989; Szmant and Gassman 1990). It has long been recognized as a generalized response to many natural and anthropogenic stresses (Glynn 1993; Brown 1997) or infectious agents (Rosemberg 2004). In 1998, the most intense and geographically diffused bleaching episode involved almost the entire tropical region (Wilkinson et al. 1999). New episodes are still occurring in several reefs around the world. In the Mediterranean Sea, no pathogenic diseases have so far been found for temperate corals, even though bleaching in the allochthonous *Oculina patagonica* has been shown to be due to the bacterium *Vibrio shiloi* (Kushmaro et al. 1996).

Several hypotheses have been proposed to explain bleaching. They include bacterial infection (Rosemberg 2004), adaptive bleaching (Buddemeier et al. 2004), oxidative/biochemical processes (Downs et al. 2002; Perez and Weis 2006), thermal stress (Lesser 2004), UV radiations (Gleason and Wellington 1993), starvation (proposed by MacKenzie, www.fisherycrisis.com/coral1.html), and the photothermal effect (suggested by Leletkin 2002, after reviewing literature on the effect of photosynthetic active radiation on zooxanthellae). A possible cyclic trend with a period of 3–4 years in the bleaching phenomena has been suggested, this cycle overlapping that of the El Niño Southern Oscillation (ENSO; Stone et al. 1999). Nevertheless, no findings are conclusive and further research is needed to clarify the causes and mechanisms of bleaching.

Some authors have proposed that coral "diseases" are opportunistic infections that are secondary to exposure to physiological stress (e.g. elevated temperature). Environmental anomalies may cause physiological stress, which leads to reduced resistance and unchecked growth of normally benign and non-pathogenic bacteria. This subsequently leads to mortality of the corals via many mechanisms, including serious infections by opportunistic pathogens (Lesser et al. 2007).

21.3.2 Other Hexacorals

In the north-western Mediterranean Sea, several populations of the zoanthid *Parazoanthus axinellae* have been suffering from a summer disease since 2000. From June to October, the polyps become covered with dense mats of the filamentous cyanobacterium *Porphyrosiphon* sp. that compromise polyp activity, causing death (Fig. 21.1b). With a loss of 80% of the *P. axinellae* surface, cover of the encrusting sponge *Crambe crambe* has increased (Cerrano et al. 2006b). A disease involving *Palythoa caribaeorum* (pathogen unknown) has been described in Brazil

(Acosta 2001), where 20% of the population of *P. caribeaorum* was affected, with necrotic zones exceeding 5% of the corals.

21.3.3 Octocorals

Alcyonacea are often infected or killed, with most occurrences along the coasts of the Caribbean and West Indies. One of the first recorded cases of mass mortality of sea fans involved *Gorgonia flabellum* in the Caribbean, where an algal infection led to almost 100% mortality (Guzman and Cortes 1984). In 1992, a similar pathogenic agent affected colonies of *Gorgonia ventalina* (Garzon-Ferreira and Zea 1992). In 1995, it was a fungus *Aspergillus sydowii*, probably transported by African dust (Weir-Brush et al. 2004), that caused severe mortality of *G. ventalina* and *G. flabellum* in the Caribbean (Nagelkerken et al. 1996, 1997). In 1998, another common Caribbean species, *Briareum asbestinum*, experienced a mass mortality tentatively attributed to a cyanobacterium belonging to the genus *Scytonema*, the presence of which was concomitant with high seawater temperatures (Harvell et al. 1999). The sudden death of sea fans in Indonesia (North Sulawesi) was, however, linked to an outbreak of caprellid amphipods (Scinto et al. 2008).

In the Mediterranean, the first episode of mass mortality of gorgonians was described for *Corallium rubrum* (Harmelin 1984) but later, in 1986, *Eunicella cavolinii* was affected (Bavestrello and Boero 1986). *P. clavata* died off along the Ligurian and Provencal coasts in 1992 (Harmelin and Marinopoulos 1994; Bavestrello et al. 1994) and, in the summers of 1999 and 2003, major events affecting many species of filter-feeders occurred in the Ligurian Sea and Gulf of Lion (Garrabou et al. 2009). The most severely affected group was the gorgonians (Fig. 21.1c, d), particularly the violet sea fan *Paramuricea clavata*. No etiological agents involved in the mass mortalities in the Mediterranean have been identified (Cerrano et al. 2000; Rodolfo-Metalpa et al. 2000; Martin et al. 2002), although the bacterium *Vibrio corallyliticus* can cause disease in *P. clavata* colonies (Bally and Garrabou 2007).

Since 2002, cold-water regions have also started to be affected. For example, every summer now, a disease affects *Eunicella verrucosa* in SW England. Although *Vibrio splendidus* has been found associated with this disease, Koch's postulates have not been satisfied to prove aetiology (Hall-Spencer et al. 2007).

There are few data on intraspecific differences in resistance to diseases causing mass mortality. In the Mediterranean, the disease of sea fans in 1999 did not similarly affect all populations. Small colonies (<10 cm) were only slightly affected but larger ones almost completely, and females more intensely than males. This caused a major change in sex ratios, with a significant male bias (3.3:1.0, n=150) a year later. This contrasts with the original ratio of 1:1 and could have unpredictable consequences on the recovery of natural populations (Cerrano et al. 2005).

Rates of tissue regeneration depend on the species and amount of damage. Recovery of populations is slow but, if left undisturbed, pre-mortality densities can be reached within a year, due to recruitment. Biomass, however, remains small for several years, mainly because new colonies are small and the remaining ones reduce their sizes because of breaking of affected branches (Cerrano et al. 2005; Linares et al. 2005). The organic skeletons of dead colonies can, however, remain in situ for several years, supporting alternative food webs (Fig. 21.1e).

21.4 Molluscs

Live or dead mollusc shells can be important ecosystem engineers (Gutièrrez et al. 2003; Cerrano et al. 2006c). For example, oyster reefs support much biodiversity. These reefs are endangered by pollution and global warming (Lanning et al. 2006). The loss of oyster reefs in Chesapeake Bay was caused by the pathogen *Perkinsus marinus*, which led to the commercial collapse of the regional oyster industry (see also Chap. 19 by Airoldi et al.).

In the Mediterranean, one of the most endangered species is the bivalve *Spondylus gaederopus*, which suffered widespread mortality in 1981, 1982 and 1983 (north-western Mediterranean; Meinesz and Mercier 1983), 2003 (French coast; Garrabou et al. 2008), 2005 (Columbrets Islands; Kersting and Templado 2006) and 2006 (Thyrrenian basin; Garrabou et al. 2008). Because there were no temperature anomalies from 1981 to 1983, Meinesz and Mercier (1983) hypothesised a viral, bacterial or fungal infection (Fig. 21.1f).

The only known post-Pleistocene case of an extinction of a marine invertebrate is the limpet *Lottia alveus*, a stenohaline species that ranged from southern Labrador to Long Island Sound. It lived only on the blades of the eelgrass *Zostera marina* (Carlton et al. 1991), and widespread disease-induced regression of the eelgrass led to the complete loss of habitat for the limpet. Other gastropods (*Littoraria flammea*, *Cerithidea fuscata*, and "*Collisella*" *edmitchelli*) may have recently become extinct, although there are doubts about this (Carlton 1993).

21.5 Echinoderms

A disease can be so widespread and aggressive that there is little recovery, even after decades. An example can be furnished by the mass mortality involving the sea urchin *Diadema antillarum* in the Caribbean in the early 1980s, which led to 98% loss (Lessios et al. 1984). The cause of this disease has not been identified (Lessios 1988). It was initially recognized by an accumulation of sediment and loss of pigment and some spines but then the affected urchins could no longer attach to the substratum and started disintegrating. Moreover, they did not seek shelter during daylight and were being preyed upon by fish that did not feed on healthy *D. antillarum*. This disease led to many Caribbean reefs changing from coral- to macro-algal-dominated. To date, a reversal of this development has been observed only in a small area in Jamaica (Edmunds and Carpenter 2001).

Other mass mortalities of sea urchins have been recorded—e.g. *Strongylocentrotus droebachiensis* on the coast of Nova Scotia in 1980–1982. Although an amoeboid protist has been tentatively identified as the pathogenic agent, temperature appears to have played an important role in triggering epizootics and determining the extent of mortality (Scheibling 1984). Three species (*Paracentrotus lividus*, *Sphaerechinus granularis*, *Arbacia lixula*) were involved in local mass mortalities in the Mediterranean in 1978–1979 (Miller 1985). On the French coast, mortality of *P. lividus* led to massive growth of epiphytes on seagrass (Boudouresque et al. 1980). In the late 1980s, a widespread harmful algal bloom (*Chrysochromulina* spp.) in the North Sea (Skagerrak and Kattegat coasts) affected several taxa and, also in this case, the most deleterious effects were observed on benthic organisms including plants (red and brown seaweed), molluscs (bivalves and snails) and echinoderms (sea stars and sea urchins).

21.6 Ascidians

After the 1997/1998 El Niño event, a mass mortality of several ascidian species was recorded in northern Bahia (Brazil). Their recovery was limited by the mortality of numerous other benthic organisms (hydroids, octocorals and bryozoans) that are a commonly exploited substratum (Kelmo et al. 2006). In the Mediterranean, *Microcosmus* spp. were involved in several episodes of mortality since the 1980s and, in 1999, *Halocynthia papillosa* and *Pyura pyura* also suffered from a thermal anomaly (Perez at al. 2000).

21.7 Extinctions and Massive Mortalities: Effects on Benthic Communities

Mass mortalities reported here are generally due to environmental stress (e.g. thermal stress) and/or pathogens. Environmental anomalies can switch on and regulate pathogen virulence. Regarding the extinction of marine organisms, the main causes are different from those leading to mass mortalities and are generally due to habitat destruction, pollution, invasive species and overexploitation (see other chapters in this book part for further details), although there is no unequivocal list for these extinct species. The main problems with identifying these causes are (1) taxonomic uncertainties, (2) little attention paid to small invertebrates and (3) natural variability of populations. Mass mortalities have highlighted the need to focus more studies on life histories of many of the species that might be affected.

Another major problem that limits our understanding of the role of diseases and recovery of affected species is the lack of long-term monitoring projects. Mass mortalities strongly modify population and community structure and functioning. There may be a decrease in the three-dimensional complexity of the structure of biogenic habitat, especially when the affected species are long-lived and lack a calcareous skeleton. This can cause loss of numerous microhabitats, affecting biodiversity and threatening interspecific relationships in the community (McCoy and Bell 1991). The disappearance of structural elements in a community can also lead to changes to the substratum and edaphic variables, altering food webs. Filter-feeders are important in contributing to the transfer of energy from the water column to the benthos. When these organisms disappear, suspended particulate organic matter, phytoplankton and zooplankton, are potentially unexploited.

In the tropics, algal turfs commonly overgrow dead corals after bleaching, limiting recovery. Elsewhere, coralline algae can grow over the dead coral and there can be a major increase in fast-growing soft corals (Stobart et al. 2005). The abundances of herbivorous fish are important in controlling the algae, and so may be crucial for the recovery of these reefs (Arthur et al. 2005). The effects of bleaching on coral fishes differ between the short and long term. Months after mortality of the coral, the biomass and numbers of fish species can increase but they subsequently decrease, with different functional groups being significantly influenced by these habitat changes. Abundances of herbivores increased as an immediate response to bleaching but then decreased because of loss of structural habitat complexity (Garpe et al. 2006). In the long term, there may be a decline in smaller fishes and an increase in larger ones, perhaps due to a time lag in response to changes in habitat complexity, as fishes are lost through natural mortality and fishing and not replaced by juveniles because of loss of juvenile habitat (Graham et al. 2007).

The mass mortality of *Diadema antillarum* in the Caribbean caused dramatic changes to the benthic algal community, and algal biomass increased by 22–439% across the reef within 16 months of the die-off of urchins. The composition of the algal species also changed dramatically, from a dominance of algal turfs and crustose algae, with few macro-algae before the mass mortality of urchins, to a reduction of turfs to 40% and an increase in macro-algae to 47% 25 months later.

In large areas of subtidal reefs in the North Mediterranean Sea, mass mortalities of sponges and sea fans have dramatically changed the landscape down to 45 m depth, although it is difficult to evaluate the effect of the almost complete loss of these organisms on the entire coralligenous assemblages. The widespread strong reduction of species of economic interest—e.g. bath sponges and coral—has impaired commercial fishery activities. Several years after the mass mortality, there are as yet no sign of recovery of shallow populations of bath sponges. The partial recover of shallow sea fans is probably due to larvae influx from deep populations (Cerrano and Bavestrello 2008). The preservation of these deep populations is crucial for the conservation of these characteristic elements of the Mediterranean landscape.

Benthic suspension feeders are very efficient in extracting and processing energy from plankton. Considering the ecological importance of benthic filter-feeders in regulating littoral food chains (Gili and Coma 1998), and the slow growth of many important species, mass mortalities may induce long-term effects, not only in benthic communities but also in the water column (Sala and Knowlton 2006). To correctly evaluate these complex dynamics, it is important to develop long-term monitoring projects of the most vulnerable species to evaluate when and where populations dip below threshold densities, which may compromise recovery and lead to local or global extinction (Levitan and McGovern 2005).

Mass mortalities determine sudden fragmentation of affected populations. In the case of populations of benthic sessile species, recruitment could be strongly compromised. For spawning organisms, the absolute distances that sperms can travel are limited, so that eggs of isolated individuals may remain unfertilized. The Allee effect is a threshold in numbers below which rates of population growth become negative, rather than positive. When recruitment of the affected species continued after the mass mortality, a possible explanation could be the occurrence of pockets of populations with sufficient density continuing to seed other areas. In this way, communities living in the mesophotic zone, between 60 and 100–120 m depth, may represent key areas for the recovery of affected species, and would need to be included in management programs. Unfortunately, still today these environments are too poorly explored and known to be adequately considered for protection.

References

Acosta A (2001) Disease in zoanthids: dynamics in space and time. Hydrobiologia 460:113–130

Antonius A (1973) New observations in coral destruction in reefs. Assoc Mar Lab Caribb Abstr 10:3

Arthur R, Done TJ, Marsh H, Harriott VJ (2005) Benthic recovery 4 years after an El Niño-induced coral mass mortality in the Lakshadweep atolls. Curr Sci 89:694–699

Bally M, Garrabou J (2007) Thermodependent bacterial pathogens and mass mortalities in temperate benthic communities: a new case of emerging disease linked to climate change. Global Change Biol 13:2078–2088

Bavestrello G, Boero F (1986) Necrosi e rigenerazione in *Eunicella cavolinii* (Anthozoa, Cnidaria) in Mar Ligure. Boll Mus Ist Biol Univ Genova 52:295–300

Bavestrello G, Bertone S, Cattaneo-Vietti R, Cerrano C, Gaino E, Zanzi D (1994) Mass mortality of *Paramuricea clavata* (Anthozoa: Cnidaria) on Portofino Promontory cliffs (Ligurian Sea). Mar Life 4:15–19

Boero F (1996) Episodic events: their relevance to ecology and evolution. PSZNI Mar Ecol 17:237–250

Boudouresque CF, Nedelec N, Sheperd SA (1980) The decline of a population of the sea urchin *Paracentrotus lividus* in the Bay of Port-Cros (Var, France). Trav Sci Parc Natl Port-Cros 6:243–251

Brown BE (1997) Coral bleaching: causes and consequences. Coral Reefs 16:129–138

Buddemeier RW, Baker AC, Fautin DG, Jacobs JR (2004) The adaptive hypothesis of bleaching. In: Rosemberg E, Loya Y (eds) Coral health and disease. Springer, Berlin Heidelberg New York, pp 427–444

Butler MJ, Hunt JH, Herrnkind WF, Childress MJ, Bertelsen R, Sharp W, Matthews T, Field JM, Marshall HG (1995) Cascading disturbances in Florida Bay, USA: cyanobacteria blooms, sponge mortality, and implications for juvenile spiny lobsters Panulirus argus. Mar Ecol Prog Ser 129:119–125

Carlton JT (1993) Neoextinctions in marine invertebrates. Am Zool 33:499–507

Carlton JT, Vermeij GJ, Lindberg DR, Carlton DA, Dudley EC (1991) The first historical extinction of a marine invertebrate in an ocean basin: the demise of eelgrass limpet *Lottia alveus*. Biol Bull 180:72–80

Carlton JT, Geller JB, Reaka-Kudla ML, Norse EA (1999) Historical extinctions in the sea. Annu Rev Ecol Syst 30:515–538
Carter HJ (1878) Parasites of the Spongida. Ann Mag Nat Hist 2:157–172
Cerrano C, Bavestrello G (2008) Medium-term effects of die-off of rocky benthos in the Ligurian Sea. What can we learn from gorgonians? Chem Ecol 24:73–82
Cerrano C, Bavestrello G, Bianchi CN, Cattaneo-Vietti R, Bava S, Morganti C, Morri C, Picco P, Sara G, Schiaparelli S, Siccardi A, Sponga F (2000) A catastrophic mass-mortality episode of gorgonians and other organisms in the Ligurian Sea (NW Mediterranean), summer 1999. Ecol Lett 3:284–293
Cerrano C, Magnino G, Sarà A, Bavestrello G, Gaino E (2001) Necrosis in a population of *Petrosia ficiformis* (Porifera, Demospongiae) in relations with environmental stress. Ital J Zool 68:131–136
Cerrano C, Arillo A, Azzini F, Calcinai B, Castellano L, Muti C, Valisano L, Zega G, Bavestrello G (2005) Gorgonian population recovery after a mass mortality event. Aquat Conserv Mar Freshw Res 15:147–157
Cerrano C, Calcinai B, Pinca S, Bavestrello G (2006a) Reef sponges as hosts of biodiversity: cases from North Sulawesi. In: Suzuki Y, Nakamori T, Hidaka M, Kayanne H, Casareto BE, Nadao K, Yamano H, Tsuchiya M(eds)Proc 10th Coral Reef Symp, Okinawa, pp 208–213
Cerrano C, Totti C, Sponga F, Bavestrello G (2006b) Summer disease in *Parazoanthus axinellae* (Schmidt, 1862) (Cnidaria, Zoanthidea). Ital J Zool 73:355–361
Cerrano C, Calcinai B, Bertolino M, Valisano L, Bavestrello G (2006c) Epibionts of the scallop *Adamussium colbecki* in the Ross Sea, Antarctica. Chem Ecol 22:235–244
Cervino JM, Winiarski-Cervino K, Polson SW, Goreau T, Smith GW (2006) Identification of bacteria associated with a disease affecting the marine sponge *Ianthella bastain* New Britain, Papua New Guinea. Mar Ecol Prog Ser 324:139–150
Downs CA, Fauth JE, Halas JC, Dustan P, Bemiss J, Woodley CM (2002) Oxidative stress and seasonal coral bleaching. Free Rad Biol Med 33:533–543
Dustan P (1977) Vitality of reef coral populations off Key Largo, Florida: recruitment and mortality. Environ Geol 2:51–58
Edmunds PJ, Carpenter RC (2001) Recovery of *Diadema antillarum* reduces macroalgal cover and increases abundance of juvenile corals on a Caribbean reef. Proc Natl Acad Sci 98:5067–5071
Epstein PR, Sherman K, Spanger-Siegfried E, Langston A, Prasad S, McKay B (1998) Marine ecosystems: emerging diseases as indicators of change. Health Ecological and Economic Dimensions (HEED), NOAA Global Change Program
Gaino E, Pronzato R (1989) Ultrastructural evidence of bacterial damage to *Spongia officinalis* fibres (Porifera, Demospongiae). Diseases Aquat Organisms 6:67–74
Garpe KC, Yahya SAS, Lindahl U, Ohman MC (2006) Long-term effects of the 1998 coral bleaching event on reef fish assemblages. Mar Ecol Prog Ser 315:237–247
Garrabou J, Perez T, Sartoretto S, Harmelin JG (2001) Mass mortality event in red coral *Corallium rubrum* populations in Provence region (France, NW Mediterranean). Mar Ecol Prog Ser 217: 263–272
Garrabou J, Coma R, Bensoussan N, Chevaldonné P, Cigliano M, Diaz D, Harmelin JG, Gambi MC, Graille R, Kersting D K, Lejeusne C, Linares C, Marschal C, Perez T, Ribes M, Romano JC, Torrents O, Zabala M, Zuberer F, Cerrano C (2009) Mass mortality in Northwestern Mediterranean rocky benthic communities: effects of the 2003 heat wave. Global Change Biol doi: 10.1111/j.1365-2486.2008.01823.x
Garzon-Ferreira J, Zea S (1992) A mass mortality of *Gorgonia ventalina* (Cnidaria: Gorgoniidae) in the Santa Marta area, Caribbean coast of Colombia. Bull Mar Sci 50:522–526
Gili JM, Coma R (1998) Benthic suspension feeders: their paramount role in littoral marine food webs. Trends Ecol Evol 13:316–321
Gleason DF, Wellington GM (1993) Ultraviolet radiation and coral bleaching. Science 365:836–838

Glynn PW (1984) Widespread coral mortality and the 1982/83 El Niño warming event. Environ Conserv 11:133–146
Glynn PW (1993) Coral reef bleaching: ecological perspectives. Coral Reefs 12:1–17
Graham NAJ, McClanahan TR, Letourneur Y, Galzin R (2007) Anthropogenic stressors, interspecific competition and ENSO effects on a Mauritian coral reef. Environ Biol Fish 78:57–69
Gutièrrez JL, Jones CG, Strayer DL, Iribarne OO (2003) Molluscs as ecosystem engineers: the role of shell production in aquatic habitats. Oikos 101:79–90
Guzman HM, Cortes J (1984) Mass death of *Gorgonia flabellum* L. (Octocorallia: Gorgoniidae) in the Caribbean coast of Costa Rica. Rev Biol Trop 32:305–308
Hall-Spencer JM, Pike J, Munn CB (2007) Diseases affected cold-water corals too: *Eunicella verrucosa* (Cnidaria: Gorgonacea) necrosis in SW England. Diseases Aquat Organisms 76:87–97
Harmelin JG (1984) Biologie du corail rouge. Paramètres de populations, croissance et mortalité naturelle. Etat des connaissances en France. FAO Fish Rep 306:99–103
Harmelin JG, Marinopoulos J (1994) Population structure and partial mortality of the gorgonian *Paramuricea clavata* in the north-western Mediterranean. Mar Life 4:5–13
Harvell CD, Kim K, Burkholder JM, Colwell RR, Epstein PR, Grimes DJ, Hofmann EE, Lipp EK, Osterhaus ADME, Overstreet RM, Porter JW, Smith GW and Vasta GR (1999) Emerging marine diseases—climate links and anthropogenic factors. Science 285:1505–1510
Harvell CD, Mitchell CE, Ward JR, Altizer S, Dobson A, Ostfeld RS, Samuel MD (2002) Climate warming and disease risks for terrestrial and marine biota. Science 296:2158–2162
Hayes RL, Goreau NI (1998) The significance of emerging diseases in the tropical coral reef ecosystems. Rev Biol Trop 46:172–185
Hayes ML, Bonaventura J, Mitchell TP, Propero JM, Shinn EA, Van Dolah F, Barber RT (2001) How are climate changes and emerging marine diseases functionally linked? Hydrobiologia 460:213–220
Hoegh-Guldberg O, Smith GJ (1989) The effect of sudden changes in temperature, light and salinity on population density and export of zooxanthellae from the reef corals *Seriatopora hystrix* and *Stylopora pistillata*. J Exp Mar Biol Ecol 129:279–303
Kelmo F, Attrill MJ, Jones MB (2006) Mass mortality of coral reef ascidians following the 1997/1998 El Niño event. Hydrobiologia 555:231–240
Kersting DK, Templado J (2006) Evento de mortandad masiva del coral *Cladocora caespitosa* (Scleractinia) en las Islas Columbretes tras del calentamiento anormal del agua en el verano de 2003. In: Abstr Vol 80 XIV SIEMB Barcelona, 12–15 September, pp 80
Kleppel GS, Dodge RE, Reese CJ (1989) Changes in pigment associated with the bleaching of stony corals. Limnol Oceanogr 34:1331–1335
Kushmaro A, Loya Y, Fine M, Rosemberg E (1996) Bacterial infection and coral bleaching. Nature 380:396
Lanning G, Flores JF, Sokolova IM (2006) Temperature-dependent stress response in oysters, *Crassostrea virginica*: pollution reduces temperature tolerance in oysters. Aquat Toxicol 79:278–287
Leletkin VA (2002) Bleaching of hermatypic corals. Russian J Mar Biol 28:532–540
Lesser MP (2004) Experimental biology of coral reef ecosystems. J Exp Mar Biol Ecol 300:217–252
Lesser MP, Bythell JC, Gates RD, Johnston RW, Hoegh-Guldberg O (2007) Are infectious diseases really killing corals? Alternative interpretations of the experimental and ecological data. J Exp Mar Biol Ecol 346:36–44
Lessios H (1988) Mass mortality of *Diadema antillarum* in the Caribbean: what have we learned? Annu Rev Ecol Syst 19:371–393
Lessios HA, Robertson DR, Cubit JD (1984) Spread of *Diadema* mass mortality throughout the Caribbean. Science 226:335–337
Levitan DR, McGovern TM (2005) The Allee effect in the sea. In: Norse EA, Crowder LB (eds) Marine conservation biology: the science of maintaining the sea's biodiversity. Island Press, Washington, DC, pp 47–57

Linares C, Coma R, Diaz D, Zabala M, Hereu B, Dantart L (2005) Immediate and delayed effects of a mass mortality event on gorgonian population dynamics and benthic community structure in the NW Mediterranean Sea. Mar Ecol Prog Ser 305:127–137

Martin Y, Bonnefort JL, Chancerell L (2002) Gorgonians mass mortality during the 1999 late summer in French Mediterranean coastal waters: the bacterial hypothesis. Water Res 36:779–782

McCoy ED, Bell SS (1991) Habitat structure: the evolution and diversification of a complex topic. In: Bell SS, McCoy ED, Mushinsky HR (eds) Habitat structure: the physical arrangement of objects in space. Chapman & Hall, London, pp 3–27

Meinesz A, Mercier D (1983) Sur les fortes mortalités de Spongylus gaederopous Linné observées sur les cotes de Méditerranée. Trav Sci Parc Natl Port-Cros 9:89–95

Miller RJ (1985) Succession in sea urchin and seaweed abundance in Nova Scotia, Canada. Mar Biol 84:275–286

Nagelkerken I, Buchan K, Smith GW, Bonair K, Bush P, Garzón-Ferreira J, Botero L, Gayle P, Heberer C, Petrovic C, Pors L, Yoshioka P (1996) Widespread disease in Caribbean sea fans: I. Spreading and general characteristics. Int Coral Reef Symp 1:679–682

Nagelkerken I, Buchan K, Smith GW, Bonair K, Bush P, Garzon-Ferreira J, Botero L, Gayle P, Harvell CD, Heberer C, Kim K, Petrovic C, Pors L, Yoshioka P (1997) Widespread disease in Caribbean sea fans: patterns of infection and tissue loss. Mar Ecol Prog Ser 160:255–263

Perez S, Weis V (2006) Nitric oxide and cnidarian bleaching: an eviction notice mediates breakdown of a symbiosis. J Exp Biol 209:2804–2810

Perez T, Garrabou J, Sartoretto S, Harmelin JG, Francour P, Vacelet J (2000) Mass mortality of marine invertebrates: an unprecedented event in the Northwestern Mediterranean. C R Acad Sci III 323:853–865

Rodolfo-Metalpa R, Bianchi CN, Peirano A, Morri C (2000) Coral mortality in NW Mediterranean. Coral Reefs 9:24

Rosemberg E (2004) The bacterial disease hypothesis of coral bleaching. In: Rosemberg E, Loya Y (eds) Coral health and disease. Springer, Berlin Heidelberg New York, pp 445–461

Sala E, Knowlton N (2006) Global marine biodiversity trends. Annu Rev Environ Resources 31:93–122

Scheibling RE (1984) Echinoids, epizootics and ecological stability in the rocky subtidal off Nova Scotia, Canada. Helgoländ Meeresunters 37:233–242

Scinto A, Bavestrello G, Boyer M, Previati M, Cerrano C (2008) Gorgonian mortality related to a caprellids massive attack in the marine park of Bunaken (North Sulawesi, Indonesia). J Mar Biol Assoc UK 88:723–727

Stobart B, Kristian T, Buckley R, Downing N, Callow M (2005) Coral recovery at Aldabra Atoll, Seychelles: five years after the 1998 bleaching event. Philos Trans R Soc Math Phys Eng Sci 363:251–255

Stone L, Huppert A, Rajagopalan B, Bhasin H, Loya Y (1999) Mass coral reef bleaching: a recent outcome of increased El Niño activity? Ecol Lett 2:325–330

Szmant AM, Gassman NJ (1990) The effects of prolonged ‘bleaching’ on the tissue biomass and reproduction of the reef coral *Montastrea annularis*. Coral Reefs 8:217–224

Vacelet J (1994). The struggle against the epidemic which is decimating Mediterranean sponges. Rome, FAO Tech Rep

Webster N (2007) Sponge disease: a global threat? Environ Microbiol 9:1363–1375

Webster NS, Negri AP, Webb RI, Hill RT (2002) A spongin-boring α-proteobacterium is the etiological agent of disease in the Great Barrier Reef sponge Rhopaloeides odorabile. Mar Ecol Prog Ser 232:305–309

Weil E (2004) Coral reef diseases in the wider Caribbean. In: Rosemberg E, Loya Y(eds) Coral health and disease. Springer, Berlin Heidelberg New York, pp 35–68

Weir-Brush JR, Garrison VH, Smith GW, Shinn EA (2004) The relationship between gorgonian coral (Cnidaria: Gorgonacea) diseases and African dust storms. Aerobiologia 20:119–126

Wilkinson C, Linden O, Cesar H, Hodgson G, Rubens J, Strong AE (1999) Ecological and socioeconomic impacts of 1998 coral mortality in the Indian Ocean: an ENSO impact and a warning of future change? Ambio 28:188–196

Willis BL, Page CA, Dinsdale EA (2004) Coral disease on the Great Barrier Reef. In: Rosemberg E, Loya Y (eds) Coral health and disease. Springer, Berlin Heidelberg New York, pp 69–104
Wulff J (2006a) Ecological interactions of marine sponges. Can J Zool 84:146–166
Wulff J (2006b) A simple model of growth form-dependent recovery from disease in coral reef sponges, and implications for monitoring. Coral Reefs 3:157–163
Zocchi E, Carpaneto A, Cerrano C, Bavestrello G, Giovine M, Bruzzone S, Guida L, Luisa F, Usai C (2001) The temperature-signalling cascade in sponges involves a heat-gated cation channel, abscisic acid and cyclic ADP-ribose. Proc Natl Acad Sci 98(26):14859–14864

Chapter 22
Biological Invasions: Insights from Marine Benthic Communities

Christopher D. McQuaid and Francisco Arenas

22.1 Introduction

Biological invasions occur when a species enters and spreads into areas beyond its natural range of distribution (Vermeij 1996). Such invasions may be among the greatest threats to global biodiversity (Bax et al. 2003), though this perception is contested (Briggs 2007). Coastal marine systems are among the most invaded systems on the planet (Carlton 1996), and invasive species can have dramatic effects on the structure and functioning of invaded ecosystems. Yet, marine invasions are much less well studied than invasions in terrestrial and freshwater communities (Grosholz 2002). Here, we review the characteristics of biological invasions on marine hard-bottom communities from the points of view of both the invasive species and the recipient community.

22.2 The Arrival of Introduced Species: Vectors and Propagule Pressure

The first stage of any invasion is the transfer of an organism to a new region. This stage consists of the uptake of an organism (or a subset of organisms) from the source species pool and its delivery to a recipient region (Ruiz and Carlton 2003). Although invasions may result from natural dispersal, human-mediated transfers appear to be the prevalent pathway (Ruiz et al. 1997), and thousands of species are moved around the world's oceans by shipping, aquaculture and the aquarium trade, scientific research, etc. Given the variety of potential dispersal mechanisms, it is difficult to determine the vector or multiple vectors involved for most introduced marine species. Nevertheless, shipping and aquaculture are recognized as the most important of these vectors (Ruiz et al. 2000; Streftaris et al. 2005; Minchin 2007).

Marine species are transported by ships in many ways. Sessile species may attach to the hull, the anchor or to solid ballast (Fofonoff et al. 2003). Since the introduction of steel-hulled vessels, ballast water pumped into vessels to provide stability includes many holoplanktonic organisms as well as the larval stages and

M. Wahl (ed.), *Marine Hard Bottom Communities*, Ecological Studies 206,
DOI: 10.1007/978-3-540-92704-4_22,

propagules of benthic organisms (Carlton and Geller 1993), so that ballast water appears to be the single largest pathway for the unintentional transfer of marine species. Carlton (1999) estimated that more than 10,000 species are transported every day in ballast water.

Aquaculture-related activities are the second most frequent vector of marine introductions (Minchin 2007). Examples of introduced species with commercial value are abundant in the literature (e.g. the Japanese oyster, *Crassostea gigas*, or the Japanese seaweed known commercially as wakame, *Undaria pinnatifida*). Additionally, aquaculture, especially oyster culture, has unintentionally introduced associated biota attached to shells or algae used for packaging. This pathway accounted for 15% of the marine invasions reported by Ruiz et al. (2000) in North America and 19% of the marine invasions in Europe (Streftaris et al. 2005).

The removal of geographical barriers through the construction of shipping canals, such as the Suez and Panama canals, also promotes marine invasions (Galil 2000). This is particularly important in the Mediterranean Sea, where introductions via the Suez Canal constitute 52% of the total number of introductions; the so-called Lessepsian introductions (Streftaris et al. 2005).

Benthic communities on hard substrata are characterized by large, long-lived sessile species that create biogenic habitats for other organisms. Despite being attached to the substratum, most of these sessile species spread through planktonic propagules. While macroalgal spores disperse passively and have a short planktonic phase, the larvae of sessile invertebrates are usually active swimmers capable of behaviour and have a longer planktonic life. Thus, benthic organisms can be delivered as adults attached to hulls, shells or other hard structures, or during the planktonic phase in ballast tanks.

We reviewed the literature for 372 introductions of 271 species of macroalgae and benthic animals (ascidians, bryozoans and crustaceans) found on hard substrata all around the world, excluding Lessepsian introductions as these are almost exclusive to the Mediterranean Sea. Not surprisingly, we found a prevalence of aquaculture-related introductions in the case of macroalgae, while shipping was the major introduction vector for the animals (Table 22.1).

The relative contributions of these vectors to organism transfer is highly variable in space and time, affecting not only the identity of the potential invaders but also

Table 22.1 Suggested introduction vectors for different aquatic organisms

	Macroalgae 190 introductions 164 species	Bryozoans 52 introductions 29 species	Crustaceans 77 introductions 50 species	Ascidians 53 introductions 28 species
Aquaculture	38%	0%	8%	13%
Shipping	26%	94%	92%	85%
Aquaculture +shipping	0%	15%	4%	2%
Other (research, aquaria, etc.)	4%	2%	0%	2%
Unknown	31%	4%	0%	0%

the frequency of introductions and the conditions of transfer, thereby modifying the propagule pressure exerted by the invading species. Propagule pressure includes both the absolute number of propagules released during each event (inoculum or propagule size) and the number of release events (frequency of introduction). Together, these two features can explain why some introductions are successful and others are not (Lockwood et al. 2005). Massive propagule pressure by established populations of an invasive species can also explain how biotic resistance of the native community can be overcome to enable rapid spread (e.g. Hollebone and Hay 2007). In marine ecosystems, propagule pressure has probably increased over the last few decades with the increase of the size, frequency and speed of vessels (Carlton 1996). Sustained propagule pressure may also have critical effects by enabling the introduction of new genetic strains and by maintaining gene flow between invasive and donor populations of the same species. Voisin et al. (2005) examined the genetic structure of introduced populations of the red alga *Undaria pinnatifida* in Europe and Australasia and found that most populations had high genetic variability, which was ascribed to recurrent introductions. Such genetic exchange can have important effects. In the case of the crab *Carcinus maenas*, repeated inoculations through ballast water seem to be responsible for high genetic diversity in populations along the eastern seaboard of North America, which has enabled rapid range extension into colder waters in Canada (Roman 2006).

22.3 What Makes a Good Invader?

Williamson and Brown (1986) originally postulated the ten-ten rule, later modified by Williamson and Fritter (1996), which suggests that 1 in 10 of imported species appear in the wild, 1 in 10 of these become established and, in turn, 1 in 10 of these become pests. While this is obviously a crude rule, it does focus attention on the fact that the vast majority of species introduced into a new environment either fail to establish as viable populations or, at least, fail to become invasive (i.e. fail to have a noticeable effect on the host environment). Grosholz and Ruiz (1996) list 20 species that have spread from a single source region to several parts of the world, some of which have been outstandingly successful in establishing in widely separated areas. This includes the mussel *Mytilus galloprovincialis*, which is invasive on every continent except Antarctica (Daguin and Borsa 2000; Branch and Steffani 2004). Given the ability of some species to be successful invaders in many places, it is tempting to try to identify what characteristics make some species more successful invaders than others. This line of research has, however, largely proven unsuccessful and there is a general consensus that invasion processes are highly idiosyncratic (Meiners et al. 2004). Regarding possible characteristics of successful invaders, there are some that seem intuitively obvious, such as high fecundity and strong competitive abilities or other ecological or behavioural properties (e.g. Paglianti and Gherardi 2004). Yet, a review of the literature has highlighted the absence of consistency in characteristics among invasive species (Colautti et al. 2006).

Nyberg and Wallentinus (2005) take a more useful approach by not attempting to describe the average traits of invasive algae but, rather, examining patterns of features that increase the likelihood of invasion. Nevertheless, some taxa are more prevalent as invaders than others. In North America, half of all marine invasives are either crustaceans or molluscs (Ruiz et al. 2000).

Thus, more useful are broad comparisons that suggest taxon-specific characteristics that may be important in determining the ability to be invasive, such as body size in marine bivalves (Roy et al. 2002) or, among fish, crevice-seeking behaviour that allows gobies and blennies to utilize ballast holes on ships (Wonham et al. 2000). At a still broader level, carnivores and top predators in coastal marine systems are thought to have especially high levels of extinction rates through human activities. Although there are many examples of predators that have been successful as invasives, especially among the crabs (e.g. Abelló and Hispano 2006; Brousseau and Goldberg 2007), it appears that most invasions are by lower trophic levels, including detritivores and deposit feeders. The combination of these two biases results in an overall change in the structure of invaded food webs (Byrnes et al. 2007).

22.4 Which Communities Are More Susceptible to Invasion?

Despite the often idiosyncratic and context-specific character of biological invasions, researchers have tried to find some generalizations that could explain the susceptibility of communities to invasion. Thus, ecologists have long appreciated the importance of predation, parasitism and competition in the outcome of invasions. Nowadays, the roles of facilitation and disturbance are also increasingly recognized. Finally, with the outburst of the biodiversity-ecosystem functioning debate, the role of diversity has been at the core of research into the susceptibility of communities for the last decade. We review each of these topics separately.

22.4.1 Biotic Resistance, Competition, Predation and Facilitation: Interactions Between Native and Invasive Species

Interactions with indigenous species may inhibit or slow invasion by exotic species, usually through competition or predation—for example, the consumption of invasive predators by indigenous ones (Harding 2003; DeRivera et al. 2005), though consumption of introduced species can improve fitness of the native predator, with negative effects on native prey (Noonburg and Byers 2005). Another example of the interaction between native and introduced species with similar niche requirements comes from the west coast of South Africa, where the intertidal mussel *Mytilus galloprovincialis* dominates the entire mussel zone at the expense of the indigenous mussel *Aulacomya ater* and of limpets (Branch and Steffani 2004). In contrast, on the south coast of the country, *Mytilus* is excluded from the lower part of the mussel zone, partly through competition for space with the indigenous *Perna perna* (Rius and McQuaid 2006). Competition with an ecologically similar native species is an obvious example of biotic resistance but indirect effects among different trophic levels have also been detected. For example, the native *Zostera marina* has a strong negative effect on the introduced bivalve *Musculista senhousia* by decreasing water flow and, thus, reducing the availability of suspended food (Reusch and Williams 1999).

The opposite can also occur when indigenous species interact with introduced species in ways that benefit the newcomer. On the south coast of South Africa, *P. perna* facilitates colonisation of the low shore by *Mytilus* because *Perna* is more tolerant of strong wave action (Zardi et al. 2006), and survival of *Mytilus* is initially improved in mixed-species mussel beds, though it is later eliminated through competition (Rius and McQuaid 2006). Facilitation of invasion by native species may explain high local abundances of the Pacific oyster, *Crassostea gigas*, in northeast Canada (Ruesink 2007), and Siddon and Witman (2004) showed that predation on sea urchins by native crabs indirectly increased abundances of an introduced ascidian.

There is also evidence that invasive species can facilitate native species (Rodriguez 2006), and the idea that multiple invasive species may facilitate each other's survival has been put forward by Simberloff and Von Holle (1999) under the catchy heading of "invasion meltdown". The idea was taken up in a number of studies, including that of Grosholz (2005) who showed that a previously rare, introduced clam (*Gemma gemma*) became much more abundant after the later introduction of the crab *Carcinus maenas*. Simberloff (2006) later emphasised the original definition of the phrase to mean that interspecific facilitation results in accelerating *rates* of new introductions. This is different from simple facilitation of one invasive species by another, such as Grosholz's example, and Simberloff concludes that there have not yet been any described cases of "full meltdown".

22.4.2 The Role of Diversity in the Susceptibility of Communities to Invasion

Elton (1958) suggested that species-rich communities should be more stable and less susceptible to invasion, the so-named biotic resistance hypothesis. Although Elton never identified the underlying mechanism, most subsequent studies proposed it to be the more complete utilization of available resources by highly diverse communities, i.e. complementary use of resources by different species (Levine and D'Antonio 1999). The key concept then is a negative relationship between diversity and the availability of resources for newly introduced species. A second possible mechanism driving the relationship between diversity and invasibility is the sampling effect

(Wardle 2001). This is defined as the increasing occurrence of suppressive or facilitative species in more diverse communities.

The debate on the role of diversity in preventing invasion has been enhanced by the conflicting results found between observational surveys and experimental manipulations, the so-called invasion paradox (Fridley et al. 2007). Observational studies carried out at large spatial scales have frequently reported positive relationships between native diversity and invader diversity (Stohlgren et al. 1999). In contrast, most of the smaller-scale experimental work supports the biotic resistance hypothesis.

Few marine studies have examined the relationship between native diversity and invasibility, and the few large-scale observational studies have found no clear patterns (Klein et al. 2005; Wasson et al. 2005). In addition, non-experimental studies at local scales can reveal opposite findings. Stachowicz et al. (2002) found a negative correlation between native and invader diversity in sessile epibenthic invertebrate communities at small scales (quadrats of 250 cm^2). The invaders were at such low abundances that this was unlikely to result from their effects on native diversity. In contrast, Dunstan and Johnson (2004) examined a natural community of sessile marine invertebrates and found an increase in invasion rates where species richness was greater. Species-rich areas were dominated by small colonies with higher mortality rates, providing bare space for new species to colonise. Furthermore, some facilitative interactions among species reinforced this effect.

Experimental studies of the diversity-invasibility relationship in hard-bottom communities have been carried out both by manipulating diversity in natural communities and by assembling new communities (i.e. synthetic assemblages). Stachowicz et al. (1999, 2002) assembled synthetic communities of sessile invertebrates with different levels of diversity and followed the survivorship of two non-native species placed into the assemblages. They found that decreasing diversity increased the survivorship and growth of the invaders because decreasing diversity resulted in more space for the invaders. Resource availability seems also to be the mechanism underlying invasion resistance in communities of benthic marine algae. Britton-Simmons (2006) and Arenas et al. (2006) showed experimentally that biotic resistance depended more on the identity of the functional groups than on their diversity, and that resource availability (light and space) was critical. Similar identity-specific effects were also detected by Ceccherelli et al. (2002) in an experiment on invasibility of Mediterranean algal assemblages by *Caulerpa taxifolia* and *C. racemosa*.

22.4.3 Disturbance and the Susceptibility to Invasion

Disturbance also alters the resistance of communities to invasion (Mack et al. 2000), probably because disturbances remove individuals, freeing resources and decreasing competition from resident species (Shea and Chesson 2002). Experimental work on macroalgae has shown that disturbance can improve invasion success probably by preventing the monopolisation of light and space by the native flora (Valentine and Johnson 2003; Sánchez and Fernández 2006).

22.5 The Effects of Invasions

Non-indigenous species can have ecological effects at a wide range of scales, from local to global, but the main concern is with those that become invasive, as this implies large-scale alterations to the original community. An important concept here comes from a review by Ricciardi and Cohen (2007) that included terrestrial and marine examples. They found that there was no link between the ability of an introduced species to spread rapidly (its invasiveness) and the likelihood that it will have strong effects on the recipient community (its impact). Perhaps the two most extreme effects that can be predicted for an invasive species are extinction of indigenous species leading to a loss of biodiversity, and the reverse, the facilitation of indigenous species, especially through habitat engineering. Many papers have recorded negative effects of invasive species on community diversity, and there is a perceived threat that this could lead to the homogenisation of the world's biota (e.g. Occhipinti-Ambrogi and Sheppard 2007), so that biological invasions are understood, both popularly and in the scientific literature, to be a major threat to global biodiversity (Mack et al. 2000; Bax et al. 2003). Although there is good evidence of large-scale extinctions due to invasive species in freshwater systems (e.g. see Kaufman 1992 on the case of Lake Victoria and Vitousek et al. 1997 for cases involving freshwater fish in North America), the situation is different in the sea, perhaps because marine species are less susceptible to extinction (Ruiz et al. 1997). Well-known examples, such as the ctenophore *Mnemiopsis* in the Black Sea (Kideys and Niermann 1994), can lead to radical changes in an entire ecosystem but there seems to be a developing realisation that, within marine systems, the threat of species extinctions may be overstated as a result of considering inappropriate scales. Gurevitch and Padilla (2004) point out that the presumed loss of native species to invasive aliens is largely untested or unproven. Briggs (2007) goes farther; he found that there is no evidence from modern events or from the fossil record that biological invasions have caused either loss of biodiversity or the extinction of native marine species.

Reise et al. (2006a) reconcile these different perspectives by looking at the results of invasions in European waters. They conclude that local extinctions of indigenous species occur as a result of biological invasions but that there is no evidence of European-scale extinctions. Galil (2007) forms similar conclusions for the Mediterranean Sea. Essentially, this comes down to defining what is meant by extinction—local or global, and will one lead inevitably to the other? In other words, will the local-scale extinctions already recorded inevitably lead to global species loss? Briggs' (2007) findings suggest not, but perhaps the pressure from non-native species is so much greater now that things may turn out differently from the past. Cassey et al. (2005) argue that this is the case, using the term "current mass invasion event" to suggest comparison with past mass extinction events, though Brown and Sax (2005) reject the idea that historic and recent, human-caused invasions are fundamentally different.

Due to its relative youth, the North Atlantic biota is generally impoverished (e.g. Vermeij 1991 for the molluscs) and, despite high numbers of invasive species, the European coastal biota shows no loss of biodiversity, while invasive species can actually expand ecosystem functioning (Reise et al. 2006b; Galil 2007). Ricciardi and Atkinson (2004) explicitly name the North, Baltic and Mediterranean seas as systems that have notably few introduced species that have caused dramatic declines in native species—what they term "high-impact" invaders. This may be a special case, because the European biota is unsaturated in terms of species (e.g. Por 1971), but it does highlight the difficulty of global predictions.

While it is clear that the effects of invasive species are highly dependent on their ecological context, there is no doubt that they can have dramatic effects on community structure and ecosystem functioning, even if this falls short of eliminating indigenous species. The direct and indirect effects of most invasive species remain unknown and difficult to predict (Ruiz et al. 1997) but Ricciardi and Atkinson (2004) suggest that, within aquatic ecosystems, invaders are more likely to displace indigenous species if they are distinctively different (a suggestion supported by Stachowitz and Tilman 2005)—for example, if they belong to a different genus. On the other hand, they found no correlation between the likelihood of strong effects and the richness of the recipient community. Some effects of invasive species are strong but nevertheless comparatively subtle or inconspicuous. These include niche shifts (e.g. northeast Atlantic littorinids; Eastwood et al. 2006), genetic changes in indigenous species (e.g. northeast Atlantic mussels; Freeman and Byers 2006) and hybridization between introduced and indigenous species (e.g. Mediterranean killifish; Goren and Galil 2005). Other effects, especially those caused by species that change environmental conditions, are profound and conspicuous. Thus, introduced species can act as ecosystem engineers, changing habitat architecture, patterns of water movement, sediment accumulation and light penetration, as well as reducing food availability in the water column by filtration (Dukes and Mooney 2004; Wallentinus and Nyberg 2007). The negative effects of such changes on the indigenous biota are obvious but so are some positive ones. In the North Sea, non-native *Sargassum muticum* provides habitats for epibiota otherwise absent in sediments, resulting in a strong effect on diversity of soft sediments but not of hard substrata, where native species perform the same function (Buschbaum et al. 2006).

Galil (2007) describes a number of effects of alien invasive species in the Mediterranean. These include the loss of biodiversity through the replacement of the native seagrass *Posidonia oceanica* by the structurally less complex, invasive alga *Caulerpa taxifolia*; increases in the rate of algal recycling by the introduction of the herbivorous rabbitfish, *Siganus* spp., where herbivorous fish were previously rare; replacement of native mussels by the introduced *Brachidontes pharaonis* and of native limpets by the introduced *Cellana rota*. Two of these examples have led to increases in densities of native predators. In the case of rabbitfish, the addition of a new component to the food web has increased prey levels for subtidal predators such as groupers, while the whelk *Stramonita haemastoma* prefers introduced to native mytilids, and has increased in abundance since *Brachidontes* became dominant.

22.6 Overview

- Invasions are idiosyncratic and context-dependent. For example, the relationship between the diversity of communities and their vulnerability to invasion is unclear and it seems that the effects of diversity are identity-specific.
- There are few general characteristics of successful invaders but most invasions are by lower trophic levels, and some taxa are particularly prevalent (e.g. molluscs and crustaceans, among the animals). The importance of the various invasion vectors differs among taxa but aquaculture is particularly important for macroalgae and shipping for animals.
- There is no clear link between the ability of a species to invade and the likelihood that it will have strong effects on the recipient community. Interactions with indigenous biota are important and can vary between facilitation and biotic resistance due to predation or competition. While invasive species can have dramatic effects on community structure and functioning, there is little evidence that they have yet led to loss of biodiversity on anything other than local scales.

References

Abelló P, Hispano C (2006) The capture of the Indo-Pacific crab *Charybdis feriata* (Linnaeus, 1758) (Brachyura: Portunidae) in the Mediterranean Sea. Aquat Invasions 1:13–16

Arenas F, Sánchez I, Hawkins SJ, Jenkins SR (2006) The invasibility of marine algal assemblages: role of functional diversity and identity. Ecology 87:2851–2861

Bax N, Williamson A, Aguero M, Gonzalez E, Geeves W (2003) Marine invasive alien species: a threat to global biodiversity. Mar Policy 27:313–323

Branch GM, Steffani CN (2004) Can we predict the effects of alien species? A case-study of the invasion of South Africa by *Mytilus galloprovincialis* (Lamarck). J Exp Mar Biol Ecol 31:189–215

Briggs JC (2007) Marine biogeography and ecology: invasions and introductions. J Biogeogr 34:193–198

Britton-Simmons KH (2006) Functional group diversity, resource pre-emption and the genesis of invasion resistance in a community of marine algae. Oikos 113:395–401

Brousseau DJ, Goldberg R (2007) Effect of predation by the invasive crab *Hemigrapsus sanguineus* on recruiting barnacles *Semibalanus balanoides* in western Long Island Sound, USA. Mar Ecol Prog Ser 339:221–228

Brown JH, Sax DF (2005) Biological invasions and scientific objectivity: Reply to Cassey et al. (2005). Austral Ecol 30:481–483

Buschbaum C, Chapman AS, Saier B (2006) How an introduced seaweed can affect epibiota diversity in different coastal systems. Mar Biol 148:743–754

Byrnes JE, Reynolds PL, Stachowicz JJ (2007) Invasions and extinctions reshape coastal marine food webs. Plos ONE 2:e295

Carlton JT (1996) Pattern, process and prediction in marine invasion ecology. Biol Conserv 78:97–106
Carlton JT (1999) The scale and ecological consequences of biological invasions in the world's oceans. In: Sandlund OT, Schei PJ, Viken A (eds) Invasive species and biodiversity management. Kluwer, Dordrecht, pp 195–212
Carlton JT, Geller JB (1993) Ecological roulette: the global transport of nonindigenous marine organisms. Science 261:78–82
Cassey P, Blackburn TM, Duncan RP, Chown SL (2005) Concerning invasive species: Reply to Brown and Sax. Austral Ecol 30:475–480
Ceccherelli G, Piazzi L, Balata D (2002) Spread of introduced *Caulerpa* species in macroalgal habitats. J Exp Mar Biol Ecol 280:1–11
Colautti RI, Grigorovich IA, MacIsaac HJ (2006) Propagule pressure: a null model for biological invasions. Biol Invasions 8:1023–1037
Daguin C, Borsa P (2000) Genetic relationships of *Mytilus galloprovincialis* Lamarck populations worldwide: evidence from nuclear-DNA markers. Geol Soc Lond Spec Publ 177:389–397
DeRivera CE, Ruiz G, Hines A, Jivoff P (2005) Biotic resistance to invasion: native predator limits abundance and distribution of an introduced crab. Ecology 86(12):3364–3376
Dukes JS, Mooney HA (2004) Disruption of ecosystem processes in western North America invasive species. Rev Chilena Hist Nat 77:411–437
Dunstan PK, Johnson CR (2004) Invasion rates increase with species richness in a marine epibenthic community by two mechanisms. Oecologia 138:285–292
Eastwood MM, Donahue MJ, Fowler AE (2006) Reconstructing past biological invasions: niche shifts in response to invasive predators and competitors. Biol Invasions 9:397–407
Elton CS (1958) The ecology of invasions by animals and plants. Methuen, London
Fofonoff PW, Ruiz GM, Steves B, Carlton JT (2003) In ships or on ships? Mechanisms of transfer and invasion for nonnative species to the coasts of North America. In: Ruiz GM, Carlton JT (eds) Invasive species: vectors and management strategies. Island Press, Washington, DC, pp 152–182
Freeman AS, Byers JE (2006) Divergent induced responses to an invasive predator in marine mussel populations. Science 313:831–833
Fridley JD, Stachowicz JJ, Naeem S, Sax DF, Seabloom EW, Smith MD, Stohlgren TJ, Tilman D, Von Holle B (2007) The invasion paradox: reconciling pattern and process in species invasions. Ecology 88:3–17
Galil BS (2000) A sea under siege—alien species in the Mediterranean. Biol Invasions 2:177–186
Galil BS (2007) Loss or gain? Invasive aliens and biodiversity in the Mediterranean Sea. Mar Pollut Bull 55:314–322
Goren M, Galil BS (2005) A review of changes in the fish assemblages of Levantine inland and marine ecosystems following the introduction of non-native fishes. J Appl Ichthyol 21:364–370
Grosholz E (2002) Ecological and evolutionary consequences of coastal invasions. Trends Ecol Evol 17:22–27
Grosholz ED (2005) Recent biological invasion may hasten invasional meltdown by accelerating historical introductions. Proc Natl Acad Sci 102:1088–1091
Grosholz ED, Ruiz GM (1996) Predicting the impact of introduced marine species: lesson from multiple invasion of the European green crab *Carcinus maenas*. Biol Conserv 78:59–66
Gurevitch J, Padilla DK (2004) Are invasive species a major cause of extinctions? Trends Ecol Evol 19:470–474
Harding JM (2003) Predation by blue crabs, *Callinectes sapidus*, on rapa whelks, *Rapana venosa*: possible natural controls for an invasive species? J Exp Mar Biol Ecol 297:161–178
Hollebone AL, Hay ME (2007) Propagule pressure of an invasive crab overwhelms native biotic resistance. Mar Ecol Prog Ser 342:191–196
Kaufman L (1992) Catastrophic change in species-rich freshwater ecosystems: the lessons of Lake Victoria. BioScience 42:846–858
Kideys AE, Niermann U (1994) Occurrence of *Mnemiopsis* along the Turkish coast. ICES J Mar Sci 51:423–427

Klein J, Ruitton S, Verlaque M, Boudouresque CF (2005) Species introductions, diversity and disturbances in marine macrophyte assemblages of the northwestern Mediterranean Sea. Mar Ecol Prog Ser 290:79–88
Levine JM, D'Antonio CM (1999) Elton revisited: a review of evidence linking diversity and invasibility. Oikos 87:15–26
Lockwood JL, Cassey P, Blackburn T (2005) The role of propagule pressure in explaining species invasions. Trends Ecol Evol 20:223–228
Mack RN, Simberloff D, Lonsdale WM, Evans H, Clout M, Bazzaz F (2000) Biotic invasions: causes, epidemiology, global consequences, and control. Ecol Appl 10:689–710
Meiners SJ, Cadenasso ML, Pickett STA (2004) Beyond biodiversity: individualistic controls of invasion in a self-assembled community. Ecol Lett 7:121–126
Minchin D (2007) Aquaculture and transport in a changing environment: overlap and links in the spread of alien biota. Mar Pollut Bull 55:302–313
Noonburg EG, Byers JE (2005) More harm than good: when invader vulnerability to predators enhances impact on native species. Ecology 86:2555–2560
Nyberg CD, Wallentinus I (2005) Can species traits be used to predict marine macroalgal introductions? Biol Invasions 7:265–279
Occhipinti-Ambrogi A, Sheppard C (2007) Marine bioinvasions: a collection of reviews. Mar Pollut Bull 55:299–301
Paglianti A, Gherardi F (2004) Combined effects of temperature and diet on growth and survival of young-of-year crayfish: a comparison between indigenous and invasive species. J Crust Biol 24:140–148
Por FD (1971) One hundred years of Suez Canal—a century of Lessepsian migration. Syst Zool 20:138–159
Reise K, Olenin S, Thieltges DW (2006a) Editorial. Helgoland Mar Res 60:75–76
Reise K, Olenin S, Thieltges DW (2006b) Are aliens threatening European aquatic coastal ecosystems? Helgoland Mar Res 60:77–83
Reusch TBH, Williams SL (1999) Macrophyte canopy structure and the success of an invasive marine bivalve. Oikos 84:398–419
Ricciardi A, Atkinson SK (2004) Distinctiveness magnifies the impact of biological invaders in aquatic ecosystems. Ecol Lett 7:781–784
Ricciardi A, Cohen J (2007) The invasiveness of an introduced species does not predict its impact. Biol Invasions 9:309–315
Rius M, McQuaid CD (2006) Wave action and competitive interaction between the invasive mussel *Mytilus galloprovincialis* and the indigenous *Perna perna* in South Africa. Mar Biol 150:69–78
Rodriguez LF (2006) Can invasive species facilitate native species? Evidence of how, when, and why these impacts occur. Biol Invasions 8:927–939
Roman J (2006) Diluting the founder effect: cryptic invasions expand a marine invader's range. Proc R Soc B 273:2453–2459
Roy K, Jalonski D, Valentine JW (2002) Body size and invasion success in marine bivalves. Ecol Lett 5:163–167
Ruesink JL (2007) Biotic resistance and facilitation of a non-native oyster on rocky shores. Mar Ecol Prog Ser 331:1–9
Ruiz GM, Carlton JT (2003) Invasion vectors: a conceptual framework for management. In: Ruiz GM, Carlton JT (eds) Invasive species: vectors and management strategies. Island Press, Washington, DC, pp 459–504
Ruiz GM, Carlton JT, Grosholz ED, Hines AH (1997) Global invasions of marine and estuarine habitats by non-indigenous species: mechanisms, extent, and consequences. Am Zool 37:621–632
Ruiz GM, Fofonoff PW, Carlton JT, Wonham MJ, Hines AH (2000) Invasion of coastal marine communities in North America: apparent patterns, processes, and biases. Annu Rev Ecol Syst 31:481–531
Sánchez I, Fernández C (2006) Resource availability and invasibility in an intertidal macroalgal assemblage. Mar Ecol Prog Ser 313:85–94

Shea S, Chesson P (2002) Community ecology theory as a framework for biological invasions. Trends Ecol Evol 17:170–176
Siddon CE, Witman JD (2004) Behavioural indirect interactions: multiple predator effects and prey switching in the rocky subtidal. Ecology 85:2938–2945
Simberloff D (2006) Invasional meltdown 6 years later: important phenomenon, unfortunate metaphor, or both? Ecol Lett 9:912–919
Simberloff D, Von Holle B (1999) Positive interactions of nonindigenous species: invasional meltdown. Biol Invasions 1:21–32
Stachowicz JJ, Tilman D (2005) Species invasions and the relationships between species diversity, community saturation, and ecosystem functioning. Community saturation, diversity, and ecosystem functioning. In: Sax DF, Stachowicz JJ, Gaines SD (eds) Species invasions. Insights into ecology, evolution and biogeography. Sinauer, Sunderland, MA, pp 42–64
Stachowicz JJ, Whitlatch RB, Osman RW (1999) Species diversity and invasion resistance in a marine ecosystem. Science 286:1577–1579
Stachowicz JJ, Fried H, Whitlatch RB, Osman RW (2002) Biodiversity, invasion resistance and marine ecosystem function: reconciling pattern and process. Ecology 83:2575–2590
Stohlgren TL, Binkley D, Chong GW, Kalhan MA, Schell LD, Bull KA, Otsuki Y, Newman G, Bashkin M, Son Y (1999) Exotic plant species invade hot spots of native plant diversity. Ecol Monogr 69:25–46
Streftaris N, Zenetos A, Papathanassiou E (2005) Globalisation in marine ecosystems: the story of non-indigenous marine species across European Seas. Oceanogr Mar Biol Annu Rev 43:419–453
Valentine JP, Johnson CR (2003) Establishment of the introduced kelp *Undaria pinnatifida* in Tasmania depends on disturbance to native algal assemblages. J Exp Mar Biol Ecol 295:63–90
Vermeij GJ (1991) Anatomy of an invasion: the trans-Arctic interchange. Paleobiology 17:281–307
Vermeij GJ (1996) An agenda for invasion biology. Biol Conserv 78:3–9
Vitousek PM, D'Antonio CM, Loope LLz, Rejmanek M, Westbrook R (1997) Introduced species: a significant component of human-induced global change. N Z J Ecol 21:1–16
Voisin M, Engel CR, Viard F (2005) Differential shuffling of native genetic diversity across introduced regions in a brown alga: aquaculture vs. maritime traffic effects. Proc Natl Acad Sci 102:5432–5437
Wallentinus I, Nyberg CD (2007) Introduced marine organisms as habitat modifiers. Mar Pollut Bull 55:323–332
Wardle DA (2001) Experimental demonstration that plant diversity reduces invasibility-evidence of a biological mechanism or a consequence of sampling effect. Oikos 95:161–170
Wasson K, Fenn K, Pearse JS (2005) Habitat differences in marine invasions of central California. Biol Invasions 7:935–948
Williamson MH, Brown KC (1986) The analysis and modelling of British invasions. Philos Trans R Soc B 314:505–522
Williamson M, Fritter A (1996) The varying success of invasions. Ecology 77:1661–1666
Wonham MJ, Carlton JT, Ruiz GM, Smith LD (2000) Fish and ships: relating dispersal frequency to success in biological invasions. Mar Biol 136:1111–1121
Zardi GI, Nicastro KR, McQuaid CD, Rius M, Porri F (2006) Hydrodynamic stress as a determinant factor in habitat segregation between the indigenous mussel *Perna perna* and the invasive *Mytilus galloprovincialis* in South Africa. Mar Biol 150:79–88

Chapter 23
Habitat Distribution and Heterogeneity in Marine Invasion Dynamics: the Importance of Hard Substrate and Artificial Structure

Gregory M. Ruiz, Amy L. Freestone, Paul W. Fofonoff, and Christina Simkanin

23.1 Introduction

Biological invasions by non-native species are common in coastal marine communities in many global regions, and the rate of documented invasions has increased dramatically in recent time (e.g., Por 1978; Cohen and Carlton 1995; Reise et al. 1999; Ruiz et al. 2000; Orensanz et al. 2002; Hewitt et al. 2004; Castilla et al. 2005; Kerckhof et al. 2007). It appears, however, that not all habitats are equally important as sites for colonization. For example, most invasions in marine systems are known from temperate latitudes (Ruiz and Hewitt 2008). Non-native populations are usually found in bays and estuaries, and relatively few non-native species have been reported from exposed outer coasts and offshore locations (Carlton 1979; Wasson et al. 2005). Furthermore, within bays and estuaries, it appears that hard bottom communities and especially artificial hard substrata, such as docks and pilings, are often foci for colonization (Cohen and Carlton 1995; Wasson et al. 2005).

Patterns of habitat utilization may have significant implications for the dynamics of marine invasions, and efforts to curtail their establishment, spread and impacts. Here, we evaluate the known distribution of non-native species in coastal habitats, using North America as a model. Our goal is to provide a summary of available information on distributions of non-native marine invertebrates and algae according to habitat, highlighting the relative importance of hard substrata. We also examine the potential role of artificial (anthropogenic) hard substrata for marine invasions, and consider some possible implications of observed distributions of habitats for invasion dynamics.

23.2 Habitat Distribution of Non-native Species in North America

Over 300 non-native species of invertebrates and algae are considered established in coastal waters of North America, including the continental United States and Canada. We have compiled information about taxonomic identity, distribution, and

M. Wahl (ed.), *Marine Hard Bottom Communities*, Ecological Studies 206,
DOI: 10.1007/978-3-540-92704-4_23, © Springer-Verlag Berlin Heidelberg 2009

characteristics of these species in a database (NEMESIS 2008), which we use here to summarize the relative importance of hard substrata as habitat for non-native species. Below, we report on the frequency of occurrence among different types of habitat for adult forms of each species (many of which have planktonic larval stages) in tidal waters of North America, spanning from full marine salinities, to oligohaline and freshwater reaches of estuaries.

Our analysis deals with those species known to have established populations in North America, based on an extensive synthesis of available literature. This approach has several implications. First, our analysis provides a minimum estimate of the number of invasions in space and time. Many additional species have been reported, but it is not clear whether they have established self-sustaining populations. Still other non-native species have certainly gone undetected (see Carlton 1996 for discussion). Second, there is the potential for bias in the literature for particular taxonomic groups, time periods, and habitats (Ruiz et al. 2000). We present a summary of data as available, to describe our current state of knowledge, and the patterns of marine invasions that emerge from the literature.

23.2.1 Importance of Hard Substrata for Marine Invasions

Most non-native marine and estuarine species that are reported and established in North America are associated with hard substrata as adults (Fig. 23.1). Of 327 such species, 46% were reported only on hard substrata, and 22% occurred on both hard substrata and soft sediments. Another 3% of species were associated with hard substrata and the water column, representing coelenterates with relatively long-lived benthic (polyp) and pelagic (medusa) life stages. In total, 71% of the non-native species occurred on hard substrata, either solely or in combination with other habitats. Of these 232 species, most (63%) were sessile or sedentary as adults. Not surprisingly, the majority of mobile species (e.g., amphipods, crabs, non-tube-building polychaetes) that utilize hard substrata were also reported in soft-sediment habitats (Fig. 23.1).

It is also noteworthy that some of the non-native parasites and commensals occur on host species that, as such, are part of hard bottom communities. A good example is the introduced parasitic copepod *Mytilicola orientalis*, which infects mussels, oysters, and other molluscan hosts along western North America (Carlton 1979). We did not include these species in our analyses, because we considered the hosts to be the primary substratum. We have also excluded these parasites and commensals from subsequent analyses, focusing only on free-living biota.

This summary underscores the relative dominance of hard substrata as habitat for known marine invasions on a continental scale. A similar pattern appears to exist at the local scale of bays and estuaries, as demonstrated by a recent analysis of Elkhorn Slough in Central California by Wasson et al. (2005). Sampling multiple habitats in this single estuary, they found that numbers of non-native species were greater for hard substrata than for soft-sediment habitats. Non-native species also accounted for a higher percentage of total individual organisms for habitats on hard

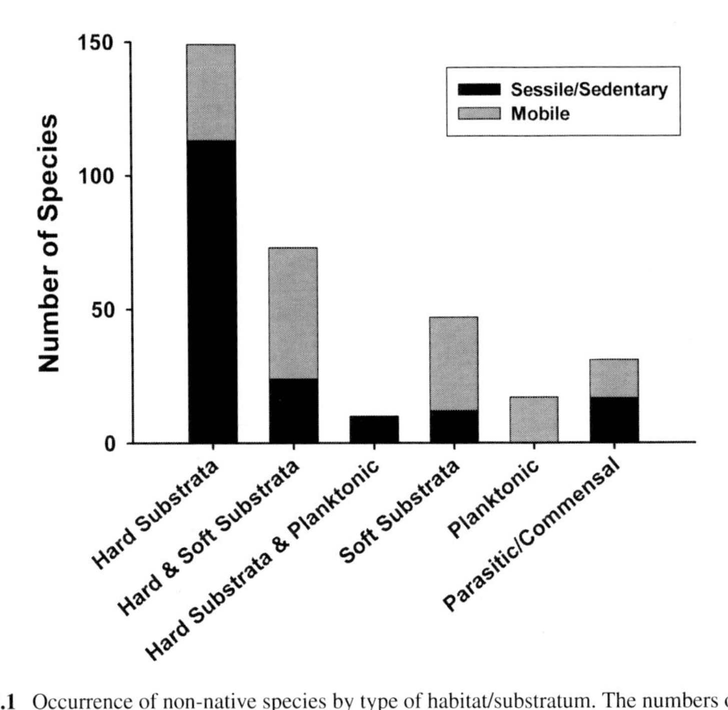

Fig. 23.1 Occurrence of non-native species by type of habitat/substratum. The numbers of established species in marine and estuarine waters of North America are shown by habitat separately for mobile and sessile/sedentary species

substrata than those in soft sediments. For both diversity and total abundance, the difference between habitats was most striking for sessile invertebrates.

These results do not imply that biological invasions are functionally less important in other habitats, in terms of impacts or dynamics, but instead that non-native species richness is simply concentrated in communities living on hard substrata (see Wasson et al. 2005 for possible explanations). Invasions can clearly have significant impacts in soft-sediment and planktonic communities, and as parasites in host assemblages (e.g., Alpine and Cloern 1992; Burreson et al. 2000; Grosholz et al. 2000). In addition, many of the species on hard substrata occur in other habitats (e.g., plankton) during various life stages, or affect processes across habitats and on broader scales.

23.2.2 Temporal Pattern of Marine Invasions on Hard Substrata

The rate of detection for marine invasions has increased dramatically in North America, as reported previously (Cohen and Carlton 1998; Ruiz et al. 2000; Wonham and Carlton 2005). The number of new invasions reported for communities on hard substrata contributes strongly to this pattern, increasing ≥47% in each 30-year interval since 1830 (Fig. 23.2). For example, the largest number of

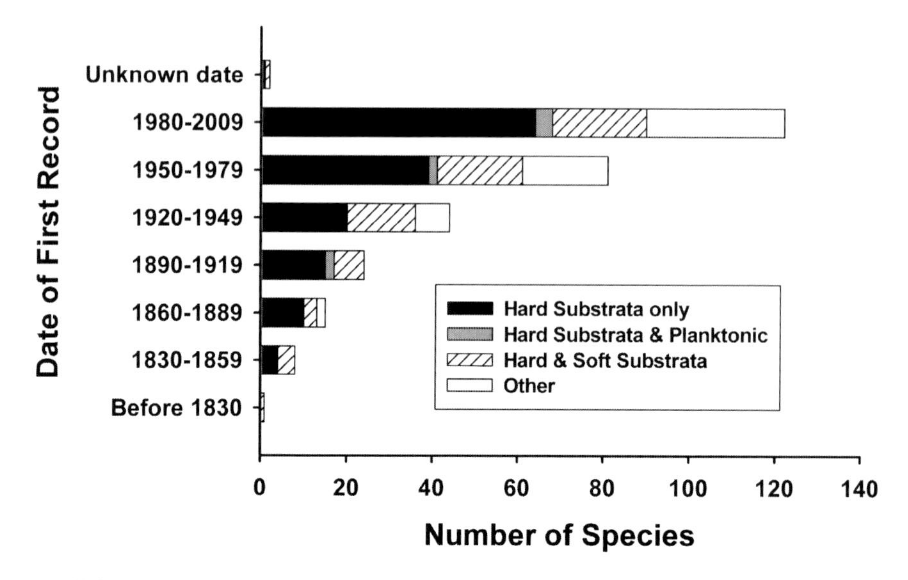

Fig. 23.2 Relative contribution of species from hard substrata to observed temporal patterns in invasion. The number of newly established marine invasions reported in North America is shown by habitat type for 30-year intervals (parasites, pathogens, and obligate commensals excluded)

invasions (122 non-native species) was reported in the last 30-year interval, from 1980–2009. Of these recent invaders, 52% are associated as adults with hard substrata alone, and another 22% occur as adults on hard substrata in combination with soft sediments or plankton. Thus, a surprising 74% of reported invasions in the past three decades appear to utilize hard substrata.

It remains challenging to estimate the precise dates of species invasions. We have reported the date of first record, including either the date of detection (if available) or the date of publication. For many species, the actual invasion may have occurred long before the first record, and significant lag times can exist before detection of a non-native population (Ruiz et al. 2000; Costello and Solow 2003). Nonetheless, the available evidence suggests that the pace of invasions has increased strongly in recent time, especially for large, conspicuous, and well-studied marine organisms such as shelled molluscs and decapod crustaceans.

23.2.3 Distribution of Non-native Species Among Bays, Estuaries and Outer Coasts

For the 232 non-native species associated with hard substrata in North America (see above), the vast majority occur at high salinity in relatively sheltered waters of bays and estuaries. Most species inhabit the polyhaline and euhaline zones, with a smaller number reported for tidal waters of lower salinity (Fig. 23.3; note that some species

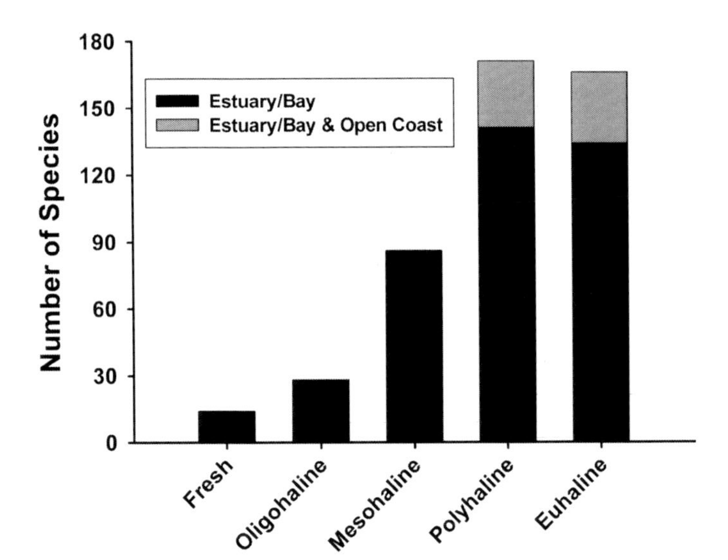

Fig. 23.3 Distribution of established non-native species from hard substrata among salinity and exposure zones in North America. (Species can occur in multiple zones, so the total number of species is not additive; parasites, pathogens, and obligate commensals excluded.)

occur in multiple salinity zones, so that the total number shown across zones is not additive). A small subset (42 species) is also reported on more exposed areas of outer coastal habitats, often near the entrance of bays or surrounding coastal harbors. Of the few non-native species known for outer coast areas, most (95%) also occur in adjacent bays and estuaries.

The disparity in number of non-native species between bays/estuaries and outer coasts (including offshore waters) has been noted previously (Carlton 1979; Ruiz et al. 1997). In a recent formal comparison, Wasson et al. (2005) reported significantly greater richness of non-native species in Elkhorn Slough versus the adjacent California coast. Fifty-eight non-native species were present in the estuary, and most were associated with hard substrata. Only eight non-native species were confirmed for the exposed outer coast; all of these species were found on hard substrata, and were also present in the estuary.

Some recent examples of non-native species along the outer coast of North America include the ascidian *Didemnum* sp., alga *Undaria pinnatifada*, amphipod *Caprella mutica*, bryozoan *Watersipora subtorqata*, and lionfish *Pterois volitans* (Silva et al. 2002; Whitfield et al. 2002; Page et al. 2006; Valentine et al. 2007). While indicating that invasions are possible here, it is also noteworthy that many invasions to outer coasts were only reported recently, and some of these species are clearly abundant and spreading over large areas. Search effort has certainly been greatest in bays, but this cannot easily explain the magnitude of difference in reported number of non-native species compared to the outer coast, since the biota for some outer coast sites (e.g., intertidal and shallow subtidal zones of rocky habitats in California) has been intensively surveyed. Thus, we surmise that most non-native species in North America occur in bays and estuaries, and colonization of more exposed outer coasts may be limited by suitable environmental conditions and/or biotic interactions.

Hard substrata along coastlines are often common, and do not appear to be limiting in regions such as western North America. However, in such regions, non-native species frequently have a discontinuous distribution, occurring in multiple bays that are separated by outer rocky coast where the species are seemingly absent, with few documented exceptions (as above). The extent to which current non-native populations can spread from bays to outer coasts appears limited, but this topic has not received much formal analysis to date, and represents a gap in our understanding of invasion risk.

One plausible explanation for the observed concentration of non-native species in bays and estuaries results from the relative magnitude of human-mediated transfer mechanisms or vectors (see Chap. 22 by McQuaid and Arenas for discussion). Bays and estuaries are often centers of commerce, including shipping, fisheries, aquaculture, and a variety of other human activities that are known to transfer non-native species (Cohen and Carlton 1995; Ruiz et al. 2000). Through history, bays have served as hubs for global trade. Consequently, it has predominantly been marine organisms inhabiting bays and estuaries that are transferred by human activities, and delivered directly to other bays and estuaries.

There can be little doubt that the supply of propagules for non-native species has been greatest to bays, compared to exposed outer coasts, but this does not explain the current distribution, and why so few species have subsequently spread to outer coastal habitats. Many non-native species have been established in high numbers for considerable time (decades to centuries), and have successfully spread coastwise among estuaries, apparently without establishing populations on outer coasts. While inter-bay spread has often been human-mediated (Fofonoff et al. 2008), we should still expect some dispersal of non-native propagules to outer coasts, creating opportunity for colonization.

Outer coast environments may simply be less susceptible to invasion by those species that have invaded to date. It is certainly possible that habitat and environmental characteristics limit many non-native species to bays and estuaries, which are the predominant source environments in the native regions (as above). An alternative possibility is that biotic interactions such as predation or competition have restricted colonization on the outer coast. There is support for the role of predators in limiting the offshore distribution of some non-native species in Long Island Sound, USA (Osman and Whitlatch 1998, 2004), but such tests are rare. Thus, the relative importance of biotic resistance to invasion versus environmental constraints (e.g., suitable habitat, physiological tolerance) is not understood.

Despite the small numbers of non-native species known from outer coasts, we may expect this to change with increases in human activities offshore. There is considerable interest in developing offshore aquaculture, harnessing wave energy, and increasing oil exploration and extraction (Bulleri 2005). One consequence may be an increased supply of propagules, including those from bays as well as outer coast areas. If there is a positive relationship between propagule supply of non-native species and establishment success, then we may expect an increase in invasions. In addition to species transfers, these activities would result in the creation of artificial structures, providing hard substrata that may be more vulnerable to colonization (see below). The effects of increasing such offshore activities on risk of invasion raise many concerns for resource management and conservation, but remain to be evaluated.

23.2.4 Role of Artificial Hard Substrata in Marine Invasions

In our initial attempt to characterize the types of hard substrata occupied by non-native species, it was clear that most occur on artificial substrata (Fig. 23.4). Of the 232 non-native species found on hard substrata (Fig. 23.1), over 200 are reported to occur on artificial structures at docks and marinas, including pier pilings, floats, bulkheads, pipes, and lines. Although the greatest richness of non-native species is reported for these structures, most of the species also occur on other hard substrata, especially rock.

We attempted to differentiate further the occurrence of non-native species on artificial rocky habitat (breakwaters, riprap) from natural rock habitat, but the distinction was often absent from the literature reports for these species. Although

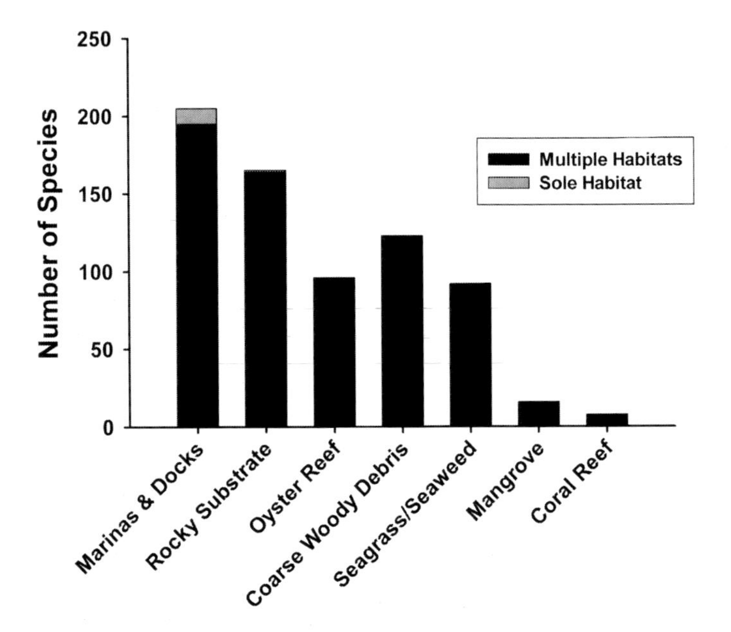

Fig. 23.4 Reported distribution of established non-native species among types of hard substratum in North America (parasites, pathogens, and obligate commensals excluded)

we observed that many of the non-native species reported on rock occupy artificial habitat, our analysis cannot readily (1) identify the use of natural rock habitat, or (2) be used to estimate frequency of occurrence exclusively for natural versus artificial rock substrata. For many regions, such characterization will require more direct field-based comparisons. Along much of the eastern USA (including the southeastern states and Gulf of Mexico), however, nearly all rocky habitat in bays is artificial (Hedgpeth 1954).

Many of the oyster reefs where non-native species are reported also represent artificial substrata in two ways. First, many of the oysters now grown in western North America are the non-native *Crassostrea gigas*, arguably an artificial habitat. Second, the oysters are often cultured with associated structures in areas where oysters were not previously present. Thus, oysters and oyster culture have provided a novel substrate that is now occupied by non-native species, but that did not exist historically at many sites.

While many different types of hard substratum are occupied by non-native taxa (including natural reefs, macrophytes, and woody debris), our coarse-level analysis suggests that anthropogenic structure may play a particularly important role in the dynamics of marine invasions. Specifically, it is possible that (1) artificial hard substrata are often the first colonized by non-native species arriving in a bay or estuary, and (2) artificial hard substrata are foci for growth and spread of many non-native populations.

Such a role for artificial substrata may result partly from their predominance in and surrounding developed shorelines, such as ports and marinas, where vectors and supply of propagules are presumably concentrated. Furthermore, the availability of natural hard substrata (especially rock) is very limited in some estuaries. This is the case for San Francisco Bay and Chesapeake Bay, each having relatively large numbers of non-native species (Cohen and Carlton 1995; Fofonoff et al. 2008). Thus, one outcome of increases in artificial hard substrata may be to increase the opportunity for colonization, whereby the probability of establishment increases as a positive function of proximity or quantity of hard substratum. While it is likely that such a positive relationship exists at some scale, the actual quantitative relationship(s) remain to be tested.

It may also be the case that artificial hard substrata are more susceptible to invasion than are natural hard substrata, due to differences in recruitment and survivorship. Few studies that formally evaluate this possibility are available for marine communities. Working on fouling communities in Australia, Glasby et al. (2007) found that richness of non-native species was significantly greater on some types of artificial substrata, especially pontoons and pilings, than on nearby rocky reefs. Using an extensive series of controlled experiments, they also demonstrated that recruitment of non-native species was greater on these artificial surfaces, compared to natural substrata. Based on these results, Glasby et al. (2007) suggested that artificial structures may provide critical beachheads, increasing success of establishment and subsequent spread of non-native species. In a study of the non-native alga *Codium fragile* spp. *tomentosoides* on the Adriatic coast of Italy, Bulleri and Airoldi (2005) provide further support for the role of artificial substrate as a unique habitat that can facilitate the spread of non-native marine taxa, creating a dispersal corridor for stepwise colonization along a coastline where other suitable habitats may be limiting.

In observed associations with artificial substrata, the relative importance of habitat selection (e.g., larval choice of substrata) by non-native species versus biotic interactions with resident communities (invasion resistance) remains to be tested. Importantly, Glasby et al. (2007) demonstrated differential recruitment among different types of new (unoccupied) substratum. Several studies have also examined effects of existing communities on invasion dynamics on single types of substratum, especially artificial substrata (e.g., Stachowicz et al. 1999, 2002; Osman and Whitlatch 2004). The differential effects of existing communities on colonization—separating the effects of substrate selection versus biotic resistance—have, however, not been examined for non-native species across types of substratum. It is generally thought that biotic resistance to invasions is greater on natural substrata, but this has not been demonstrated in hard bottom marine communities. Challenge experiments offer one approach to quantitatively compare the susceptibility of communities on artificial versus natural hard substrata, controlling for actual propagule supply (e.g., Clark and Johnston 2005).

In general, we hypothesize that artificial substrata operate as focal points for invasion of hard bottom communities by (1) facilitating initial invasions through greater availability of resources (both in total area, and decreased biotic resistance) over those on natural substrata, and (2) providing seed populations that persist and spread regionally, increase in abundance, and (as a result) can spread to natural substrata. This follows directly from the conceptual framework of Glasby et al. (2007). The qualitative or observed association with artificial habitat provides some additional support for such facilitation, especially given the relatively large proportion of artificial habitats available in some areas. Nonetheless, the functional roles of different habitats for persistence and spread of non-native species, and whether artificial substrata frequently serve as "stepping stones" to natural habitats remain poorly resolved at present.

Regardless of the nature of invasion dynamics across habitats, the large numbers of species associated with artificial habitats, with approximately 90% of non-native taxa inhabiting hard substrata being reported from docks and marinas alone (Fig. 23.4), indicate that these are particularly useful sites for efficient monitoring and detection of new invasions. As foci for invasions, artificial substrata represent high-value sites for resource managers interested in tracking or responding to new incursions, in addition to researchers interested in the biology and ecology of the organisms. Importantly, such artificial substrata are often readily accessible from shore, and are highly suitable for repeated, quantitative assessments to rigorously examine spatial and temporal patterns of invasion (see Ruiz and Hewitt 2002 for discussion).

23.3 Integrating Substratum Heterogeneity and Spatial Scale

Non-native species exhibit strong habitat-specific distributions in marine systems. In North America, most known non-native species are associated with hard substrata and high salinities. Field observations indicate that within hard bottom habitats, invaders are more common on artificial than on natural substrata (see Sect. 23.2.4). Moreover, most invasions are reported from bays and estuaries, with non-native species rarely occurring along more exposed outer coasts. Thus, non-native species occur in habitat patches, having implications at multiple spatial scales.

At small or neighborhood scales, such as study plots or quadrats, it appears that biotic interactions may limit coexistence of native and non-native species (Elton 1958; Stachowicz 1999; Fridley et al. 2007). Community structure and invasion dynamics may, however, differ greatly across different types of substratum. Within hard bottom habitats in particular, substrata vary from natural rock (of multiple types) to artificial pilings, floating docks, riprap, and concrete. These provide qualitatively different habitats, and some may be more conducive to establishment of non-native taxa.

There is a growing perception that artificial substrata are foci for marine invasions, and may be especially important as beachheads for establishment, facilitating persistence and spread. A few studies indicate that artificial substrata are particularly prone to colonization by invaders, but these are very restricted geographically and taxonomically. The relative susceptibilities of various types of hard substratum are largely unexplored. Such measurements would advance understanding of invasion risk, and potentially help inform management options to reduce high-risk habitats.

At landscape scales, the effect of habitat on marine invasions has also received little attention to date. Yet, the configuration of habitat patches, and also environmental heterogeneity can affect invasion processes at multiple spatial and temporal scales (Melbourne et al. 2007). Understanding dynamics at larger scales across many different types of habitat, such as within a bay, requires knowledge not only about the dynamics of individual patches of habitat, but also of the degree of connectivity among patches, and the effects of existing environmental heterogeneity. All of these components are scale-dependent, varying among organisms as a function of life-history and other attributes.

By utilizing a landscape approach, it is possible to evaluate the density, size, and spatial arrangement of types of habitat (patches) that influence the establishment and spread of non-native taxa. Modeling the population dynamics of selected species, with estimates of dispersal and neighborhood interactions across patches, would be very informative and yield interesting predictions to test empirically. It would be especially fruitful to identify critical thresholds for persistence and spread, as current knowledge of the quantitative effects of propagule delivery on success of establishment is very limited (Ruiz and Carlton 2003; Lockwood et al. 2005).

Overall, the processes that control the establishment, persistence, and spread of marine invaders are poorly understood. Integrative models to examine and predict invasions in coastal ecosystems are largely lacking, especially to evaluate scale-dependent processes. A meta-community framework may provide one useful approach to explicitly examine the invasion dynamics of communities that are linked at various spatial scales (e.g., Leibold et al. 2004; Holyoak et al. 2005). In our view, such integration is needed to understand underlying invasion processes, and develop effective management strategies aimed at reducing the establishment and spread of non-native species.

Acknowledgements We wish to thank Martin Wahl for the opportunity to contribute this chapter. The development of this paper benefited from discussions with Gail Ashton, Jim Carlton, Chad

Hewitt, Whitman Miller, Dan Minchin, and Rick Osman. This research was supported by the National Sea Grant Program, Smithsonian Institution and U.S. Coast Guard.

References

Alpine AE, Cloern JE (1992) Trophic interactions and direct physical effects control phytoplankton biomass and production in an estuary. Limnol Oceanogr 37:946–955

Bulleri F (2005) The introduction of artificial structures on marine soft- and hard-bottoms: ecological implications of epibiota. Environ Conserv 32:101–102

Bulleri F, Airoldi L (2005) Artificial marine structures facilitate the spread of a nonindigenous green alga, *Codium fragile* spp. *tomentosoides*, in the north Adriatic Sea. J Appl Ecol 42:1063–1072

Burreson EM, Stokes NA, Friedman CS (2000) Increased virulence in an introduced pathogen: *Haplosporidium nelsoni* (MSX) in the eastern oyster *Crassostrea virginica*. J Aquat Anim Health 12:1–8

Carlton JT (1979) History, biogeography, and ecology of the introduced marine and estuarine invertebrates of the Pacific coast of North America. PhD Dissertation, University of California, Davis, CA

Carlton JT (1996) Biological invasions and cryptogenic species. Ecology 77:1653–1655

Castilla JC, Uribe M, Bahamonde N, Clarke M, Desqueyroux-Faúndez R, Kong I, Moyano H, Rozbaczylo N, Santilices B, Valdovinos C, Zavala P (2005) Down under the southeastern Pacific: marine non-indigenous species in Chile. Biol Invasions 7:213–232

Clark GF, Johnston EL (2005) Manipulating larval supply in the field: a controlled study of marine invisibility. Mar Ecol Prog Ser 298:9–19

Cohen AN, Carlton JT (1995) Biological study: non-indigenous aquatic species in a United States estuary: a case study of the biological invasions of the San Francisco Bay and delta. US Fisheries and Wildlife and National Sea Grant College Program Rep NTIS PB96-166525, Springfield, VI

Cohen AN, Carlton JT (1998) Accelerating invasion rate in a highly invaded estuary. Science 279:555–558

Costello CJ, Solow AR (2003) On the pattern of discovery of introduced species. Proc Natl Acad Sci USA 100:3321–3323

Elton CS (1958) The ecology of invasions by animals and plants. Methuen, London

Fofonoff PW, Ruiz GM, Hines AH, Steves BP, Carlton JT (2008) Four centuries of estuarine biological invasions in the Chesapeake Bay region. In:Rilov G, Crooks J (eds) Marine bioinvasions: ecology, conservation, and management perspectives. Springer, Berlin Heidelberg New York (in press)

Fridley JD, Stachowicz JJ, Naeem S, Sax DF, Seabloom EW, Smith MD, Stohlgren TJ, Tilman D, Von Holle B (2007) The invasion paradox: reconciling pattern and process in species invasions. Ecology 88:3–17

Glasby TM, Connell SD, Holloway MG, Hewitt CL (2007) Nonindigenous biota on artificial structures: could habitat creation facilitate biological invasions? Mar Biol 151:887–895

Grosholz ED, Ruiz GM, Dean CA, Shirley KA, Maron JL, Connors PG (2000) The impacts of a nonindigenous marine predator on multiple trophic levels. Ecology 81:1206–1224

Hedgpeth JW (1954) Bottom communities of the Gulf of Mexico. In: Galstoff PS (ed) Gulf of Mexico. Its origin, waters, and marine life. Fishery Bull 89. Fish Bull Fish Wildl Serv 55: 203–126

Hewitt CL, Campbell ML, Thresher RE, Martin RB, Boyd S, Cohen BF, Currie DR, Gomon MF, Keogh MJ, Lewis JA, Lockett MM, Mays N, McArthur MA, O'Hara TD, Poore GDB, Ross DJ, Storey MJ, Watson JE, Wilson RS (2004) Introduced and cryptogenic species in Port Phillip Bay, Victoria, Australia. Mar Biol 144:183–202

Holyoak M, Leibold MA, Holt RD (2005) Metacommunities: spatial dynamics and ecological communities. The University of Chicago Press, Chicago, IL

Kerckhof FJ, Haelters J, Gollasch S (2007) Alien species in the marine and brackish ecosystem: the situation in Belgian waters. Aquat Invasions 2:243–257

Leibold MA, Holyoak M, Mouquet N, Amarasekare P, Chase JM, Hoopes MF, Holt RD, Shurin JB, Law R, Tilman D, Loreau M, Gonzalez A (2004) The metacommunity concept: a framework for multi-scale community ecology. Ecol Lett 7:601–613

Lockwood JL, Cassey P, Blackburn T (2005) The role of propagule pressure in explaining species invasions. Trends Ecol Evol 20:223–228

Melbourne BA, Cornell HV, Davies KF, Dugaw CJ, Elmendorf S, Freestone AL, Hall RJ, Harrison S, Hastings A, Holland M, Holyoak M, Lambrinos J, Moore K, Yokomizo H (2007) Invasion in a heterogeneous world: resistance, coexistence or hostile takeover? Ecol Lett 10:77–94

NEMESIS (2008) National Exotic Marine and Estuarine Information System. http://invasions.si.edu/nemesis/

Oresanz JM, Schwindt E, Pastorino G, Bortolus A, Casas G, Darrigran G, Elías R, López Gappa JJ, Obenat S, Pascual M, Penchaszadeh P, Piriz ML, Scarabino F, Spivak ED, Vallarino EA (2002) No longer the pristine confines of the world ocean: a survey of exotic marine species in the southwestern Atlantic. Biol Invasions 4:115–143

Osman RW, Whitlatch RB (1998) Local control of recruitment in an epifaunal community and the consequences to colonization processes. Hydrobiologia 375/376:113–123

Osman RW, Whitlatch RB (2004) The control of the development of a marine benthic community by predation on recruits. J Exp Mar Biol Ecol 311:117–145

Page HM, Dugan JE, Culver CS, Hoesterey JC (2006) Exotic invertebrate species on offshore oil platforms. Mar Ecol Prog Ser 325:101–107

Por FD (1978) Lessepsian migration: the influx of Red Sea biota into the Mediterranean by way of the Suez Canal. Springer, Berlin Heidelberg New York

Reise KS, Gollasch S, Wolff WJ (1999) Introduced marine species of the North Sea coasts. Helgoländ Meeresunters 52:219–234

Ruiz GM, Carlton JT (2003) Invasion vectors: a conceptual framework for management. In:Ruiz GM, Carlton JT (eds) Invasive species: vectors and management strategies. Island Press, Washington, DC, pp 459–504

Ruiz GM, Hewitt CL (2002) Toward understanding patterns of coastal marine invasions: a prospectus. In:Leppakoski E, Olenin S, Gollasch S (eds) Invasive aquatic species of Europe. Kluwer, Dordrecht, pp 529–547

Ruiz GM, Hewitt CL (2008) Latitudinal patterns of biological invasions in marine ecosystems: a polar perspective. In:Krupnik I, Lang MA, Miller SE (eds) Smithsonian at the poles: contributions to international polar year science. Smithsonian Institution Scholarly Press, Washington, DC, pp 347–358

Ruiz GM, Carlton JT, Grosholz ED, Hines AH (1997) Global invasions of marine and estuarine habitats by non-indigenous species: mechanisms, extent, and consequences. Am Zool 37:619–630

Ruiz GM, Fofonoff PW, Carlton JT, Wonham MJ, Hines AH (2000) Invasion of coastal marine communities in North America: apparent patterns, processes, and biases. Annu Rev Ecol Syst 31:481–531

Silva PC, Woodfield RA, Cohen AN, Harris LH, Goddard JH (2002) First report of the Asian kelp *Undaria pinnatifida* in the northeastern Pacific Ocean. Biol Invasions 4:333–338

Stachowicz JJ, Whitlatch RB, Osman RW (1999) Species diversity and invasion resistance in a marine ecosystem. Science 286:1577–1579

Stachowicz JJ, Fried H, Osman RW, Whitlatch RB (2002) Biodiversity, invasion resistance, and marine ecosystem function. Ecology 83:2575–2590

Valentine PC, Collie JS, Reid RN, Asch, RG, Guida VG, Blackwood DS (2007) The occurrence of the colonial ascidian *Didemnum* sp. on Georges Bank gravel habitat—ecological observations and potential effects on groundfish and scallop fisheries. J Exp Mar Biol Ecol 342:179–181

Wasson K, Fenn K, Pearse JS (2005) Habitat differences in marine invasions of central California. Biol Invasions 7:935–946

Whitfield P, Gardner T, Vives SP, Gilligan MR, Courtenay WR Jr, Ray GC, Hare JA (2002) Biological invasion of the Indo-Pacific lionfish *Pterois volitans*, along the Atlantic coast of North America. Mar Ecol Prog Ser 234:289–297

Wonham M, Carlton JT (2005) Cool-temperate marine invasions at local and regional scales: the Northeast Pacific Ocean as a model system. Biol Invasions 7:369–392

Chapter 24
Rehabilitation of Habitat and the Value of Artificial Reefs

Paris J. Goodsell and M. Gee Chapman

24.1 Introduction

There are many controversies regarding goals, aims and assessment of restoration, including the term itself. Here, we do not debate the term "restoration" versus "rehabilitation" but note that the former implies recreating a previous condition. This is problematic because the choice of such conditions is subjective and the conditions usually unknown (Chapman 1999). Nevertheless, "restoration" is widely used in the literature, public debate and policies. Here, we use the more objective term "rehabilitation" to cover activities that attempt to improve degraded conditions by removing impacts, actively managing habitat or species, or constructing new habitat in mitigation for irreversible or future disturbances.

Disturbances are generally pulses (short-lived, e.g. chemical spills) or presses (sustained, e.g. land reclamation), although cumulative impacts also occur (see Chap. 17 by Clynick et al.). Pulse or press perturbations can be followed by either pulse or press responses (Glasby and Underwood 1996), with different species responding to a given perturbation in different ways. For example, building a marina may cause a short-term response from fish during construction but a different, long-term response if they use the marina as habitat. The marina itself is a press perturbation (a permanent change to the environment) until, and perhaps even after, it is removed.

Generally, rehabilitation is applicable only to press or cumulative pulse disturbances. If a disturbance is a pulse and populations recover naturally, then there is no need for intervention. If degradation is in response to a press disturbance, then the need for rehabilitation and how to achieve it are usually clear—e.g. removing obstructions to tidal flow. There is a problem if there has been a press response to a short-term pulse perturbation because all one sees is the long-term response, with no understanding of what caused it. In fact, the immediate cause may no longer be present. For example, an abnormal disturbance may remove species from an area and populations not recover because of other circumstances (Underwood 1996). Without knowledge of what caused the loss, one would necessarily and erroneously focus on current suspected causes for their demise.

M. Wahl (ed.), *Marine Hard Bottom Communities*, Ecological Studies 206,
DOI: 10.1007/978-3-540-92704-4_24,

24.2 Rehabilitation of Marine Habitats

Most of the theory about rehabilitation comes from terrestrial landscapes (e.g. Cairns 1988). It typically involves revegetation—removal of exotic vegetation or planting of native vegetation. Simultaneously, other impediments to recovery may be removed and there may be (but often is not) continued management (e.g. removal of weeds).

A search of published studies of rehabilitation of marine habitats showed that most were done in soft sediments, i.e. saltmarsh, mangroves, seagrasses. It is not immediately clear why this should be the case, although managers may see the possibility of adapting current practices and theory into marine habitats that are dominated by large plants. Many coastal habitats are also used for recreation; rehabilitating these habitats is focused on preventing loss of or augmenting public amenity, rather than benefiting biota (Martin et al. 2005). Nonetheless, hard substrata are neither less impacted nor at less risk of impact than are soft sediments.

At the simplest level, a disturbance may simply be removed—e.g. decommissioning of oilrigs when resources are depleted (Quigel and Thornton 1989)—although one may argue that this level of intervention is not rehabilitation. At the other endpoint, new natural or artificial habitats can intentionally or unintentionally be created (Clynick et al. 2007). Although most artificial reefs are built to enhance fish stocks, rather than replace habitat, rehabilitation can include artificial reefs. The overall aim of rehabilitation is to create habitat for a diversity of native flora and fauna; most programmes rely on natural colonization and, often, only few species use such habitats.

24.2.1 Removal of Obstructions to Natural Recovery

There are few examples where the removal of disturbance with no further intervention is explicitly considered under the banner of rehabilitation because many of these decisions are based on anthropocentric goals. Restrictions on fishing are almost always to "restore" a fishery so that it can be further exploited. Ecological goals behind imposition of changes to fishing effort and gear that impact soft sediments (e.g. Watling and Norse 1998) are less common for hard substrata, where the impacts of fishing on the habitat are less clear-cut (Dayton et al. 1995). A notable exception is harvesting the intertidal whelk *Concholepas concholepas* in Chile. Restricted harvesting since 1982 had the primary aim of maintaining the valuable fishery but concurrently enabled much research on ecological processes on shores protected from harvesting (e.g. Manríquez and Castilla 2001). Limiting fishing is discussed further in Chapter 25 by Goodsell and Underwood.

Attempts to reduce contamination are also primarily to minimise danger to human health, rather than to protect biota. Indeed, diversion of sewage from inshore to offshore can be implemented without evidence of impacts on the biota (Chapman et al. 1995). In contrast, banning TBT in marine paints was imposed because of its strong association with abnormalities in molluscs (Stewart et al. 1992).

Rehabilitation involving removal of particular species is generally concentrated on introduced or invasive species (see Chap. 22 by McQuaid and Arenas). On some coral reefs, disease, bleaching and overfishing have resulted in persistent changes and domination by macro-algae (Hughes 1994). McClanahan et al. (2001) experimentally removed algae in areas where fishing was restricted, to investigate whether this would increase cover of coral. The number of fish increased where algae were removed but no substantial increase in coral was evident. Increased numbers of sea urchins, perhaps due to reductions of large fish, can have substantial influence on the recovery of algae and experimental removals of urchins can result in recruitment of large kelps and other canopy-forming algae (Leinaas and Christie 1996). This does not, however, always happen (Prince 1995), indicating the complexity of interactions that can affect recovery of depleted populations.

Built structures, such as pilings and wharves, are a common press disturbance on urbanized shorelines (Chapman and Bulleri 2003). They are often removed for ecological, rather than for anthropogenic reasons; yet, there have been few quantitative studies about the ecological effects of removing infrastructure. Toft (2007) described removing a seawall in Puget Sound, USA, in order to restore the habitat for Chinook salmon and its prey (an explicit ecological goal), but inadequate replication makes it impossible to unambiguously test for rehabilitation. Obsolete oilrigs can be towed inshore to create artificial habitat but, again, there has been little assessment of the ecological impact of removing rigs (Quigel and Thornton 1989) or on their effectiveness as artificial habitat (see Sect. 24.2.3).

24.2.2 Adding Biota or Structure to Existing Habitat

Rehabilitation can involve adding specific components of habitat or biota to degraded areas. This can be problematic if one tries to change a site with the aim of returning conditions to some prior state. Natural populations fluctuate through time from place to place, often asynchronously; this includes their disappearance from some sites and appearance at others, naturally. One must be clear that, by reintroducing a species into an area, one is not reversing what is a natural change. There are similar problems about knowing what densities should be introduced. In addition, one must consider other ongoing disturbances and changes to the environment. If a species can no longer survive in degraded areas because of larger-scale conditions that local rehabilitation cannot "mend", it would be futile to keep attempting to replace such populations. This is, to some extent, overcome if current "reference", rather than previous conditions are defined as the goal.

Seldom are species, other than large plants, added to habitat in rehabilitation programmes, so that rehabilitation is usually done in habitats such as seagrass (e.g. Sheridan 2004), mangroves (e.g. Field 1998) and saltmarshes (e.g. Zedler 1992). Adding plants to hard substrata is difficult, especially where areas are wave-exposed, because most plants attach to (and are not embedded in) the substratum. In the Northern Hemisphere, populations of kelp have been depleted by disturbances such as harvesting, storms and increased herbivory (Dayton 1985). Transplanting large macro-algae to subtidal and intertidal areas is possible (e.g. Hernández-Carmona et al. 2000) but time-consuming and costly and usually limited to large-scale projects. Transplanted plants may be vulnerable to grazers (Schiel and Foster 1992) and some species do not grow large after transplantation (Correa et al. 2006). Attempts to restore *Macrocystis pyrifera* in California using artificial reefs and transplants produced viable populations for only a few years, after which they were lost (Deysher et al. 2002).

It may be easier to increase recruitment of the target plants. Dayton et al. (1984) transplanted fertile spores that colonized surrounding habitat; success will, however, probably be limited to particular environmental conditions (Schiel and Foster 1992). Hernández-Carmona et al. (2000) successfully increased recruitment of kelp by caging reproductive blades to provide a supply of spores. They argued that transplanting adults was more productive because the understorey algae inhibited growth of spores. Many environmental factors need to be appropriate to maintain viable populations of adult transplants—e.g. water-depth, substratum, wave-exposure and grazers (Schiel and Foster 1992; Deysher et al. 2002). Although spores can be easily cultured, the potential impacts of removing plants from one area in order to replant these in another area need justification. Recruitment has been artificially enhanced by the addition of cues that induce/enhance settlement of larvae: conspecifics (e.g. mussels; Tamburri et al. 2007), chemical cues (e.g. corals; Morse and Morse 1996) and microhabitat (e.g. abalone; Daume et al. 1999). These techniques are still being trialled and little is known of their success in large-scale rehabilitation.

It is unnecessary to add new components of habitat if lesser intervention for a short period will enable natural recovery. Raymundo et al. (2007) showed that preventing movement of coral rubble accelerated the natural recovery of reefs because this stabilized pieces of rubble that then grew onto the substratum. Recruitment and development of coral may, however, be limited if the substratum is smothered by sediment from associated disturbances (Clark and Edwards 1994). Most tests of adding different abiotic components of habitat to hard substrata to enhance rehabilitation involve deployment of artificial reefs—e.g. providing cavities to reefs (Hixon and Beets 1993).

24.2.3 Providing Novel Habitat

This usually comes under the banner of artificial reefs and is primarily subtidal. Artificial reefs have been constructed from floating structures, sunken vehicles or boats, or purposely designed. The primary goal is to create habitat to increase populations of harvestable taxa (Bombace 1989), with little attempt to provide habitat for all biota. Most projects are designed to be cheap and use available material (sometimes justifying disposal of "waste" under the banner of "habitat creation"; Van Treeck and Schuhmacher 1999). The extent to which such habitats are suitable for all biota is seldom assessed, usually only in terms of food for fish (Bohnsack and Sutherland 1985).

Artificial reefs may be associated with large populations of some species but there is debate as to whether they actually enhance populations or, rather, simply concentrate individuals into a small area (Lindberg 1997). The speed at which reefs are colonized by fish indicates that attraction may be a primary method of enhancement. It is difficult to resolve this issue because of complications of tracking larvae and identifying small changes in abundances over large surrounding areas. Svane and Petersen (2001) identified the evidence needed to demonstrate increased production of fish resources on artificial reefs and the lack of studies that have demonstrated it.

Nevertheless, fish rapidly colonize and successfully use artificial habitats built of many materials and with varying complexity (Baine 2001). Complexity can influence rates of colonization and the types or sizes of fish that use reefs (Hixon and Beets 1993) but, generally, fish are adaptable and readily use a range of structures as habitat (Clynick et al. 2007). There is little concern about the effects (on surrounding biota) of deploying such structures or enhancing populations of fish, although fish foraging outside artificial reefs can eliminate benthic organisms over large areas (Davis and Van Blaricom 1982).

Use of artificial reefs by invertebrates or algae has seldom been evaluated—e.g. Baine's (2001) review of 249 studies of artificial reefs showed that only four examined the effects of the structure on invertebrates. Invertebrate and algal assemblages on artificial reefs can differ from those on natural reefs (Clark and Edwards 1999); such differences can persist for many years (Perkol-Finkel and Benayahu 2005). Carter et al. (1985) showed that algae and invertebrates recruiting to constructed boulder reefs remained at an early stage of succession for 3 years, perhaps due to large amounts of foraging by fish that rapidly colonized the reefs.

24.2.4 Constructing Biotic Habitat

This must initially involve the removal of disturbance(s) and then replacement of the degraded habitat by adding gametes, juveniles or adult individuals in situ. Again, much of this type of rehabilitation is done subtidally.

Oyster-reefs (*Crassostrea virginica*) in Chesapeake Bay, USA have been the subject of many rehabilitation efforts. Stocks of oysters have been severely reduced by harvesting, habitat-destruction, water-quality and disease (see Chap. 19 by Airoldi et al.). Early attempts at rehabilitation included out-planting spat and transferring juveniles from productive areas to restock the fishery, but continued disturbances limited recovery (Lenihan and Peterson 1998). Disease-free, hatchery-reared spat or juveniles were later deployed on shells or artificial structures, and the development of natural biofilms was expected to induce settlement (Coen and Luckenbach

2000). Few rigorous experiments have examined the effects that rehabilitation of these reefs may have on other biota, although reefs created with seeded oysters have greater structural complexity and more macrofauna than are found on natural reefs (Rodney and Paynter 2006).

Coral reefs are vulnerable to many forms of anthropogenic damage—dredging, fishing, runoff and climate change (Jaap 2000). Natural recovery can take decades because of slow growth rates (Jaap 2000), leading to numerous attempts to actively rehabilitate degraded reefs. Unstable rubble, created when reefs are disturbed and concentrated predation on small populations of recruits can inhibit recovery (Clark and Edwards 1994). Lack of progeny and/or cues for settlement in areas where populations have been severely depleted, have led to attempts to transplant whole corals (Clark and Edwards 1994) or fragments (Harriott and Fisk 1988; Van Treeck and Schuhmacher 1999) onto artificial reefs or stabilized rubble. This is successful in many cases, with good survival as long as the coral is supported above sediment. Success depends on the species, technique and environmental conditions. Attempts to transplant coral larvae (Richmond 1995) or to "paint" the substratum with cues to enhance settlement (Morse and Morse 1996) have been done but the long-term success of these methods is unknown.

24.3 Evaluating Success of Rehabilitation

To evaluate rehabilitation, changes to assemblages or species need to be tested against background measures, which are inevitably spatially and temporally variable. There has been little thought about sampling designs to best measure rehabilitation, compared to those that measure environmental impacts (Chapman 1999). The need to measure impacts on organisms that are naturally very patchy in time and space has led to robust analytical techniques (Underwood 1994). Similar procedures are needed to measure rehabilitation, with more emphasis on recreating natural variability (Simenstad and Thom 1996).

Before any data are collected, there must be clear aims. In the simplest case, suppose rehabilitation were aimed to increase the numbers of a particular species. If successful, there should be an increase at the rehabilitated site after the start of the work but this change must not occur also at degraded sites that have not been rehabilitated (Chapman 1999)—otherwise, any change cannot be attributed to rehabilitation. For success, there should not only be a change but also it should reach a desired endpoint, as represented by reference sites. Before rehabilitation, the target site should be similar to other degraded control sites and, afterwards, similar to external reference sites (Fig. 24.1).

This is complicated by the fact that interpretation of whether means differ is affected by the size of their associated variances. Much variability among replicates (common to most marine organisms) is associated with a large measure of error around the mean, unless there are sufficient replicates. The only way to reduce this error is to collect more replicates. If standard errors are large, then

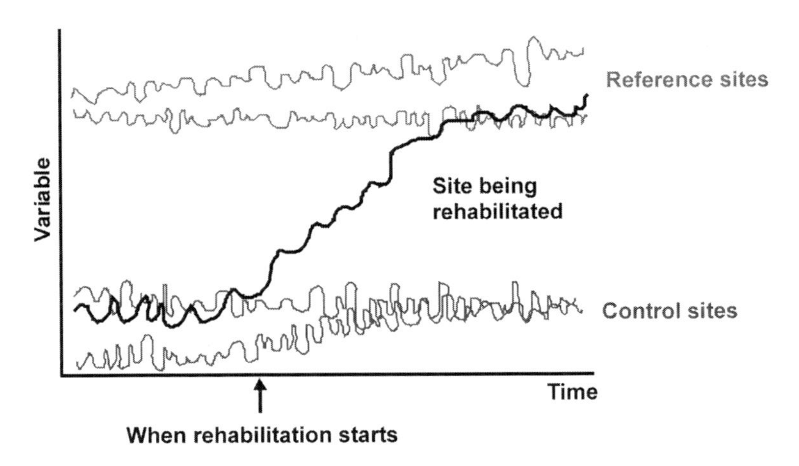

Fig. 24.1 Success of rehabilitation is best measured against multiple control and reference sites

there may be no statistically significant difference between a target site and reference sites, even when populations have not recovered (Fig. 24.2a). Similarly, minimal sampling may indicate that the target site is not statistically different from control sites (i.e. rehabilitation is not successful), even when there has been an improvement (Fig. 24.2b).

A robust approach emphasizes the importance of reducing the error around the mean to ensure better interpretation of rehabilitation projects—e.g. bio-equivalence (McDonald and Erickson 1994). One must decide in advance what level of conditions at reference sites will be acceptable as a measure of success—e.g. rehabilitation is adequate if the target site ultimately contains at least 80% of the species at reference sites (Fig. 24.2c). This test emphasizes the need to reduce error, i.e. have a good and reliable estimate of the mean, before rehabilitation can be considered successful.

This also requires knowledge (or a justifiable decision) about what values can be defined to represent adequate rehabilitation. There may be current data from which to estimate this but there is probably insufficient information for many measures to develop well-informed criteria. Questions that need addressing are: (1) is recovery of all species needed? (2) If not, which species (or functional groups) matter and why? (3) Are there particular functional groups that must be present and does it matter which species are found in each group? (4) Do we need to ensure that populations recover to natural abundances, or will smaller populations suffice? (5) Are natural abundances of a few species more important than unnatural abundances of more species? These are difficult questions but they need to be addressed urgently and analytical methods needs to be developed for assemblages, rather than single species.

Not all rehabilitation has control and reference conditions (e.g. newly created habitat in new areas). It is then important to understand what are the consequences of the lack of some of this information on interpreting whether rehabilitation has

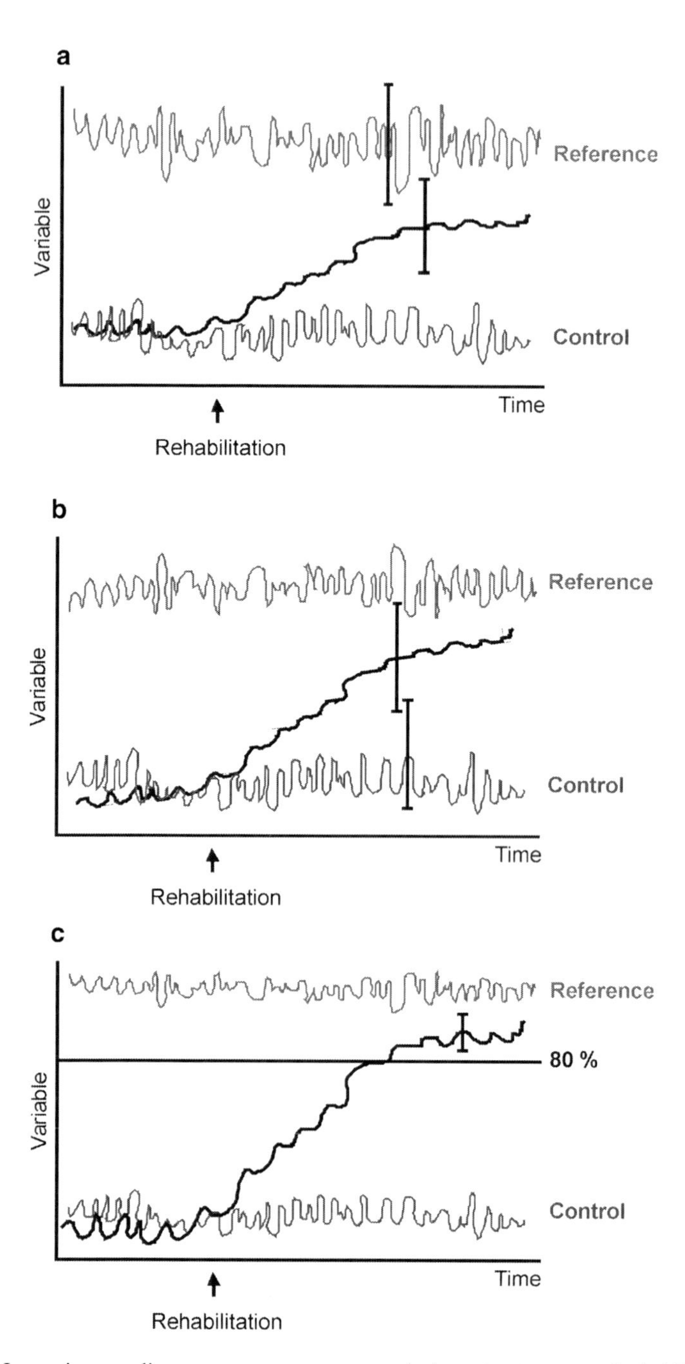

Fig. 24.2 Imprecise sampling can cause erroneous conclusions about success of rehabilitation. In **a**, large errors around mean measures at a reference site and the rehabilitated site cause non-significant differences, which would indicate that rehabilitation was successful, despite the large difference between the two means. Similarly, in **b** large errors around the means for a control site and rehabilitated site could show that they were not statistically different, leading to the conclusion that rehabilitation was not successful, despite large changes in the rehabilitated site. In **c**, small errors will enable one to determine whether conditions are now within pre-determined criteria (e.g. 80% of reference conditions)

been successful (Chapman 1999). For example, without controls, any change in the target site cannot unambiguously be attributed to the rehabilitation work.

24.4 Conclusions

Success of rehabilitation can be determined only when goals have been defined, be they societal (e.g. recreational or aesthetic) or ecological. They must be explicit—otherwise, it is not clear what must be measured. If the goal is to restore "ecologically functioning habitat" (e.g. Cairns 1988), then appropriate functions (e.g. rates of production) must be measured, rather than the structure of the assemblage. Relationships between ecological structure and function are not clear; which species are essential and which are functionally redundant? (Clarke and Warwick 1998; Tilman et al. 2006). Studies are necessary to understand these relationships, as focus may need to be on species essential for ecologically functioning rehabilitated sites, rather than the full complement of species.

Different aspects of structure and function vary at different temporal and spatial scales (e.g. Archambault and Bourget 1996; Underwood and Chapman 1998). Disturbances affect different aspects of assemblage structure in different ways and at different scales (Fairweather 1988). It is crucial to understand the scales at which different structural and functional variables are affected by disturbance before one can measure recovery from disturbances when they are removed.

To restore ecological functions, natural temporal and spatial variability must be restored, and more long-term studies to assess the relationships between variability and ecological structure and function are necessary (Hawkins et al. 2002). For example, changes to intertidal assemblages following the use of dispersants on the Torrey Canyon oil spill lasted up to 15 years (Southward et al. 2005). Yet, many assessments of rehabilitation are done with sporadic measures over very short time-scales.

The need to develop effective strategies assessing success of rehabilitation is becoming increasingly urgent. This will benefit from considering recent techniques that can successfully measure impacts and relevant temporal and spatial scales. To enhance success of rehabilitation, we must consider the complexity of natural changes as a background to changes in response to disturbance and recovery.

References

Archambault P, Bourget E (1996) Scales of coastal heterogeneity and benthic intertidal species richness, diversity and abundance. Mar Ecol Prog Ser 136:11–121

Baine M (2001) Artificial reefs: a review of their design, application, management and performance. Ocean Coast Manag 44:241–259

Bohnsack JA, Sutherland DL (1985) Artificial reef research: a review with recommendations for future priorities. Bull Mar Sci 37:11–39

Bombace G (1989) Artificial reefs in the Mediterranean Sea. Bull Mar Sci 44:1023–1032

Cairns J (1988) Rehabilitating damaged ecosystems. CRC Press, Boca Raton, FL
Carter JW, Carpenter AL, Foster MS, Jessee WN (1985) Benthic succession on an artificial reef designed to support a kelp-reef community. Bull Mar Sci 37:86–113
Chapman MG (1999) Improving sampling designs for measuring restoration in aquatic habitats. J Aquat Ecosyst Stress Recovery 6:235–251
Chapman MG, Bulleri F (2003) Intertidal seawalls—new features of landscape in intertidal environments. Landsc Urban Plan 62:159–172
Chapman MG, Underwood AJ, Skilleter GA (1995) Variability at different spatial scales between a subtidal assemblage exposed to the discharge of sewage and two control assemblages. J Exp Mar Biol Ecol 189:103–122
Clark S, Edwards AJ (1994) Use of artificial reef structures to rehabilitate reef flats degraded by coral mining in the Maldives. Bull Mar Sci 55:724–744
Clark S, Edwards AJ (1999) An evaluation of artificial reef structures as tools for marine habitat rehabilitation in the Maldives. Aquat Conserv Mar Freshw Ecosyst 9:5–21
Clarke KR, Warwick RM (1998) Quantifying structural redundancy in ecological communities. Oecologia 113:278–289
Clynick BG, Chapman MG, Underwood AJ (2007) Effects of epibiota on assemblages of fish associated with urban structures. Mar Ecol Prog Ser 332:201–210
Coen LD, Luckenbach MW (2000) Developing success criteria and goals for evaluating oyster reef restoration: ecological function or resource exploitation. Ecol Eng 15:323–343
Correa JA, Lagos NA, Medina MH, Castilla JC, Cerda M, Ramirez M, Martinez E, Faugeron S, Andrade S, Pinto R, Contreras L (2006) Experimental transplants of the large kelp *Lessonia nigrescens* (Phaeophyceae) in high-energy wave exposed rocky intertidal habitats of northern Chile: experimental, restoration and management applications. J Exp Mar Biol Ecol 335:13–18
Daume S, Brand-Gardner S, Woelkerling WJ (1999) Preferential settlement of abalone larvae: diatom films vs. non-geniculate coralline red algae. Aquaculture 174:243–254
Davis N, Van Blaricom GR (1982) Man-made structures on marine sediments: effects on adjacent benthic communities. Mar Biol 70:295–303
Dayton PK (1985) Ecology of kelp communities. Annu Rev Ecol Syst 16:215–245
Dayton PK, Currie V, Gerrodette T, Keller BD, Rosenthal R, Ven Tresca D (1984) Patch dynamics and stability of some Californian kelp communities. Ecol Monogr 54:253–289
Dayton PK, Thrush SF, Agardy TM, Hofman RJ (1995) Environmental effects of marine fishing. Aquat Conserv Mar Freshw Ecosyst 5:205–232
Deysher LE, Dean TA, Grove RS, Jahn A (2002) Design considerations for an artificial reef to grow giant kelp (*Macrocystis pyrifera*) in Southern California. ICES J Mar Sci 59:S201–207
Fairweather PG (1988) Sewage and the biota on seashores: assessment of impact in relation to natural variability. Environ Monit Assess 14:197–210
Field CD (1998) Rehabilitation of mangrove ecosystems: an overview. Mar Pollut Bull 37:383–392
Glasby TM, Underwood AJ (1996) Sampling to differentiate between pulse and press perturbations. Environ Monit Assess 42:241–252
Harriott VJ, Fisk DA (1988) Coral transplantation as a reef management option. Proc 6th Int Coral Reef Symp 2:375–379
Hawkins SJ, Gibbs PE, Pope ND, Burt GR, Chesman BS, Bray S, Proud SV, Spence SK, Southward AJ, Southward GA, Langston WJ (2002) Recovery of polluted ecosystems: the case for long-term studies. Mar Environ Res 54:215–222
Hernández-Carmona G, Garcia O, Robledo D, Foster M (2000) Restoration techniques for *Macrocystis pyrifera* (Phaeophyceae) populations at the southern limit of their distribution in Mexico. Bot Mar 43:273–284
Hixon MA, Beets JP (1993) Predation, prey refuges, and the structure of coral-reef fish assemblages. Ecol Monogr 63:77–101
Hughes TP (1994) Catastrophes, phase shifts, and large-scale degradation of a Caribbean coral reef. Science 265:1547–1551

Jaap WC (2000) Coral reef restoration. Ecol Eng 15:345–364
Leinaas HP, Christie H (1996) Effects of removing sea urchins (*Strongylocentrotus droebachiensis*): stability of the barren state and succession of kelp forest recovery in the east Atlantic. Oecologia 105:524–536
Lenihan HS, Peterson CH (1998) How habitat degradation through fishery disturbance enhances impacts of hypoxia on oyster reefs. Ecol Appl 8:128–140
Lindberg WJ (1997) Can science resolve the attraction-production issue? Fisheries 22:10–13
Manríquez PH, Castilla JC (2001) Significance of marine protected areas in central Chile as seeding grounds for the gastropod *Concholepas concholepas*. Mar Ecol Prog Ser 215:201–211
Martin D, Bertasi F, Colangelo MA, de Vries M, Frost M, Hawkins SJ, Macpherson E, Moschella PS, Satta MP, Thompson RC, Ceccherelli VU (2005) Ecological impact of coastal defence structures on sediment and mobile fauna: evaluating and forecasting consequences of unavoidable modifications of native habitats. Coast Eng 52:1027–1051
McClanahan TR, McField M, Huitric M, Bergman K, Sala E, Nystrom M, Nordemar I, Elfwing T, Muthiga NA (2001) Responses of algae, corals and fish to the reduction of macroalgae in fished and unfished patch reefs of Glovers Reef Atoll, Belize. Coral Reefs 19:367–379
McDonald LL, Erickson WP (1994) Testing for bio-equivalence in field studies: has a disturbed site been adequately reclaimed? In: Fletcher DJ, Manly BF (eds) Statistics and environmental monitoring. University of Otago Press, Dunedin, pp 183–197
Morse ANC, Morse DE (1996) Flypapers for coral and other planktonic larvae. Bioscience 46:254–262
Perkol-Finkel S, Benayahu Y (2005) Recruitment of benthic organisms onto a planned artificial reef: shifts in community structure one decade post-deployment. Mar Environ Res 59:79–99
Prince J (1995) Limited effects of the sea-urchin *Echinometra mathaei* (Deblainville) on the recruitment of benthic algae and macroinvertebrates into intertidal rock platforms at Rottnest Island, Western Australia. J Exp Mar Biol Ecol 186:237–258
Quigel JC, Thornton WL (1989) Rigs to reefs—a case history. Bull Mar Sci 44:799–806
Raymundo LJ, Maypa AP, Gomez ED (2007) Can dynamite blasted reefs recover? A novel, low-tech approach to stimulating natural recovery of fish and coral populations. Mar Pollut Bull 54:1009–1019
Richmond R (1995) Coral reef health: concerns, approaches and needs. In: Crosby M (ed) Proc Coral Reef Symp. Practical, Reliable, Low Cost Monitoring Methods for Assessing the Biota and Habitat Conditions of Coral Reefs. US EPA & NOAA, Maryland, pp 25–28
Rodney WS, Paynter KT (2006) Comparisons of macrofaunal assemblages on restored and non-restored oyster reefs in mesohaline regions of Chesapeake Bay in Maryland. J Exp Mar Biol Ecol 335:39–51
Schiel DR, Foster MS (1992) Restoring kelp forest. In: Thayer GW (ed) Restoring the Nation's marine environment. Maryland Sea Grant, College station, MD, pp 279–342
Sheridan P (2004) Comparison of restored and natural seagrass beds near Corpus Christi, Texas. Estuaries 27:781–792
Simenstad CA, Thom RM (1996) Functional equivalency trajectories of the restored Gog-Le-Hi-Te estuarine wetland. Ecol Appl 6:38–56
Southward AJ, Langmead O, Hardman-Mountford NJ, Aiken J, Boalch GT, Dando PR, Genner MJ, Joint I, Kendall MA, Halliday NC, Harris RP, Leaper R, Mieszkowska N, Pingree RD, Richardson AJ, Sims DW, Smith T, Walne AW, Hawkins SJ (2005) Long-term oceanographic and ecological research in the western English Channel. Adv Mar Biol 47:1–105
Stewart C, Demora SJ, Jones MRL, Miller MC (1992) Imposex in New-Zealand neogastropods. Mar Pollut Bull 24:204–209
Svane I, Petersen JK (2001) On the problems of epibioses, fouling and artificial reefs, a review. Mar Ecol 22:169–188
Tamburri MN, Zimmer RK, Zimmer CA (2007) Mechanisms reconciling gregarious larval settlement with adult cannibalism. Ecol Monogr 77:255–268
Tilman D, Reich PB, Knops JMH (2006) Biodiversity and ecosystem stability in a decade-long grassland experiment. Nature 441:629–632

Toft J (2007) Benthic macroinvertebrate monitoring at Seahurst Park 2006, post-construction of seawall removal. Report for the City of Burien, School of Aquatic & Fishery Science, University of Washington, Washington, DC

Underwood AJ (1994) On beyond BACI: sampling designs that might reliably detect environmental disturbances. Ecol Appl 4:3–15

Underwood AJ (1996) Spatial patterns of variance in density of intertidal populations. In: Floyd RB, Sheppard AW, De Barro PJ (eds) Frontiers of population ecology. CSIRO, Victoria, pp 369–389

Underwood AJ, Chapman MG (1998) Variation in algal assemblages on wave-exposed rocky shores in New South Wales. Mar Freshw Res 49:241–254

Van Treeck P, Schuhmacher H (1999) Artificial reefs created by electrolysis and coral transplantation: an approach ensuring the compatibility of environmental protection and diving tourism. Estuar Coast Shelf Sci 49:75–81

Watling L, Norse EA (1998) Special section: effects of mobile fishing gear on marine benthos. Conserv Biol 12:1178–1179

Zedler JB (1992) Restoring cordgrass meadows in southern California. In: Thayer GW (ed) Restoring the Nation's marine environment. Maryland Sea Grant, College Station, MD, pp 7–51

Chapter 25
Protection of Biota and the Value of Marine Protected Areas

Paris J. Goodsell and A.J. Underwood

25.1 Introduction

Threats to marine habitats are many. They vary in severity, intensity, extent and their consequences for biota. They include loss of habitat, over-harvesting, contamination, invasions, debris, modification of coastal processes and climatic change.

Here, only protection of habitats or assemblages that are considered or known to be threatened is discussed. "Threatened" is not used in its legal sense (i.e. that a species or habitat has been listed by law as threatened) but refers to species, assemblages or habitats that are affected by some human activity.

In this chapter, the threats and processes that necessitate protection are considered, including establishment of marine reserves. Some of the ways reserves are planned and how well they achieve their goals are also briefly considered.

25.2 Protection Outside Reserves

Appropriate protection of marine ecology must vary according to understanding of the particular threats and risks. For example, strategies to protect resources at risk from large-scale disturbance (such as over-harvesting, large-scale contamination) will necessarily be different from those dealing with small-scale disturbances (such as point-source contamination). Disturbances and contamination are generally considered to be of two broad types, "presses" and "pulses" (Bender et al. 1984; see Chaps. 17 by Clynick et al. and 24 by Goodsell and Chapman). Ecological responses to press disturbances (long-term alterations—e.g. the development of a coastal port) are generally considered to be long-term, continuous changes to a new mean level of relevant ecological variables (e.g. disruption of alongshore currents that transport larvae). As a result, there is a long-term impact "pressing" the abundances of, for example, fish to new but decreased numbers. In contrast, a pulse disturbance is generally short-term (e.g. contamination to local sediments during construction of a port, ending when construction is completed). Such disturbances are generally thought to cause short-term responses by the biota, which recover after the disturbance ends.

M. Wahl (ed.), *Marine Hard Bottom Communities*, Ecological Studies 206,
DOI: 10.1007/978-3-540-92704-4_25, © Springer-Verlag Berlin Heidelberg 2009

The dichotomy is an oversimplification because either type of disturbance can cause either type of response (Glasby and Underwood 1996). Nevertheless, it is important to understand the type of disturbance if there are plans to protect biota or to create reserves to prevent further ecological damage. Press responses are generally easier to detect and interpret because they are alterations of average values of density, size, reproduction, etc. Pulse responses are more difficult because they change the time-courses (temporal variance) of ecological variables and, if a pulse disturbance has caused a press response, it is no longer present to be detected as the cause of some identified impact.

Similarly, severe deleterious effects (e.g. loss of habitat) are much simpler to identify and interpret than are more subtle influences. Contamination (or disturbances) causing pollution (or impacts) to numerous species are often easier to detect than are processes threatening one or only a few species. Obviously, widespread contamination or disturbance is easier to regulate or ameliorate. Areas to declare as marine reserves are much easier to identify if the threatening processes are localized, because these can be avoided much more easily than is the case for larger-scale problems.

Most protection is done by legislation to restrict, regulate or prohibit destruction or modification of habitat, disturbances—e.g. discharge of contaminants, harvesting and other human activities. Note that it is not actually possible to manage the *ecology* itself (to sustain biodiversity or to control ecological functions). Management, including protection, is management of people's behaviour.

25.2.1 Contaminants

Contaminants are potentially toxic chemicals that may or may not cause pollution. Some theoretically toxic chemicals are contaminants (i.e. present in unnaturally large concentrations) but do not cause pollution. There is no ecological response (no direct or indirect mortality, impaired growth or reproduction), because organisms do not take up the chemicals and detoxify them if they are absorbed, or because they are "inert" to their potential effects.

It is always necessary to demonstrate that any contaminant does, in fact, have an adverse effect before it can be considered a pollutant. For example, antifouling paints containing the endocrine-suppressant TBT have been banned worldwide since 2003 (Ellis and Pattisina 1990). This was done to protect marine organisms susceptible to its effects. Previous research had shown that a large number of gastropods developed imposex (sterilised females) in areas contaminated by TBT, even in small concentrations (Ellis and Pattisina 1990). Substantial changes in the abundance and distribution of the gastropods and, in some cases, localized extinctions were recorded (Hawkins et al. 1994).

Contaminants causing pollution can be point-source, in which case protection most likely requires local-scale legislation to restrict discharges to "safe" levels. Alternatively, they can be large-scale, in which case protection and compliance are much more difficult. In either case, it is probably common that adequate protection comes substantially too late for adequate conservation because pollution is often identified only after it has occurred.

Protection or management of discharges is often based on standards of water quality. In the ocean, however, most contaminants sink to the sediments or are taken up by organisms; "monitoring" water quality is unlikely to measure any ecologically meaningful impact. For example, it is quite widespread that regulatory authorities, local governmental agencies or community groups monitor concentrations of nitrogen and phosphorus in seawater—particularly in estuaries or coastal lagoons. There are numerous issues about the reliability of the actual measurements and problems of adequacy of spatial and temporal aspects of the design of sampling that can make the data unrepresentative, uninformative or uninterpretable. Over and above these, the purpose of monitoring needs to be logical. For example, monitoring is often motivated by perceived problems of excessive algal blooms that can develop in response to increased nutrients in the water. In many cases, however, the excessive algal growth removes the nutrients (which are taken up into the algae), so that measures of water quality do not actually change. In cases like this, it is obviously more effective to measure the algae per se, instead of the non-responsive indicator of water quality.

Furthermore, the true toxicity of contaminants is largely unknown because of substantial complexity and variability in the dosage/concentration, bio-accumulation, detoxification in bodies, dispersion or concentration via circulation, interactive effects with other contaminants, species-specific responses, etc. (Bryan and Langston 1992). It is obviously difficult to protect against effects about which little is known.

Prohibition or regulation of many ecologically harmful contaminants lags far behind realization of their ecological effects (Thompson et al. 2002). Protection of marine organisms may therefore be inhibited by lack of adequate experimental evidence to establish causal links between a disturbance and any adverse ecological consequences. Lack of reliable data, in turn, inhibits legislative processes to restrict, regulate or prohibit disturbances.

25.2.2 Harvesting

Commercial or amateur exploitation of resources is a major disturbance to many organisms in hard-bottom habitats, resulting in changes in their sizes, distributions or abundances. Organisms are often protected from over-harvesting by limits to the amounts and sizes of organisms that can be collected. In some cases, harvesting is banned, often by declaring reserves (see below).

Habitat can be destroyed by some techniques of collection (e.g. chain-trawling) and trampling (during recreational harvesting). Protection from the disturbances associated with harvesting is done by banning certain fishing equipment or restricting access. The latter is often also done by declaring protected areas.

25.3 Reserves as Protection—Principles

Establishing protected areas has, for more than a century, been a strategy to conserve biodiversity. The general aim of protected areas is conservation "for present and future human use, the diversity and integrity of biotic communities of plants and animals within natural ecosystems" (UNESCO 1974). The goals of protected areas are blurry at best, as made clear by the number of unclear words in the preceding sentence. For example, some areas have been chosen because they have large biodiversity, but which are not necessarily threatened or at risk. Areas have also been chosen that were at risk, but not necessarily representative of a diverse suite of flora and fauna. This is unfortunate, given that ad hoc selection of reserves sites does nothing to achieve the goal of conserving biodiversity (Sullivan and Shaffer 1975).

Nonetheless, two major functions are generally accepted: (1) to protect areas representative of, or containing, many endemic or representative species and (2) to insure the long-term survival of species and ecological functions within these areas (Margules et al. 1988).

Protecting organisms by reserving areas of habitat and restricting human access, harvesting, development or modification of such areas has a theoretical basis in terrestrial ecology (Kunin 1997). Designing reserves and evaluating the success of terrestrial reserves has been based largely on theories of island biogeography (MacArthur and Wilson 1967), followed by those of landscape ecology (e.g. Forman and Godron 1986) and metapopulation ecology (Hanski 1997). In the 1980s, research on shape, quality, size, etc. of reserves gave rise to debates about which were better for the ultimate goal of conserving diversity and ecological function (e.g. the SLOSS, single large or several small, debate; Diamond 1975).

It is, however, now recognised that a landscape approach is needed to create networks of reserves that are potentially connected, such as via corridors (Margules et al. 1988). Despite many terrestrial reserves and protected areas being designated some time ago, many have suffered altered ecological processes, invasion by exotic taxa and extinction of native taxa (Parks and Harcourt 2002). Indeed, with increasing urbanisation or human domination of areas surrounding reserves, the failure of many of these to conserve and protect biota and ecological functions may be due to lack of awareness of the effects of the surrounding land and/or the configuration of many reserves within a landscape (Hansen and DeFries 2007).

Concepts of reserve design developed in terrestrial habitats may be beneficial for the development of marine reserves, which has a shorter history than does that of terrestrial reserves (Day and Roff 2000). There are, however, fundamental differences between marine and terrestrial systems (reviewed by Carr et al. 2003). The most obvious is water, which has physical and chemical properties different from those of air (see Chap. 1 by Wahl). Also, phyletic biodiversity is greater in marine than in terrestrial habitats. The latter, however, tend to have greater diversity of species (Gray 1997). Large size, circular shapes and connectivity by close distances between reserves or corridors have been identified as beneficial for terrestrial reserves (Kunin 1997). These criteria may, however, not all be useful for maintaining biodiversity and ecological functions in marine habitats.

Because many marine organisms disperse via planktonic larvae, regional-scale processes have been considered more likely to affect local populations in marine than in terrestrial systems (Palmer et al. 1996). Reproduction and recruitment can depend on processes occurring elsewhere than in the reserve. Of course, this depends on the life-history characteristics of species; those with pelagic adult stages (e.g. fish) may not benefit from protected areas, whereas species that rely on spawning sites, nursery areas, etc., or those that are relatively sedentary as adults or have limited dispersal of propagules (e.g. algae) may benefit.

Differences in connectivity are argued to be a major difference between terrestrial and marine reserves because connectivity is assumed to be of less importance for maintaining dispersive marine populations. There is, however, accumulating evidence that marine organisms are affected by the distance between remnant habitats, the suitability of matrix habitat and the presence of corridors (e.g. Robbins and Bell 1994). "Corridors" for many marine habitats are likely to be currents that transport propagules among isolated populations (Carr et al. 2003).

25.4 Reserves as Protection—Practice

Because marine reserves have been developed for different purposes (e.g. increased fisheries, reduced human impact, restoring and maintaining biodiversity, protecting endangered taxa), there are several kinds of protected areas. Confusingly, these are ambiguously and differentially defined among different countries. Protected areas range from "no-take" areas (completely closed to any methods of extraction), to those where only recreational fishing is allowed, subject to catch limits, and to areas where commercial fishing is allowed at certain times (temporary closures) or in certain areas of the reserve. There are attempts to develop a unifying set of classifications and a representative system of protected areas, at least within nations—e.g. Australia and New Zealand (e.g. ANZECC 1999; Day et al. 2007). Chape et al. (2003) catalogued the world's protected areas. It is estimated that only 0.5% of the surface area of the oceans is protected in some form.

Reserves are usually developed with long-term goals in mind, so that responses of populations inside (and processes that govern recruitment and dispersal outside) reserves should be estimated. For example, episodic climatic changes can have a marked effect on populations in reserves if dispersal events outside are changed (i.e. by El Niño events) or mass mortality events occur (Allison et al. 1998). Moreover, current conditions in reserves may not persist during long-term climatic changes such as those predicted for the future (Soto 2001). For example, it is recognised that species and assemblages will redistribute themselves as an effect of increased temperature (Frank et al. 1990). The current location of reserves may not represent, nor maintain, levels of biodiversity or ecological functions. This is particularly important in marine (cf. terrestrial) reserves, given the shorter life spans of many primary producers (e.g. algae); so, responses will likely be manifested within decades, rather than hundreds of years, as is predicted for terrestrial systems (Carr et al. 2003). Possible ecological scenarios in marine reserves following climatic change were reviewed by Soto (2001).

Reserves are unlikely to protect against all human impacts, primarily because boundaries are diffuse. Allison et al. (1998) provided guidelines to integrate protection outside, in addition to inside reserves. A widespread strategy, for terrestrial and marine habitats, is a network of reserves encompassing a range of environmental conditions, populations and habitats and sufficiently large and connected to insure against stochastic and long-term changes (Margules et al. 1988; Roberts et al. 2003).

Some reserves do provide examples of good outcomes. For example, Leigh Marine (Goat Island) Reserve in north-eastern New Zealand was designated in the late 1970s. Macro-algae are the primary occupiers of space in the area and support many invertebrate grazers (e.g. urchins and gastropods) and fish. Throughout the 1960s, large areas of urchin barrens replaced forests of macro-algae, probably due to decreased predation on urchins (Schiel 1982). Since the establishment of the reserve, predators have increased, urchins have declined and forests of kelp have expanded (Babcock et al. 1999). Some mobile species are likely to move outside a reserve. For example, Kelly and MacDiarmid (2003) showed that 20% of lobsters (*Jasus edwardsii*) surviving to maturity inside the reserve emigrated and potentially became part of the fishery. Other mobile species (for example, snapper) are larger, denser and more fecund in the reserve (and other reserves) than in comparable fished areas (Willis et al. 2003).

The Estación Costera de Investigaciones Marinas (ECIM) in central Chile has an intertidal protected area established for scientific purposes in 1982, which was fenced to prevent human perturbation. The rocky shore in this area was dominated by large beds of the mussel *Perumytilus purpuratu*, kelp *Lessonia nigrecsens* and the harvested gastropod, *Concholepas concholepas*. Restriction of human harvesting resulted in a large increase in density of the predatory gastropod and a persistent switch of the dominant space-occupiers from mussels to barnacles (*Jehlius cirratus* and *Chthamalus scabrosus*). This is a particularly interesting case because, whilst there was an initial increase in diversity after the exclusion of humans, barnacles persisted in sufficient numbers to exclude other organisms, resulting in less diversity in the protected area (Castilla 1999).

25.5 What Happens Outside Reserves?

One of the major issues influencing political and social will to develop marine reserves is the so-called spill-over effect on harvested resources. In several parts of the world, there has been a major change in emphasis over the role of marine reserves. The earlier view was dominated by the notion that reserves were desirable because they would enhance sustainability of habitats and diversity of species. Refuges or areas protected from some or all human interference would provide benefit, largely of an ecocentric kind—i.e. for the greater good of ecological functions or of biological species (see Fairweather 1993 for a discussion of different ethical positions about such issues).

In contrast, now a major view is that marine reserves are part of the toolkit for managing the allocation and sustainability of harvestable resources, mostly fisheries, i.e. an anthropocentric view. It is considered that "selling" reserves on this basis gains more public support, despite denial of access to some areas previously available for activities.

The basis for a reserve being valuable for management of resources is twofold. First, as discussed by Alcala and Russ (1990), reserves can be areas closed against specific activities, such as fishing, for a defined period and then opened again, in a "rolling" manner through space. The principle is that recovery of fished stocks in the closed reserve enables good fishing when the area is reopened. Different areas are closed and recovering while harvesting takes place in the former reserve.

In contrast, the other strategy involves "spill-over". Species in the reserve increase in numbers or sizes or other biological variables that enhance reproductive output. They have dispersive propagules, or the adults move out of areas of increased density and biomass. As a result, recruitment of juveniles over relatively large distances, or adults over smaller distances, enhances the stocks of exploited species outside the reserves (see earlier discussion of the Leigh Reserve). For this to work in practice, at the very least the enhanced harvestable stock outside the reserve must equal or exceed the stock lost in the protected area. So, in general, if areas of size a are protected, out of a total area of fishery size A, the loss of stock is related to a/A. The populations in the reserved areas (a) must then increase productivity at least sufficiently to replace the lost stock, by migration of animals from the reserve into the fished areas. To achieve any increased productivity of the harvested resources, the spill-over from the reserve to the harvested areas must be greater than the amount of harvest lost from the area of the reserve that is no longer exploited.

Modelling to assess the conditions under which stocks will be maintained or enhanced by spill-over from a reserve indicates that this approach can be successful, depending on the intensity of harvesting, the reproductive biology and dispersal of the organisms, and the proportion of the total area that is reserved (Gaylord et al. 2005). There are, however, numerous complications, including fishing being focused along the edges of reserves and aspects of the ecology of the assemblages in reserves. For example, if predators of commercially exploited species increase in numbers, sizes, activity, etc. in a reserve, any increased production would be consumed locally and could not disperse into exploited areas. Whether or not a reserve will have the predicted outcomes for other areas will depend on the direct and indirect ecological interactions in the reserve. Determining whether predicted goals are achieved must therefore depend on assessing the outcome (i.e. testing the predictions—hypotheses) by using the reserve(s) as experimental manipulations of management (e.g. Underwood 1995).

25.6 Assessing Effectiveness of Marine Reserves

It is important to gather relevant data in appropriate ways to determine whether a marine reserve (or other form of protection against threats) is effectively achieving its aims. This is the case whatever the nature of analytical procedures to be used (e.g. frequentists or Bayesian) or whether there are considered good reasons for using design-based or model-based ways of determining what is happening. There is insufficient space to consider these issues in detail. There are, however, several critical elements in achieving efficient and reliable understanding of how well a reserve is performing. The details of many aspects of this are covered in Chapter 24 by Goodsell and Chapman, in the context of estimating the effectiveness of restoration of damaged habitat. Certain elements of an appropriate research programme are common to assessments of reserves. They include:

1. having clearly defined goals. What is the reserve for? What risks and threats are being removed or reduced? What are the mechanisms of protection being put in place and how (i.e. by what series of processes) are these supposed to be removing or reducing threats? What habitats, assemblages, species are of primary concern? Are the issues solely about maintaining (or increasing) numbers of species, or particular species? Do the issues concern maintaining the sustainability of the ecological systems? If so, must they include maintaining patchiness in space and time (i.e. the spatial and temporal variance), rather than only average conditions?
2. ensuring that the goals are translated into or interpreted as defined hypotheses. So, if the goal is to maintain (at least) the current diversity of seabirds in a reserve, the hypothesis is clear. Where the goal is to maintain the turnover of nutrients in a reserve, the hypotheses are likely to be much more complex.
3. having identified explicit predictions, appropriate sampling and experimental designs must be formulated (as discussed in Chap. 4.8). Where possible or appropriate, designs should include quantitative targets (appropriate reference areas). If it is necessary to be sure that the protection created by the reserve is actually causing the defined outcomes, there should be relevant control areas (see Chap. 24 by Goodsell and Chapman). Relevant designs must also include the appropriate timescales of sampling.

Once a suitable sampling design and the associated appropriate focus of analysis have been determined, it is necessary to do the sampling, as planned. Reducing the number of samples or the time period over which sampling is made usually reduces the capacity of the study to determine anything or, at the very least, makes it less powerful (less likely to determine anything well).

25.7 Conclusions

Decisions about which areas to protect are difficult without detailed information about the distribution and abundance of biota. It is necessary to understand which species (or levels of diversity) are necessary to maintain ecological functions, an aspect that has been and still is hotly debated (e.g. Wardle et al. 2000). What surrogates can or should be used to estimate biodiversity in a potential reserve is always difficult to decide, unless it is possible to sample all taxa in an area. Some taxa may be more important for survival and functioning of many other species (Simberloff 1998) and their protection should be a target.

Despite their goals, many marine reserves conflict with development and economic gain. As a result, they become compromised and are no longer representative of diversity or become inadequate to maintain survival of ecological functions. To be successful, there may need to be a suite of connected reserves representing a range of environmental conditions, habitats and biota (Roberts et al. 2003). In particular, this is predicted to insure against stochastic or long-term shifts in the distribution and abundance of marine organisms (Soto 2001; Carr et al. 2003). One strategy is to choose different types of reserves, each of which achieves some percentage of protection of habitat or other ecological criteria, such as sufficient connectivity and area to sustain ecological function (Roberts et al. 2003).

The development and planning of marine reserves involve economic, social and political, in addition to ecological issues. Ultimate goals involve protecting marine taxa—so, ecology must have an important role. Ecological literature from the 1970s onwards should assist with the initial design of reserves to ensure that they are representative of biodiversity and are sufficiently large or connected to maintain ecological processes. Moreover, ecological studies can assess the effectiveness of reserves to achieve their desired goals.

Estimating an adequate range of biodiversity in any potential reserve and understanding the processes that maintain ecological function and diversity are difficult tasks. It is time for increased empirical research into roles, resources, etc. of marine reserves, rather than continuing to emphasize theoretical and modelling approaches to their design.

Acknowledgements The preparation of this paper was supported by funds from the Australian Research Council through its Special Centres' programme.

References

Alcala AC, Russ GR (1990) A direct test of the effects of protective management on abundance and yield of tropical marine resources. J Cons Int Explor Mer 46:40–47

Allison GW, Lubchenco J, Carr MH (1998) Marine reserves are necessary but not sufficient for marine conservation. Ecol Appl 8:S79–S92

ANZECC (1999) Guidelines for establishing the national representative system of marine protected areas. Australian and New Zealand Environment and Conservation Council, Task Force on Marine Protected Areas, Environment Australia, Canberra

Babcock RC, Kelly S, Shears NT, Walker JW, Willis TJ (1999) Changes in community structure in temperate marine reserves. Mar Ecol Prog Ser 189:125–134

Bender EA, Case TJ, Gilpin ME (1984) Perturbation experiments in community ecology: theory and practice. Ecology 65:1–13

Bryan GW, Langston WJ (1992) Bioavailability, accumulation and effects of heavy metals in sediments with special reference to United Kingdom estuaries: a review. Environ Pollut 76:89–131

Carr MH, Neigel JE, Estes JA, Andelman S, Warner RR, Largier JL (2003) Comparing marine and terrestrial ecosystems: implications for the design of coastal marine reserves. Ecol Appl 13:S90–S107

Castilla JC (1999) Coastal marine communities: trends and perspectives from human-exclusion experiments. Trends Ecol Evol 14:280–283

Chape S, Blyth S, Fish L, Fox P, Spalding M (compilers) (2003) 2003 United Nations list of protected areas. IUCN, Gland; Cambridge and UNEP World Conservation Monitoring Centre, Cambridge

Day JC, Roff JC (2000) Planning for representative marine protected areas: a framework for Canada's oceans. World Wildlife Fund, Toronto

Day JC, Senior J, Monk S, Neal W (eds) (2007) Proceedings of the First International Marine Protected Areas Congress. IMPAC1 2005, Victoria

Diamond JM (1975) The island dilemma: lessons of modern biogeographic studies for the design of natural reserves. Biol Conserv 7:129–146

Ellis DV, Pattisina LA (1990) Widespread neogastropod imposex—a biological indicator of global TBT contamination. Mar Pollut Bull 21:248–253

Fairweather PG (1993) Links between ecology and ecophilosophy, ethics and the requirements of environmental management. Aust J Ecol 18:3–19

Forman RTT, Godron M (1986) Landscape ecology. Wiley, New York

Frank KT, Perry RI, Drinkwater KF (1990) Predicted response of northwest Atlantic fish stocks to CO2 induced climate change. Trans Am Fish Soc 119:353–365

Gaylord B, Gaines SD, Siegel DA, Carr MH (2005) Marine reserves exploit population structure and life history in potentially improving fisheries yields. Ecol Appl 15:2180–2191

Glasby TM, Underwood AJ (1996) Sampling to differentiate between pulse and press perturbations. Environ Monit Assess 42:241–252

Gray SJ (1997) Marine biodiversity: patterns, threats and conservation needs. Biodivers Conserv 6:153–175

Hansen AJ, DeFries R (2007) Ecological mechanisms linking protected areas to surrounding lands. Ecol Appl 17:974–988

Hanski I (1997) Metapopulation dynamics: from concepts and observations to predictive models. In: Hanksi I (ed) Metapopulation biology: ecology, genetics and evolution. Academic Press, San Diego, CA, pp 69–92

Hawkins SJ, Proud SV, Spence SK, Southward AJ (1994) From the individual to the community and beyond: water quality, stress indicators and key species in coastal ecosystems. In: Sutcliffe DW (ed) Water quality and stress indicators in marine and freshwater ecosystems: linking levels of organisation (individuals, populations, communities). Freshwater Biological Association, Ambleside, pp 35–62

Kelly S, MacDiarmid AB (2003) Movement patterns of mature spiny lobsters, *Jasus edwardsii*, from a marine reserve. N Z J Mar Freshw Res 37:149–158

Kunin WE (1997) Sample shape, spatial scale and species counts: implications for reserve design. Biol Conserv 82:369–377

MacArthur RH, Wilson E (1967) The equilibrium theory of island biogeography. Princeton University Press, Princeton, NJ

Margules CR, Nicholls AO, Pressey RL (1988) Selecting networks of reserves to maximise biological diversity. Biol Conserv 43:63–76
Palmer MA, Allan JD, Butman CA (1996) Dispersal as a regional process affecting the local dynamics of marine and stream benthic invertebrates. Trends Ecol Evol 11:322–326
Parks SA, Harcourt AH (2002) Reserve size, local human density, and mammalian extinctions in US protected areas. Conserv Biol 16:800–808
Robbins BD, Bell SS (1994) Seagrass landscapes: a terrestrial approach to the marine subtidal environment. Trends Ecol Evol 9:301–304
Roberts CM, Branch G, Bustamante RH, Castilla JC, Dugan J, Halpern BS, Lafferty KD, Leslie H, Lubchenco J, McArdle D, Ruckelshaus M, Warner RR (2003) Application of ecological criteria in selecting marine reserves and developing reserve networks. Ecol Appl 13:S215–S228
Schiel DR (1982) Selective feeding by the echinoid, Evechinus chloroticus, and the removal of plants from subtidal algal stands in northern New Zealand. Oecologia 54:379–388
Simberloff D (1998) Flagships, umbrellas, and keystones: is single-species management passé in the landscape era? Biol Conserv 83:247–257
Soto CG (2001) The potential impacts of global climate change on marine protected areas. Rev Fish Biol Fish 11:181–195
Sullivan AL, Shaffer ML (1975) Biogeography of the megazoo. Science 189:13–17
Thompson RC, Crowe TP, Hawkins SJ (2002) Rocky intertidal communities: past environmental changes, present status and predictions for the next 25 years. Environ Conserv 29:168–191
Underwood AJ (1995) Ecological research and (and research into) environmental management. Ecol Appl 5:232–247
UNESCO (1974) Final report. Task Force on Criteria and Guidelines for the Choice and Establishment of Biosphere Reserves, UNESCO Man and the Biosphere Programme, Paris
Wardle DA, Huston MA, Grime JP, Berendse F, Garnier E, Lauenroth WK, Setala H, Wilson SD (2000) Biodiversity and ecosystem function: an issue in ecology. Bull Ecol Soc Am 81:235–239
Willis TJ, Millar RB, Babcock RC (2003) Protection of exploited fish in temperate regions: high density and biomass of snapper Pagrus auratus (Sparidae) in northern New Zealand marine reserves. J Appl Ecol 40:214–227

Part V
Role of Diversity

Coordinated by Tasman P. Crowe, Heather E. Sugden, and Stephen J. Hawkins

Introduction

Tasman P. Crowe, Heather E. Sugden, and Stephen J. Hawkins

Much of this book is dedicated to patterns of community structure and the processes that cause these. In Part V, the consequences of those patterns for the functioning of the ecosystems of which the communities are a part are considered. This topic, the so-called biodiversity-ecosystem function debate, has been the focus of extensive and at times controversial research over the past decade or so. Comparatively little of this work has been done in marine ecosystems, even less on marine hard substrata. This is perhaps surprising, given the history of influential ecological research on marine hard substrata and their suitability for research in this field. More recently, marine researchers have begun to engage in addressing the issue, partly in response to growing concern about the potential loss of ecosystem goods and services to society that may result from changes in marine biodiversity.

An overview of current theory, recent findings and future directions is provided by Gamfeldt and Bracken in Chapter 26. In Chapter 27, Crowe and Russell elaborate on a specific issue that marine ecosystems may be of particular value in resolving: relative utility of taxonomic versus trait-based (i.e. 'functional') classifications of biodiversity in predicting ecosystem functioning. Chapter 28 focuses on the long-standing diversity-stability debate. In his contribution, Benedetti-Cecchi presents a new theoretical framework and an empirical test of its main predictions using data from Mediterranean rocky shores. Chapter 29, by Sugden et al., addresses the intangible, aesthetic value of rocky shores and reefs and discusses some of the ethical issues arising from our attempts to understand them.

Conclusion

While most researchers now recognise a link between biodiversity and ecosystem functioning, Gamfeldt and Bracken presented evidence from marine systems that specific species often have disproportionate effects (Chap. 26). They stressed, however, that such effects are more likely to be detected by current experimental proto-

M. Wahl (ed.), *Marine Hard Bottom Communities*, Ecological Studies 206,
DOI: 10.1007/978-3-540-92704-4_V, © Springer-Verlag Berlin Heidelberg 2009

cols than are straightforward relationships between numbers of species and ecosystem performance. In concluding, they emphasised the need to take account of large-scale variation, temporal change, realistic extinction scenarios and multiple ecosystem functions in attempting to predict effects of loss of species or genotypes.

In Chapter 27, Crowe and Russell outlined the potential value of characterising the functional roles of species to help predict consequences of biodiversity loss. Researchers have yet to agree upon the most appropriate way of classifying and quantifying functionality. A range of approaches are available and their evaluation should involve a combination of observation and experiment and modelling.

Analyses by Benedetti-Cecchi revealed the important roles of statistical averaging, overyielding and asynchronous fluctuations in abundance of individual taxa in reducing variability in the overall abundance of organisms in assemblages (Chap. 28). He concluded that loss of diversity does not necessarily imply lower stability of communities but emphasised the need to focus on dynamics of individual species and their interactions, particularly when addressing conservation challenges.

Sugden et al. described a range of ways in which humans derive enjoyment from coastal environments (Chap. 29). They concluded, however, that researchers have potential to damage marine ecosystems if their work is not conducted with considerable care. They point out the codes of practise that have emerged in some situations, to ensure that scientists engage with managers and stakeholders so that their science helps, rather than hinders the conservation of marine biodiversity and does not interfere in the enjoyment of the seashore by the general public.

In all, it is clear that ecosystems based on marine hard substrata provide a valuable opportunity to develop and test ecological ideas of wide relevance and influence. They also constitute a resource that is highly valued by society and fully deserving of careful stewardship.

Chapter 26
The Role of Biodiversity for the Functioning of Rocky Reef Communities

Lars Gamfeldt and Matthew E.S. Bracken

26.1 Introduction

A major focus of ecological research has been on understanding the processes that determine the diversity of organisms at a particular location (Hutchinson 1959). Many important classic studies elucidating these local-scale processes have been conducted in marine hard-bottom communities (e.g. Kitching and Ebling 1961; Paine 1966; Menge and Sutherland 1976; Connell 1978; Lubchenco 1978; Sousa 1979; Fletcher 1987). These studies and subsequent work have shown that diversity in rocky intertidal and subtidal systems is regulated by a combination of abiotic (e.g. disturbance, productivity, nutrient availability) and biotic (e.g. competition, consumption) processes (Worm et al. 2002; Nielsen 2003; Graham 2004; Bracken and Nielsen 2004; see Part II of this book). However, the organisms inhabiting these ecosystems are not only influenced by the performance of the systems—they also, in part, determine it. Growing evidence from a variety of systems indicates that organisms, by virtue of their roles in mediating ecosystem performance, influence the rates of biogeochemical processes, stability, and other attributes of the ecosystems in which they live (Naeem 2002; Kinzig et al. 2002; Cardinale et al. 2006).

Because of the unique roles that different organisms play in mediating ecosystem performance, there is a growing awareness that the diversity of organisms in a system can influence that system's performance (reviewed in Hooper et al. 2005). Recent studies have illustrated several key ways in which organisms inhabiting temperate and tropical reefs—and, by extension, the diversity of those organisms—influence the performance of marine ecosystems (reviewed in Worm et al. 2006). Below, we review studies from hard-bottom marine ecosystems and provide suggestions for future research into the functional roles of marine biodiversity in temperate and tropical hard-bottom communities. Accumulated evidence and theoretical reasoning suggest that the diversity of benthic communities may be an important property for the functioning of these ecosystems.

M. Wahl (ed.), *Marine Hard Bottom Communities*, Ecological Studies 206,
DOI: 10.1007/978-3-540-92704-4_26, © Springer-Verlag Berlin Heidelberg 2009

26.2 How and Why Biodiversity Can Be Linked to Ecosystem Performance

Several general mechanisms have been identified that link the diversity of organisms in an ecosystem with the performance of that system. The first, the "selection effect" or "sampling effect", arises because of the increased likelihood that a diverse community contains a species that performs well with respect to the ecosystem function being measured (Huston 1997). These species may be highly productive, in which case the selection effect results in communities with high productivity that rely on only one or a few species. Species may also dominate assemblages because they are the superior competitors even though they are not highly productive (Bruno et al. 2006). In such a case, the selection effect results in low-productivity communities. A second mechanism, complementarity, occurs when species partition limiting resources, so that resource use is enhanced when more species are present (Loreau and Hector 2001). Finally, facilitation is associated with enhanced performance due to positive interactions between species (Cardinale et al. 2002).

Research in marine ecosystems suggests that "species-identity effects"—key contributions by one or a few species—are often responsible for enhanced performance in high-diversity assemblages (O'Connor and Crowe 2005; Bruno et al. 2005, 2006; but see Byrnes et al. 2006), and a recent cross-system meta-analysis by Cardinale et al. (2006) suggests that these effects commonly link diversity and ecosystem performance in most ecosystems worldwide. Why might identity effects be important in marine systems? Two key ecological terms—the "keystone species" concept (sensu Paine 1969) and the roles of "foundation species" (sensu Dayton 1972)—were coined based on benthic marine research. These ideas highlight the importance of one or a few species (e.g. consumers or providers of habitat) in ultimately determining the functioning of hard-bottom marine ecosystems (Schiel 2006). Furthermore, the importance of density-dependence in these systems means that experiments must consider not only the number of species but also the identities of those species and their densities (O'Connor and Crowe 2005). This is especially important in experiments that manipulate diversity, as the interpretations and results of often-used replacement-series designs are critically dependent on the density of organisms (Jolliffe 2000). Benedetti-Cecchi (2004, 2006) has developed a detailed experimental and statistical framework for distinguishing between the effects of the number, identity and density of species, and these distinctions need to be incorporated into future experimental designs. It is important to note, however, that the design of most experiments hitherto has been relatively limited in both temporal and spatial scale, and researchers have often aimed to minimize environmental heterogeneity. These factors actually make studies more likely to detect species-specific effects, rather than effects of diversity per se (Cardinale et al. 2004; Ives et al. 2005). Spatial and temporal heterogeneity, and the ways in which organisms partition those spatial and temporal resources, are important mechanisms underlying diversity-function relationships (see Chap. 28 by Benedetti-Cecchi).

26.3 Roles of Species in Mediating Ecosystem Performance

The diversity of organisms can potentially influence both bottom-up (e.g. nutrient uptake and production) and top-down (e.g. grazing and predation) processes, as well as other key processes such as larval settlement and survival (Fig. 26.1). For example, when one organism consumes another organism, energy and structural and functional materials (e.g. carbon, nitrogen and phosphorus) are moved from one trophic level to the next. One important function involving consumers in benthic

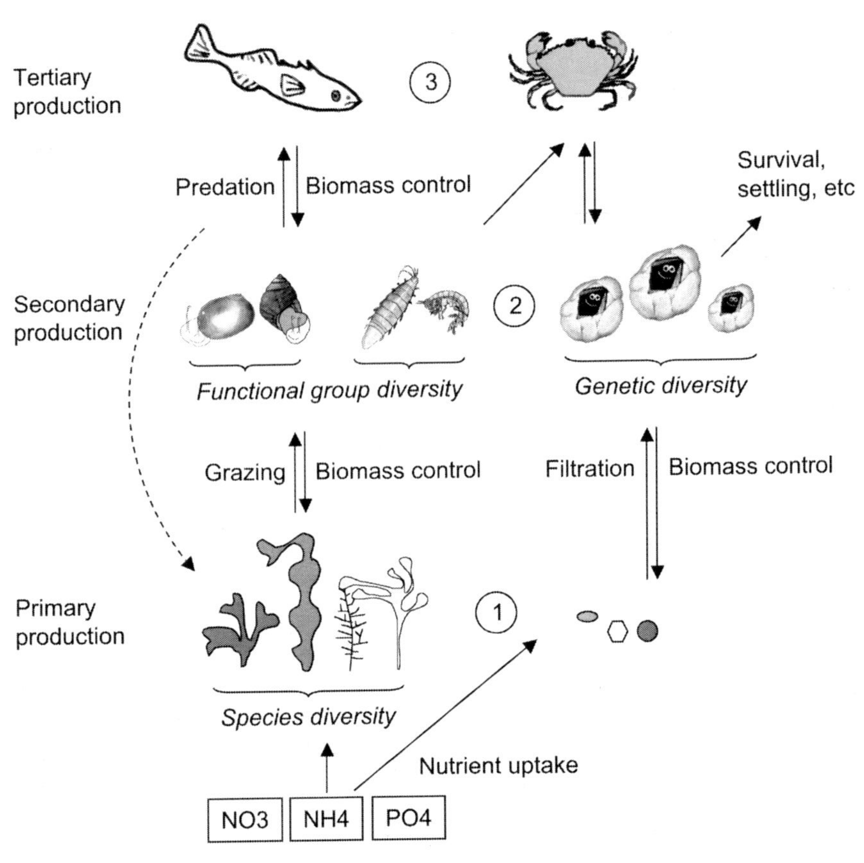

Fig. 26.1 Hypothetical benthic marine food web showing different aspects of diversity and functions. The figure presents three different levels of biodiversity (genetic, species and functional group) at three different trophic levels: *1* primary producers (macroalgae and phytoplankton), *2* primary consumers (grazers and filter feeders) and *3* predators (e.g. fish and crabs). *Arrows* indicate important ecosystem processes that can be affected by biodiversity. Arrows pointing upwards mark processes associated with energy transfer up the food web (exceptions are functions such as settlement and survival), and arrows pointing downwards relate to processes of biocontrol of biomass at lower trophic levels. The *dashed arrow* is representative of the many possible indirect interactions among species

marine systems is the role that various filter-feeding taxa play in capturing particulate organic matter (which is otherwise unavailable to many benthic taxa) and transforming it into a consumable form. The well-known role that mussels and barnacles play as basal organisms (sensu Pimm 1982) in hard-bottom ecosystems is an example of this type of benthic-pelagic coupling (e.g. Menge et al. 1997). Consumers also serve as "chemical transformers", organisms that change materials from one form into another form. For example, by consuming particulate organic nitrogen and excreting ammonium as a metabolic waste product, invertebrates mediate the supply of nitrogen to marine algae (Bracken and Nielsen 2004; Bracken et al. 2007a). Another important consequence of benthic diversity may be to limit the success of invading algae and invertebrates (e.g. Stachowicz et al. 1999). Furthermore, many benthic primary producers, especially large seaweeds (Graham 2004; Schiel 2006), also serve as foundation species (sensu Dayton 1972), providing physical structure and habitat. These "ecosystem engineers" (Jones et al. 1994), which also include bivalves (Witman 1985; Suchanek 1992), tunicates (Castilla et al. 2004), and other sessile invertebrates, modify the environment and thereby influence the functioning of marine ecosystems. The physical structure provided by foundation species commonly enhances the diversity and abundance of associated taxa (Bracken et al. 2007b), partly by expanding the realized niche of these organisms (Bruno et al. 2003).

26.4 Biodiversity and Primary Production

Most marine diversity-function research has focused on consumers, whereas most terrestrial work has been on producers, making it difficult to adequately compare the mechanisms linking diversity and performance in different systems and to come up with a general framework for understanding the consequences of declining diversity (Giller et al. 2004). This discrepancy is unfortunate, because mechanisms linking algal diversity and productivity in marine systems do not necessarily overlap with those identified in terrestrial habitats. Unlike higher plants, seaweeds have no roots, take up nutrients directly through their tissues, are surrounded by water and nutrients, rather than air, and compete mainly for space in a two-dimensional arena. Several key mechanisms often used to explain diversity-function relationships on land do not apply in marine systems. For example, in hard-bottom aquatic environments, water movement makes nutrient stratification less apparent, and there are no belowground interactions like those in the soil. Furthermore, competition for space involves not only algae but also sessile animals such as mussels, barnacles and corals.

Using a set of both outdoor mesocosm- and short-term field experiments, Bruno et al. (2005) investigated the role of macroalgal richness and identity for primary productivity in North Carolina, USA. They found that species-identity effects, i.e. dominant roles of certain species, were much more important than richness effects in terms of biomass accumulation. Bruno et al. (2006) used a similar approach for coral reef algae and obtained similar results. They concluded that both complementarity and facilitation were generally present but that their positive effects on biomass production were cancelled out by negative selection effects. These results suggest that algal species that are not highly productive are competitively superior, as would be expected if species experience trade-offs among different traits such as growth, reproduction and defence.

The experiments used by Bruno et al. (2005, 2006) were run for relatively short periods of time (2.5–5weeks), and emerging evidence suggests that the results of diversity–performance experiments are dependent on the length of time over which an experiment is maintained (Stachowicz et al. 2008a; see also Chap. 16 by Molis and da Gama). Many factors likely contribute to enhanced performance in high-diversity experimental units, and it may take relatively long periods of time (months to years) for the effects of some of those factors (e.g. seasonal buffering of stressful events, facilitation of recruitment) to become apparent. For example, in an experiment on the coast of northern California, USA, Stachowicz et al. (2008a) removed seaweeds from multi-species assemblages to create monocultures of each of the four common mid-intertidal seaweed species. They also removed an equivalent amount of biomass from polycultures containing all four of those species. They found that cover (an excellent surrogate for biomass in this system) in the four-species polycultures was higher than that in any of the component monocultures. Furthermore, polyculture cover was no different from that in unmanipulated controls, despite the biomass removed from the polycultures. However, these effects took approximately a year to become apparent. Maintaining experiments for longer periods of time under field conditions will benefit our understanding of both the consequences of declining marine diversity and the mechanisms underlying diversity-mediated changes in performance.

Why might we expect to see enhanced primary production in more diverse producer assemblages? One mechanism, often suggested in terrestrial studies (Kinzig et al. 2002), is that more diverse producer assemblages are more effective at using limiting nutrients (Fig. 26.1). However, the relationship between seaweed diversity and nutrient use remains largely unexplored. Bracken and Stachowicz (2006) evaluated the relationship between intertidal algal diversity and nutrient uptake and found that more diverse seaweed assemblages are more effective at utilizing limiting nutrients because some algal species are more effective at using nitrate as a nitrogen source, whereas others are better at using ammonium. When both nitrogen forms are present, as they often are in benthic marine systems, total nitrogen use by diverse seaweed assemblages is therefore higher than predicted based on the uptake rates of the component species. Another potential mechanism is that different seaweeds prefer different microhabitats. Some find refuge from grazing in cracks and crevices while others are unpalatable or able to withstand high grazing pressure. Some species are tolerant to desiccation whereas others require a constant and moist environment. Environmental heterogeneity might be a key factor in explaining why diverse producer assemblages experience enhanced primary production (Stachowicz et al. 2008a).

26.5 The Role of Consumer Diversity

Whereas it is widely known that consumers can have dramatic effects on algal communities, both directly (Hay 1981; Hughes and Connell 1999) and via trophic cascades (Estes et al. 1998), many of the studies that have evaluated effects of producer diversity on productivity and resource acquisition (e.g. Bruno et al. 2005, 2006; Bracken and Stachowicz 2006) have deliberately excluded grazers. Thus, while these experiments may help us understand some of the mechanisms underlying effects of producer diversity, the ultimate fate of production will depend on higher trophic levels (Paine 2002) and the numerous indirect effects that are associated with complex food webs (Menge 1995; Fig. 26.1, dashed arrow). At higher trophic levels in the food web, the chance that a species will go ecologically extinct increases (Duffy 2003; Petchey et al. 2004) and, while documented extinctions of primary producers are rare, they appear to be more common for consumer species (Duffy 2003; Byrnes et al. 2007). For example, overexploitation of fish and shellfish in the oceans has resulted in dramatic shifts in many ocean ecosystems (Dayton et al. 1998; Jackson et al. 2001; Bellwood et al. 2004), often with significant effects on vital ecosystem processes (Worm et al. 2006; Myers et al. 2007). This is likely to skew ecosystem structure on rocky bottoms, especially as species additions by invasions are dominated by species lower in the food web (Byrnes et al. 2007).

Differences in feeding structures and strategies (e.g. Steneck and Watling 1982) suggest that high diversity of consumers could facilitate efficient use of the available resource space (Duffy 2002). High consumer diversity may play roles in both top-down control and in energy transfer up the food chain, partly due to its interaction with prey diversity (Gamfeldt et al. 2005a; Fig. 26.1). Partitioning of algal resources has been shown for mesograzers inhabiting algal communities (Fletcher 1987; Lotze and Worm 2000; Råberg and Kautsky 2007). Similarly, the disappearance of both herbivorous fishes and urchins on coral reefs resulted in a shift from coral to algal dominance (Hughes and Connell 1999). Grazing by urchins and fish was important for controlling the growth of macroalgae, and the loss of both these components of diversity had dramatic consequences. In contrast, field manipulations of grazer diversity in rock pools in Ireland did not show any effects of consumer richness on the algal community level (O'Connor and Crowe 2005).

Observational data indicate that predator abundance can have strong effects on the structure of hard-bottom communities (Estes et al. 1998; Halpern et al. 2006), and that predator species richness per se may be strongly linked to functional diversity (Micheli and Halpern 2005). Laboratory experimental manipulations of the richness of predators from macroalgal communities have rendered mixed results. Using a set of crabs, Griffin et al. (2008) showed that crab diversity was important for the consumption of prey, due to feeding complementarity, but only at high predator density. Byrnes et al. (2006) showed that a more diverse predator assemblage is more effective at controlling herbivores (via a combination of behavioural changes and consumption), leading to enhanced algal production. Bruno and O'Connor (2005) also found evidence for predator richness effects but only in the absence of omnivores. In the presence of omnivorous fish (which also consumed the algae in their experiments), there were no diversity effects. Together with results from terrestrial environments, results in marine systems seem to show that predator diversity can have important effects but the exact form of these effects depends strongly on feeding preferences (Duffy et al. 2007). One gap in our understanding of the role of consumer diversity concerns the potential link between consumer diversity and consumer productivity. Most work has focused only on consumer effects on their prey (but see Fox 2004; Gamfeldt et al. 2005a). This is unfortunate, since the production of consumer biomass is an important ecosystem process that ultimately drives fisheries production.

26.6 The Role of Within-Species Diversity

Many marine ecosystems like seagrass meadows, kelp forests, salt marshes and deep-sea coral reefs are dominated by, or dependent on, one or a few species (see our discussion of foundation species above). Even though these dominant species are unlikely to go globally extinct, habitat destruction and fragmentation, as well as local population declines are predicted to alter the genetic structure of populations through processes such as genetic drift, inbreeding and reduced gene flow among populations (although this is not always the case; see Young et al. 1996). This may reduce the ability of species to adapt to changes in the environment (Frankham 2005). Furthermore, whereas consumer species high in the food web are not numerically dominant, they can play disproportional roles in regulating community structure. These consumers may decline in numbers as an effect of human exploitation, release of toxic chemicals, and habitat destruction (Jackson et al. 2001; Myers et al. 2007). Changes in the intraspecific diversity of these species are likely to influence ecosystem processes.

However, whereas the importance of individual species for ecosystems is widely agreed upon, there has been little focus on the importance of altered changes in within-species diversity for ecosystem processes. There is evidence for important effects of within-species diversity in both terrestrial (Crutsinger et al. 2006) and marine (Hughes and Stachowicz 2004; Reusch et al. 2005) angiosperms. Increasing genetic diversity of these plants enhanced productivity or their resistance to stressors such as consumption and high water temperatures. For Pacific sockeye salmon, the diversity of different populations is critical for sustaining total salmon production (Hilborn et al. 2003). Only two studies, however, have examined hard-bottom organisms. Gamfeldt et al. (2005b) found that within-species diversity of the barnacle *Balanus improvisus* Darwin was important for larval settlement, and Gamfeldt and Källström (2007) showed that increasing the diversity of amphipod populations enhanced predictability in population survival (Fig. 26.1). Even though current evidence is scarce, these two experiments, together with a recent study on honey-bee colonies elegantly showing that genotypic diversity can be important for bee colony productivity (Mattila and Seeley 2007), suggest that within-species diversity of animals can influence community structuring and ecosystem functioning. Furthermore, since the results of experiments on both producers and consumers are similar, the importance of genetic diversity for ecosystem functioning may be a general phenomenon of ecosystems.

26.7 Conclusions and Outlook

Dominant genotypes and species have often been identified as the most important drivers of community and ecosystem performance (Cardinale et al. 2006), especially in marine ecosystems (O'Connor and Crowe 2005; Bruno et al. 2005, 2006). As described in this chapter, however, diversity per se is often an important factor as well. Due to logistical constraints, both laboratory and field manipulations of biodiversity have been limited in scale (both temporal and spatial) and heterogeneity. This has probably biased experiments to detect species-identity effects (Ives et al. 2005; Stachowicz et al. 2008b). Both theory and more recent experimental results suggest that increasing our scales of inference will make effects of diversity more prominent (Cardinale et al. 2004; Stachowicz et al. 2008a). An interesting and as yet unexamined aspect of diversity-function research is the incorporation of habitat heterogeneity and landscape perspectives (Bengtsson et al. 2002). Just as diversity of species may be important, so may be a diversity of different habitats. Furthermore, most studies have hitherto looked only at individual functions. Emerging evidence suggests that different species maximize different ecosystem functions (Hooper and Vitousek 1998; Duffy et al. 2003; Bracken and Stachowicz 2006; Gamfeldt et al. 2008). This "multivariate complementarity" means that even when species-identity effects are more important than complementarity for any one performance measure, diversity can enhance overall performance when multiple functions are considered simultaneously. Future work should explicitly evaluate multiple ecosystem functions simultaneously.

An important task of future research is to explore the interactive effects of diversity loss across multiple trophic levels. All benthic habitats contain multiple trophic levels and only by studying the effects of diversity loss of one level on other levels can we gain a more comprehensive understanding of the consequences for whole systems. Recent evidence from experiments shows clearly that the effects of diversity loss of both algae and grazers are dependent on the presence and diversity of adjacent trophic levels (Gamfeldt et al. 2005a; Bruno et al. 2008; Douglass et al. 2008).

Whereas most experiments linking diversity and performance have used randomly selected species from a region to construct gradients of diversity, we know that diversity does not change randomly in natural ecosystems. Instead, benthic marine diversity is influenced by a variety of factors, including consumers (Paine 1966; Lubchenco 1978), disturbance (Sousa 1979) and nutrient availability (Bracken and Nielsen 2004). Recent work evaluating the consequences of non-random diversity changes (Solan et al. 2004)—e.g. due to overexploitation of resources by humans (Myers et al. 2007)—and explicitly comparing these to random diversity gradients (e.g. Bracken et al. 2008) is improving the realism of our assessments of the consequences of declines in marine biodiversity.

Because many species in marine systems are broadcast spawners with far-dispersing larvae, and because many of the consumers in marine habitats are highly mobile, it is likely that the conclusions of diversity-function research—much of which has, to date, been conducted in closed mesocosms—might change when experiments are carried out under more "open" field conditions (Giller et al. 2004). Results from marine microcosm studies that have included connectivity between local patches in their experimental designs have found that migration of small grazers have quite strong effects on grazing rates and the structure of algal communities (France and Duffy 2006; Matthiessen et al. 2007). Migration of species is an important and ubiquitous feature of marine systems, and it is reasonable to hypothesize that migration of both grazers and highly mobile predators will have notable effects on local (patch) dynamics and community structure.

Given the rate of anthropogenic global change, and the potential effects of those changes on species' interactions in marine habitats (e.g. Sanford 1999), the roles of species in mediating community and ecosystem performance may change radically over the next century. The marine environment is likely to change both locally and globally due to factors such as environmental warming and acidification (see Part IV of this book). This means that today's winners may be tomorrow's losers. Such scenarios call for precautionary actions in terms of protecting biodiversity. Models and experiments that incorporate global-change scenarios will make diversity-function research a more predictive field and will help us understand the future consequences of changing biodiversity.

Acknowledgements Many thanks to Jarrett Byrnes for comments that helped to improve this chapter. LG was supported by a postdoctoral grant from the Swedish Research Council Formas (2006-1173), and MESB was supported by the National Science Foundation (OCE-0549944 to Susan Williams and MESB).

References

Bellwood DR, Hughes TP, Folke C, Nyström M (2004) Confronting the coral reef crisis. Nature 429:827–833

Benedetti-Cecchi L (2004) Increasing accuracy of causal inference in experimental analyses of biodiversity. Funct Ecol 18:761–768

Benedetti-Cecchi L (2006) Understanding the consequences of changing biodiversity on rocky shores: how much have we learned from past experiments? J Exp Mar Biol Ecol 338:193–204

Bengtsson J, Engelhardt K, Giller PS, Hobbie S, Lawrence D, Levine JM, Vilá M, Wolters V (2002) Slippin' and slidin' between the scales: the scaling components of biodiversity-ecosystem functioning relations. In: Loreau M, Naeem S, Inchausti P (eds) Biodiversity and ecosystem functioning synthesis and perspectives. Oxford University Press, Oxford, pp 209–220

Bracken MES, Nielsen KJ (2004) Diversity of intertidal macroalgae increases with nutrient loading by invertebrates. Ecology 85:2828–2836

Bracken MES, Stachowicz JJ (2006) Seaweed diversity enhances nitrogen uptake via complementary use of nitrate and ammonium. Ecology 87:2397–2403

Bracken MES, Gonzalez-Dorantes CA, Stachowicz JJ (2007a) Whole-community mutualism: associated invertebrates facilitate a dominant habitat-forming seaweed. Ecology 88:2211–2219

Bracken MES, Rogers-Bennett L, Bracken BE (2007b) Species diversity and foundation species: potential indicators of fisheries yields and marine ecosystem functioning. Calif Coop Oceanic Fish Invest Rep 48:82–91

Bracken MES, Friberg SE, Gonzalez-Dorantes CA, Williams SL (2008) Functional consequences of realistic biodiversity changes in a marine ecosystem. Proc Natl Acad Sci USA 105:924–928

Bruno JF, O'Connor MI (2005) Cascading effects of predator diversity and omnivory in a marine food web. Ecol Lett 8:1048–1056

Bruno JF, Stachowicz JJ, Bertness MD (2003) Inclusion of facilitation into ecological theory. Trends Ecol Evol 18:119–125

Bruno JF, Boyer KE, Duffy JE, Lee SC, Kertesz JS (2005) Effects of macroalgal species identity and richness on primary production in benthic marine communities. Ecol Lett 8:1165–1174

Bruno JF, Lee SC, Kertesz JS, Carpenter RC, Long ZT, Emmett Duffy J (2006) Partitioning the effects of algal species identity and richness on benthic marine primary production. Oikos 115:170–178

Bruno JF, Boyer KE, Duffy JE, Lee SC (2008) Relative and interactive effects of plant and grazer richness in a benthic marine community. Ecology 89:2518–2528

Byrnes JE, Stachowicz JJ, Hultgren KM, Randall Hughes A, Olyarnik SV, Thornber CS (2006) Predator diversity strengthens trophic cascades in kelp forests by modifying herbivore behaviour. Ecol Lett 9:61–71

Byrnes JE, Reynolds PL, Stachowicz JJ (2007) Invasions and extinctions reshape coastal marine food webs. PLoS One 2:e295

Cardinale BJ, Palmer MA, Collins SL (2002) Species diversity enhances ecosystem functioning through interspecific facilitation. Nature 415:426–429

Cardinale BJ, Ives AR, Inchausti P (2004) Effects of species diversity on the primary productivity of ecosystems: extending our spatial and temporal scales of inference. Oikos 104:437–450

Cardinale BJ, Srivastava DS, Emmett Duffy J, Wright JP, Downing AL, Sankaran M, Jouseau C (2006) Effects of biodiversity on the functioning of trophic groups and ecosystems. Nature 443:989–992

Castilla JC, Lagos NA, Cerda M (2004) Marine ecosystem engineering by the alien ascidian *Pyura praeputialis* on a mid-intertidal rocky shore. Mar Ecol Prog Ser 268:119–130

Connell JH (1978) Diversity in tropical rain forests and coral reefs—high diversity of trees and corals is maintained only in a non-equilibrium state. Science 199:1302–1310

Crutsinger GM, Collins MD, Fordyce JA, Gompert Z, Nice CC, Sanders NJ (2006) Plant genotypic diversity predicts community structure and governs an ecosystem process. Science 313:966–968

Dayton PK (1972) Toward an understanding of community resilience and the potential effects of enrichment to the benthos at McMurdo Sound, Antarctica. In: Parker BC (ed) Proc Colloquium Conservation Problems in Antarctica. Allen Press, Lawrence, KA, pp 81–96

Dayton PK, Tegner MJ, Edwards PB, Riser KL (1998) Sliding baselines, ghosts, and reduced expectations in kelp forest communities. Ecol Appl 8:309–322

Douglass JG, Duffy JE, Bruno JF (2008) Herbivore and predator diversity interactively affect ecosystem properties in an experimental marine community. Ecol Lett 11:598–608

Duffy JE (2002) Biodiversity and ecosystem function: the consumer connection. Oikos 99:201–219

Duffy JE (2003) Biodiversity loss, trophic skew and ecosystem functioning. Ecol Lett 6:680–687

Duffy JE, Richardson JP, Canuel EA (2003) Grazer diversity effects on ecosystem functioning in seagrass beds. Ecol Lett 6:637–645

Duffy JE, Cardinale BJ, France KE, McIntyre PB, Thebault E, Loreau M (2007) The functional role of biodiversity in ecosystems: incorporating trophic complexity. Ecol Lett 10:522–538

Estes JA, Tinker MT, Williams TM, Doak DF (1998) Killer whale predation on sea otters linking oceanic and nearshore ecosystems. Science 282:473–476

Fletcher WJ (1987) Interactions among subtidal Australian sea-urchins, gastropods, and algae—effects of experimental removals. Ecol Monogr 57:89–109

Fox JW (2004) Effects of algal and herbivore diversity on the partitioning of biomass within and among trophic levels. Ecology 85:549–559
France KE, Duffy JE (2006) Diversity and dispersal interactively affect predictability of ecosystem function. Nature 441:1139–1143
Frankham R (2005) Conservation biology—ecosystem recovery enhanced by genotypic diversity. Heredity 95:183–183
Gamfeldt L, Källström B (2007) Increasing intraspecific diversity increases predictability in population survival in the face of perturbations. Oikos 116:700–705
Gamfeldt L, Hillebrand H, Jonsson PR (2005a) Species richness changes across two trophic levels simultaneously affect prey and consumer biomass. Ecol Lett 8:696–703
Gamfeldt L, Wallén J, Jonsson PR, Berntsson K, Havenhand J (2005b) Intraspecific diversity enhances settling success in a marine invertebrate. Ecology 86:3219–3224
Gamfeldt L, Hillebrand H, Jonsson PR (2008) Multiple functions increase the importance of biodiversity for overall ecosystem functioning. Ecology 89:1223–1231
Giller PS, Hillebrand H, Berninger U-G, Gessner MO, Hawkins S, Inchausti P, Inglis C, Leslie H, Malmqvist B, Monaghan MT, Morin PJ, O'Mullan G (2004) Biodiversity effects on ecosystem functioning: emerging issues and their experimental test in aquatic environments. Oikos 104:423–436
Graham MH (2004) Effects of local deforestation on the diversity and structure of Southern California giant kelp forest food webs. Ecosystems 7:341–357
Griffin J, de la Haye K, Hawkins S, Thompson R, Jenkins S (2008) Predator diversity effects and ecosystem functioning: density modifies the effect of resource partitioning. Ecology 89:298–305
Halpern BS, Cottenie K, Broitman BR (2006) Strong top-down control in Southern California kelp forest ecosystems. Science 312:1230–1232
Hay ME (1981) Herbivory, algal distribution, and the maintenance of between-habitat diversity on a tropical fringing reef. Am Nat 118:520–540
Hilborn R, Quinn TP, Schindler DE, Rogers DE (2003) Biocomplexity and fisheries sustainability. Proc Natl Acad Sci USA 100:6564–6568
Hooper DU, Vitousek PM (1998) Effects of plant composition and diversity on nutrient cycling. Ecol Monogr 68:121–149
Hooper DU, Chapin FS, Ewel JJ, Hector A, Inchausti P, Lavorel S, Lawton JH, Lodge DM, Loreau M, Naeem S, Schmid B, Setälä H, Symstad AJ, Vandermeer J, Wardle DA (2005) Effects of biodiversity on ecosystem functioning: a concensus of current knowledge. Ecol Monogr 75:3–35
Hughes TP, Connell JH (1999) Multiple stressors on coral reefs: a long term perspective. Limnol Oceanogr 44:932–940
Hughes AR, Stachowicz JJ (2004) Genetic diversity enhances the resistance of a seagrass ecosystem to disturbance. Proc Natl Acad Sci USA 101:8998–9002
Huston MA (1997) Hidden treatments in ecological experiments: re-evaluating the ecosystem function of biodiversity. Oecologia 110:449–460
Hutchinson GE (1959) Homage to Santa Rosalia or why are there so many kinds of animals? Am Nat 93:145–159
Ives AR, Cardinale BJ, Snyder WE (2005) A synthesis of subdisciplines: predator-prey interactions, and biodiversity and ecosystem functioning. Ecol Lett 8:102–116
Jackson JBC, Kirby MX, Berger WH, Bjorndal KA, Botsford LW, Bourque BJ, Bradbury RH, Cooke R, Erlandson J, Estes JA, Hughes TP, Kidwell S, Lange CB, Lenihan HS, Pandolfi JM, Peterson CH, Steneck RS, Tegner MJ, Warner RR (2001) Historical overfishing and the recent collapse of coastal ecosystems. Science 293:629–638
Jolliffe PA (2000) The replacement series. J Ecol 88:371–385
Jones CG, Lawton JH, Shachak M (1994) Organisms as ecosystem engineers. Oikos 69:373–386
Kinzig AP, Pacala SW, Tilman D (2002) The functional consequences of biodiversity. *Princeton* University Press, Princeton, NJ
Kitching JA, Ebling FJ (1961) The ecology of Lough Ine: XI. The control of algae by Paracentrotus *lividus* (Echinoidea). J Animal Ecol 30:373–383
Loreau M, Hector A (2001) Partitioning selection and complementarity in biodiversity experiments. Nature 412:72–75

Lotze HK, Worm B (2000) Variable and complementary effects of herbivores on different life stages of bloom-forming macroalgae. Mar Ecol Prog Ser 200:167–175
Lubchenco J (1978) Plant species diversity in a marine intertidal community: importance of herbivore food preference and algal competitive abilities. Am Nat 112:23–39
Matthiessen B, Gamfeldt L, Jonsson PR, Hillebrand H (2007) Effects of grazer richness and composition on algal biomass in a closed and open marine system. Ecology 88:178–187
Mattila HR, Seeley TD (2007) Genetic diversity in honey bee colonies enhances productivity and fitness. Science 317:362–364
Menge BA (1995) Indirect effects in marine rocky intertidal interaction webs: patterns and importance. Ecol Monogr 65:21–74
Menge BA, Sutherland JP (1976) Species-diversity gradients—synthesis of roles of predation, competition, and temporal heterogeneity. Am Nat 110:351–369
Menge BA, Daley BA, Wheeler PA, Dahlhoff E, Sanford E, Strub PT (1997) Benthic-pelagic links and rocky intertidal communities: bottom-up effects on top-down control? Proc Natl Acad Sci USA 94:14530–14535
Micheli F, Halpern BS (2005) Low functional redundancy in coastal marine assemblages. Ecol Lett 8:391–400
Myers RA, Baum JK, Shepherd TD, Powers SP, Peterson CH (2007) Cascading effects of the loss of apex predatory sharks from a coastal ocean. Science 315:1846–1850
Naeem S (2002) Ecosystem consequences of biodiversity loss: the evolution of a paradigm. Ecology 83:1537–1552
Nielsen KJ (2003) Nutrient loading and consumers: agents of change in open-coast macrophyte assemblages. Proc Natl Acad Sci USA 100:7660–7665
O'Connor NE, Crowe TP (2005) Biodiversity loss and ecosystem functioning: distinguishing between number and identity of species. Ecology 86:1783–1796
Paine RT (1966) Food web complexity and species diversity. Am Nat 100:65–75
Paine RT (1969) A note on trophic complexity and species diversity. Am Nat 103:91–93
Paine RT (2002) Trophic control of production in a rocky intertidal community. Science 296:736–739
Petchey OL, Downing AL, Mittelbach GG, Persson L, Steiner CF, Warren PH, Woodward G (2004) Species loss and the structure and functioning of multitrophic aquatic systems. Oikos 104:467–478
Pimm SL (1982) Food webs. Chapman and Hall, London
Råberg S, Kautsky L (2007) Consumers affect prey biomass and diversity through resource partitioning. Ecology 88:2468–2473
Reusch TBH, Ehlers A, Hammerli A, Worm B (2005) Ecosystem recovery after climatic extremes enhanced by genotypic diversity. Proc Natl Acad Sci USA 102:2826–2831
Sanford E (1999) Regulation of keystone predation by small changes in ocean temperature. Science 283:2095–2097
Schiel DR (2006) Rivets or bolts? When single species count in the function of temperate rocky reef communities. J Exp Mar Biol Ecol 338:233–252
Solan M, Cardinale BJ, Downing AL, Engelhardt KAM, Ruesink JL, Srivastava DS (2004) Extinction and ecosystem function in the marine benthos. Science 306:1177–1180
Sousa WP (1979) Disturbance in marine intertidal boulder fields: the nonequilibrium maintenance of species diversity. Ecology 60:1225–1239
Stachowicz JJ, Whitlatch RB, Osman RW (1999) Species diversity and invasion resistance in a marine ecosystem. Science 286:1577–1579
Stachowicz JJ, Graham M, Bracken MES, Szoboszlai AI (2008a) Diversity enhances cover and stability of seaweed assemblages: the importance of environmental heterogeneity and experimental duration. Ecology 89:3008–3019
Stachowicz JJ, Best RJ, Bracken MES, Graham MH (2008b) Complementarity in marine biodiversity manipulations: reconciling divergent evidence from field and mesocosm experiments. Proc Natl Acad Sci USA 105:18842–18847

Steneck RS, Watling L (1982) Feeding capabilities and limitation of herbivorous mollusks—a functional-group approach. Mar Biol 68:299–319

Suchanek TH (1992) Extreme biodiversity in the marine-environment—mussel bed communities of *Mytilus californianus*. Northw Environ J 8:150–152

Witman JD (1985) Refuges, biological disturbance, and rocky subtidal community structure in New England. Ecol Monogr 55:421–445

Worm B, Lotze HK, Hillebrand H, Sommer U (2002) Consumer versus resource control of species diversity and ecosystem functioning. Nature 417:848–851

Worm B, Barbier EB, Beaumont N, Duffy JE, Folke C, Halpern BS, Jackson JBC, Lotze HK, Micheli F, Palumbi SR, Sala E, Selkoe KA, Stachowicz JJ, Watson R (2006) Impacts of biodiversity loss on ocean ecosystem services. Science 314:787–790

Young A, Boyle T, Brown T (1996) The population genetic consequences of habitat fragmentation for plants. Trends Ecol Evol 11:413–418

Chapter 27
Functional and Taxonomic Perspectives of Marine Biodiversity: Functional Diversity and Ecosystem Processes

Tasman P. Crowe and Roly Russell

27.1 Introduction

The importance of biological diversity for the functioning of ecosystems has become the focus of a substantial body of literature, drawing on research in a wide variety of ecosystems. As discussed by Gamfeldt and Bracken (Chap. 26), much debate still surrounds the topic, yet some general conclusions are emerging. Here, we turn our attention to how this relationship changes given classifications of diversity based on functional differences, rather than only taxonomic differences. Little empirical research on this topic has been done in marine hard bottom communities relative to terrestrial communities, especially grasslands (Emmerson and Huxham 2002; Hooper et al. 2005; Stachowicz et al. 2007). This is perhaps surprising, given the history of influential ecological research on marine hard substrata (Connell 1974; Paine 1977). It is even more surprising when one considers the remarkable functional and phylogenetic diversity of organisms occurring on hard substrata in marine environments. Although terrestrial ecosystems often have larger numbers of species (particularly arthropods, especially beetles), marine systems have far higher diversity at higher taxonomic levels (May 1994; Vincent and Clarke 1995; Ormond 1996). In particular, 28 phyla of animals occur in the sea, 13 exclusively so, compared to only 11 that occur on land, only one exclusively so (Ray and Grassle 1991). This makes marine hard substrata of potentially great utility in addressing an outstanding and emerging issue in the biodiversity-ecosystem function (BEF) debate: the relative utility of taxonomic versus trait-based (i.e. 'functional') classifications of biodiversity. In this chapter, we will explore the complexities of defining these different types of biological diversity, and go on to discuss the relative utility and relationships between them, focusing on examples from marine hard substrata.

27.2 Defining Diversity

Before discussing the implications of functional diversity, we must explicitly define biological diversity. We adopt the definition used by the Convention on Biological Diversity (CBD 1992), which we paraphrase to "the variety of biological form and

M. Wahl (ed.), *Marine Hard Bottom Communities*, Ecological Studies 206,
DOI: 10.1007/978-3-540-92704-4_27, © Springer-Verlag Berlin Heidelberg 2009

function". Given the cumbersome breadth of this definition, biodiversity is widely conceptualised at three levels of organisation for operational purposes: genetic, organismal and ecosystem (or habitat) level (CBD 1992). Genetic diversity generally arises within populations of a single species, within which this 'variety in form and function' involves differences such as unique alleles or different levels of heterozygosity. A more genetically diverse population could have greater diversity at any level of genetic organisation: anything from nucleic acid differences in base sequences (even non-coding sequences) to unique protein metabolism. Organismal diversity encompasses a wide range of classifications and will be the main focus of this chapter. It includes the most intuitive level of diversity—that at the species level—but also encompasses higher taxomonic levels such as classes and phyla, includes subspecies differences such as populations or ontogenetic shifts, and also includes classifications into functional groups and trophic levels. Organismal diversity thus refers to groups of individual organisms (generally of different species) that are diverse in terms of form or function: Lake Victoria cichlids and Great Barrier Reef coral ecosystems would both be examples of organismally diverse systems both in a taxonomic (e.g. number of species) and functional sense (e.g. the number of functions performed by different types of fish). Habitat-level diversity refers to variety of habitat types in an area. Five km^2 of abyssal plain, for example, would be considered less diverse in this respect than 5 km^2 of the coastal zone, which might encompass rocky reefs, sandy beaches and mudflats in both the intertidal and subtidal—a greater variety in habitat form and function.

In part because of the relative ease of empirical operationalisation at a species level, rather than other levels, the vast majority of biodiversity-ecosystem function research has focused on species-level diversity (see Chap. 26 by Gamfeldt and Bracken). It has emerged from this research, however, that the number of species in an ecosystem is often less important to its functioning than are their identities or the numbers of functional groups, particularly in marine systems (e.g. Gamfeldt and Bracken, Chap. 26; Stachowicz et al. 2007)—that is to say, in many cases a few species play key functional roles and their addition or loss has much greater influence than that of other species (e.g. Solan et al. 2006). This has led to the suggestion that understanding the role of diversity in ecosystem processes and predicting effects of loss of species on ecosystem functioning may best be achieved by taking account of their functional roles and some indication of the functional diversity of a system.

Given our desire to study the roles of biological 'form and function', we must derive some methods of quantifying diversity. At each level and each 'mode' of diversity, there is a balance between predictive power, convenience and generality that produces a useful measure of diversity. At the most fundamental organismal level, diversity is manifest in species identities (via the variation in form and function encapsulated in the variation that has evolved between species) but is most often quantified by proxy via the number of species (species richness). Likewise, variation in traits is a key element of biological diversity but there is an essentially infinite number of potential characteristics of organisms that could feasibly be identified as relevant traits. Thus, a decision must be made—at least in some part a subjective decision—to determine what traits to include in such an analysis—once again, based on a balance of predictive power, convenience and generality. Given the necessity for making functional diversity operational and the complexities in doing so, in the following section we turn our attention to this challenge and approaches adopted by various schemes to identify functional diversity.

27.3 Operational Characterisation of Functional Diversity

Examinations of the role of functional diversity are hampered by the lack of a universal system for classifying functional groups or guilds (Simberloff and Dayan 1991; Sullivan and Zedler 1999; Díaz and Cabido 2001; Lavorel and Garnier 2001; Blondel 2003, Wright et al. 2006). Effective operationalisation of functional diversity would be of benefit for two primary reasons. Firstly, functional diversity ought to be a more effective predictor of ecosystem functioning than is taxonomic diversity; an effective classification scheme could be a cost-effective way of predicting effects of loss (or restoration) of particular taxa on ecosystem functioning and could have valuable applications in conservation and management (Micheli and Halpern 2005). Secondly, functional classifications can enable meaningful comparisons of the roles of biodiversity in different ecosystems as they transcend taxonomic differences (Micheli and Halpern 2005; Bremner et al. 2006). Functional classification can also improve mechanistic understanding of community assembly (Micheli and Halpern 2005). In 1972, Robert MacArthur predicted that research statements such as 'for organisms of type A in environment of structure B, such and such relationships will hold' would capture most of the creative energy of future ecologists: effective operationalisation of functional diversity would clarify the *types* of organisms that are pertinent to community ecology.

In broad terms, functional classifications may include aspects of demography (body size, longevity, etc.), feeding strategies, morphology (e.g. Steneck and Dethier 1994), responses to environmental or resources changes such as disturbance (Lavorel et al. 1997; McIntyre et al. 1999; Diaz et al. 2001; Lavorel and Garnier 2002), adaptive strategies, or effects on ecosystem function through biogeochemical cycles (Lavorel and Garnier 2001). Naeem and Wright (2003) emphasise the importance of distinguishing between traits that define a species' responses to environmental drivers and those that define its functionality in a system. Species can be grossly divided on the basis of trophic levels (primary producer, primary consumer, decomposer, etc.) or other broad roles, such as autogenic or allogenic ecosystem engineers (Jones et al. 1994). Within each broad category, species can be further subdivided into functional groups on the basis of morphological, dietary or other characteristics. For example, intertidal primary producers can be further divided into microalgae, crustose algae, foliose algae, articulated calcareous algae, etc. (Dring 1982; Steneck and Dethier 1994; Arenas et al. 2006).

These groupings may produce analytical problems relative to classically used taxonomic divisions, given issues such as species that perform multiple functions (and thus belong to multiple functional groups) or species that temporally switch functional roles (e.g. the abalone *Haliotis roei* sometimes feeds on drift algae and sometimes grazes; Scheibling 1994).

27.3.1 Trophic Position

Trophic levels have been singled out by a number of authors as a key criterion for classification, with a number of calls for research that explicitly recognises differences in consequences of losing species from a single trophic level versus multiple trophic levels (Rafaelli et al. 2002; Duffy et al. 2007). Indeed, most distinctions of functional differences are relevant only once trophic position has been specified. Duffy et al. (2007) advocate a two-tier classification, distinguishing between diversity within trophic levels ('horizontal diversity') and diversity between them ('vertical diversity'). The identity of the trophic level from which species are lost can also strongly influence the consequences of their loss. It has been argued that higher trophic levels are generally less diverse than lower trophic levels (i.e. contain less redundancy) and also contain species that are inherently vulnerable to extinction (Dobson et al. 2006). Loss of species from higher trophic levels may therefore be more likely and more significant than loss of species from lower trophic levels (see also Coleman and Williams 2002).

As would be predicted, trophic position has been demonstrated to not be a reliable independent indicator of the influence of a species on a community, in that some species have inordinate effects on system processes when compared to other species in the same trophic position (e.g. Chalcraft and Resetarits 2003; Bellwood et al. 2003). Thus, although trophic position is a fundamental and informative component of functional grouping systems, it is not sufficiently predictive for most purposes.

27.3.2 Ad-hoc Groupings Based on Individual Characteristics

Simberloff and Dayan (1991) reviewed many aspects of the techniques used to classify organisms into different guilds, noting that most researchers do not use quantitative methods for classification but, rather, "taxonomy plus intuition". Indeed, Root's seminal work on the topic immediately recognised that guilds (equally applicable to functional groups in this chapter) were subjective: "As with the genus in taxonomy, the limits that circumscribe the membership of any guild must be somewhat arbitrary" (Root 1967, p. 335, quoted in Simberloff and Dayan 1991). For marine rocky bottom species, two such ad-hoc grouping systems have been commonly utilized to group taxa at a resolution different from that of species. For marine producers, there are long-standing attempts to develop a 'functional grouping' classification that had predictive power for organismal responses to environmental gradients (e.g. Littler and Littler 1980). Likewise, for marine consumer communities, people have utilized feeding-mode type groupings to refine a grouping system that is more ecologically informative than taxonomic affiliation alone (e.g. Steneck and Watling 1982; Micheli and Halpern 2005).

For marine algae, morphological features form the backbone of most historical systems to group these species: Littler and Littler (1980) produced a system of classifying algae into functional groups, which was further refined by Steneck and Dethier (1994) to identify the following groupings: (1) single cell microalgae, (2) filamentous algae, (3) foliose algae, (3.5) corticated foliose algae, (4) corticated macrophytes, (5) leathery macrophytes, (6) articulated calcareous algae and (7) crustose algae. These groupings have been shown to correspond well with some particular ecophysiological responses (for example, response to herbivory: Steneck and Watling 1982, or to dessication: Ji and Tanaka 2002; Johansson and Snoeijs 2002) but it is also clear that these grouping schemes, based on morphological characters alone, are not a panacea for the troubles of functional grouping systems. Padilla and Allen (2000) call for a revision of our functional-form classification systems of macroalgae, in preference for classification systems in which functional groups are uniquely defined based on the particular function to be predicted, rather than morphological form for attempted broad ecological utility.

Marine consumers have been divided into functional groups based on feeding mode—a readily apparent and important characteristic of the ecological niche of these taxa. Distinctions between groups such as grazers, filter feeders and deposit feeders have seen much attention in the ecological literature for many years, and are often used to give some ecologically relevant grouping scheme to communities of consumers.

27.3.3 Classifications Based on Multiple Traits

Plants and algae vary substantially in their contribution to ecosystem functioning, both as primary producers and as structural elements. Although these contributions are the result of a large number of character gradients between species, plant biologists have found that multivariate analyses of a small number of readily measurable morphological traits or indices derived from such traits capture a great deal of information about functionality. Seed size, plant height and specific leaf area were singled out by Westoby (1998) and have been repeatedly shown to correspond to plant functioning under different environmental conditions, and responses to disturbance and to biotic interactions (Lavorel and Garnier 2001).

Investigations of unique ecosystem functions may best be served by unique classifications of functionality (as recommended by Padilla and Allen 2000 for marine algal groupings). Sullivan and Zedler (1999) found that a classification of saltmarsh plants based on morphological and metabolic criteria did not predict a classification subsequently derived from experimental evidence of a range of ecosystem functions of the species. Although the loss of any given species may not have an effect on any one ecosystem function (e.g. productivity), this does not necessarily imply that it is not critical for other functions (e.g. nutrient retention). Ideally, therefore, it would be of value to consider more than one function at a time, an approach currently lacking in much of the relevant research (but see Naeem et al. 1994; Zedler et al. 2001).

A promising way forward is offered by biological traits analysis, which captures all biological information relevant to a wide range of functions (Bremner et al. 2003, 2006). Traits selected by Bremner et al. (2003) include individual/colony size, adult longevity, reproductive technique, relative adult mobility, body form and feeding habit. Each of these traits is assigned a number of subcategories, based on the taxa present. For example, adult longevity could have three categories: <2 years, 2–5 years and >5 years; adult life habit could have four categories: sessile, swim, crawl, crevice dweller. The use of a "fuzzy coding" scheme effectively solves the problem of species varying their functioning under different circumstances (Chevenet et al. 1994; Bremner et al. 2003). A score of 0–3 can be assigned for each trait by a particular species and this score can be subdivided among categories if the species varies in its expression of that trait. For example, if *H. roei* (see above) feeds on drift algae <30% of the time and grazes for the remainder, it would score 1 for drift feeding, 2 for grazing and 0 for any other feeding strategies in the classification scheme. Biological traits analysis characterises all relevant traits in all taxa present and expresses them as a matrix of functionality in which values assigned for particular taxa are weighted by their abundance. Such matrices can be analysed using exactly the same methods as multivariate matrices of species' abundances or biomass, and are sensitive to functionally relevant changes in assemblage structure (Bremner et al. 2003). Biological traits composition has been shown to be more stable than taxonomic composition and is responsive to human impacts (Bremner et al. 2003), enabling the possible functional consequences of assemblage-level impacts to be characterised.

27.3.4 Generalisable Quantifications of Functional/Trait Diversity

There are two primary drawbacks of the multidimensional *grouping* methods. First, most of these methods impose a lack of generality—that is, when one identifies certain functional groups based on a particular suite of traits, the final product is a list of functional groups particular to that location, time and system. Once a different context is approached with differing types of important functions, new focal species and identification of different 'key' traits will result in identification of quite different functional groupings; in essence, functional groupings are often too specific to a particular function or too general to be of significant ecological insight (cf. Padilla and Allen 2000). Second, part of the concern with most functional grouping methods is that community functional diversity involves not only counting of the number of different types present but a sense of the potential trait-space occupied by the community as well, and few measures have been proposed to adequately incorporate this 'volumetric' aspect of functional diversity. There have recently been many attempts to increase the generality of these methods, to produce a metric proxy of functional diversity that is continuous, robust to analytic or systemic changes, and more similar to the convenience and generality of species richness (and related diversity indices), the proxy for species diversity.

In attempts to overcome these problems, Petchey and Gaston (2002) proposed a method of calculating functional diversity (FD) based on community occupancy of trait-space. Their method involves (1) obtaining a matrix of functional traits as above (Bremner et al. 2003), (2) converting this to a distance matrix, (3) running a cluster analysis on this distance matrix and producing a dendrogram of functional state-space and then (4) the total branch length of a community (defined as FD) is used as a measure of functional diversity. FD relates well to taxonomic diversity in a range of situations (Petchey and Gaston 2002). It has been widely used since its development but continues to be debated and refined (e.g. Petchey et al. 2004; Mouchet et al. 2008). It retains the drawback that it is sensitive to the number and identity of traits specified in the matrix. As with any type of analysis using clustering to compute functional groupings, there is also substantial variation in the outcome depending on the clustering method (e.g. hierarchical, non-hierarchical, agglomerative, divisive, etc.), as well as the linkage method used (e.g. nearest neighbour, group means, etc.; Mouchet et al. 2008). Since clustering analysis necessitates computation of intergroup distances, it seems as though a logical next step would be to refine a similar method of estimating functional diversity that omits the clustering step and utilizes cumulative distances between species in a community.

One information-rich approach currently being developed utilizes the methodology of convex hulls. If one imagines a multidimensional hypervolume in which each axis is an organismal trait, functional diversity would ideally be quantified by (1) the number of points in this space, (2) the overall volume encompassed by the community, representing the 'trait limits' of the community, and (3) the density of points within this volume, representing the 'occupancy' density of the trait-space. The statistical techniques being refined for ecology using convex hulls may help considerably in quantifying the range (or volume) occupied by elements in trait-space (see, e.g. Cornwell et al. 2006). Although very few examples exist in the ecological literature utilizing this type of approach for functional diversity, there are statistical packages available that make it feasible to explore further (e.g. '*chplot*' in the R package; Quickhull algorithm for Matlab; QHull program of Barber and Huhdanpaa; see Cornwell et al. 2006).

27.4 How to Test the Validity and Value of Particular Methods/Groupings

Given the wealth of approaches and techniques for operationalising functional diversity that are available, we now turn our attention to how ecologists have tested the relevance, or predictability, of these approaches. All functional classification schemes have limitations: all are dependent on the number of functional traits or groups considered; all involve a decision about how many traits or groups to include, which is to some extent arbitrary (Micheli and Halpern 2005). It has been argued that, in the context of the objectives defined above, some traits are more important than others and some system of weighting should be used (Petchy and Gaston 2002). However, given imperfect knowledge of the relationship between traits and functioning, such weightings would inevitably introduce additional subjectivity. Given the limitations inherent in functional classification schemes, research is needed to assess their effectiveness in fulfilling particular objectives (Micheli and Halpern 2005). There are a number of approaches to testing validity of functional classifications (see also Chap. 1 by Wahl for an example). Each will be discussed briefly below.

27.4.1 Correlational Approaches

At a fundamental level, there will inevitably be some degree of correlation between taxonomic and functional diversity (Naeem 2002). The strength and shape of the relationship will vary among schemes for classifying functional diversity and will depend primarily on the level of functional redundancy (i.e. numbers of species fulfilling a given functional role) under a given scheme for a given system (Fig. 27.1). Compared to functional diversity, taxonomic diversity is generally better known or more readily derived. However, if the relationship between taxonomic and functional diversity is not strong, then taxonomic diversity will not be a good predictor of ecosystem functioning (Naeem 2002; Naeem and Wright 2003).

Micheli and Halpern (2005) found a good correlation between taxonomic and functional richness for temperate rocky reefs in the Channel Islands of California and also in a dataset drawn from 31 studies of fish assemblages in and near to marine protected areas (MPAs). They used a classification scheme based on broad dietary categories—e.g. for the temperate reefs, categories included ‘herbivorous invertebrates’, ‘sessile planktivorous invertebrates’, ‘mobile small-invertebrate-eating fish’, etc. Each category included one to 13 species. In analysing fish assemblages in and near MPAs, three traits were considered: trophic group, size and mobility. The effectiveness of MPAs was analysed for functional diversity as classified using different combinations of these three traits (with increasing numbers of functional groups defined by classifications involving increasing numbers of traits). Although the amount of variation explained increased with increasing numbers of functional groups, all regressions were significant and positive. For this system, it may therefore be possible to capture essential functional information with comparatively crude classification schemes. However, the analysis also showed that increasing the detail of functional classification should improve its predictive power, emphasising the arbitrary nature of functional classification schemes and analyses based on them. Micheli and Halpern (2005) also showed that small changes in species richness can

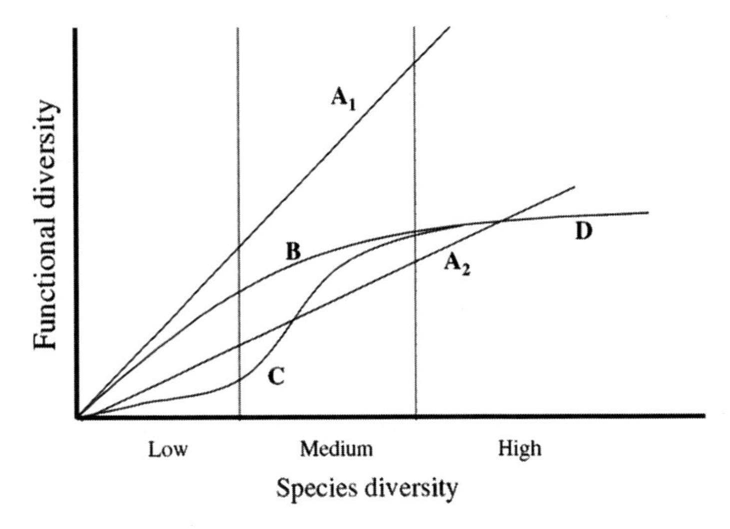

Fig. 27.1 Schematic of possible relationships between functional and species diversity within three levels of diversity at which communities have been arbitrarily delimited: low, medium and high diversity. In scenario A, functional diversity increases linearly with increasing species diversity, regardless of diversity level (although it would likely reach a plateau at high-diversity levels). In scenario A_1, each species plays a unique functional role (slope=1), whereas multiple species have similar functional traits in A_2 (slope<1). In scenario B, functional diversity increases at declining rates with increased species diversity and reaches an asymptote at high-diversity levels (D). Scenario C depicts high functional redundancy at low species diversity followed by rapid increase at intermediate species diversity, until functional diversity asymptotes at high-diversity levels (D). Figure reproduced with permission from Micheli and Halpern (2005)

result in significant changes in functional diversity—in cases where functional roles are fulfilled by one or few taxa.

27.4.2 Experimental Approaches

Although there have been a number of studies testing the importance of functional diversity, and a large number of studies testing the importance of species diversity (Hooper et al. 2005; see Chap. 26 by Gamfeldt and Bracken), there have been none to date that manipulate both levels of diversity simultaneously and can therefore directly compare their relative importance in a given context (but see Naeem 2002 for an example adapting existing data). Designs to do this have been developed for direct comparisons and experiments have been undertaken using laboratory rock pools (S. Jenkins et al., unpublished data; T. Crowe and S. Nicol, unpublished data) but results are not yet available.

All experimental tests of relationships between diversity and ecosystem functioning must necessarily be complex. A detailed review of experimental design for BEF research is beyond the scope of this chapter. Many of the issues have been reviewed, with particular reference to marine hard substrata, by Allison (1999) and Benedetti-Cecchi (2004). It is worth noting, however, that hard substrata are particularly suitable for the field-based removal experiments that have been called for by some authors (Díaz and Cabido 2001).

27.4.3 Modelling Approaches

Given the complexities and grandiose scales of appropriate experimental designs, and the intrinsic shortcomings of plausible correlational approaches, modelling can provide an important alternative approach to exploring the roles of different modes of diversity. Simple models incorporating the heterogeneity of resource-use strategies generally indicate that more functionally diverse communities tend to be more efficient at capture and processing of resources (e.g. Hulot et al. 2000). Solan et al. (2004) used data to parameterize models of different extinction regimes and how they would influence the functioning of the community. They demonstrated that the functional differences between species matter, clearly manifest in one species (the brittlestar *Aphiura filiformis*) that had a disproportionately large impact on this ecosystem process. Modelling that has been done with variously defined functional groups (rather than functional diversity) tends to show more mixed results (Hulot et al. 2000). Very little modelling has tackled the topic of the relative role of functional versus taxonomic diversity in influencing ecosystem functioning, probably due to the need for more robust empirical data (observational or experimental) to provide a solid foundation for this type of modelling. Without more evidence, there is no a priori reason to predict that taxonomic diversity would be a better predictor of ecosystem processing or functioning than is functional diversity.

27.5 Evidence from Hard Substrata Regarding Sensitivity of Systems to Changes in Functional Diversity

Taxa inhabiting marine hard substrata have been classified into functional groups to test their importance to a range of ecosystem processes. Some of the insights this approach has yielded are discussed below (and see Allison et al. 1996 and Crowe 2005, who reviewed relevant research not necessarily targeted at the BEF debate).

An important recent focus has been based on simple functional classifications into trophic levels. As a progression from the initial emphasis on a single trophic level (primary producers), BEF researchers have been asking whether changes in diversity at one trophic level affect the functioning of others and stressing the need to consider multi-trophic extinctions (Raffaelli et al. 2002; Duffy et al. 2007; Stachowicz et al. 2007). Marine hard substrata have been used as a model system for a number of tests of these ideas and these are reviewed by Gamfeldt and Bracken in the preceding chapter (and see Stachowicz et al. 2007).

The role of functional diversity in resisting invasion has been the focus of a number of studies using marine hard substrata. Arenas et al. (2006) manipulated numbers and identities of functional groups of intertidal algae to test effects on invasibility of rock pools. They found that the identities of the functional groups present were more important than the functional richness per se. Although Britton-Simmons (2006) draws some unsupported inferences about interactive effects, his study also supports the hypothesis that different functional groups of native algae vary in the extent to which they prevent recruitment by a non-native alga and the mechanism by which they influence the process (Britton-Simmons 2006). Stachowicz and colleagues (Stachowicz et al. 1999, 2002; Stachowicz and Byrnes 2006) have also completed an influential body of work in this area, using subtidal fouling communities as the model system. Ingenious experiments found clear negative relationships between species richness and invasion. Temporal fluctuations in space occupation by individual taxa provided opportunities for invaders if those taxa were part of low-diversity communities, whereas diverse communities appeared to be buffered against such fluctuations in space availability. Species could be classified into the same functional group (sessile suspension feeders) and were essentially interchangeable in their effects; the presence of multiple species within this functional group enables space to remain occupied even if one species declines—a good illustration of the value of 'biological insurance' (Yachi and Loreau 1999).

In other systems, different species within functional groups have been shown to have different effects. For example, in manipulating diversity of grazing limpets, Beovich and Quinn (1992) found that only loss of *Siphonaria diemenensis* would lead to major effects on foliose algae (e.g. *Scytosiphon lomentaria*); removal of *Cellana tramoserica* had little effect. This difference was attributed partly to differences in radula morphology and feeding biology, such that only *Siphonaria* could graze mature foliose algae. Similarly, O'Connor and Crowe (2005) showed that loss of limpets (*Patella ulyssiponensis*) had far greater influence on macroalgal cover than did the loss of one or more other species of grazing gastropod. Loss of limpets could be compensated for by increased biomass of other grazers, but only in the short term. In this case, biological insurance was comparatively ineffective—the influence of the functional group depended largely on one of its members.

Micheli and Halpern (2005) used data from marine hard substrata specifically to assess the value of functional classification in aiding prediction of effects of species loss (see Chap. 6 by Kotta and Witman). They found good relationships between species and functional diversity (regardless of functional classification scheme), again suggesting low functional redundancy in coastal marine assemblages.

Allison et al. (1996) derived a framework for predicting effects of loss of species based on a broad classification of interaction strengths into weak, diffuse and strong (Menge et al. 1994). This idea was followed up by Berlow et al. (1999) who formalised the quantification of interaction strengths. More recently, Navarrette and Berlow (2006) used empirical evidence from a rocky shore to infer a link between resilience to environmental stochasticity and variation in interaction strength.

27.6 The Relative Importance of Functional and Taxonomic Diversity: Summary of Current Knowledge and Suggestions for the Future

Biological diversity matters for ecosystem processes. Regardless of scale or focal unit, it seems that biodiversity affects how systems function. The relative roles of different levels of biological diversity, however, remain much less clear. In this review, we have focused on functional and taxonomic diversity. Research at the genetic and habitat levels is also underway and has been discussed elsewhere (e.g. Gamfeldt and Bracken, Chap. 26; Hawkins 2004).

After early research (primarily in grasslands) studying the role of species richness in ecosystem functioning, it was shown that functional groups or species identities may often predict ecosystem functioning (primarily productivity) more effectively than does species richness. This point should be implicit in this field of research: it seems clear that if all species were functionally homogenous—or, alternatively, carried wholly identical suites of traits—then species would not be unique, and diversity ('variety in form and function', as defined above) would not exist. As such, the mechanism by which species diversity can matter to ecosystem functioning is necessarily through variety in organismal form and function (i.e. traits) and the interactions between these traits. It is because interspecific individuals are more different than conspecific individuals that species richness can matter.

Classifying species according to their function can have a number of benefits (see Sect. 27.3). However, a number of issues arise in applying this approach. A recurrent issue is the extent to which individual members of a functional group vary in the strength of their influence. Stachowitz et al. (2007) claimed that "the magnitude of the identity or composition effects was roughly 10 times stronger than that of richness" in studies of the effect of algal richness and identity on community productivity. Such effects seem prevalent in a wide range of marine ecosystems and lend support to the argument that basic research focused at the level of individual species is still needed to underpin our understanding of their functional roles.

Strong differences among species within a functional group may be partially attributable to the scheme used to classify them. There is a wide range of approaches to operationalising and classifying diversity according to its functional traits (see Sect. 27.3). Development of a widely accepted scheme for classifying functional diversity would be a very significant step forward. A critical issue to resolve is the reduction of subjectivity or arbitrariness in selecting traits and character states for those traits. Indeed, some authors have questioned the validity of creating groupings at all (Wright et al. 2006). Petchey and Gaston's (2002) metric of functional diversity (FD) distils multivariate information about functionality into a generalisable index but, in taking the steps from multivariate matrix to index, becomes limited in the same way as diversity indices, in that detailed information about the functional composition of the community is lost. Biological traits analysis (BTA) also dispenses with groupings and offers an effective way of capturing variation in functional capacity among assemblages (Bremner et al. 2003). Being based directly on matrices of functional traits weighted by abundance, BTA also lends itself to assessing the specific functional consequences of changes in assemblage structure and may usefully be applied directly to environmental management. Nevertheless, the results remain dependent upon the choice of traits with which the matrix is populated. Naeem and Wright (2003) argue that, in order for BEF science to serve as a predictive tool, effect and response traits should be selected on the basis of the function of interest and the key drivers of biodiversity change in the system. Perhaps the optimal solution in the search for a more general scheme is to seek a small subset of traits of which the relationship to a wide range of functions is or can be well established. In plant science, for example, seed size, plant height and specific leaf area have been shown repeatedly to relate to a range of aspects of functioning (Westoby 1998; Lavorel and Garnier 2001). In animal science, body size has also been singled out as a key trait (e.g. Emmerson and Raffaelli 2004; Jonsson et al. 2005).

Another key difficulty with the functional group approach is that species vary in their functioning under different environmental circumstances and in relation to different ecosystem functions. Operationally, this is dealt with quite effectively using the fuzzy coding approach (Bremner et al. 2006). Conceptually, however, there remains a pressing need to develop a broader framework for assessing the roles of individual species or functional groups and of their diversity. To encompass a range of environmental contexts, experiments need either to manipulate environmental variables explicitly and/or to run in the field over extended periods (O'Connor and Crowe 2005). More experiments that simultaneously assess more than one ecosystem function are also required (Gamfeldt et al. 2008). As has been pointed out by others (Raffaelli et al. 2002; Naeem 2006; Stachowicz et al. 2007), small-scale field and laboratory experimentation cannot alone drive the development of a broader framework. The complexity involved requires a multidisciplinary approach, including observation, experimentation and modelling (Duffy et al. 2007). General theory must, by definition, emerge from and apply to a wide range of ecosystems. Marine hard substrata continue to provide an ideal test bed for many of the key ideas in this debate. Closer links between marine and terrestrial research in the field would be of benefit to all.

References

Allison GW (1999) The implications of experimental design for biodiversity manipulations. Am Nat 153:26–45

Allison GW, Menge BA, Lubchenco J, Navarrete SA (1996) Predictability and uncertainty in community regulation: consequences of reduced consumer diversity in coastal rocky ecosystems. In: Mooney H, Cushman JH, Medina E, Sala O, Schulze ED (eds) Functional roles of biodiversity: a global perspective. Wiley, New York, pp 371–392

Arenas F, Sanchez I, Hawkins SJ, Jenkins SR (2006) The invasibility of marine algal assemblages: role of functional diversity and identity. Ecology 87(11):2851–2861

Bellwood DR, Hoey AS, Choat JH (2003) Limited functional redundancy in high diversity systems: resilience and ecosystem function in coral reefs. Ecol Lett 6:281–285

Benedetti-Cecchi L (2004) Increasing accuracy of causal inference in experimental analyses of biodiversity. Funct Ecol 18:761–768
Beovich EK, Quinn GP (1992) The grazing effect of limpets on the macroalgal community of a rocky intertidal shore. Aust J Ecol 17:75–82
Berlow E, Navarrete SA, Briggs CJ, Power ME, Menge BA (1999) Quantifying variation in the strengths of species interactions. Ecology 80:2206–2224
Blondel J (2003) Guilds or functional groups: does it matter? Oikos 100:223–231
Bremner J, Rogers SI, Frid CLJ (2003) Assessing functional diversity in marine benthic ecosystems: a comparison of approaches. Mar Ecol Prog Ser 254:11–25
Bremner J, Rogers SI, Frid CLJ (2006) Matching biological traits to environmental conditions in marine benthic ecosystems. J Mar Syst 60(3/4):302–316
Britton-Simmons KH (2006) Functional group diversity, resource preemption and the genesis of invasion resistance in a community of marine algae. Oikos 113(3):395–401
CBD (1992) Convention on Biological Diversity. United Nations, http://www.cbd.int/convention/convention
Chalcraft DR, Resetarits WJ Jr (2003) Mapping functional similarity of predators on the basis of trait similarities. Am Nat 162(4):390–402
Charvet S, Statzner B, Usseglio-Polatera P, Dumont B (2000) Traits of benthic macroinvertebrates in semi-natural French streams: an initial application to biomonitoring in Europe. Freshw Biol 43:277–296
Chevenet F, Doledec S, Chessel D (1994) A fuzzy coding approach for the analysis of long-term ecological data. Freshw Biol 31:295–309
Coleman FC, Williams SL (2002) Overexploiting marine ecosystem engineers: potential consequences for biodiversity. Trends Ecol Evol 17:40–44
Connell JH (1974) Ecology: field experiments in marine ecology. In: Mariscal RN (ed) Experimental marine biology. Academic Press, New York, pp 131–138
Cornwell WK, Schwilk DW, Ackerly DD (2006) A trait-based test for habitat filtering: convex hull volume. Ecology 87(6):1465–1471
Crowe TP (2005) What do species do in intertidal ecosystems? In: Wilson JG (ed) The intertidal ecosystem: the value of Ireland's shores. Royal Irish Academy, Dublin, pp 115–133
Díaz S, Cabido M (2001) Vive la différence: plant functional diversity matters to ecosystem processes. Trends Ecol Evol 16(11):646–655
Diaz S, Noy-Meir I, Cabido M (2001) Can grazing response of herbaceous plants be predicted from simple vegetative traits? J Appl Ecol 38(3):497–508
Díaz S, Symstad AJ, Chapin FS III, Wardle DA, Huenneke LF (2003) Functional diversity revealed by removal experiments. Trends Ecol Evol 18(3):140–146
Dobson A, Lodge D, Alder J, Cumming GS, Keymer J, McGlade J, Mooney H, Rusak JA, Sala O, Wolters V, Wall D, Winfree R, Xenopoulos MA (2006) Habitat loss, trophic collapse, and the decline of ecosystem services. Ecology 87(8):1915–1924
Dring MJ (1982) The biology of marine plants. Edward Arnold, London
Duffy JE, Cardinale BJ, France KE, McIntyre PB, Thebault E, Loreau M (2007) The functional role of biodiversity in ecosystems: incorporating trophic complexity. Ecol Lett 10(6):522–538
Emmerson M, Huxham M (2002) How can marine ecology contribute to the biodiversity-ecosystem functioning debate? In: Loreau M, Naeem S, Inchausti P (eds) Biodiversity and ecosystem functioning: synthesis and perspectives. Oxford University Press, Oxford, pp 139–146
Emmerson MC, Raffaelli D (2004) Predator-prey body size, interaction strength and the stability of a real food web. J Anim Ecol 73(3):399–409
Gamfeldt L, Hillebrand H, Jonsson PR (2008) Multiple functions increase the importance of biodiversity for overall ecosystem functioning. Ecology (in press)
Hawkins SJ (2004) Scaling up: the role of species and habitat patches in the functioning of coastal ecosystems. Aquat Conserv Mar Freshw Ecosyst 14(3):217–219
Hooper DU, Chapin FS, Ewel JJ, Hector A, Inchausti P, Lavorel S, Lawton JH, Lodge DM, Loreau M, Naeem S, Schmid B, Setlala H, Symstad AJ, Vandermeer J, Wardle DA (2005) Effects of biodiversity on ecosystem functioning: a consensus of current knowledge. Ecol Monogr 75(1):3–35

Hulot FD, Lacroix G, Lescher-Moutoue F, Loreau M (2000) Functional diversity governs ecosystem response to nutrient enrichment. Nature 405:340–344

Ji Y, Tanaka J (2002) Effect of desiccation on the photosynthesis of seaweeds from the intertidal zone in Honshu, Japan. Phycol Res 50(2):145–153

Johansson G, Snoeijs P (2002) Macroalgal photosynthetic responses to light in relation to thallus morphology and depth zonation. Mar Ecol Prog Ser 244:63–72

Jones CG, Lawton JH, Shachak M (1994) Organisms as ecosystem engineers. Oikos 69:373–386

Jonsson T, Cohen JE, Carpenter SR (2005) Food webs, body size, and species abundance in ecological community description. Adv Ecol Res 36:1–84

Lavorel S, Garnier E (2001) Aardvark to Zyzyxia-functional groups across kingdoms. New Phytol 149:360–364

Lavorel S, Garnier E (2002) Predicting changes in community composition and ecosystem functioning from plant traits: Revisiting the Holy Grail. Funct Ecol 16(5):545–556

Lavorel S, McIntyre S, Landsberg J, Forbes D (1997) Plant functional classifications: from general groups to specific groups based on response to disturbance. Trends Ecol Evol 12:474–478

Littler MM, Littler DS (1980) The evolution of thallus form and survival strategies in benthic marine macroalgae: field and laboratory tests of a functional form model. Am Nat 116:25–44

MacArthur RH (1972) Geographical ecology: patterns in the distribution of species. Harper & Row, New York

May RM (1994) Biological diversity: differences between land and sea. Philos Trans R Soc Lond B 343:105–111

McIntyre S, Lavorel S, Landsberg J, Forbes TDA (1999) Disturbance response in vegetation: towards a global perspective on functional traits. J Veg Sci 10(5):621–630

Menge BA, Berlow EL, Blanchette CA, Navarrete SA, Yamada SB (1994) The keystone species concept: variation in interaction strength in a rocky intertidal habitat. Ecol Monogr 64:249–286

Micheli F, Halpern BS (2005) Low functional redundancy in coastal marine assemblages. Ecol Lett 8(4):391–400

Mouchet M, Guilhaumon F, Villeger S, Mason NWH, Tomasini JA, Mouillot D (2008) Towards a consensus for calculating dendrogram-based functional diversity indices. Oikos 117(5):794–800

Naeem S (2002) Disentangling the impacts of diversity on ecosystem functioning in combinatorial experiments. Ecology 83(10):2925–2935

Naeem S (2006) Expanding scales in biodiversity-based research: challenges and solutions for marine systems. Mar Ecol Prog Ser 311:273–283

Naeem S, Wright JP (2003) Disentangling biodiversity effects on ecosystem functioning: deriving solutions to a seemingly insurmountable problem. Ecol Lett 6:567–579

Naeem S, Thompson LJ, Lawler SP, Lawton JH, Woodfin RM (1994) Declining biodiversity can alter the performance of ecosystems. Nature 368:734–737

Navarrete SA, Berlow EL (2006) Variable interaction strengths stabilize marine community pattern. Ecol Lett 9(5):526–536

O'Connor NE, Crowe TP (2005) Biodiversity and ecosystem functioning: distinguishing between effects of the number of species and their identities Ecology 86(7):1783–1796

Ormond RFG (1996) Marine biodiversity: causes and consequences. J Mar Biol Assoc UK 76:151–152

Padilla DK, Allen BJ (2000) Paradigm lost: reconsidering functional form and group hypotheses in marine ecology. J Exp Mar Biol Ecol 250(1/2):207–221

Paine RT (1977) Controlled manipulations in the marine intertidal zone, and their contributions to ecological theory. In: Goulden CE (ed) Changing scenes in natural sciences, 1776–1976. Academy of Natural Sciences, Philadelphia, PA, pp 245–270

Petchey OL, Gaston KJ (2002) Functional diversity (FD), species richness and community composition. Ecol Lett 5:402–411

Petchey OL, Hector A, Gaston KJ (2004) How do different measures of functional diversity perform? Ecology 85(3):847–857

Raffaelli D, van der Putten WH, Persson L, Wardle DA, Petchey OL, Koricheva J, van der Heijden M, Mikola J, Kennedy T (2002) Multi-trophic dynamics and ecosystem processes. In: Naeem S,

Loreau M, nchausti I P (eds) Biodiversity and ecosystem functioning: synthesis and perspectives. Oxford University Press, Oxford, pp 147–154
Ray GC, Grassle JF (1991) Marine biological diversity. Bioscience 41:453–457
Scheibling RE (1994) Molluscan grazing and macroalgal zonation on a rocky intertidal platform at Perth, Western Australia. Aust J Ecol 19:141–149
Simberloff D, Dayan T (1991) The guild concept and the structure of ecological communities. Annu Rev Ecol Syst 22:115–143
Solan M, Cardinale B, Downing AL, Engelhardt KAM, Ruesink JL, Srivastava DS (2004) Extinction and ecosystem function in the marine benthos. Science 306:1177–1180
Solan M, Raffaelli DG, Paterson DM, White PCL, Pierce GJ (2006) Marine biodiversity and ecosystem functioning: empirical approaches and future research needs. Mar Ecol Prog Ser 311:175–178
Stachowicz JJ, Byrnes JE (2006) Species diversity, invasion success, and ecosystem functioning: disentangling the influence of resource competition, facilitation, and extrinsic factors. Mar Ecol Prog Ser 311:251–262
Stachowicz JJ, Whitlatch RB, Osman RW (1999) Species diversity and invasion resistance in a marine ecosystem. Science 286(5444):1577–1579
Stachowicz JJ, Fried H, Osman RW, Whitlatch RB (2002) Biodiversity, invasion resistance, and marine ecosystem function: reconciling pattern and process. Ecology 83(9):2575–2590
Stachowicz JJ, Bruno JF, Duffy JE (2007) Understanding the effects of marine biodiversity on communities and ecosystems. Annu Rev Ecol Syst 38:739–766
Steneck RS, Dethier MN (1994) A functional group approach to the structure of algal-dominated communities. Oikos 69:476–498
Steneck RS, Watling L (1982) Feeding capabilities and limitation of herbivorous molluscs: a functional group approach. Mar Biol 68:299–319
Sullivan G, Zedler JB (1999) Functional redundancy among tidal marsh halophytes: a test. Oikos 84(2):246–260
Vincent A, Clarke A (1995) Diversity in the marine environment. Trends Ecol Evol10:55–56
Westoby M (1998) A leaf-height-seed plant-ecology strategy scheme. Plant Soil 199:213–227
Wright JP, Naeem S, Hector A, Lehman C, Reich PB, Schmid B, Tilman D (2006) Conventional functional classification schemes underestimate the relationship with ecosystem functioning. Ecol Lett 9:111–120
Yachi S, Loreau M (1999) Biodiversity and ecosystem productivity in a fluctuating environment: the insurance hypothesis. Proc Natl Acad Sci USA 96(4):1463–1468
Zedler JB, Callaway JC, Sullivan G (2001) Declining biodiversity: why species matter and how their functions might be restored in Californian tidal marshes. Bioscience 51(12):1005–1017

Chapter 28
Mechanisms Underpinning Diversity–Stability Relationships in Hard Bottom Assemblages

Lisandro Benedetti-Cecchi

28.1 Introduction

The relationship between diversity and stability is an important ecological paradigm that has stimulated considerable theoretical and empirical research in ecology. As it is often the case with core principles, this relationship has also stimulated considerable debate. The hypothesis that more diverse assemblages are also more stable (i.e. less variable in space and time and more resistant to invasion and disturbance) was advanced by a number of authors in the 1950s (Odum 1953; McArthur 1955; Elton 1958). These intuitive ideas were challenged by May (1973), who showed that species diversity reduced the stability of individual populations in mathematical simulations where species interactions were governed by random coefficients. Later, Yodzis (1981) showed the opposite, using more realistic values of species interaction strengths. In a review of the topic, Pimm (1984) argued that the lack of consistent results among studies on stability could be a consequence of the vague definition of this term. This author distinguished several definitions of stability to which theoretical and empirical analyses had referred (including mathematical stability, resilience, persistence, resistance and variability), and concluded that these analyses often focused on different hypotheses about stability. Pimm (1984) went further and emphasized the need to identify appropriate measurable quantities to reflect the different definitions of stability and to enable their theoretical and empirical scrutiny.

The contribution of empirical ecologists to this debate was minimal at early stages. A notable exception was the work of McNaughton on herbivore grazing in the Serengeti (McNaughton 1977). This author excluded large grazers from one area with diverse vegetation and one area with less diverse vegetation, and compared the resulting patterns in plant biomass with those occurring in unmanipulated areas. Results showed that herbivores had a lower impact on the more diverse assemblage, providing support to Elton's hypothesis.

Increasing concern with the consequences of loss of biodiversity stimulated a research programme on biodiversity and ecosystem functioning that also renewed the interest of empirical ecologists in the diversity–stability debate (McCann 2000; Cottingham et al. 2001). By manipulating nutrient availability in grassland plots,

M. Wahl (ed.), *Marine Hard Bottom Communities*, Ecological Studies 206,
DOI: 10.1007/978-3-540-92704-4_28, © Springer-Verlag Berlin Heidelberg 2009

Tilman (1996) obtained a gradient in species richness that was used to determine the extent to which diversity increased resistance to physical disturbance (a major drought event). Tilman found a negative relationship between temporal variability of total biomass and species richness in grassland plots, whilst the variability of individual species was positively related to richness. These results refocused attention on the important point already made by May (1973), but then overlooked, that aggregate variables such as total species biomass and abundance measured at the assemblage level may be more stable than the corresponding variables measured at the population level. These findings also reconciled the outcomes of May's simulations that focused on the stability of individual populations and the original idea of stability, which applied to entire assemblages (e.g. Elton 1958).

An important contribution towards the understanding of diversity-stability relationships at different levels of biological organization was made by Doak et al. (1998). These authors used probability theory to show that a positive diversity-stability relationship can be expected for aggregate variables simply as a statistical consequence of averaging across multiple independent variables (the abundances of individual populations). This effect was termed 'statistical averaging' by Doak et al. (1998) and 'portfolio effect' by Tilman et al. (1998), to recall the same stabilizing mechanism that operates when a fixed financial investment is diversified across a portfolio of stocks. Indeed, this mechanism had already been known to ecologists for a long time as 'spreading the risk', after den Boer (1968), who should be credited more often for his contribution.

A synthetic theory is emerging where statistical averaging, biological interactions and differential responses of species to environmental fluctuations are recognized as three non-mutually exclusive mechanisms regulating diversity–stability relationships (McCann 2000; Cottingham et al. 2001). In addition to the work of Doak et al. (1998) and Tilman and co-workers (Tilman et al. 1998; Tilman 1999; Lehman and Tilman 2000), the 'insurance hypothesis' proposed by Yachi and Loreau (1999) contributed to the theoretical foundations of this theory. This hypothesis ascribes a dual role to diversity: a negative (stabilizing) effect on temporal variance of ecosystem processes (e.g. productivity) and a positive effect on temporal means. These effects are driven largely by the amount of dissimilarity in the response of species to environmental fluctuations, which is a function of species diversity.

Empirical studies have started to disentangle the contribution of the three mechanisms of stability invoked by the theory, on the basis of long-term observational data and experiments (e.g. Ives et al. 1999, 2000, 2003; Petchey et al. 2002; Gonzalez and Descamps-Julien 2004; Vasseur and Gaedke 2007). Most empirical studies have used plankton assemblages and grasslands as model systems, whilst investigations on rocky shore assemblages have contributed only little to this topic. Apart from few notable exceptions on resistance to disturbance and invasion (Stachowicz et al. 1999, 2002; Allison 2004; Arenas et al. 2006), most studies on variation in populations and assemblages of rocky shores were not designed to investigate diversity–stability relationships explicitly (reviewed in Benedetti-Cecchi 2006). While manipulating the biodiversity of hard bottom assemblages is challenging (but possible; see also O'Connor and Crowe 2005; Bruno et al. 2005; O'Connor and Bruno 2007; Stachowicz et al. 2007), descriptive studies may be readily available to examine diversity-stability relationships, albeit on a correlative basis. In the following sections I illustrate the basic principles that underpin the synthetic theory on diversity and stability, and provide a test of this theory using data on temporal change in abundance of algae and invertebrates from Mediterranean rocky shores. In the remainder of the paper, I will refer to diversity simply as the number of species (or taxa) in an assemblage, and to stability as the inverse of temporal variability for response variables measured at the population and assemblage levels. This definition seems to correspond to the way early proponents thought of stability (Lehman and Tilman 2000).

28.2 Measures of Stability

Traditionally, stability is assessed from estimates of temporal variance in chosen response variables (e.g. biomass or abundance) that reflect the properties of individual populations or entire assemblages. There has been some discussion on how temporal variances can be best estimated from real data. Gaston and McArdle (1994) provided a detailed account of the measures of (population) variability that have been used in the literature. Most studies of stability have employed either the standard deviation of log-transformed values, $S[\log(X)]$, where X is a random variable measured over time, or the coefficient of variation, $CV = S/\bar{X}$ (Pimm 1984, 1991; Doak et al. 1998; Tilman 1999; Lehman and Tilman 2000; Cottingham et al. 2001). Both measures quantify temporal variation on a proportional scale—i.e. relative to the mean of the response variable. The former measure is problematic when there are zeros in the data. The common procedure is to transform the data as $\log(X+1)$, which introduces a bias in estimation whenever there are low values of the response variable (McArdle and Gaston 1990; Gaston and McArdle 1994). The CV does not require a transformation of the data and, therefore, it is free of this particular problem. This is probably a main reason why the CV has become a common measure of variability and its inverse, $\bar{X}/S$, has been proposed as a measure of stability (Lehman and Tilman 2000).

The rationale for expressing variation on a proportional scale is to give the same value for the measure of variability to a population that varies, say, from ten to 20 individuals, as to another that varies from 40 to 80 individuals. Expressing variation on a common scale facilitates comparisons among populations, and it is a desirable property of measures of variability when these are used to assess the stability of aggregate response variables (as in the procedures described in Sect. 28.3). The use of a proportional scale may, however, not be justified on ecological grounds. The functional consequences (e.g. in terms of productivity or for species interactions) of a change in density from ten to 20 individuals may be fundamentally different from that for a change in density from 40 to 80 individuals. Clearly, the proportional scale does not account for the ecology of individual species, and any measure of variability based on this scale treats species as equivalent on a per capita basis. This is often a necessary assumption, however, because we do not know the ecology of every single species in an assemblage.

Another argument against the use of the CV as a measure of variability is that it actually confounds changes in variance with changes in means of response variables (Underwood 1996; Underwood and Chapman 2000). In other words, the CV is not a pure measure of variability, and any change can be ascribed both to the variance and to the mean of response variables. Underwood and Chapman (2000) have proposed to use pure estimates of variance of populations and to determine the extent to which changes in variance reflect changes in the mean by examining variance-to-mean relationships. This seems the most logical approach when examining the variability of population-level variables and when the research focuses on the ecology of individual species or taxa.

A final issue that needs to be considered about measures of variability is spatial confounding. If a site is sampled through time with independent replicate samples at each time, the variance among sample means combines spatial and temporal variation and cannot be considered a pure measure of temporal variability. This problem has been detailed clearly by other authors (e.g. Stewart-Oaten et al. 1995) and will not be pursued further here. It is sufficient to say that obtaining a pure measure of temporal variability from a sampling design that involves both spatial and temporal sources of variation is a problem of estimating the appropriate variance component.

28.3 Three Mechanisms Relating Stability to Diversity

In this section I provide an overview of the mechanisms underlying diversity–stability relationships as discussed in the general ecological literature (Doak et al. 1998; Tilman et al. 1998; Tilman 1999; Lehman and Tilman 2000; Cottingham et al. 2001).

28.3.1 The Statistical Averaging (Portfolio) Effect

Doak et al. (1998) started from very simple assumptions (then partially relaxed) to show that the stability of an aggregate response variable is expected to increase with increasing number of species, simply on the basis of a probabilistic argument. These authors based their explanation on the comparison of the coefficient of variation for an assemblage with N species with that of a hypothetical assemblage composed of a single species. The simplest scenario assumed that species are random and independent, each with mean abundance $x_s = M/N$, where M is the mean of the total abundance of the assemblage over a period of time. M is assumed constant regardless of N, implying that the mean abundance of the individual species decreases as diversity increases. This scenario also assumes that species have the same variances (and, hence, the same coefficient of variation) defined as σ_s^2 / N^2, where σ_s^2 is the variance of a species living by itself. Given these assumptions, the mean and the variance of total assemblage abundance are $m_{ass} = Nx_s$, which is equal to M as originally defined, and $\sigma_{ass}^2 = N\sigma_s^2/N^2$ respectively. Therefore, the coefficient of variation for the total abundance of the assemblage is $CV_{ass}=100\sigma_s/(N^{1/2}M)$. Because $100\sigma_s/N$ is the coefficient of variation for an assemblage with a single species, CV_s (recall that M is assumed independent of N), the coefficient of variation for the entire assemblage can be written as

$$CV_{ass} = CV_s N^{-1/2} \tag{28.1}$$

The inverse relationship between N and CV_{ass} indicates that the variability of the assemblage is expected to decrease as the number of species increases, regardless of species interactions and environmental context. Then, Doak et al. (1998) relaxed the assumption of equal species abundances and derived an expression for CV_{ass} that depended on the degree of species dominance. Because species may now have different means and variances, the coefficient of variation for the entire assemblage becomes $CV_{ass} = 100(\sigma_1^2 + \sigma_2^2 + \ldots \sigma_N^2)^{1/2} / (x_1 + x_2 + \ldots x_N)$, where $\sigma_1^2 + \sigma_2^2 + \ldots \sigma_N^2$ and $x_1+x_2+\ldots x_n$ are the variances and the means of individual species respectively. Assuming again that the coefficient of variation is the same for all species in the assemblage and equal to CV_s, Doak et al. (1998) derived the following relation:

$$CV_{ass} = CV_s (x_1^2 + x_2^2 + \ldots x_N^2)^{1/2} / (x_1 + x_2 + \ldots x_N) \tag{28.2}$$

Equation (28.2) was further modified to include the effect of species dominance. To do this, Doak et al. (1998) assumed an exponential rank–abundance distribution—i.e. if species are ranked in decreasing order of abundance, then the abundance of the ith species becomes $x_i = x_1 \exp[-a(i-1)]$, where x_1 indicates the most abundant species and a is the coefficient determining the rate at which mean abundance decreases with increasing rank. After some manipulation (see also Lhomme and Winkel 2002), Eq. (28.2) becomes

$$CV_{ass} = CV_s \left[\frac{(1-e^{-a})(1+e^{-aN})}{(1+e^{-a})(1-e^{-aN})} \right]^{1/2} \tag{28.3}$$

Equation (28.3) shows that CV_{ass} is still a decreasing function of N and that the strength of this relationship diminishes as species dominance increases (large values of a). Thus, the statistical averaging effect becomes less important as unevenness in species abundance increases.

Commenting on these ideas, Tilman et al. (1998) provided a very simple and intuitive description of the statistical averaging effect by recalling that if x is a random variable, then $\mathrm{Var}(x)/n = \mathrm{Var}(x)/n^2$. Hence, if the mean abundance of species i, x_i, is reduced by half, its variance is reduced to one-fourth. Under the assumption that the mean abundance of the entire assemblage, M, is constant and independent of N, as in Doak et al. (1998), the mean abundance of the ith species, x_i, must decrease as species richness increases. It follows that also the variance of the ith species, σ_i^2, must decrease (more than proportionally) with increasing species richness, so that the contribution of the variances of individual species to total variability decreases, leading to greater stability at the assemblage level.

The work of Tilman et al. (1998) showed that the scaling relationship between the mean and the variance is a key factor regulating diversity–stability relationships. By assuming that all species have equal coefficient of variation, Doak et al. (1998) implicitly assumed that the variance in abundance of species i scales as the square of its abundance—i.e. $\sigma_i^2 = cx_i^2$, where c is a constant. In this case, the coefficient of variation for an assemblage containing a single species, $CV_s = 100\sigma_s/M$, becomes $CV_s = 100(cM^2)^{1/2}/M$, which is equal to $100(c)^{1/2}$. Therefore, Eq. (28.1) becomes $CV_{ass}=100(c/N)^{1/2}$, recovering the negative relationship between variability and diversity of Eq. (28.1).

Tilman et al. (1998) generalized this case assuming that the variance in the abundance of the ith individual species depends on its mean abundance as $\sigma_i^2 = cx_i^z$, where z is the scaling coefficient determining the strength of the relationship between the mean and the variance. In this case $CV_s=100c^{1/2}M^{(z-2)1/2}$. In an assemblage with N independent species, each with abundance M/N, the variance in abundance of each species would be cM^z/N^z. The variance in total abundance would be $N(cM^z/N^z$, which is cM^z/N^{1-z}. Thus, the coefficient of variation for an assemblage containing N species becomes $CV_{ass} = 100c^{1/2}M^{(z-2)/2}N^{(1-z)/2}$. To investigate the effect of diversity on stability, Tilman et al. (1998) compared the variability of an assemblage with N species to that of an assemblage consisting of a single species:

$$CV_{ass}/CV_s = N^{(1-z)/2} \tag{28.4}$$

Equation (28.4) shows that diversity can have different effects of stability (lower values of CV_{ass}/CV_s imply greater stability), depending on z. Specifically, stability can increase, remain unchanged or decrease as the number of species in an assemblage increases, for $z > 1$, $z = 1$ and $z < 1$ respectively. Thus, the case described by Doak et al. (1998) is a special case for $z = 2$. The main result of the extension due to Tilman et al. (1998) is that while CV_{ass} will inevitably be lower than CV_s (implying that the total abundance of organisms in an assemblage will always be more stable than the abundance of individual species), increasing diversity will not inevitably lead to greater stability, the relationship depending on the scaling parameter z.

28.3.2 The Covariance Effect

The analytical treatment of the diversity–stability relationship presented above assumed that the abundances of the N species were uncorrelated—i.e. species fluctuated independently in time. This assumption is clearly unrealistic: negative covariances, indicating asynchronous temporal fluctuations, are expected for competing species (when one competitor increases in abundance, the other decreases) or when species respond differently to environmental factors. Positive covariances, underscoring synchronous fluctuations in species abundance, are expected if species respond in similar ways to environmental change and/or as a consequence of indirect (Connell 1983) and positive (F. Bulleri, personal communication) species interactions. Thus, negative or positive covariances may modify the predictions made by Eqs. (28.1–28.4). The problem is now how to incorporate these covariance effects in the diversity–stability relationship.

It is known from probability theory (Feller 1950) that the variance of a sum of random variables is equal to the sum of the variances plus twice the sum of the covariances. Thus, for an assemblage of N species, total abundance (i.e. $x_1+x_2+\ldots x_N$) will have a variance of (Schluter 1984)

$$\sigma_{ass}^2 = \sum_{i=1}^{N} \sigma_i^2 + 2\sum_{i=1}^{N-1} \sum_{j=i+1}^{N} \mathrm{cov}(i,j) \qquad (28.5)$$

where σ_i^2 is the variance of the ith species and cov(i, j) is the covariance between species i and species j. Equation (28.5) shows that the variance in total abundance is the sum of all terms in the full $N \times N$ covariance matrix. For the diversity–stability relationship to hold, it is necessary that the summed variances and/or the summed covariances decline as the number of species in the assemblage increases. Negative covariances will contribute to reduce variability in total abundance, whilst positive covariances will have the opposite effect.

How does Eq. (28.1) relate to diversity and stability? Vasseur and Gaedke (2007) have noted that under the assumption of species independence (zero covariance), equal mean abundance and equal coefficient of variation (the most restrictive assumptions made by Doak et al. 1998), $\sigma_{ass} / \sum_{i=1}^{N} \sigma_i = \mathrm{CV}_{ass} / \mathrm{CV}_s$. These authors have called the ratio $\sigma_{ass}^2 / \sum_{i=1}^{N} \sigma_i^2$ the explained variance ratio (EVR), which is equal to $1/N$ for independent population dynamics (from Eq. 28.1), recovering the negative relationship between diversity and variability. If the assumption of equal species abundance is relaxed, then using Eq. (28.3) one arrives at what Vasseur and Gaedke (2007) called the threshold EVR (EVRt), which is equal to $\left[\frac{(1-e^{-a})(1+e^{-aN})}{(1+e^{-a})(1-e^{-aN})}\right]$. EVRt is the expected reduction in variance of total abundance with respect to variance in abundance of individual species due solely to the statistical averaging effect, assuming an exponential rank-abundance species distribution with $z = 2$. EVRt marks the threshold from synchronous to compensatory species dynamics, and the difference between EVR and EVRt (ΔEVR) can be used to assess whether the covariance effect contributes to increase (positive difference values) or decrease (negative difference values) variability in total species abundance (Vasseur and Gaedke 2007).

28.3.3 *Overyielding*

This term was introduced by terrestrial plant ecologists to indicate the case in which the average biomass of a species (usually the most productive species) increases with diversity as a consequence of a better use of resources (Trenbath and Harper 1973). In the context of the diversity–stability debate, Overyielding indicates the positive effect of diversity on the average total biomass of an assemblage, as indicated by Tilman (1999). This author related the total biomass of an assemblage to diversity by expressing the abundance of the *i*th species in the assemblage as $x_i = m/N^k$, where *m* is a constant. The overyielding parameter, *k*, measures how the mean abundance of individual species changes with diversity. If $k < 1$, the abundance of each species decreases less than proportionally to diversity, and this causes an increase in mean total abundance with diversity, indicating overyielding by all species. If $k = 1$, the mean abundance of single species decreases proportionally to diversity, and mean total abundance is constant and unrelated to diversity. If $k > 1$, mean total abundance decreases as diversity increases, indicating underyielding by all species. By expressing stability as the inverse of the coefficient of variation, Tilman (1999) related the ratio of assemblage to population stability (S_{ass}/S_1) to the parameters *k* and *z* using the general scaling relationship between the mean and the variance, and assuming species independence in Eq. (28.5), as $S_{ass}/S_1 = N^{1-k-(1-kz)/2}$. This shows that overyielding ($k < 1$) has a stabilizing effect by reducing the variance in total abundance relative to the variance in the abundance of individual species. An equivalent formula can be derived to relate *k* and *z* to EVR:

$$\mathrm{EVR} = N^{(1-zk)-2(1-k)} \tag{28.6}$$

28.4 Diversity–Stability Relationships in Assemblages of Rocky Shores

I now examine the relative importance of the different components of the diversity–stability relationship for assemblages of algae and invertebrates of rocky shores in the northwest Mediterranean. I based these analyses on data from 45 shores (stretches of coast 20–50m long) that have been sampled 7–10 times over periods of 2–3years between 1998 and 2002. Two types of assemblages were sampled on each shore: midshore assemblages located at 0.15–0.25m above the mean low level water (MLLW), and lowshore assemblages located at 0.0–0.1m with respect to the MLLW. Organisms were sampled visually in quadrats of

20 × 20cm, and abundances were expressed either as counts of occupied units of space for sessile organisms or as number of individuals for mobile species. Each quadrat consisted of 100 units of space obtained by dividing the surface into 25 sub-quadrats of 2 × 2cm, and assigning to each sessile organism an integer score ranging from 0 (absence) to 4 (when the sub-quadrat was entirely filled by the target organism; Dethier et al. 1993). Whilst the abundance of individual taxa could not exceed 100% (only few taxa in few occasions occupied the entire quadrat), total abundances were not constrained to 100 because assemblages were multilayered.

Of the 45 shores examined, 33 were sampled according to a hierarchical design consisting of two random sites (stretches of coast about 2m long) nested within each time×assemblage combination, and five replicate quadrats distributed randomly at each site. The remaining 12 shores were sampled with eight replicate quadrats distributed randomly at each time of sampling in each assemblage. In either case, care was taken to ensure that assemblages were sampled independently in space and time by design.

Estimates of temporal variance in total abundance and in the abundance of individual taxa (species or morphological groups) were obtained for each shore and assemblage separately (90 cases in total), using a generalized linear mixed-effect model fitted by restricted maximum likelihood (REML). Depending on sampling design, the model of analysis was either a two-factor hierarchical model, with time and site nested within time as random factors, or a one-factor model with time as the main effect. Because the methods to assess diversity–stability relationships discussed in this paper require comparisons of temporal variances across taxa, it was desirable to express temporal variability on a proportional scale. This was achieved by specifying a quasi-Poisson distribution for the error structure in combination with the log-link function (McCullagh and Nelder 1989). The quasi-Poisson distribution was preferred over Poisson errors because it allows the dispersion parameter to vary and, therefore, can model overdispersion, which is a common feature for organisms living on rocky shores. Analyses were done using the function lmer in R 2.6.

The distribution of the explained variance ratios (EVRs) clearly indicated that temporal fluctuations in summed abundances were much lower than the sums of the fluctuations in abundance of individual taxa (Fig. 28.1a). This effect decreased slightly when analyses were restricted to the most abundant taxa (Fig. 28.1b, c). The assumption of an exponentially decaying rank–abundance distribution was realistic for the assemblages investigated (Fig. 28.2), so Eq. (28.3) was used to estimate the EVRt values, which reflected the influence of the statistical averaging effect on EVR. EVRt values ranged from 0.090 to 0.681, with a median value of 0.189, indicating that the statistical averaging effect alone was responsible for about a three- up to a nine-fold reduction in variance of total abundance, compared to the variance in abundance of individual taxa. In addition to statistical averaging, asynchronous dynamics of individual taxa contributed to stabilize total abundance, as indicated by the prevalence of negative covariance effects (estimated as ΔEVR, the differences between EVR and EVRt) in Fig. 28.3a. Restricting the analyses to the most abundant taxa—which resulted in increased equitability—had the effect of enhancing the relative importance of statistical averaging and of decreasing the

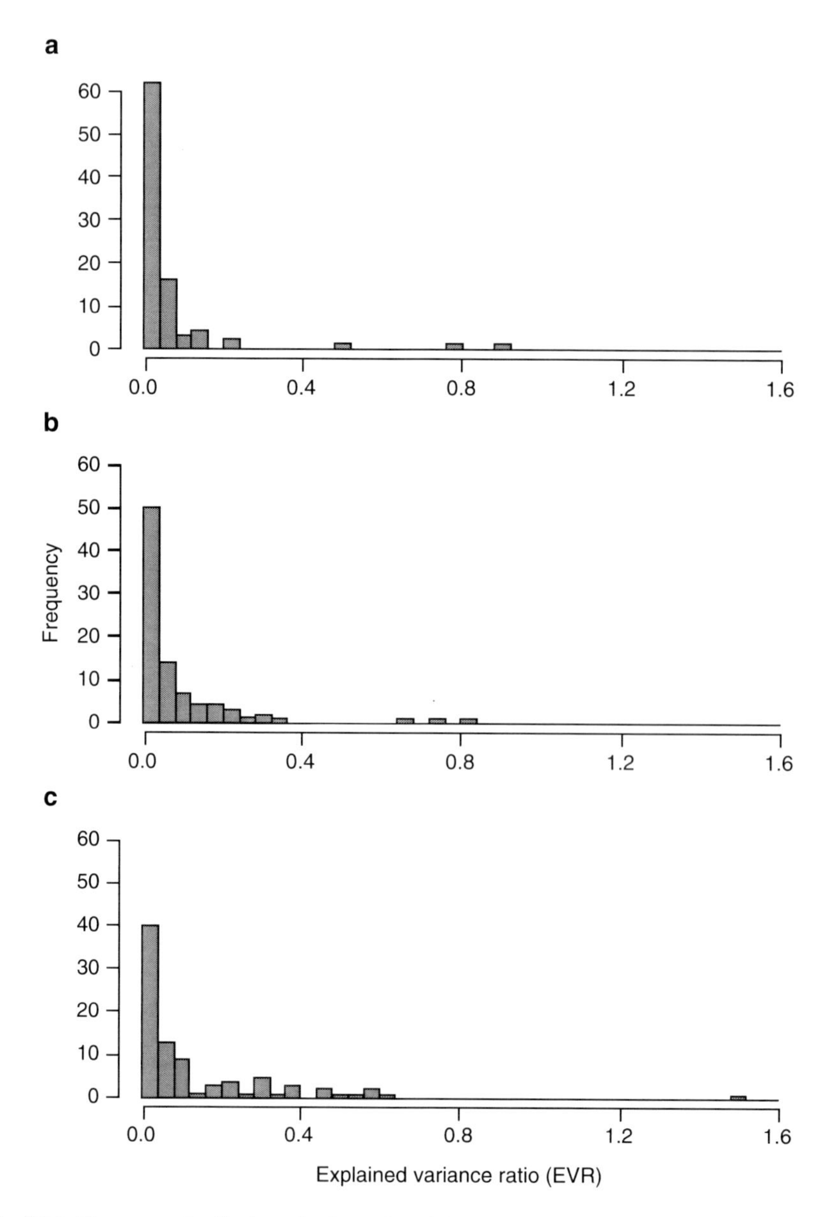

Fig. 28.1 Frequency distribution of values of explained variance ratio (EVR) for **a** the full set of taxa, **b** taxa with relative abundance greater than 5% and **c** taxa with relative abundance greater than 10%

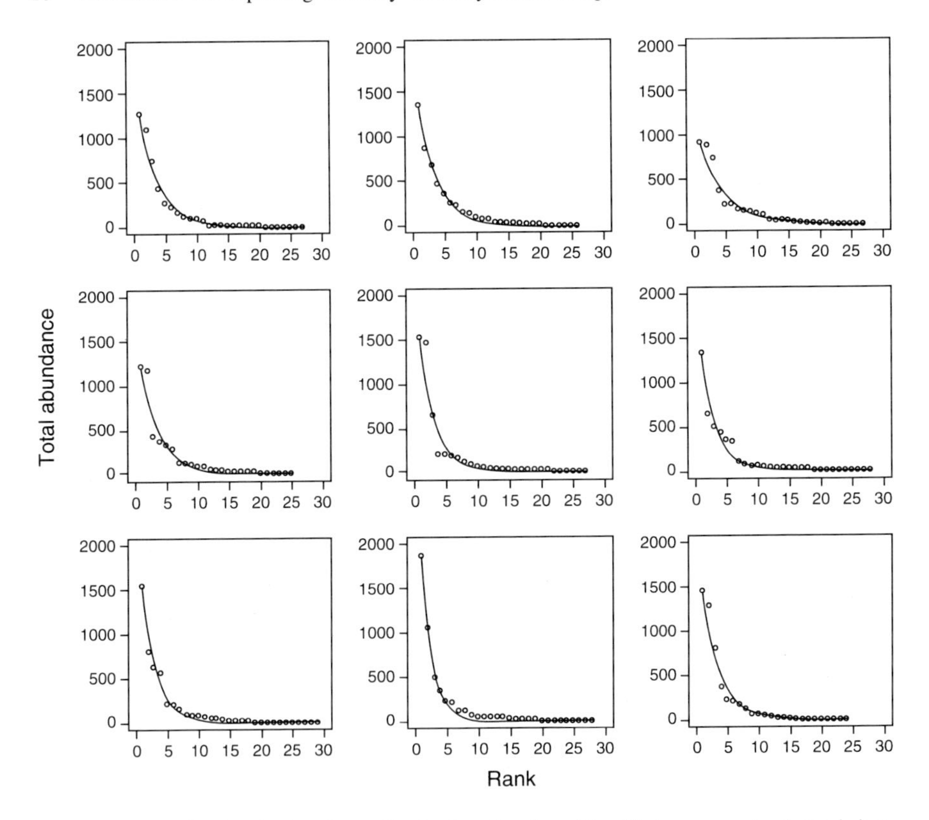

Fig. 28.2 Example of rank–abundance distributions for nine of the 45 shores examined. Total abundance is the sum of abundances observed for each taxon on a shore over the course of the study

covariance effect, as expected from theory (Fig. 28.3b, c, Eq. 28.3). Negative ΔEVR occurred significantly more often than expected by chance in all the analyses corresponding to the data reported in Fig. 28.3 (two-tailed binomial tests: $P < 0.001$ in all cases).

The derivation of the EVRt values assumed a scaling relationship between mean and variance for individual taxa of $z = 2$. The data showed a global value of (± s.e.) of 1.5 (± 0.05), which is lower than the assumed value. This implies that the statistical averaging effect was underestimated in the analyses. Lhomme and Winkel (2002) provided an expression equivalent to Eq. (28.3) where the scaling parameter is allowed to vary. Using this more complex expression to derive EVR and EVRt did not change the qualitative nature of the results (data not shown).

Finally, one can use the estimates of z in combination with Eq. (28.6) to examine whether overyielding can be expected for these assemblages. The overyielding coefficient, k, was negative in some occasions, with non-sense values ($k \ll 0$) occurring at two sites, probably as a consequence of the sparse abundance of organisms and the large proportion of unoccupied space characterizing those sites. If outliers

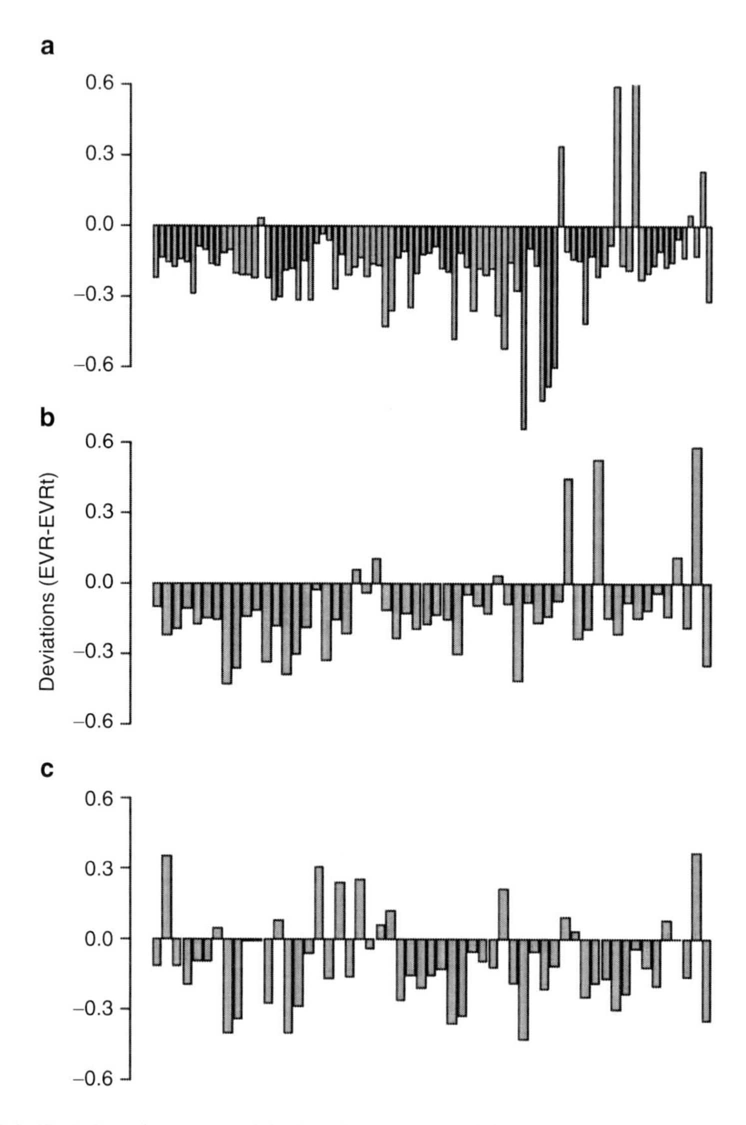

Fig. 28.3 Deviations between explained variance ratios and thresholds (ΔEVR) for **a** the full set of taxa, **b** taxa with relative abundance greater than 5% and **c** taxa with relative abundance greater than 10%

were removed, then k ranged between –0.907 and 3.487 with a median value of 0.682 (compared to a median of 0.485 using all data). This suggested that at most sites diversity had a strong stabilizing effect on temporal variance in total abundance through overyielding.

28.5 Discussion

Concern over increasing rates of species loss from natural environments has spurred a considerable amount of research on diversity–stability relationships in the last 10 years. We now have a well-developed theory to predict how changes in diversity affect the temporal stability of aggregate response variables, such as total abundance, through changes in the dynamics of individual species abundances, including their interactions and responses to environmental fluctuations. Statistical averaging, species covariances and overyielding are the key mechanisms underlying these relationships. The next relevant step for marine ecologists is to engage in the critical tests of the theory.

Increasing diversity may have two opposite effects on the variability of individual species in an assemblage and, therefore, on their contribution to stability in total abundance. All else being equal, increasing diversity is expected to reduce the mean abundance of individual species and, therefore, their variance, if means and variances are positively related (the statistical averaging effect). Increasing diversity may, however, also increase the chance that strong interacting species occur in the assemblage, enhancing the intensity of species interactions and making the abundance of individual species more variable. Theory ascribes to the scaling relationship between the mean and the variance a key role in determining how individual species respond to increasing diversity (Doak et al. 1998; Tilman et al. 1998). Variability at the species level is expected to decrease with increasing diversity for values of $z > 2$ (where z determines the strength of the relationship between the mean and the variance; see Eq. 28.4), making statistical averaging an important stabilizing mechanism at the level of the assemblage when the abundances of these species are summed. In contrast, the variability in individual species abundance is expected to increase with diversity for $z < 2$, reducing the importance of statistical averaging as a stabilizing mechanism. Heterogeneity in species rank–abundance distributions (large values of a in Eq. 28.3) further decreases the importance of statistical averaging (Doak et al. 1998; Vasseur and Gaedke 2007).

The nature and intensity of biological interactions also play important roles in determining the relationship between diversity and stability at the species level. Empirical studies have provided conflicting evidence regarding this link. Vogt et al. (2006), for example, found a decrease in population variability with diversity in multitrophic meiofaunal assemblages of rock pools, despite a mean value of 1.91 for the z coefficient, close to the critical value of $z = 2$ for which temporal variability at the species level should be independent of diversity. This finding suggested that complex species interactions in multitrophic assemblages may introduce uncertainty in theoretical predictions. Other studies have reported either an increase (Tilman 1996), a decrease (Ives et al. 1999, 2000; Valone and Hoffman 2003) or no effect (McGrady-Steed and Morin 2000) of diversity on variability in abundance of individual species.

Biological interactions and differential responses of species to environmental fluctuations may affect variability in total abundance through the covariance effect. Negative covariances (compensatory dynamics) are expected to reduce variability in total abundance by introducing a negative term in Eq. (28.5). To date, the importance of compensatory dynamics as a stabilizing mechanism of aggregate properties of natural assemblages has been examined in a limited number of systems, and no general consensus has emerged (Micheli et al. 1999; Keitt and Fisher 2006; Houlahan et al. 2007).

Increasing efficiency in use of resources with increasing diversity is also expected to reduce variability in total abundance through overyielding (Tilman 1999). In the presence of overyielding, the abundance of individual species does not decline with increasing diversity, leading to greater stability at the assemblage level. This effect also depends on the scaling mean–variance relationship, and disentangling the contribution parameters z and k in Eq. (28.6) can be arduous from observational data. The large number of experiments reporting positive diversity–productivity relationships provide evidence that overyielding is common in some systems like grasslands, where most of these experiments have been done (Hooper et al. 2005). However, lack of consensus on the specific mechanism(s) underlying positive diversity–productivity relationships (essentially, the heated debate over sampling versus complementary effects—e.g. Hooper et al. 2005) precludes a clear explanation of how diversity drives stability through overyielding.

It is clear from the preceding discussion that no general consensus has emerged on the relative importance of the different mechanisms linking diversity to stability. To date, these issues have been examined in a limited number of assemblages, mostly herbaceous plants and freshwater plankton. Using a dataset that generated 90 independent estimates of the different components of the diversity–stability relationships, I have shown that all the three mechanisms invoked by theory—statistical averaging, compensatory dynamics and overyielding—contributed to reduce fluctuations in total abundance of assemblages of algae and invertebrates of rocky shores. Although a detailed analysis of the relationship between the metrics describing assemblage-to-species (taxa) variability—i.e. EVR, EVRt and ΔEVR—and diversity was beyond the scope of the present paper, some lines of evidence indicated that loss of diversity might have contrasting effects on the stability of total abundance for the assemblages examined. A value of $z > 1$ and the low values of EVRt led to the prediction that loss of diversity would reduce the stabilizing effect of statistical averaging. Furthermore, decreasing diversity would alter the web of interactions that likely accounted for compensatory dynamics in these assemblages, reducing the influence of the covariance effect on stability (Benedetti-Cecchi 2000). When analyses focused on the most abundant species, however, the statistical averaging effect became stronger due to reduced dominance (data not shown), and the covariance effect decreased. Thus, a decline in diversity may not necessarily result in less stability if it is driven by the loss of rare species, which are those most prone to local extinction.

Of course, there are also a number of caveats that must be taken into account when approaching diversity–stability relationships using the framework described here. Assessing the stability properties of aggregate measures like total abundance or biomass may be irrelevant to conservation issues, if these properties can occur under very different combinations of species abundances and identities, including unnatural or unwanted ones. In this case, a detailed account of changes in the structure (species composition and abundance) of assemblages would be needed. Furthermore, whilst aggregate measures can provide convenient response variables to assess the functioning of an ecological system as a whole, it is the ecology of the component species that ultimately we need to understand to preserve biodiversity. Thus, examining diversity–stability relationships at the species level has merit per se, regardless of the implications these links may have for the stability of total abundance or other aggregate response variables.

With these considerations in mind, I believe the synthetic theory emerging from the diversity–stability debate offers a unique opportunity to integrate research at different levels of ecological organization. The data presented in this paper indicated that the mechanisms commonly invoked to explain stability–diversity relationships in terrestrial and freshwater systems may also be important for marine benthic assemblages. Hopefully, marine ecologists will take the challenge of contributing to this intriguing research program with additional observational and experimental studies.

Acknowledgements This study was supported by grants from the University of Pisa, the Census of Marine Life Programme NaGISA and the MARBEF (Marine Biodiversity and Ecosystem Functioning) Network of Excellence funded in the European Community's Sixth Framework Programme. This is contribution number 8041 of MARBEF. I thank I. Bertocci, F. Bulleri, E. Maggi and L. Tamburello for commenting on an early version of the manuscript.

References

Allison G (2004) The influence of species diversity and stress intensity on community resistance and resilience. Ecol Monogr 74:117–134

Arenas F, Sánchez I, Hawkins SJ, Jenkins SR (2006) The invasibility of marine algal assemblages: role of functional diversity and identity. Ecology 87:2851–2861

Benedetti-Cecchi L (2000) Variance in ecological consumer-resource interactions. Nature 407:370–374

Benedetti-Cecchi L (2006) Understanding the consequences of changing biodiversity on rocky shores: how much have we learned from past experiments? J Exp Mar Biol Ecol 338:193–204

Bruno JF, Boyer KE, Duffy JE, Lee SC, Kertesz JS (2005) Effects of macroalgal species identity and richness on primary production in benthic marine communities. Ecol Lett 8:1165–1174

Connell JH (1983) On the prevalence and relative importance of interspecific competition: evidence from field experiments. Am Nat 122:661–696

Cottingham KL, Brown BL, Lennon JT (2001) Biodiversity may regulate the temporal variability of ecological systems. Ecol Lett 4:72–85

den Boer PJ (1968) Spreading of risk and stabilization of animal numbers. Acta Biotheoretica 18:165–194

Dethier MN, Graham ES, Cohen S, Tear LM (1993) Visual versus random-point percent cover estimations: "objective" is not always better. Mar Ecol Prog Ser 96:93–100

Doak DF, Bigger D, Harding EK, Marvier MA, O'Malley RE, Thomson D (1998) The statistical inevitability of stability-diversity relationships in community ecology. Am Nat 151:264–276

Elton CS (1958) The ecology of invasions by animals and plants. Wiley, London

Feller W (1950) An introduction to probability theory and its applications. Vol I. Wiley, New York
Gaston KJ, McArdle BH (1994) The temporal variability of animal abundances: measures, methods and patterns. Philos Trans R Soc Lond B 345:335–358
Gonzalez A, Descamps-Julien B (2004) Populations and community variability in randomly fluctuating environments. Oikos 106:105–116
Hooper DU, Chapin FS III, Ewell JJ, Hector A, Inchausti P, Lavorel S, Lawton JH, Lodge DM, Loreau M, Naeem S, Schmid B, Setälä H, Symstad AJ, Vandermeer J, Wardle DA (2005) Effects of biodiversity on ecosystem functioning: a consensus of current knowledge. Ecol Monogr 75:3–35
Houlahan JE, Currie DJ, Cottenie K, Cumming GS, Ernest SKM, Findlay CS, Fuhlendorf SD, Gaedke U, Legendre P, Magnuson JJ, McArdle BH, Muldavin EH, Noble D, Russell R, Stevens RD, Willis TJ, Woiwod IP, Wondzell SM (2007) Compensatory dynamics are rare in natural ecological communities. PNAS 104:3273–3277
Ives AR, Gross K, Klug JL (1999) Stability and variability in competitive communities. Science 286:542–544
Ives AR, Klug JL, Gross K (2000) Stability and species richness in complex communities. Ecol Lett 3:399–411
Ives AR, Dennis B, Cottingham KL, Carpenter SR (2003) Estimating community stability and ecological interactions from time-series data. Ecol Monogr 73:301–330
Keitt TH, Fisher J (2006) Detection of scale-specific community dynamics using wavelets. Ecology 87:2895–2904
Lehman C, Tilman D (2000) Biodiversity, stability and productivity in competitive communities. Am Nat 156:534–552
Lhomme JP, Winkel T (2002) Diversity-stability relationships in community ecology: re-examination of the portfolio effect. Theor Popul Biol 62:271–279
May RM (1973) Stability and complexity in model ecosystems, 2nd edn. Princeton University Press, Princeton, NJ
McArdle BH, Gaston KJ (1990) Comparing population variabilities. Oikos 64:610–612
McArthur RH (1955) Fluctuations of animal populations and a measure of community stability. Ecology 36:533–536
McCann KS (2000) The diversity-stability debate. Nature 405:228–233
McCullagh P, Nelder JA (1989) Generalized linear models, 2nd edn. Chapman & Hall, London
McGrady-Steed J, Morin PJ (2000) Biodiversity, density compensation, and the dynamics of populations and functional groups. Ecology 81:361–373
McNaughton SJ (1977) Diversity and stability of ecological communities: a comment on the role of empiricism in ecology. Am Nat 111:515–525
Micheli F, Cottingham KL, Bascompte J, Bjornstad O, Eckert J, Fisher J, Keitt T, Kendall B, Klug J, Rusak J (1999) The dual nature of community variability. Oikos 85:161–169
O'Connor NE, Bruno JF (2007) Predatory fish loss affects the structure and functioning of a model marine food web. Oikos 116:2027–2038
O'Connor NE, Crowe TP (2005) Biodiversity loss and ecosystem functioning: distinguishing between number and identity of species. Ecology 86:1783–1796
Odum EP (1953) Fundamentals of ecology. Saunders, Philadelphia, PA
Petchey OL, Casey T, Jiang L, McPearson PT, Price J (2002) Species richness, environmental fluctuations, and temporal change in total community biomass. Oikos 99:231–240
Pimm SL (1984) The complexity and stability of ecosystems. Nature 307:321–326
Pimm SL (1991) The balance of nature? University of Chicago Press, Chicago, IL
Schluter D (1984) A variance test for detecting species associations, with some example applications. Ecology 65:998–1005
Stachowicz JJ, Whitlatch RB, Osman RW (1999) Species diversity and invasion resistance in a marine ecosystem. Science 286:1577–1579
Stachowicz JJ, Fried H, Osman RW, Whitlatch RB (2002) Biodiversity invasion resistance, and marine ecosystem function: reconciling pattern and process. Ecology 83:2575–2590

Stachowicz JJ, Bruno JF, Duffy JE (2007) Understanding the effects of marine biodiversity on communities and ecosystems. Annu Rev Ecol Evol Syst 38:739–766

Stewart-Oaten A, Murdoch WW, Walde SJ (1995) Estimation of temporal variability in populations. Am Nat 146:519–535

Tilman D (1996) Biodiversity: population versus ecosystem stability. Ecology 77:350–363

Tilman D (1999) The ecological consequences of changes in biodiversity: a search for general principles. Ecology 80:1455–1474

Tilman D, Lehman CL, Bristow E (1998) Diversity-stability relationships: statistical inevitability or ecological consequence? Am Nat 151:277–282

Trenbath BR, Harper JL (1973) Neighbour effects in the genus Avena. I. Comparison of crop species. J Appl Ecol 10:379–400

Underwood AJ (1996) Spatial patterns of variance in density of intertidal populations. In: Floyd RB, Sheppard AW, De Barro PJ (eds) Frontiers of population ecology. CSIRO, Melbourne, pp 369–389

Underwood AJ, Chapman MG (2000) Variation in abundances of intertidal populations: consequences of extremities of environment. Hydrobiologia 426:25–36

Valone TJ, Hoffman CD (2003) A mechanistic examination of diversity-stability relationships in annual plant communities. Oikos 103:519–527

Vasseur DA, Gaedke U (2007) Spectral analysis unmasks synchronous and compensatory dynamics in plankton communities. Ecology 88:2058–2071

Vogt RJ, Romanuk TN, Kolasa J (2006) Species richness-variability relationships in multi-trophic aquatic microcosms. Oikos 113:55–66

Yachi S, Loreau M (1999) Biodiversity and ecosystem productivity in a fluctuating environment: the insurance hypothesis. PNAS 96:1463–1468

Yodzis P (1981) The stability of real ecosystems. Nature 289:674–676

Chapter 29
The Aesthetic Value of Littoral Hard Substrata and Consideration of Ethical Frameworks for Their Investigation and Conservation

Heather E. Sugden, A.J. Underwood, and Stephen J. Hawkins

29.1 Introduction

Rocky coastlines provide impressive vistas. Above and below the waves, hard substrata provide habitat for a wide diversity of beautiful plants and animals, appreciated by walkers on the seashore and divers alike. The recreational and aesthetic value of rocky intertidal areas encompasses a large proportion of their ecosystem services, along with their provisioning and regulatory services such as nutrient regulation and supply of food (Dayton et al. 2005). Through increasing human populations and the modification of coastlines, overexploitation and pollution, these important coastal systems are becoming increasingly degraded and vulnerable, reducing their appeal and compromising the value of the goods and services they provide (Dayton et al. 2005).

Advances in technology and easier access to the seashore inevitably lead to growing public and scientific attention and usage. Whilst drawing attention to the value and aesthetic appeal of biodiversity on rocky substrata can promote their conservation, greater usage for scientific study and recreation can also damage coastal biodiversity. Nevertheless, by bringing attention to the common threats which face these increasingly popular areas, abuses such as the deposition of raw sewage directly onto shores and the dumping of waste materials can be reduced, and the fascinating flora and fauna which inhabit these areas can be protected.

In this chapter, we first consider the aesthetics of assemblages on hard substrata and impacts reducing their appeal. We then discuss some of the ethical issues involved in their use for research and the general public. Codes of practice currently in use are given, as are case studies of how ethical considerations can enable the sustainable usage of shores.

M. Wahl (ed.), *Marine Hard Bottom Communities*, Ecological Studies 206,
DOI: 10.1007/978-3-540-92704-4_29, © Springer-Verlag Berlin Heidelberg 2009

29.2 Aesthetics

29.2.1 Rocky Shores

There has been a long history of recreational use of the seashore. In the 19th century, natural history of the seashore became popular and fashionable amongst a new wave of tourists taking seaside holidays. This enthusiasm was reflected in numerous books on the seashore and its natural history in the UK and elsewhere (e.g. Gosse 1854, 1855, and his many imitators). Much attention was given to rock pools and to the collection of specimens, some bound for public or private aquaria. Instructions were given on where to go and how to collect. Pressed seaweeds were highly valued and became a genteel pursuit for many ladies. The wealthier visitors to the seashore often drew on the help of the locals, such as "stout backed quarrymen" (Kingsley 1855). Some shores suffered from over-collecting and, by the end of the Victorian era, writers were lamenting the damage done to seashores in the vicinity of fashionable resorts in Britain.

The enthusiasm for the seashore persists to this day. In many countries, recreational activity is primarily limited to gathering food or collecting bait (Kingsford et al. 1991). "Rock pooling" is still a feature of seaside holidays worldwide and the UK Royal Commission on Environmental Pollution, in its report on the *Torrey Canyon* oil spill, commented that one of the delights of a seaside holiday had been severely affected by this major pollution incident. Artificial sea-defences are now proliferating on many coastlines due to rising sea levels and stormier seas threatening coastal urban areas and infrastructure such as roads and railways (Airoldi et al. 2005). The incorporation of design features which maximise biodiversity has been advocated (Chapman 2003; Moschella et al. 2005). Many people value the sea life of artificial shores as well as natural coastlines. In recent years, rock pools have been incorporated into artificial habitats, such as seawalls and breakwaters (Chapman 2007; Moreira et al. 2007). This can often include life which has completely colonised artificial habitats such as former dock basins (Allen et al. 1995), and which can become oases of marine life in urban areas and contribute to inner city renewal schemes (Hawkins et al. 1999).

29.2.2 Diving

A second wave of recreational appreciation of the biodiversity on hard substrata arose via snorkelling or SCUBA diving. Whilst the early history of recreational diving involved those interested in wrecks, spear-fishing or collecting food, the advent of underwater cameras led to much interest in photographing all forms of marine life. There is a plethora of books for divers and non-divers, based on underwater photographs which display to a new audience the beauty of marine life from shallow waters down to the limits of recreational diving (e.g. Naylor 2005).

The enduring popularity of natural history broadcasts means that startling underwater images have entered our living rooms via TV, due to pioneers such as Jacques Cousteau and Hans and Lottie Hass and, more recently, David Attenborough and Sylvia Earle. The Internet has increased accessibility to the underwater world, and websites such as *MarLIN* (www.marlin.ac.uk) have pioneered the virtual dive. Submersibles have extended the range and depth of hard substrata brought to the public attention, including wrecked liners such as the *Titanic* and sunken battleships (e.g. *HMS Hood*), through the work of Ballard and others (Ballard 1995).

The enthusiasm shown by divers from a variety of different backgrounds has also led to huge support for conservation initiatives such as *Seasearch* in the UK (www.seasearch.org), a national project in which volunteer sports divers can become involved in recording marine life to provide information to protect marine habitats.

29.2.3 Impacts on Aesthetic Value

The aesthetic appeal of rocky shores can be reduced by many human impacts. These include stranding of ephemeral algae as a result of eutrophication (Karez et al. 2004; Worm and Lotze 2006), sewage pollution from untreated effluents and outfalls (Fraschetti et al. 2006), oil pollution and unsightly flotsam and jetsam—particularly the burgeoning quantity of plastic debris and rubbish in recent decades (Thompson et al. 2004, 2005).

On a much larger scale, disasters such as the *Torrey Canyon* event of 1967 have much more significant and severe effects. In this instance, 31,000,000 gallons of crude oil spread along 120 miles of English coast and 50 miles of French coast, devastating all marine and bird life (Smith 1968). It was the largest environmental disaster known at the time, and was amplified by the toxic detergents which were used in an attempt to disperse the oil slick (Southward and Southward 1978; Hawkins and Southward 1992; Raffaelli and Hawkins 1996). The combined effects of the oil and use of dispersants meant that the effects were evident long after the oil spill as such—indeed, the ramifications were still apparent up to 15 years later (Hawkins et al. 2002). Subsequent major spills such as the *Exon Valdez* (1989), *MV Braer* (1993), *Sea Empress* (1996) and *Prestige* (2002) have been treated more sensitively, but initial aesthetic impacts are still large.

Eutrophication can cause problems in many marine systems (e.g. the Adriatic and Baltic seas). Worldwide, sewage is being increasingly treated at least to the primary screening level coupled with longer outfall pipes (Raffaelli and Hawkins 1996). Ecological problems caused by sewage pollution are now much rarer in Western Europe, the United States, Australia and New Zealand. Eutrophication is also being combated by managing agricultural runoff at the catchment scale, but it can lead to unsightly "green" or "brown" tides (Phillips 2006). Whilst the ecological damage is often only slight, sewage-derived solids and algal blooms are often unsightly and extremely messy.

The last 30 years has seen an explosion in the use of plastics, particularly in packaging materials, but also in fishing gear. Plastic debris litters the strandline and is common on rocky shores. As well as being unsightly, it is suspected to have toxic effects when broken down, especially when ingested by deposit-feeding animals on soft shores (Handy and Shaw 2007).

29.3 Ethics

29.3.1 A Brief Background

Ethical issues about marine habitats are somewhat confusing. Any consideration of the ethics of "rights" of nature is intimately confused by the fact that only humans discuss ethics but humans are part of nature. The critical areas which must be considered in any discussion of ethics and rocky habitats are twofold. First, we must ensure an appropriate mix between exploitation (e.g. harvesting of resources such as fish) and conservation. Second, there must be properly developed ethical frameworks for using the organisms and the habitats themselves in ecological studies and, particularly, in experiments.

Traditionally, it has been tacitly assumed that scientific research and ethical frameworks were separate issues, not to be considered together. This was well summarized by Rolston (1975) who concluded that "natural laws" are based on moral and ethical principles for dealing with nature. They are considered to be quite different from the theories dealt with by scientists. These are descriptive and supposedly free from moral or other constraints.

More realistic views now tend to be in the ascendant, particularly the realization that development of scientific theories is in no way free from value-ladenness, personality–cultism, ambition, greed and, of course, ethical and moral constraints.

Recent decades have seen the coming together, at least in principle, of philosophers, ethicists and natural scientists in preliminary attempts to construct a more realistic framework. This has partially been driven by environmentalism—the charter of thought processes and political activism which advocates conservation and preservation of habitats and species because they have intrinsic values. Thus, development of ecophilosohies (Davis 1989) requires the development of environmental ethics which are informed and influenced by the data, information and insights gained from nature by its investigators. A useful review and introduction for ethics in relation to environmental impacts can be found in Fairweather (1993).

Such ecologically influenced views should, according to Rodman (1983), include intrinsic values in the ways the values of ecology are described, assessed and acted upon. Rodman (1983) also considered that the thought processes of ecological thinking should include the interactions and relationships among components of environments. This, at the very least, has long been the major topic investigated by ecologists. The development of more ethical, philosophical approaches would benefit by being better informed about the science! Finally, more ecosensitive ethics should attempt to prevent interference with natural processes, thereby contributing to sustaining ecological systems (Rodman 1983). This does, however, beget serious issues at the core of ethical environmental behaviour, particularly the fundamental chasm between ecocentric and anthropocentric viewpoints. This is a notoriously difficult area. At one end of the possibilities is an idealized nature—the nature which existed before modern man ruined it. This form of fundamentalist ecocentrism does not actually place values on ecological principles and processes. Instead, it undervalues the role of modern humans, who exist even though ethical considerations would rather they did not. It also completely ignores the existence of hysteresis in humans dealing with nature. Thus, whatever the negativities of past destruction of habitats, it is much easier to destroy than to recreate these. So, modern, potentially more ethical behaviour will still have to be implemented in a damaged world.

Finally, such ecocentrism ignores the activities of humans long before the modern era. Thus, in espousing the ideal that some habitat should be restored to the conditions prevailing before modern man arrived (in Australia, as recently as 1788), the previous at least 40,000 years of aboriginal activities are somehow irrelevant. In many parts of the world, early humans undoubtedly destroyed many species (e.g. large mammals in North America; Martin and Klein 1984) and repeatedly modified and destroyed large areas of habitat because they used fire as a tool in hunting. There is no obvious demarcation—in terms of ethical principles—between damage done by people of different eras (despite some being quaintly considered to have been "noble savages" and others only "savages").

Ecocentric thinking can be quite dangerous when the real basis for thought is not unequivocally clear. Thus, there is a very grey area in what is popularly called nimby-ism (i.e. "not in my backyard") where local environmental activists clamour to prevent some activity because it will interfere with nature. In reality, however, what is to be conserved is rather the local natural amenity, for the continued (but exclusive) use by locals—including the activists. This is in essence anthropocentrism. It is, in fact, quite difficult to be truly ecocentric in ones thinking but, fortunately, greater attention is now being paid to the intrinsic values of ecological systems.

The other major issue for ecologists who became enmeshed in social controversies about topics involving conservation, pollution, impacts and resource harvesting is the nature of advocacy (Soulé 1985). The desire to "do the right thing" or, sometimes, simply "to be seen to be doing the right thing" can be a dangerous attitude. Thus, ecologists espousing some specific course of action (and, thus, potentially altering decision-making) need to be particularly careful that their advice or commentaries are neutral and based on objective analyses of any opposing points of view. In addition, advocacy by ecologists can "erode the objectivity … of the scientific process" (Wiens 1996), particularly when not considering fairly any evidence arguing against what is being advocated. Yet, some areas of ecology—for example, "conservation biology"—seek to be distinguished from other areas of ecology by encouraging and praising advocacy (Noss 1999).

Although advocacy can be problematic, if not downright dishonest, there are also fundamental arguments in favour of it. For example, Shrader-Frechette (1996) favoured advocacy because, without it, ecologists may be acting to preserve an unsustainable status quo. Obviously, individual ecologists need to determine their own position but all should be able to defend it, if challenged, by demonstrating how the ethical requirement of attempting to be objective is being maintained.

Given a need to be carefully introspective about the nature of ethics and its interaction with the ethics of nature and ecology, it is worthwhile to consider some of the relevant issues for ecologists working on hard substrata.

29.3.2 Experimental Ecology

Manipulative field experiments are in essence controlled disturbances. They usually involve removal or adjustments in density of species, transplants and modification of the environment. Thus, inevitably experiments alter the state of rocky shore assemblages and thus have impacts (Hawkins 1999). Intertidal organisms are thought to be robust to many external factors and disturbances, due to the extreme and often highly disturbed environments in which they live, their great dispersal and short-lived nature (Denny 2006). This is, however, not always the case. Disturbances naturally create patches in these areas, producing mosaic landscapes (Jones 1948; Underwood and Jernakoff 1981, 1984). Nevertheless, altering the frequency, intensity or prevalent type of disturbance can have severe consequences for the species living there and, in extreme cases, cause shifts to alternate phases where the original community undergoes such damage that it is replaced by an entirely different composition of species, possibly affecting the functioning of that ecosystem due to changes in the identities of species (Petraitis and Dudgeon 1999).

Rocky intertidal areas provide ideal sites for experimental ecology (Dayton 1971). They are easily accessible, whilst the assemblages of flora and fauna inhabiting these areas are relatively short-lived. Fundamental work on rocky shores included pioneering work on competition (Connell 1961) and grazing (Jones 1948; Dayton 1971), predator-prey interactions (Paine 1966; Menge and Lubchenco 1981), disturbance ecology (Connell 1978; Sousa 1979), as well as spatial and temporal variability (Underwood and Chapman 1998) and patch dynamics (Burrows and Hawkins 1998). Rocky shores provide ideal species and assemblages for monitoring responses to climate (Southward et al. 1995; Helmuth et al. 2006) and, more recently, are becoming the focus of attempts to unravel the complexity associated with ecosystem functioning and biodiversity (see Chap. 26 by Gamfeldt and Bracken).

29.3.3 Recoverability

Human pressure on rocky shores can affect the recoverability of certain species at small to large scales. On a small scale, there are numerous investigations and examples of the effects of trampling in rocky shore environments, which affects species compositions and results in losses of larger branching algae and mussels (Beauchamp and Gowing 1982; Povey and Keough 1991; Brosnon and Crumrine 1994; Pinn and Rodgers 2005). The effects are, however, thought to be relatively short-lived for some specific species—for example, algae-barnacle assemblages can recover in as little as 1 year (Brosnan and Crumrine 1994) and, when trampling is prevented, assemblages have recovered over time to their previous abundances (Schiel and Taylor 1999). Care must, however, be exercised when undertaking sampling, especially when visiting frequently sampled sites.

29.3.4 Slow Recovery or Non-reversible Manipulations

Planning or undertaking irreversible manipulations such as modifications of hard substratum to change habitat complexity or create new habitat such as rock pools (e.g. Underwood and Skilleter 1996) require careful consideration. In most cases, biodiversity will be enhanced and, from an ecological perspective, the changes could be considered as acceptable. Other scientific stakeholders such as earth scientists may be concerned by the acceleration of geomorphological processes which naturally would take 10s to 100s of years (Hall et al. 2008).

When removing species, a precautionary approach must be taken with species known to be slow re-colonisers, such as some canopy-forming species (e.g. *Ascophyllum*, up to 15 years, Jenkins and Hawkins 2003; *Hormisira*, Underwood 1998) or those creating biogenic structures through slow growth (calcareous algae, Kuffner et al. 2007). Poorly dispersing species may also be slow to re-colonise an area (e.g. *Nucella lapillus*, Day and Bayne 1988; some littorinid gastropods, Kyle and Boulding 2000). If plots are small, the potential for recovery will be maximised. There has been some debate about the existence of alternative stable states with respect to *Ascophyllum*-dominated assemblages in New England (Petraitis and Dudgeon 1999, 2004; Bertness et al. 2002). Should such states occur, then removals pushing a system into another state should be avoided, particularly if the extent of the assemblage manipulated is large.

29.3.5 Biogeographic Studies and Non-native Species

Experimental manipulations are vital to the understanding of mechanisms determining the structure of rocky intertidal benthic communities, and transplantation experiments are often common. For example, experiments are crucial to investigate the influences of topography (Underwood 2004), to test the behaviour of common species (Rajasekharan and Crowe 2007), and to examine changes in morphology (Kelaher et al. 2003), as well as the development of transplantation techniques to re-establish species which have been lost due to various anthropogenic disturbances (e.g. Correa et al. 2006). Most experiments using transplantation techniques do not pose a threat to the natural system, due to careful consideration of the species used and transplantation usually occurring within shores.

The geographic distribution and abundance of rocky shore species are governed by processes such as recruitment, survival, reproduction, the dispersive capabilities of individuals into the pool of species in an area (Gaston 1996; Witman et al. 2004; Airoldi et al. 2005) and interactions with other species. With global climate change, significant range expansions and reductions are becoming apparent in rocky shore habitats (e.g. Mieszkowska et al. 2007). The distribution of "native" species will be dependent not only on the life history and competitive abilities of species but also on how global climates continue to change. It is tempting to use experimental transplants to understand what sets biogeographic limits (e.g. Fischer-Piette 1955). However, most scientists would accept that it is unethical to do such experiments, because they could lead to the spread of species and actually interfere with the processes under investigation.

29.3.6 Genetic Considerations

The maintenance of genetic diversity is an essential component of the conservation of global biodiversity (see Chap. 26 by Gamfeldt and Bracken) and continued ecosystem functioning (see Chap. 27 by Crowe and Russell). In a changing world with new selection pressures, a loss in genetic variation may be associated with a loss in the ability of individuals and species to adapt and persist within communities. Experimental transplantation may lead to genetic homogenization.

Although not found on hard substrata, the eelgrass *Zostera marina* is an example of a species playing a critical role in many estuarine ecosystems and acting as a driver of biotic and abiotic processes (Johannesson and André 2006). Several studies have shown that genetic diversity in eelgrass communities is linked to ecosystem functioning and evolution, and is analogous to species diversity in this respect (Hughes and Stachowicz 2004; Reusch et al. 2005). In an attempt to reverse the widespread loss of *Z. marina* and other seagrasses which has occurred worldwide, many restoration attempts have been made involving the transplantation of vegetation from nearby sites (Williams and Davis 1996). Subsequently, restored eelgrass meadows often exhibit reduced genotypic diversity, with implications for continued ecosystem functioning (Williams 2001; Rhode and Duffy 2004). Hughes and Stachowicz (2004) have advocated the use of a diversity of genotypes in restoration transplants, thereby improving the persistence of restored meadows in the face of increased climatic changes.

Such considerations may be relevant also for hard substrata species, particularly animals with direct development such as dogwhelks (*Nucella* spp. from egg capsules) and littorinids (ovoviviparous, viviparous from egg capsules) which can have small-scale genetic variability and morphological variation due to localised selection in response to sharp environmental gradients and predation pressure on the shore, sometimes reinforced by sexual selection (e.g. in *Littorina saxatilis*,

Johannesson and Johannesson 1996; in dogwhelks, e.g. Colson and Hughes 2007). Transplantation must be undertaken over small scales, with care to prevent the underlying genetic structure of the population from being affected, as this could compromise future research in addition to having unknown consequences for the survival of individuals and the population biology of the species concerned. Dogwhelks have also suffered from local extinctions due to pollution by TBT, and recovery has been slow (Hawkins et al. 2003; Colson and Hughes 2007). Again, it would be tempting to try to speed the recovery of populations via transplants but care would need to be exercised due to small-scale genetic differentiation (Hawkins et al. 1999).

Box 1 Ethics

Case Study: Wembury Point—Geological Site of Special Scientific Interest (SSSI)

Historical Background

Wembury Point in South Devon is designated a Geological SSSI, providing a range of diverse habitats including rocky shores, slate reefs, rock platforms and cliff faces. These provide important sites for nesting seabirds, as well as supporting a vast array of marine plants and animals. Since the 1940s the 56-ha coastal estate has been protected from public pressure, when it was acquired by the Ministry of Defence (MoD). After decommissioning of the area, there was intense pressure for development due to its close proximity to Plymouth. In an unprecedented step, Natural England put pressure on the MoD to sell the land to a conservation organisation before being put on the open market—ultimately, the National Trust purchased Wembury. Since 1981 (www.devonwildlifetrust.org), Wembury Point and the surrounding areas have enjoyed a range of local and national forms of protection, including the following designations:

- Site of Special Scientific Interest
- Voluntary Marine Conservation Area
- Special Area of Conservation
- Area of Outstanding Natural Beauty.

The Voluntary Marine Conservation Area was founded due to the important wildlife of the area, as well as to raise public awareness about the marine environment and the threats facing its conservation. It stretches from Gara Point to Fort Bovisand, and covers both intertidal and shallow water reefs. It is extremely popular with educational groups from schools and universities and is an important area for scientific research, with long-term datasets stretching back well over 50 years (Southward et al. 1995, 2005). Pressure from a wide variety of users can harm the very area the reserve was set up to protect.

(continued)

Box 1 (continued)

Natural England, in collaboration with the Marine Biological Association of the UK, The National Trust and The University of Plymouth, established the following code of practice for the sustainable use of the area in order to help prevent any harm.

Code of Practice

The following activities are considered acceptable:

1. Inspection, description, photographing that does not involve disturbance or where disturbance is transitory. For example, turning boulders but replacing them in the same place without damage to organisms, small scale digging of sediment and backfilling the hole.
2. Collection of single specimens of organisms by individual researchers. For example, when required for laboratory identification or demonstration.
3. Collecting multiple individuals of locally abundant common species. For example, when required for research or education.

If your proposed activity does not correspond to any of the above you must consult WVMCA staff

The following activities require consultation, consent or a licence:

1. Any removal of habitat (e.g. rocks, sediment) or change to habitats (e.g. rockpool creation) that may cause a detrimental change to the shore.
2. Removal of any species that is rare or threatened. For guidance, see Biological Action Plan species lists or the Nationally Important Marine Features list (available in the near future).
3. Any activities that involve the use of power tools or the introduction of nutrients, narcotics, poisons or alien (non-native) species.
4. Any visits by groups for courses, workshops or classes.
5. Any activities whatsoever above lowest astronomical tide on the Great or Little Mewstone.

29.4 Conclusions

Rocky shores above and below the water continue to be an area of fascination and attraction for many people and, with advances in technology, our ability not only to explore the underwater world but also to disseminate images and information via digital media will continue to increase.

The problems highlighted in this chapter have serious implications for the continued aesthetic value of coastlines which are enjoyed by many different user groups. In addition to this, the ethical considerations discussed are vital in any activity to protect the marine environment so that it can be sustainably used by all. Specific guides and regulations have been put in place for areas which are considered to be of high value in terms of their biodiversity, the services they provide, their aesthetic appeal or their geological history, in order to achieve their conservation (Box 1). More attention to the ethical responsibilities involved in any management of people's use and enjoyment of coastal habitats and species will undoubtedly aid the future sustainability of these assets.

References

Airoldi L, Abbiati M, Beck MW, Hawkins SJ, Jonsson PR, Martin D, Moschella PS, Sundelof A, Thompson RC, Aberg P (2005) An ecological perspective on the deployment and design of low-crested and other hard coastal defence structures. Coast Eng 52:1073–1087

Allen JR, Wilkinson SB, Hawkins SJ (1995) Redeveloped docks as artificial lagoons: the development of brackish-water communities and potential for conservation of lagoonal species. Aquat Conserv Mar Freshw Ecosyst 5:299–309

Ballard RD (1995) The discovery of the Titanic. Grand Central Publishing, London

Beauchamp KA, Gowing MM (1982) A quantitative assessment of human trampling effects on a rocky intertidal community. Mar Environ Res 7:279–293

Bertness MD, Trussell GC, Ewanchuk PJ, Silliman BR (2002) Do alternate stable community states exist in the Gulf of Maine rocky intertidal zone? Ecology 83:3434–3448

Brosnan DM, Crumrine LL (1994) Effects of human trampling on marine rocky shore communities. J Exp Mar Biol Ecol 177:79–97

Burrows MT, Hawkins SJ (1998) Modelling patch dynamics on rocky shores using deterministic cellular automata. Mar Ecol Prog Ser 167:1–13

Chapman MG (2003) Paucity of mobile species on constructed seawalls: effects of urbanization on biodiversity. Mar Ecol Prog Ser 264:21–29

Chapman MG (2007) Colonization of novel habitat: tests of generality of patterns in a diverse invertebrate assemblage. J Exp Mar Biol Ecol 348:97–110

Colson I, Hughes RN (2007) Contrasted patterns of genetic variation in the dogwhelk *Nucella lapillus* along two putative post-glacial expansion routes. Mar Ecol Prog Ser 343:183–191

Connell JH (1961) The influence of interspecific competition and other factors on the distribution of the barnacle *Chthamalus stellatus*. Ecology 42:710–723

Connell JH (1978) Diversity in tropical rainforests and coral reefs. Science 199:1302–1310

Correa JA, Lagos NA, Medina MH, Castilla JC, Cerda M, Ramírez M, Martínez E, Faugeron S, Andrade S, Pinto R, Contreras L (2006) Experimental transplants of the large kelp *Lessonia nigrescens* (Phaeophyceae) in high-energy wave exposed rocky intertidal habitats of northern Chile: experimental, restoration and management applications. J Exp Mar Biol Ecol 335:13–18

Davis DE (1989) Ecophilosophy: a field guide to the literature. Miles, San Pedro

Day AJ, Bayne BL (1988) Allozyme variation in populations of the dog-whelk *Nucella lapillus* (Prosobranchia: Muricacea) from the South West peninsula of England. Mar Biol 99:93–100

Dayton PK (1971) Competition, disturbance and community organization: the provision and subsequent utilization of space in a rocky intertidal community. Ecol Monogr 41:351–389

Dayton P, Curran S, Kitchingman A, Wilson M, Catenazzi A, Restrepo J, Birkeland C, Blaber S, Saifullah S, Branch G, Boersma D, Nixon S, Dugan P, Davidson N, Vörösmarty C (2005) Coastal systems. In: Baker J, Moreno Casasola P, Lugo A, Suárez Rodríguez A, Dan L, Tang L (eds) Ecosystems and Human Well-being: Current State and Trends, vol 1. Millennium Ecosystem Assessment, Intergovernmental Panel on Climate Change. Island Press, Washington, DC

Denny MW (2006) Ocean waves, nearshore ecology, and natural selection. Aquat Ecol 40:439–461

Fairweather PG (1993) Links between ecology and ecophilosophy, ethics and the requirements of environmental management. Aust J Ecol 18:3–19
Fischer-Piette E (1955) Répartition, le long des cotes septentrionales de l'Espagne, des principales espèces peuplant les rochers intercotidaux. Annu Inst Oceanogr Monaco 31:37–124
Fraschetti S, Gambi C, Giangrande A, Musco L, Terlizzi A, Donavaro R (2006) Structural and functional responses of meiofauna rocky shore assemblages to sewage pollution. Mar Pollut Bull 52:540–548
Gaston KJ (1996) Biodiversity—latitudinal gradients. Prog Phys Geogr 20:466–476
Gosse PH (1854) The aquarium: an unveiling of the wonders of the deep. London
Gosse PH (1855) A handbook to the marine aquarium: containing instructions for constructing, stocking, and maintaining a tank, and for collecting plants and animals. Oxford University, London
Hall AM, Hansom JD, Jarvis J (2008) Patterns and rates of erosion produced by high energy wave processes on hard rock headlands: the Grind of the Navir, Shetland, Scotland. Mar Geol 248:28–46
Handy RD, Shaw BJ (2007) Toxic effects of nanoparticles and nanomaterials: implications for public health, risk assessment and the public perception of nanotechnology. Health Risk Soc 9:25–144
Hawkins SJ (1999) Experimental ecology and coastal conservation: conflicts on rocky shores. Aquat Conserv Mar Freshw Ecosyst 9:565–572
Hawkins SJ, Southward AJ (1992) The Torrey Canyon oil spill: recovery of rocky shore communities. In: Thayer GW (ed) Restoring the Nation's environment. Maryland Sea Grant, *College Station*, MD, pp 583–631
Hawkins SJ, Allen JR, Bray S (1999) Restoration of temperate marine and coastal ecosystems: nudging nature. Aquat Conserv Mar Freshw Ecosyst 9:23–46
Hawkins SJ, Gibbs PE, Pope ND, Burt GR, Chesman BS, Bray S, Proud SV, Spence SK, Southward AJ, Langston WJ (2002) Recovery of polluted ecosystems: the case for long-term studies. Mar Environ Res 54:215–222
Hawkins SJ, Southward AJ, Genner MJ (2003) Detection of environmental change in a marine ecosystem—evidence from the western English Channel. Sci Total Environ 310:245–256
Helmuth B, Mieszkowska N, Moore P, Hawkins SJ (2006) Living on the edge of two changing worlds: forecasting the responses of rocky intertidal ecosystems to climate change. Annu Rev Ecol Evol Syst 37:373–404
Hughes AR, Stachowicz JJ (2004) Genetic diversity enhances the resistance of a seagrass ecosystem to disturbance. Proc Natl Acad Sci USA 101:8998–9002
Jenkins SR, Hawkins SJ (2003) Barnacle larval supply to sheltered rocky shores: a limiting factor? Hydrobiologia 503:143–151
Johannesson K, André C (2006) Life on the margins: genetic isolation and diversity loss in a peripheral marine ecosystem, the Baltic Sea. Mol Ecol 15:2013–2029
Johannesson B, Johannesson K (1996) Population differences in behaviour and morphology in the snail *Littorina saxatilis:* phenotypic plasticity or genetic differentiation? J Zool 240:475–493
Jones NS (1948) Observations and experiments on the biology of *Patella vulgata* at Port St Mary, Isle of Man. Proc Trans Liverpool Biol Soc 56:50–77
Karez R, Engelbert S, Kraufvelin P, Pedersen MF, Sommer U (2004) Biomass response and changes in composition of ephemeral macroalgal assemblages along an experimental gradient of nutrient enrichment. Aquat Biol 78:103–117
Kelaher BP, Underwood AJ, Chapman MG (2003) Experimental transplantations of coralline algal turf to demonstrate causes of differences in macrofauna at different tidal heights. J Exp Mar Biol Ecol 282:23–41
Kingsford MJ, Underwood AJ, Kennelly SJ (1991) Humans as predators on rocky reefs in New-South-Wales, Australia. Mar Ecol Prog Ser 72:1–14
Kingsley C (1855) Glaucus; or, the wonders of the sea-shore. Macmillan, Cambridge
Kuffner IB, Anderson AJ, Jokiel PL, Rodgers KS, MacKenzie FT (2007) Decreased abundance of crustose coralline algae due to ocean acidification. Nature Geosci 1:114–117
Kyle CJ, Boulding EG (2000) Comparative population genetic structure of marine gastropods (*Littorina* spp) with and without pelagic larval dispersal. Mar Biol 137:835–845

Martin PS, Klein RG (eds) (1984) Quaternary extinctions. University of Arizona Press, Tucson, AR
Menge BA, Lubchenco J (1981) Community organization in temperate and tropical rocky intertidal habitats: prey refuges in relation to consumer pressure gradients. Ecol Monogr 51:429–450
Mieszkowska N, Hawkins SJ, Burrows MT, Kendall MA (2007) Long-term changes in the geographic distribution and population structures of *Osilinus lineatus* (Gastropoda: Trochidae) in Britain and Ireland. J Mar Biol Assoc UK 87:537–545
Moreira J, Chapman MG, Underwood AJ (2007) Maintenance of chitons on seawalls using crevices on sandstone blocks as habitat in Sydney Harbour, Australia. J Exp Mar Biol Ecol 347:134–143
Moschella PS, Abbiati M, Aberg P, Airoldi L, Anderson JM, Bacchiocchi F, Bulleri F, Dinesen GE, Frost M, Gacia E, Granhag L, Jonsson PR, Satta MP, Sundelof A, Thompson RC, Hawkins SJ (2005) Low-crested coastal defence structures as artificial habitats for marine life: using ecological criteria in design. Coast Eng 52:1053–1071
Naylor P (2005) Great British marine animals, 2nd edition. Sound Diving, London
Noss R (1999) Is there a special conservation biology? Ecography 22:113–122
Paine RT (1966) Food web complexity and species diversity. Am Nat 100:65–75
Petraitis PS, Dudgeon SR (1999) Experimental evidence for the origin of alternative communities on rocky intertidal shores. Oikos 84:239–245
Petraitis PS, Dudgeon SR (2004) Do alternative stable community states exist in the Gulf of Maine rocky intertidal zone? Comm Ecol 85:1160–1165
Phillips PA (2006) Drifting blooms of the endemic filamentous brown alga *Hincksia sordida* at Noosa on the subtropical east Australian coast. Mar Pollut Bull 52:962–968
Pinn EH, Rodgers M (2005) The influence of visitors on intertidal biodiversity. J Mar Biol Assoc UK 85:263–268
Povey A, Keough MJ (1991) Effects of trampling on plant and animal populations on rocky shores. Oikos 61:355–368
Raffaelli D, Hawkins SJ (1996) Intertidal ecology. Kluwer, Dordrecht
Rajasekharan M, Crowe TP (2007) Intrinsic differences in dispersal between populations of gastropods separated by a few metres: evidence from reciprocal experimental transplantation. J Exp Mar Biol Ecol 341:264–273
Reusch TBH, Ehlers A, Hämmerli A, Worm B (2005) Ecosystem recovery after climatic extremes enhanced by genotypic diversity. Proc Natl Acad Sci USA 102:2826–2831
Rhode JM, Duffy JE (2004) Relationships between bed age, bed size, and genetic structure in Chesapeake Bay (Virginia, USA) eelgrass (*Zostera marina* L.). Conserv Genet 5:661–671
Rodman J (1983) Ecological sensibility. In: Scherer D, Attig T (eds) Ethics and the environment. Prentice Hall, Upper Saddle River, NJ, pp 88–92
Rolston H (1975) Is there an ecological ethic? Ethics 85:93–109
Schiel DR, Taylor DI (1999) Effects of trampling on a rocky intertidal algal assemblage in southern New Zealand. J Exp Mar Biol Ecol 235:213–235
Schrader-Frechette K (1996) Throwing out the bathwater of positivism, keeping the baby of objectivity: relativism and advocacy in conservation biology. Conserv Biol 10:912–914
Smith JE (1968) 'Torrey Canyon' pollution and marine life: a report by the Plymouth Laboratory of the Marine Biological Association of the UK. Cambridge University Press, Cambridge
Soulé ME (1985) What is conservation biology? Biol Sci 35:727–734
Sousa WP (1979) Disturbance in marine intertidal boulder fields: the nonequilibrium maintenance of species diversity. Ecology 60:1225–1239
Southward AJ, Southward EC (1978) Recolonization of rocky shores in Cornwall after the use of toxic dispersants to clean up the Torrey Canyon spill. J Fish Res Board Can 35:682–705
Southward AJ, Hawkins SJ, Burrows MT (1995) Seventy years of changes in distribution and abundance of zooplankton and intertidal organisms in the western English channel in relation to rising sea temperature. J Therm Biol 20:127–155
Southward AJ, Langmead O, Hardman-Mountford NJ, Aiken J, Boalch GT, Dando PR, Genner MJ, Joint I, Kendall MA, Halliday NC, Harris RP, Leaper R, Mieszkowska N, Pingree RD, Richardson AJ, Sims DW, Smith T, Walne AW, Hawkins SJ (2005) Long-term oceanographic and ecological research in the Western English Channel. Adv Mar Biol 47:1–105

Thompson RC, Olsen Y, Mitchell RP, Davis A, Rowland SJ, John AWG, McGonogle D, Russell AE (2004) Lost at sea: where is all the plastic? Science 304:838–838
Thompson RC, Moore C, Andrady A, Gregory M, Takda H, Weisburg S (2005) New directions in plastic debris. Science 310:1117–1117
Underwood AJ (1998) Grazing and disturbance: an experimental analysis of patchiness in recovery from a severe storm by the intertidal alga *Hormosira banksii* on rocky shores in New South Wales. J Exp Mar Biol Ecol 231:291–306
Underwood AJ (2004) Landing on one's foot: small-scale topographic features of habitat and the dispersion of juvenile intertidal gastropods. Mar Ecol Prog Ser 268:173–182
Underwood AJ, Chapman MG (1998) A method for analysing spatial scales of variation in composition of assemblages. Oecologia 117:570–578
Underwood AJ, Jernakoff A (1981) Effects of interactions between algae and grazing gastropods on the structure of a low-shore inter-tidal algal community. Oecologia 48:221–233
Underwood AJ, Jernakoff A (1984) The effects of tidal height, wave-exposure, seasonality and rock-pools on grazing and the distribution of intertidal macroalgae in New South-Wales. J Exp Mar Biol Ecol 75:71–96
Underwood AJ, Skilleter GA (1996) Effects of patch-size on the structure of assemblages in rock-pools. J Exp Mar Biol Ecol 197:63–90
Wiens JA (1996) Oil, seabirds, and science. Biol Sci 46:587–597
Williams SL (2001) Reduced genetic diversity in eelgrass transplantations affects both population growth and individual fitness. Ecol Appl 11:1472–1488
Williams SL, Davis CA (1996) Population genetic analyses of transplanted eelgrass (*Zostera marina*) beds reveal reduced genetic diversity in southern California. Restor Ecol 4:163–180
Witman JD, Etter RJ, Smith F (2004) The relationship between regional and local species diversity in marine benthic communities: a global perspective. Proc Natl Acad Sci USA 101:5664–5669
Worm B, Lotze HH (2006) Effects of eutrophication, grazing, and algal blooms on rocky shores. Limnol Oceanogr 51(1):569–579

Part VI
Appropriate Research Methods

Chapter 30
Field and Research Methods in Marine Ecology

A.J. Underwood and Angus C. Jackson

30.1 Field Methods in Marine Ecology

Methods used to study the ecology of hard substrata are as diverse and varied as the organisms living in these habitats. Some techniques, e.g. counting organisms in quadrats, have remained unchanged, others continue to evolve rapidly. It is beyond the scope of this chapter to describe these in detail. Here, we briefly outline some methods used to sample organisms on hard substrata and refer to comprehensive sources of information. We also provide a few examples of methods that have recently entered the literature.

30.1.1 Sampling Organisms and Habitats

In contrast to sedimentary habitats where removal of samples is the norm, most organisms or habitats with hard substrata are sampled by direct observation or some form of remote-sensing. If presence/absence of taxa is all that is required or if organisms are rare, over-dispersed or inhabit difficult to access habitats, timed searches may be appropriate. Most biota from hard substrata are, however, epibenthic and can be sampled directly using sampling units placed on the surface. Density or abundance are usually estimated by counts or measures of percentage covers of organisms at points, along transects or in defined areas such as quadrats. Often, a point-intercept technique is used where the presence/absence of an organism is recorded under a set of points. Not all organisms on hard substrata can be sampled like this. If organisms provide biogenic structure that is inhabited by others, then destructive sampling may be required (e.g. cores of coralline algal turf, removal of kelp holdfasts or suction sampling). Many of these general methods for sampling and reasons for using them are found in Kingsford and Battershill (1998) and Sutherland (2006).

M. Wahl (ed.), *Marine Hard Bottom Communities*, Ecological Studies 206,
DOI: 10.1007/978-3-540-92704-4_30, © Springer-Verlag Berlin Heidelberg 2009

Sampling is, however, being increasingly done by image capture (e.g. by cameras, videos, spectrometers or radar) from satellite, airborne, eye-level, sea-surface or submerged platforms. Online facilities, such as Google Earth, have substantially increased the availability of aerial photographs. Techniques used for remote-sensing coastal substrata to depths of 30 m are in Robinson et al. (1996). An overview of remote-sensing of algae is in Guillaumont et al. (1997). Image capture has many benefits including rapid collection, low cost, pixel by pixel resolution, access to information outside the visible spectrum (e.g. infrared), a permanent record of the sample and, because they are non-interventionist, repeated samples can be collected through time. There are, however, limitations. For example, canopies of algae overlie and obscure organisms attached to the primary substratum. The resolution or two-dimensionality of images may prevent reliable identification of organisms. Images captured when capture devices are not vertically above the field-of-view also have problems with perspective and scale. A recent solution (Automated Benthic Image-Scaling System, ABISS; Pilgrim et al. 2000) uses the relative positions of laser markers to quantify these distortions and has served to identify accurately the structure of patches in benthic habitats (Parry et al. 2003).

30.1.2 Plankton

Many benthic species settle from the plankton. Although not strictly part of hard-substratum assemblages, it is often useful to measure the number and types of potential recruits in the plankton. The numerous ways of doing this depend on the habitat being sampled and the organisms of interest (reviewed by Harris et al. 2000). Common techniques include towed nets, tethered traps or nets and light traps. For intertidal studies, special traps have been used to capture plankton from breaking waves (Todd et al. 2006). Increases in processing power have enabled not only automated plankton counting by towed devices but also recognition of some types of plankton from images (Embleton et al. 2003).

30.1.3 Settlement of Organisms

Settlement is notoriously difficult to measure and some early measure of recruitment is usually the closest that can be achieved (Schiel 2004). Settlement can be measured on natural substrata cleared of algae, mussel beds, etc., or on artificial substrata. Substrata such as artificial turf and pan scourers have long been used in marine ecology and enable settlement in areas that are of standard size, material, topography, etc. They are easy to deploy and recover and can be transplanted between locations or habitats. For example, matrices of artificial turf that differed in density and/or length or differed in composition and/or spatial arrangement of units or arrangement of units have been used to measure how the structure of habitat influences colonisation (Matias et al. 2007).

30.1.4 Measuring Behaviour

Measuring behaviour of organisms on hard substrata can be problematic. In intertidal areas, when access is easy during low tide, the animals are often quiescent. Displacements of animals between tides are often measured using triangulation. Some clever, longer-term experiments have been done that detect when mobile animals regularly returning to a particular point are "home". Movements of animals have also been recorded continuously (even at night) when tagged with tiny lights (Davies et al. 2006).

Feeding may be observed directly (consumption is observed or noises of feeding are recorded) or inferred by some indirect mechanism (e.g. drill holes of predatory whelks, rasp marks of gastropod radulae in wax discs, rates of fish biting).

30.1.5 Measuring Physical and Chemical Variables

The ecology of hard substrata is not just about organisms; it often requires measurements of physical and chemical variables. At large scales, physical variables such as waves, currents and surface temperatures can be measured effectively by sensors on satellites. There is also a range of equipment for measurement at smaller scales (e.g. wave-rider buoys, pressure sensors, upward-looking sonar devices). Acoustic Doppler velocimetry has dramatically changed how waves and currents are measured in the field, although the equipment (e.g. Triton-ADV, Sontek/YSI) is expensive. Some of these variables are well understood and useful predictions can be made (e.g. the effect of wind on waves). Denny (1993) and Vogel (1994) provided very readable overviews of the theory behind many physical aspects of life in fluid media. Ideally, measurements should be made at intervals over which they are likely to change (often very rapidly) and at scales appropriate to the organisms that experience these forces (usually cm). At these frequencies and scales, how variables change is much less well understood. For example, theory does not accurately predict the forces exerted by breaking waves on topographically complex surfaces. It is only recently, with decreases in sensor size and large increases in the memory capacity of data-loggers, that data can be collected at ecologically suitable frequencies, durations and scales. Sensors and loggers not much bigger than a watch battery are now routinely used to collect large amounts of data on temperature and humidity (e.g. I-buttons; Maxim, California) and force transducers have been used to make small loggers that directly record the forces on aquatic organisms (Boller and Carrington 2006).

Analysis of the chemical constituents of seawater is also well understood. The influence of nutrients and contaminants on assemblages living on hard substrata is important. Increasing automation enables samples to be processed more rapidly and, for many variables, to be recorded in situ. Many recommended protocols govern the collection of samples (e.g. JGOFS 1994) and a comprehensive review of methods has been given by Crompton (2006).

30.1.6 Data Handling

Common sins of surveying are: believing the results without consideration of biases or inaccuracies and not storing information where it can be retrieved in the future (Sutherland 2006). Inaccuracies in data often occur during transcription or input of data. Independent verification or checking of data is an important (yet seldom considered) step to eliminate typographical or interpretive errors. Safe storage of data is relatively straightforward, since digital storage capacity has increased exponentially over recent years; efficient backup systems are available. Electronic storage of documents is now cheaper than paper, which is fortunate because image-intensive methods require large amounts of storage space.

Making the data available is a greater challenge. Publication in the scientific literature provides interpretative information about data but the data as such are often not available. Many researchers are reluctant to make available their data, even after publication, because data are commercially valuable (e.g. fisheries data) or because of issues about intellectual copyright. This situation is improving with the advent of European marine databases (e.g. at http://mda.vliz.be) and organisations such as the UK Data Archive for Seabed Species and Habitats (DASSH). DASSH aims to safeguard data, to make data accessible and to provide digital facilities to archive data.

30.2 Experimental and Sampling Designs

30.2.1 Why Do We Need Experiments?

Much of ecology, particularly experimental ecology, is about trying to explain phenomena, i.e. observations about ecological patterns. The important issue about this endeavour is that there are usually, if not always, many potential explanations that might explain a given set of observations. For example, observing that a certain species of bryozoan is found only below a certain depth might be explained by processes that "attract" propagules there (e.g. cues from the substratum, positive responses to shade, etc.) or by processes that prevent it from being in more shallow areas (e.g. larvae cannot settle or the bryozoans cannot survive where light or wave action are too intense in shallower regions; seaweeds grow abundantly in shallower regions, causing competitive exclusion of the bryozoans, etc.).

Confronted with numerous possible explanatory mechanisms, some effort must be made to try to determine which might have reality. Several possible processes are used to achieve such a goal. One, which is still quite common in ecological literature, is simply to assert that a particular mechanism *must* be correct. Readers can find examples of this for themselves. More commonly, inductive methods are used—other examples of similar observations have been explained by competitive overgrowth by algae, so the present example is also explained by competition. This sort of argument has long been established (Popper 1968) to be completely spurious.

Slightly more sophisticated arguments involve the use of correlative information to demonstrate that some process must be the correct explanation. For example, if there is a positive correlation between numbers of barnacles and forces due to waves from place to place on a rocky shore, it is tempting to assume that wave action is responsible for the observed pattern. It may be—if wave forces *are* controlling numbers of barnacles, then there should be the observed correlation.

It may be, instead, that wave action is indirectly responsible—where waves are weak, predatory whelks eat barnacles, reducing their numbers (e.g. Menge 1978). Where waves are strong, whelks are ineffective, so barnacles are numerous. In this case, predation and how it is modified by waves, is the relevant process.

Alternatively, wave action may have no role at all and the observed pattern may have been established by processes operating during settlement of the barnacles, which was originally equally dense everywhere. During a subsequent period of calm weather, many barnacles died of desiccation and this was more intense on normally wave-exposed areas, where barnacles settle at higher levels when waves are large. During calm periods, these higher animals would be subject to greater desiccation and greater mortality, leading to the observed pattern.

Given the problems of trying to deduce processes solely on the basis of observations, it is not surprising that experimental manipulations of ecological processes to test and discriminate among competing alternative explanatory models has enabled widespread progress in ecology (see reviews by Hairston 1989; Underwood 1997; Mead 1988; Resetarits and Bernardo 1998).

30.2.2 What Are Experiments?

Experiments are structured tests of the different predictions that can be made from different explanatory models. The purpose of any experiment is to distinguish among different models by demonstrating that some of these fail because their predictions are not evident under the conditions created in the experiment. The notion behind this concept is that of falsification (Popper 1968). This is the logical argument that it is usually impossible and always very difficult to demonstrate the "correctness" or validity of a predictive hypothesis. There is no room here to consider all the issues (see summary in Underwood 1990) and the criticisms of this approach to experimental science (e.g. Mayo 1996).

In principle, the logical structure of falsification depends on the following idea. If some hypothesis (or prediction), *p*, implies that certain data, *q*, should be found in an experiment, then finding *q* does not enable one to conclude that the hypothesis, *p*, is correct. In logical form, if *p* implies *q* and *q* happens, it is not logically correct to assume that *p* was the cause. Any other process that results in *q* could also be the correct process.

In contrast, if the hypothesis, *p*, predicts data *q*, and *q* does *not* happen, it is logically sound to conclude that *p* was not a correct prediction. *p* was therefore falsified and the explanatory process from which *p* was predicted is also therefore falsified.

Experiments do not have to be manipulative, i.e. it is not necessary to alter something in order to test hypotheses, although this is common and often the best way to do a valid test. Some hypotheses are about patterns and are often called *mensurative* experiments (see Hurlbert 1984). These can be tested by examining the pattern under the appropriate conditions and determining whether it conforms to or rejects the prediction being tested. What matters is that an experiment requires collection of information about a prediction. This requires, first, that the prediction (or set of contrasting predictions) be very clear before the experiment is done and that the experiment creates or uses the conditions under which the prediction is made.

Note also that hypotheses are about the ecological variables relevant to the observations and to the processes considered in the models. So, hypotheses can be about means of variables, their variances, rates of change, frequencies of being at certain values, etc., etc., as required.

As an example of the distinction between mensurative and manipulative experiments, consider the case of sizes (diameters) of adult barnacles, which have been observed to vary from place to place. There are several possible reasons for this variation (e.g. Jeffery and Underwood 2001). To keep things simple, assume that a useful model is that size is a function of density. Where barnacles are numerous, they impede each others' growth and the adults are relatively small (Barnes and Powell 1950). Where they are sparse, adults grow larger. From this model, it can be predicted that there should be a negative correlation between mean size of adults and their density in different patches of shore. A mensurative experiment would consist of sampling numbers and sizes in different patches. A lack of correlation (or a positive correlation) would be contrary to the hypothesis and would refute the model.

A manipulative experiment would find patches of similarly numerous, small barnacles and thin them out at random to create several replicate patches of a given density, for a range of densities. When the barnacles have grown to adult size, their densities and sizes would be sampled. This enables a more safely interpretable test because it ensures that the patches being examined were all similar initially, in terms of the numbers of barnacles. It will, however, not be a useful experiment if mortality during the period of growth is density-dependent, so that the final densities are all very similar.

Note that measuring the final size of the barnacles in either type of experiment is only directly a measure of growth (which the hypothesis was about) if the starting sizes and time to reach the final size were similar in all the areas measured. This cannot be known for the correlative, mensurative study, so final size may not reflect very well the actual growth rate (i.e. animals in uncrowded conditions may grow at the same rate but for longer than those in sparse areas, resulting in different sizes but not for the reason proposed). The manipulative approach is superior because initial sizes and actual rates of growth can be measured properly.

30.2.3 Why Are Statistical Procedures Necessary?

Having determined what is to be known (i.e. the data required by the relevant hypotheses) and the methods appropriate to obtain the data, it is not always straightforward to identify patterns in the data to compare with what is predicted by the hypotheses. All sorts of variability and uncertainty will cloud the clarity of the data.

Intrinsic variability is inevitable for virtually all ecological variables. For example, the densities of organisms (or the concentrations of contaminants in the water, or the amount of light reaching a surface) will vary from place to place and time to time. Some of the variation will be systematic—for example, there will be fewer young, individual kelp at increasing distance from the outer edge of a bed of adult kelp, because spores have to travel further to reach greater distances. Much of the variability will, in contrast, be due to random sources of difference. So, even though there may be a consistent average pattern of decrease of kelp away from the edge, the actual numbers of young kelp will not be exactly those expected in the pattern. Some patches of adults will produce spores that travel further; some patches will produce spores that are carried very different distances because of variability in the local water currents; some areas will have few or no spores at some distances because they were consumed by grazers.

In the light of this, comparisons of data to test how well they fit the various hypotheses must be tests that take into account the variability of the data. These are statistical (or probabilistic) tests. They are of several types, notably Bayesian and frequentist. For a good account of Bayesian procedures for ecologists, see *Ecological Applications* (1996). For relevant information about some appropriate frequentist tests for ecological problems, see Underwood (1997).

30.2.4 Experimental and Sampling Designs

There are several important reasons why sampling and experiments must be very carefully designed. The major ones are:

1. to discriminate amongst competing models for observed phenomena, it is crucial to gather data that can actually discriminate amongst the different hypotheses (or predictions) from the models.

For example, one model might predict that rates of grazing by species A when on its own will be twice that when in areas with species B. A second model might predict that the rate of A alone will be three times that when with B. Only a carefully designed experiment, with appropriate replication and power (Cohen 1977), will enable these to be distinguished.

2. it is absolutely required that all care be taken to ensure that any logical inference from the experiment is valid, i.e. not erroneous because of artefacts due to experimental procedures and not confounded (confused with other possible conclusions) because of other possible explanations not falsified during the experiment.
   For example, an experiment is done to test the prediction that growth of barnacles will be slower where waves are strong. So, newly-settled barnacles in an area with strong and another area with weak wave-action are measured as they grow. If growth is faster in the former, so that the null hypothesis of no difference is rejected, the experimental data seem to support the model. It is, however, also possible that the difference in growth is due to other features of the two areas, unrelated to wave-action. For example, there may be more food in the wave-exposed area because it is in a region of upwelling. The experiment is unreplicated (see extensive discussion of ecological examples in Hurlbert 1984). To conclude that growth differed because of wave-action, it is necessary to show that the difference was greater than would be found by sampling several (replicated) areas of similar weak waves and several areas with strong waves. It would probably be better to manipulate waves to reduce (or, even, increase) wave-force but that is clearly difficult and, in fact, usually impossible.
3. all statistical procedures have important assumptions that must be met in order to arrive at valid conclusions.

An example of a statistical procedure having assumptions is the distribution-free (non-parametric) Kruskal-Wallis test, which is often used to compare the mean values of a series of experimental treatments. Many ecologists use it instead of alternatives (such as analysis of variance, which is the parametric test) because they believe it has no assumptions. Analysis of variance assumes that the data are independent, that they are normally distributed and that the variances of data in every treatment are equal. The Kruskal-Wallis test assumes that the data are independent and that the variances are equal in every treatment; it also assumes that the data come from a continuous distribution in every treatment.

To ensure that experimental data will fit the assumptions of statistical procedures requires care in the design (see numerous examples in Underwood 1997, and Winer et al. 1991).

30.2.5 Some Major Issues with Experimental Designs

Having decided to use experimental methods and understanding that it is important to be careful about designing the experiment (or sampling), it is sensible to understand some of the key components of design. Note that it really does not matter what framework of decision-making or statistical thought you wish to use—the problems for experimental work are mostly common.

The spatial or temporal scale at which an experiment is done can often cause problems. For example, suppose it is hypothesized that beds of mussels will recover quickly after being cleared by a storm, provided that grazing urchins are kept out of the area. The study must be long enough for this to be possible. If mussels breed and recruit successfully only every few years, then a study over 2 years may not be long enough for recruitment to occur. There will be no rapid recovery in experimental areas without urchins—regardless of the validity of the model from which the prediction was derived.

If appropriate spatial and temporal scales are not used in experiments, then the results cannot have any relevance to the model—the hypotheses are at some scale not being tested.

One area of considerable difficulty in ecological experiments is the use of laboratory conditions to test hypotheses. Under laboratory conditions, major ecological processes are missing (for example, wave-action at realistic spatial scales and forces; predators; recruitment of propagules from other populations; scouring by sand across the substratum). It is often possible in a laboratory experiment to determine that some process can influence an ecological pattern. If, in contrast, the anticipated effect is not evident, that may mean that the process does not have the hypothesized effect, or that the conditions in the laboratory were insufficient for it to be evident. Even where the process *can* have the predicted effect in a laboratory, that is a very different conclusion from knowing that it *does* have an effect in the field, where it is supposed to be operating.

A third problem is that of so-called natural "experiments". Consider again the experiment on mussels. It is decided to compare recruitment in some areas that naturally have urchins with what happens in areas naturally without urchins. If recovery is faster in the former, it may have nothing to do with the absence of urchins. Mussels may naturally recruit more into patches of habitat where urchins cannot live. Choosing areas that are naturally different in the presence or absence of urchins will confound the experiment. Any difference in recruitment of mussels may be unrelated to urchins. Manipulating the urchins, by removing them from areas where they occur, can solve this problem, because the replicate areas with or without urchins are similar at least in terms of urchins.

Another major issue in ecological experiments is that of controls for experimental artefacts. Consider that is has been proposed that urchins kill newly settling mussels, resulting in less rapid recovery of mussel beds after storms. Urchins are removed from replicate experimental plots and kept out by installing a fence around each plot. Unmanipulated areas are simply marked, without removing urchins. Suppose that, at the end of the experiment, there are more mussels in the urchin-free areas, rejecting the null hypothesis of no difference and being interpreted as supporting the model.

In this case, however, an alternative (and confounded) explanation is that the presence of the fences (and not the associated absence of urchins) influenced recruitment (by altering water-flow, transport of sediment, etc.). To be able to interpret the experimental data in relation to the hypothesis and model requires a control for any influence of fences that is independent of the fences causing urchins to be absent. Constructing a realistic control treatment (e.g. a fence that contains a natural density of urchins; a partial fence that would influence water flow but let urchins in and out) is usually very difficult. Urchins kept inside a fence or moving in and out of a partial fence may not behave normally (thus influencing their effects on mussels).

One major joy of doing ecological experiments is that all the issues about experimental design that have to be solved to enable logically valid conclusions to be drawn require deep thought about the logic, the data, the statistical procedures and, above all, the natural history of the animals and plants so that relevant scales, necessary controls and appropriate procedures will be incorporated in the experiments.

Acknowledgements The preparation of this paper was supported by funds from the Australian Research Council through its Special Centres' programme.

References

Barnes H, Powell HT (1950) The development, general morphology and subsequent elimination of barnacle populations, *Balanus crenatus* and *Balanus balanoides*, after a heavy initial set. J Anim Ecol 32:107–127

Boller ML, Carrington E (2006) In situ measurements of hydrodynamic forces imposed on *Chondrus crispus* Stackhouse. J Exp Mar Biol Ecol 337:159–170

Cohen J (1977) Statistical power analysis for the behavioural sciences. Academic Press, New York

Crompton TR (2006) Analysis of seawater: a guide for the analytical and environmental chemist. Springer, Berlin Heidelberg New York

Davies MS, Edwards M, Williams GA (2006) Movement patterns of the limpet *Cellana grata* (Gould) observed over a continuous period through a changing tidal regime. Mar Biol 149:77–787

Denny MW (1993) Air and water: the biology and physics of life's media. Princeton Universitsy Press, Princeton, NJ

Ecological Applications (1996) Volume 6, Ecological Society of America, Washington, DC

Embleton KV, Gibson CE, Heaney SI (2003) Automated counting of phytoplankton by pattern recognition: a comparison with a manual counting method. J Plankton Res 25:669–681

Guillaumont B, Bajjouk T, Talec P (1997) Seaweed and remote sensing: a critical review of sensors and data processing. In: Round FE, Chapman DJ (eds) Progress in Phycological Research, vol 12. Biopress, Bristol, pp 213–282

Hairston NG (1989) Ecological experiments: purpose, design and execution. Cambridge University Press, Cambridge

Harris R, Wiebe P, Lenz J, Skjoldal H-R, Huntley M (2000) ICES zooplankton methodology manual. Academic Press, London

Hurlbert SJ (1984) Pseudoreplication and the design of ecological field experiments. Ecol Monogr 54:187–211

Jeffery CJ, Underwood AJ (2001) Longevity determines sizes of an adult intertidal barnacle. J Exp Mar Biol Ecol 256:85–97

JGOFS (1994) Protocols for the joint global ocean flux study (JGOFS): core measurements. IOC Manuals and Guides no 29

Kingsford MJ, Battershill CN (1998) Studying temperate marine environments: a handbook for ecologists. Canterbury University Press, Christchurch

Matias MG, Underwood AJ, Coleman RA (2007) Interactions of components of habitats alter composition and variability of assemblages. J Anim Ecol 76:986–994
Mayo DG (1996) Error and the growth of experimental knowledge. University of Chicago Press, Chicago, IL
Mead R (1988) The design of experiments: statistical principles for practical applications. Cambridge University Press, Cambridge
Menge BA (1978) Predation intensity in a rocky intertidal community: relation between predator foraging activity and environmental harshness. Oecologia 34:1–16
Parry DM, Kendall MA, Pilgrim DA, Jones MB (2003) Identification of patch structure within marine benthic landscapes using a remotely operated vehicle. J Exp Mar Biol Ecol 285:497–511
Pilgrim DA, Parry DM, Jones MB, Kendall MA (2000) ROV image scaling with laser spot patterns. Underwater Tech 24:93–103
Popper KR (1968) The logic of scientific discovery. Hutchinson, London
Resetarits WJ, Bernardo J (1998) Experimental ecology: issues and perspectives. Oxford University Press, Oxford
Robinson CLK, Hay DE, Booth J, Truscott J (1996) Standard methods for sampling resources and habitats in coastal subtidal regions of British Columbia. Part 1. Review of mapping with preliminary recommendations. Part 2. Review of sampling with preliminary recommendations. BC Ministry of Agriculture, Food and Fisheries, Victoria
Schiel DR (2004) The structure and replenishment of rocky shore intertidal communities and biogeographic comparisons. J Exp Mar Biol Ecol 300:309–342
Sutherland WJ (2006) Ecological census techniques. Cambridge University Press, Cambridge
Todd CD, Phelan PJC, Weinmann BE, Gude AR, Andrews C, Paterson DM, Lonergan ME, Miron G (2006) Improvements to a passive trap for quantifying barnacle larval supply to semi-exposed rocky shores. J Exp Mar Biol Ecol 332:135–150
Underwood AJ (1990) Experiments in ecology and management: their logics, functions and interpretations. Aust J Ecol 15:365–389
Underwood AJ (1997) Experiments in ecology: their logical design and interpretation using analysis of variance.. Cambridge University Press, Cambridge
Vogel S (1994) Life in moving fluids, 2nd edn. Princeton University Press, Princeton, NJ
Winer BJ, Brown DR, Michels KM (1991) Statistical principles in experimental design, 3rd edn. McGraw-Hill, New York

Index

J

K

L

M